15th Topical on Refinery Processing 2012

Topical Conference on the 2012 AIChE Spring Meeting and 8th Global Congress on Process Safety

Houston, Texas, USA
1-5 April 2012

ISBN: 978-1-62276-567-6

Printed from e-media with permission by:

Curran Associates, Inc.
57 Morehouse Lane
Red Hook, NY 12571

Some format issues inherent in the e-media version may also appear in this print version.

Printed by Curran Associates, Inc. (2012)

For permission requests, please contact AIChE
at the address below.

AIChE
3 Park Avenue
New York, NY 10016-5991

Phone: (203) 702-7660
Fax: (203) 775-5177

www.aiche.org

Additional copies of this publication are available from:

Curran Associates, Inc.
57 Morehouse Lane
Red Hook, NY 12571 USA
Phone: 845-758-0400
Fax: 845-758-2634
Email: curran@proceedings.com
Web: www.proceedings.com

TABLE OF CONTENTS

() Future Fuels and Chemicals: Grand Challenges and Opportunities

Joseph B. Powell, Shell Projects & Technology, Houston, TX

To insure a Sustainable Future, CO_2emissions must be reduced despite strong future growth in demand. "Easy oil" from conventional production will be less abundant, though coal and natural gas, as well as unconventional hydrocarbon reserves will remain in strong supply for decades to come. Carbon footprint, cost, and convenience will be critical drivers for future energy choices. Basis recent advances in horizontal drilling and hydraulic fracturing, enhanced production of tight gas, shale gas, and coal bed methane will enable natural gas to be both an affordable and abundant option for meeting short and long-term sustainable energy goals. Sustainable production of biofuels, with consideration of well-to-wheels carbon intensity and other resource impacts, will play a major role in reducing carbon intensity in the transportation sector. Carbon capture and sequestration along with renewable energy sources and enhanced energy efficiency will play increasing roles in reducing future greenhouse gas emissions across the diversity of resources that will be deployed to meet local and global stakeholder needs for energy and security. Advantaged routes to chemicals will be impacted by new feedstocks and performance requirements for fuels, providing opportunities for innovation in gas to chemicals, biobased chemicals, and selective conversions associated with hydrocarbon portfolio balancing. Significant time and investment will be required to bring forth new alternative energy systems to materiality, driving the development of new paradigms for efficient de-risking and deployment of technology.

This presentation will address opportunities for the chemical engineering profession to differentiate and excel in sustainably meeting future needs in fuels and chemicals.

(29a) The Economics of Process Measurements

Douglas C. White, Emerson Process Management, Houston, TX

What is the economic value of a new measuring device in a process plant – a new analyzer, a more accurate flow measurement, an additional temperature or pressure measurement, etc.? Excluding the minimum set of measurements required to safely operate the plant and meet regulatory requirements, how can additional measurements be economically justified? Engineers are often presented with this issue, both for proposed new plants and existing sites. More measurements generally reduce the uncertainty in evaluating the past performance, the current state and the expected future performance of the plant. The reduction in uncertainty should then result in improved decisions. There are a hierarchy of decisions in plants that have different time scales and different magnitudes of possible financial impact. In some cases the decisions are algorithmic, i.e. a control loop. In others they are more complex, involving manual processing. Quantifying the possible improvement in future decisions and their possible impact on plant profitability is used to economically value the measurement. The value can be compared with the cost to determine an expected return on investment. In this paper a framework is presented to quantitatively address this subject and examples from the process industries are used to illustrate the procedures developed.

(29b) Applying Offline Tools to Reduce On-Site Control Implementation Effort

Kenneth V. Allsford[1], Bhaskar V Iyer[2]and Aric Tomlins[2], (1)Honeywell Process Solutions, Honeywell, Houston, TX, (2)Honeywell Process Solutions, Honeywell International, Houston, TX

Process control engineers are continuously challenged to reduce the effort to commission new advanced regulatory control strategies and/or linear/nonlinear multivariable control applications. Reducing on-site controller identification and tuning and the associated "interference" with process operations is especially desired – provided the commissioned application(s) meet(s) performance expectations. This paper will present a refining practitioner's perspective on applying tools to address this challenge.

For multivariable control applications on a number of refining units, it is becoming established practice to run automated closed-loop step testing tools to hone in on the control models that will enable good performance for the tuned/commissioned controller. However, this closed-loop step testing approach requires satisfactory “seed models” to enable control to be maintained during the step testing. For advanced regulatory control, similarly knowledge of approximate relevant process dynamics (effectively also “seed models”) – coupled with an appropriate tuning philosophy – enables good initial estimates for PID and other tuning parameters. As such the focus of this paper is on capturing seed models that can be used for either of these purposes. Approaches, referred to here as “Offline Tools”, that enable reduction of actual plant testing are reviewed.

A general resource in all instances is that (large) engineering organizations have the capacity to maintain, explicitly or implicitly, relevant data libraries – seed models for multivariable controllers or tuning parameters, possibly with associated identified process dynamics, for advanced regulatory control strategies.

For greenfield or significantly modified brownfield sites, operating data may not be available initially. However, operator training (or other (dynamic) process engineering) simulators may be available and their use for generating seed models and/or initial tuning for advanced regulatory control strategies will be reviewed. Process engineering model assumptions that impact the use of the models for these purposes will be highlighted. Of note is that with appropriate dynamic process engineering models, controller performance across a wide range of operations can conceptually be evaluated relatively quickly.

For actual operating plants, a range of operating data exists from the descriptive provided by operations and engineering staff through the very granular stored in historians. Steady state and/or dynamic models may also exist and the operating data can be used to identify the usefulness of such models, potentially initiating a discussion on making such models “current”. A range of approaches (“offline tools”) to utilize the available offline data to estimate seed models exists; it will be discussed that the optimal approach is typically some combination of these tools selected for the particular circumstances.

Useful Tips for Fired Heater Optimization

Nikki Bishop, Emerson Process Management, Atlanta, GA
Barbara Hamilton, Emerson Process Management, Garrison, NY

Introduction

Fired Heaters are used extensively throughout refineries for heating, vaporization and thermal cracking of various process fluids. These heaters are essential for refinery operation. Figure 1 below shows a typical process fired heater.

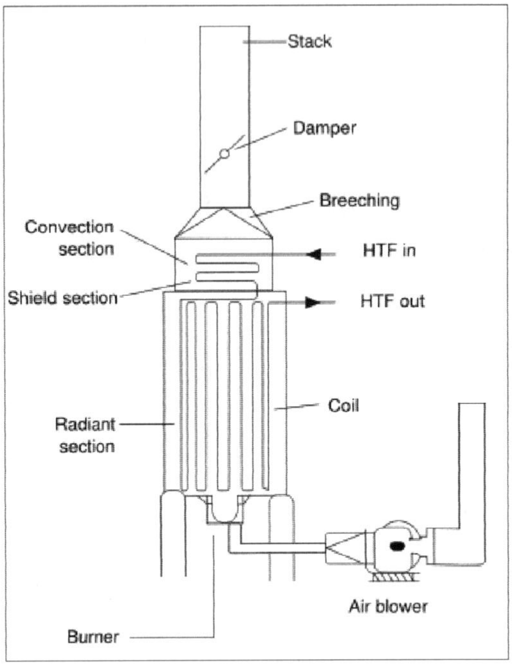

Figure 1. Typical Process Fired Heater

The main objective for fired heater optimization is to safely control to a target combined coil outlet temperature while minimizing energy costs, emissions, and overall variability. To achieve this objective, the heater must be able to safely operate near constraints and demonstrate stable and robust performance. A holistic approach that examines the field devices, basic control strategy and loop tuning is recommended to establish a solid foundation. Advanced process control methods coupled with a solid combustion control strategy provides an effective means of safely operating near constraints. Even further, pre-engineered application solutions offer great benefits in the ease of implementation and support.

Back to Basics: The Instruments

Just as a building is only as good as its foundation, a process control system is only as good as the instruments which provide the measurement data and final control. It's important to address and mitigate process and mechanical issues. Necessary measurement and actuator improvements are made in order to reduce control loop variability. Reducing variability on key loops enables the heater to operate closer to the constraint conditions, which increases throughput, efficiency, and safety. For example, it is essential to tightly control the air and fuel flows for safe and efficient combustion.

Figure 2 below shows typical fired heater measurements.

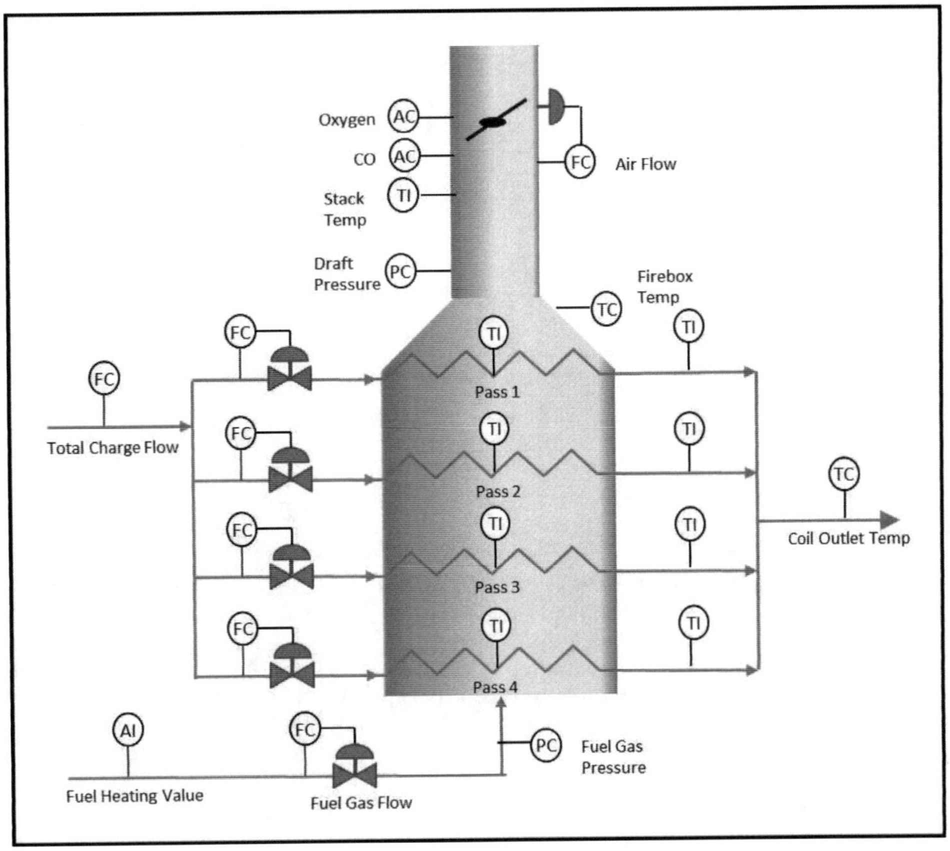

Figure 2. Typical Measurements

Beyond Base Regulatory: The Full Solution

Many fired heaters are instrumented with only the minimum instrumentation necessary for operation. While this strategy allows for operation of the heaters, it lacks the sophistication necessary for optimal performance. A full solution, including a burner management system (BMS) for safety and the necessary instruments, final control elements and damper actuator, is the ideal recipe for optimal heater performance.

The Ultimate Control Strategy

The ultimate goal for a fired heater is to heat a process fluid to a desired temperature. Maintaining a constant outlet temperature is critical for the process. Variations in outlet temperature introduce variability in the overall process. Since the optimum operation will almost always be near constraints (e.g. maximum tube temperatures, minimum excess air), variation in the process requires the operator to stay further away from the actual limit to provide the necessary buffer or safety margin to handle any unexpected process upsets. As a result, manufacturers are not always able to achieve the most efficient operation of their assets.

Figure 4 below shows a representation of the impact of variability on cost.

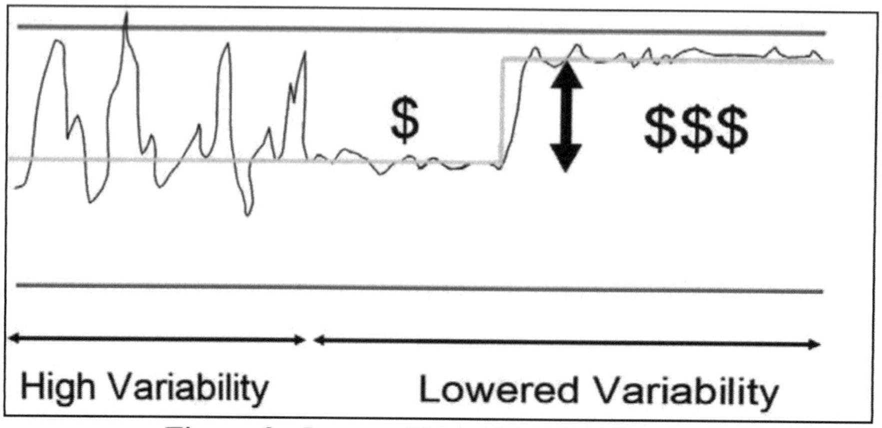

Figure 3. Process Variability Cost Impact

Reducing variability means ensuring a robust and stable control strategy is in place. Implementing a fully automatic, regulatory control system ensures that coil outlet temperature is maintained at the desired setpoint, while simultaneously balancing pass temperatures and air flow. Pass balancing adjusts the flows through each tube such that no pass runs hotter than the others, reducing hot spots within the coils, which translates to longer run lengths and the potential to operate at a higher severity.

Bringing the Heat: Combustion Controls

As previously noted, the ultimate goal for a fired heater is to safely heat a process fluid to a desired temperature. The required energy is provided by combustion of a fuel to heat the coils. The efficiency of the combustion process determines the efficiency of the heater. A "single-knob" combustion control strategy allows for safe, efficient firing at any desired charge rate. With "single-knob" control, the desired coil outlet temperature is maintained while operating within mechanical and thermodynamic heater constraints at all times.

A proven strategy for safe combustion is to employ an air and fuel "cross-limiting" solution. Cross-limiting is the traditional approach of increasing air flow first on an increasing fuel demand and decreasing fuel first on a decreasing demand. This strategy prevents unsafe, fuel-rich environments from being introduced by the automation system. Typically this is done by empirically determining the air, fuel, and stack oxygen levels at various demand points. A linear relationship is typically used between demand and fuel flow. The relationship between demand and air flow follows the empirically derived curve such that the oxygen target at each demand point ensures complete combustion.

Typically, the lowest stack oxygen level that does not produce carbon monoxide (CO) in the stack is best for both efficiency and emissions compliance. CO in the stack is an indication that the fuel is not completely burned.

Advanced combustion controls allow for maximum capture of the available heat in the fuel with less variability. This is important even if the fuel is a refinery waste gas that is "free" because it reduces greenhouse gas emissions, makes the fuel available for other uses like boilers or co-gen plants, and allows for more throughput in a capacity-constrained situation. It is impossible to safely drive stack oxygen levels down without robust combustion controls. When air and fuel are not well coordinated, not only is efficiency affected, but a fuel-rich environment risks an explosion if furnace oxygen levels get too low.

Figure 5 below shows the relationship between excess air and available heat.

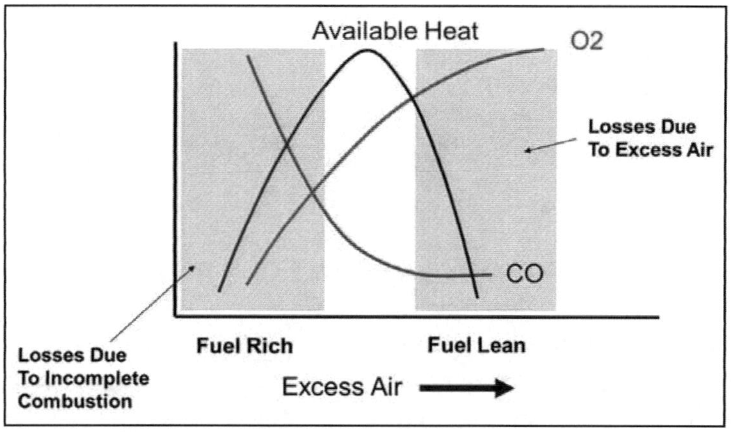

Figure 4. Stack O_2 vs. Available Heat

The economic savings resulting from operating a furnace at a lower excess air can easily be calculated based on the extra energy used to heat the extra air from ambient to the stack exhaust conditions. The stack O_2 concentration is an indication of how much excess air there is. If all the O_2 is perfectly consumed, the O_2 concentration will be 0. This, however, is not safe, so most furnaces operate with O_2 concentrations of 1% or higher. Decreasing the O_2 concentration in the stack from the current level to a lower value results in energy savings that can be substantial, depending on the starting point and the cost of fuel.

Figure 6 illustrates the relationship between reduced stack oxygen, efficiency and energy savings.

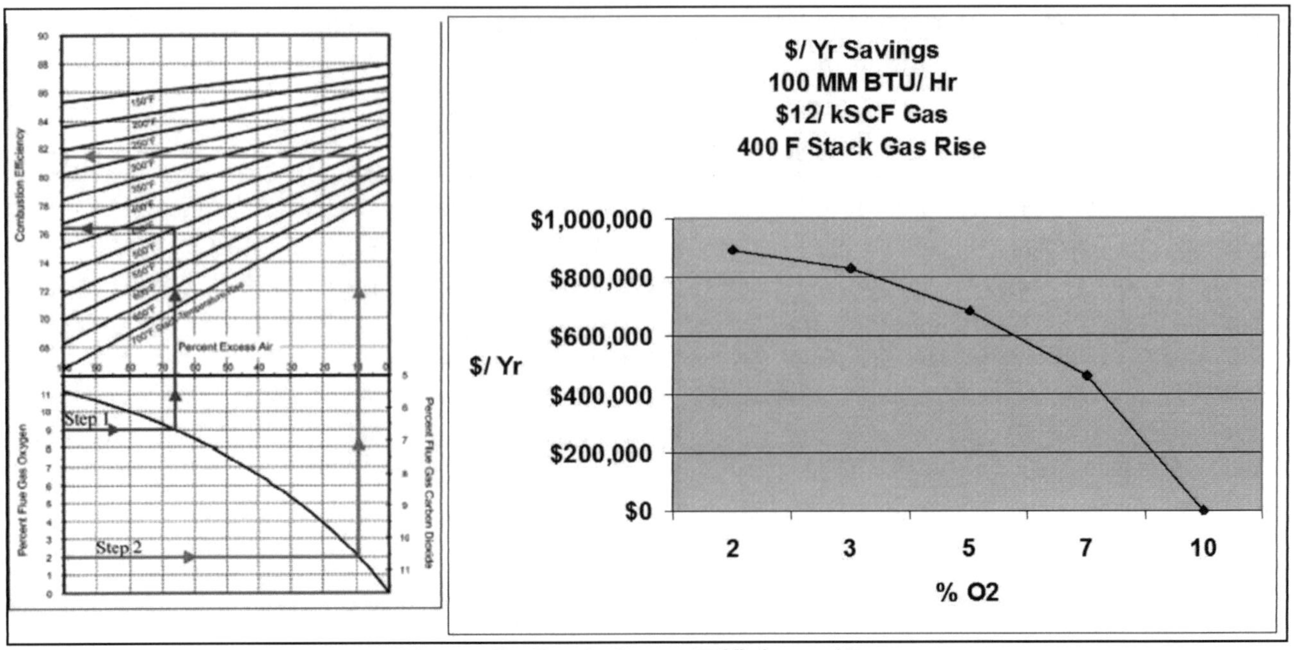

Figure 5. Stack O_2 vs. Efficiency/Cost

Advanced Process Control

To achieve optimal heater operation, the heater must be able to safely operate near constraints and demonstrate stable and robust performance. This means balancing pass temperatures while simultaneously ensuring complete combustion and minimizing energy. Model predictive control (MPC) that is native to the regulatory control processor provides a high speed, redundant, robust control solution.

MPC uses relationships between controlled, manipulated, disturbance, and constraint variables to create dynamic process models that are used with an optimizer to determine the best set of process changes that control all variables to target. Known relationships between variables can be input directly into the MPC control module. Alternatively, an automated "bump test" sequence is used to establish the process models. Constraint violations (such as skin temperatures, fuel pressures, and air/fuel cross limits) are anticipated (predicted) and manipulated variables adjusted accordingly.

A model predictive control strategy also reduces process variability, because for optimal operation, passes are balanced not to achieve equal flow, but to achieve equal temperature rise across all the tubes. Pass balancing based upon temperature rise creates process conditions that favor more efficient overall heat transfer and balance the coke build up, allowing the unit to run longer between de-coking shutdowns.

Figure 7 illustrates an actual example on an ethylene cracking furnace where MPC is used to control the temperature deviation between the passes. This chart demonstrates the significant reduction in pass outlet variability when the model predictive controller is online.

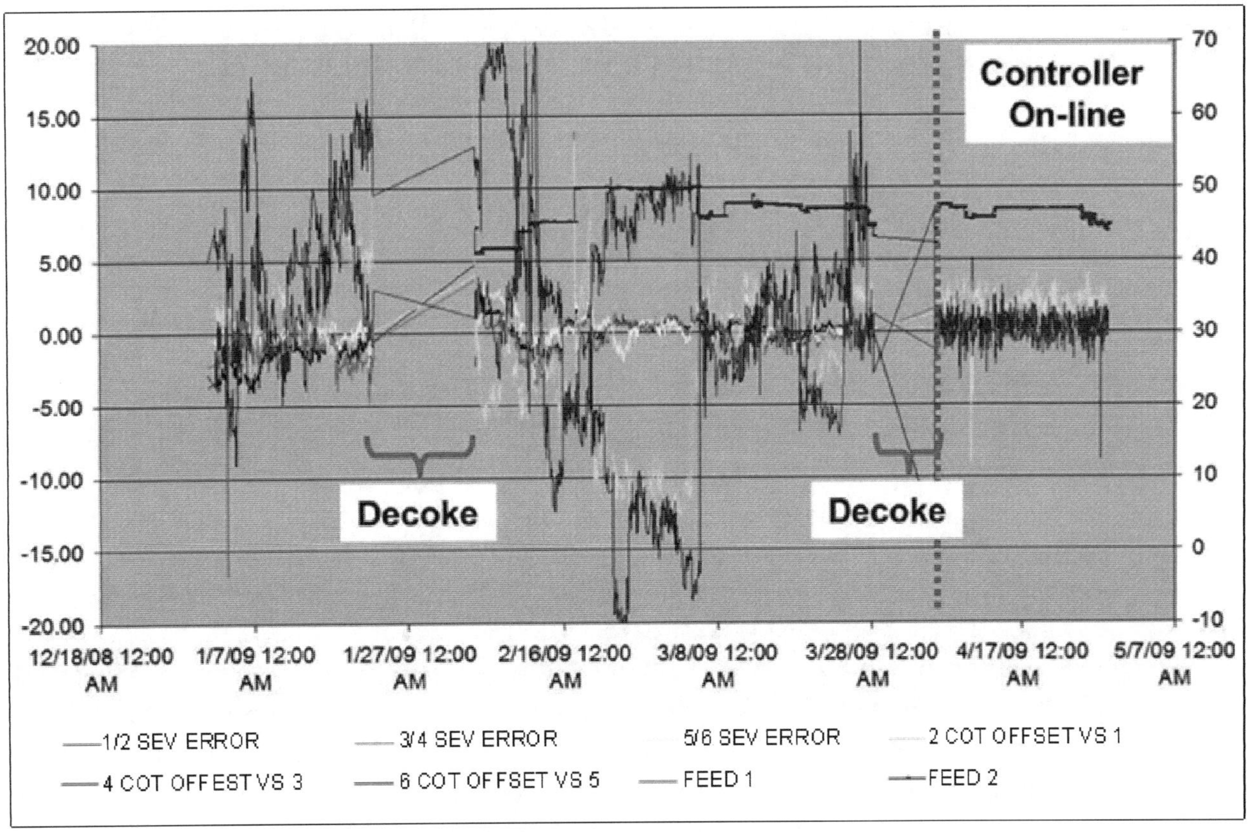

Figure 6. MPC Controller Online Variability

MPC Put to Good Use

One multi-national oil company successfully implemented an advanced process control strategy on its fired heaters as part of a larger APC project on the unit and achieved remarkable business results. Two fired heaters on the atmospheric crude unit were operated in parallel, with poor efficiency. Due to various process and equipment reasons, these heaters operated only with significant manual intervention. Process interactions made conventional, automatic control less than ideal.

The solution to the problem was an MPC strategy which allowed for safe and stable operation of both furnaces simultaneously. These furnaces can now operate near their constraints in spite of significant fuel gas, feed rate, crude composition and feed temperature variations. Implementation of this advanced process control strategy on the heaters resulted in an annual fuel savings of more than $100,000. The heaters now require less manual actions and can manage load changes automatically in closed-loop control. Excess stack oxygen is reduced and efficiency increased in both heaters.

Abqaiq Plant Virtual Plant RTO Implementation

Dr. Kuangyao Brian Peng and Saleh H. AL-Musallam,

Saudi Aramco, Abqaiq, Saudi Arabia

Introduction

Abqaiq Plant is the largest worldwide crude stabilization plant with $200MM energy consumption per year. Abqaiq NGL plant shares 30% of total energy consumption. High NGL recovery rate has been the objective because of undesired loss of hydrocarbon liquid condensate in the offgas pipeline. Recent efforts for process improvements including commissioning of TEG dehydration unit and compression train upgrades have been implemented to enhance NGL recovery and to improve the availability of desired cryogenic cooling for C2/C3 separation.

Abqaiq Natural gas liquid process includes three major processes: compressing, high pressure stripping and deethanizing. Simultaneous optimization for all three major processes under the constraints of NGL product spec. requires fundamental understanding and a robust simulation tool. It has been a dream and a challenge for process and control engineers to develop such a complete model which can map the plant operating conditions in situ and perform a complete plant optimization online.

ROMeo (Rigorous Online Modeling with Equation-based Optimization), a widely used real time optimization (RTO) platform, can be used to achieve our goal. A real time optimizer can take the variations in upstream units such as changes in crude feed flow combinations and stabilization depth in real time, and calculates the most profitable mode of operating condition by using the rigorous plant model with energy consumption equations and process and product constraint equations. Then the calculated optimal set points are submitted to the DCS control room to ramp the plant process.

Abqaiq Plant NGL engineering department has successfully implemented and tested an in-house developed RTO NGL plant model. A potential increase of $200K to $400K revenue increase for each adjustment has been predicted under the current operating mode and unit operation constraints. In addition to the quantifiable revenue advantages, this presentation will also highlight other substantial advantages of the system such as measurement health monitoring, next generation process KPIs, and energy optimization.

Background of NGL process optimization initiatives

To meet the future energy demand and sustainability requirements in the Kingdom of Saudi Arabia, Abqaiq Plant starts several energy efficiency programs for all process units. A rough study with Saudi Aramco P&CSD for NGL recovery enhancement has shown great potentials with online optimization. As a result of this study, by simply adjusting the operating conditions of HP-Stripper columns could increase NGL recovery rate by 3 to 5 MBD. Further study with relaxing product spec. of individual columns and maintain the final product spec has shown an increase of NGL production up to 9 MBD.

A joint effort among corporate optimization specialists and plant operation engineers took off in 2010. A road map was proposed to implement ROMeo RTO at Abqaiq NGL plant by the end of year 2012. The road map includes three major milestones: rigorous model development, PI integration and offline optimization.

Process description

Continuous feed gas from oil stabilization units is fed into the NGL plant with a flow rate of 550MMSCFD at 3psig and 140F. Feed gas is compressed with two stage compressors to 450psig and is cooled by cooling tower for the outlet temperature at 110F-125F.

Feed compositions are studied with the available historical data. The major variation of feed compositions come from crude oil mix containing neat condensate, Arabia Light and Arabia Extra light. As a baseline model, average compositions over the past few years are used for model stability. Typical compositional data for hydrocarbon contains C3+ above 70%. Table 1 shows the average compositional data for a typical feed.

	N2	C1	CO2	C2	H2S	C3	IC4	NC4	IC5	NC5	C6	C7	C8
Overall	0.019706	0.0785301	0.028414	0.224106	0.043989	0.342631	0.056587	0.132501	0.032532	0.032833	0.006334	0.0018	3.4E-05

Table 1

Abqaiq NGL plant contains five major high pressure stripper columns and one deethanizer column. The average operation condition of high pressure column is 145F and 400 psig. Column efficiencies vary from 68% to 85% because of different tray designs. The five high pressure stripping columns were designed with 22 bubble trays for separation of C2 and C3+. Each column equips with steam kettle reboilers and temperature controller at the 7 th stage of the column. The column temperature is the key handle for NGL product specification.

The deethanizer chiller is designed to operate at 65F and the reflux condenser is 60F. Currently, operation requires all columns product spec at 0.4 vol% which is a conservative approach. Overstripping of NGL product to offgas line would raise the liquid condensation in the pipeline, reduction of valuable NGL product recovery and waste of energy.

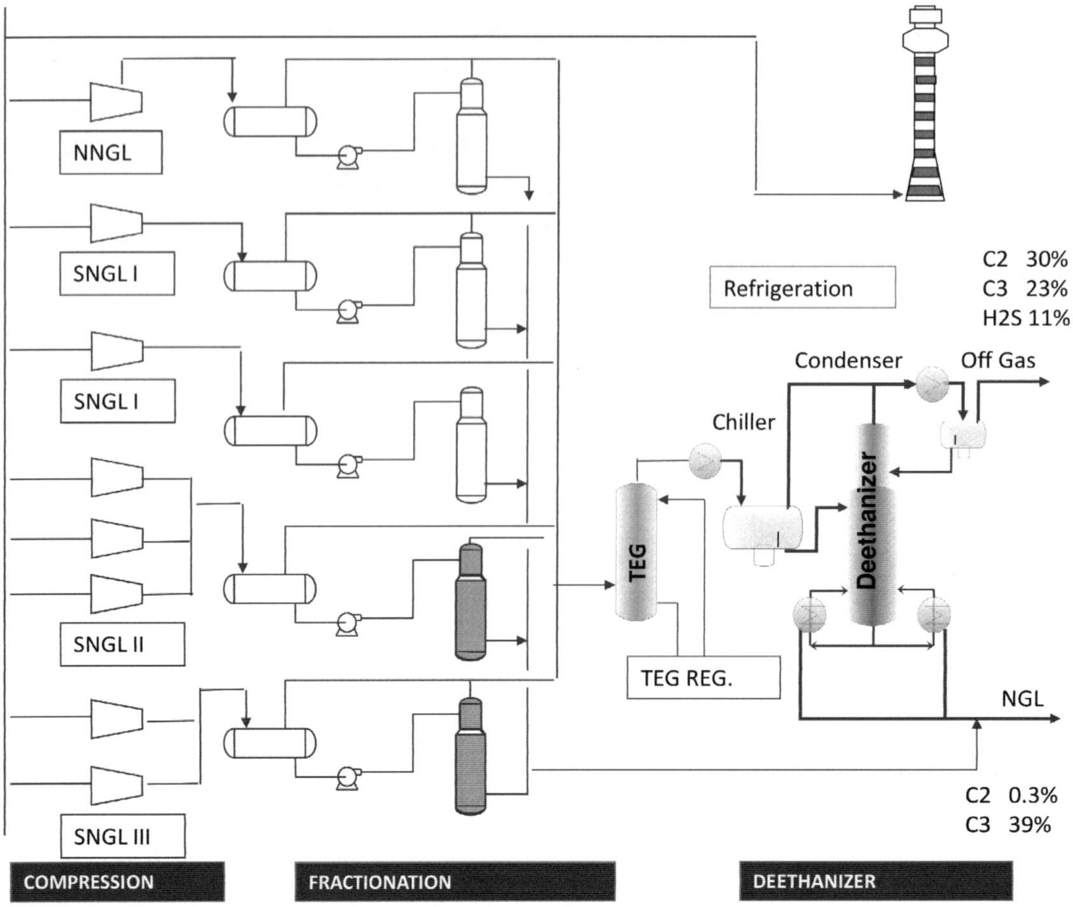

Figure 1: A simplified NGL process flow diagram.

Model development and implementation

Rigorous process model is developed based on the current process conditions in the field. Due to time constraint, the preliminary process model does not integrate with detail performance data of compressors. However, because the phase I of this program is to maximize NGL recovery and to minimize the impact of liquid condensation, the performance functions are not required at this stage.

One of the challenges of the team is to identify all the piping routes of cross feed lines. The cross feed flow rates have to match the field instrumentations. The on-off setting of process units is relied on the macro programs that were developed in house with the engineering support and training from Invensys. A total number of 180 major PI-tags were linked to PI historian for data reconciliation.

Economic variables for the sources include 60# and 625# steam, propane and feed gas. For the product steams, two offgas streams and one liquid NGL stream are included. The environmental variables include the release of CO_2 to atmosphere and the amount of sour water steam to downstream facility.

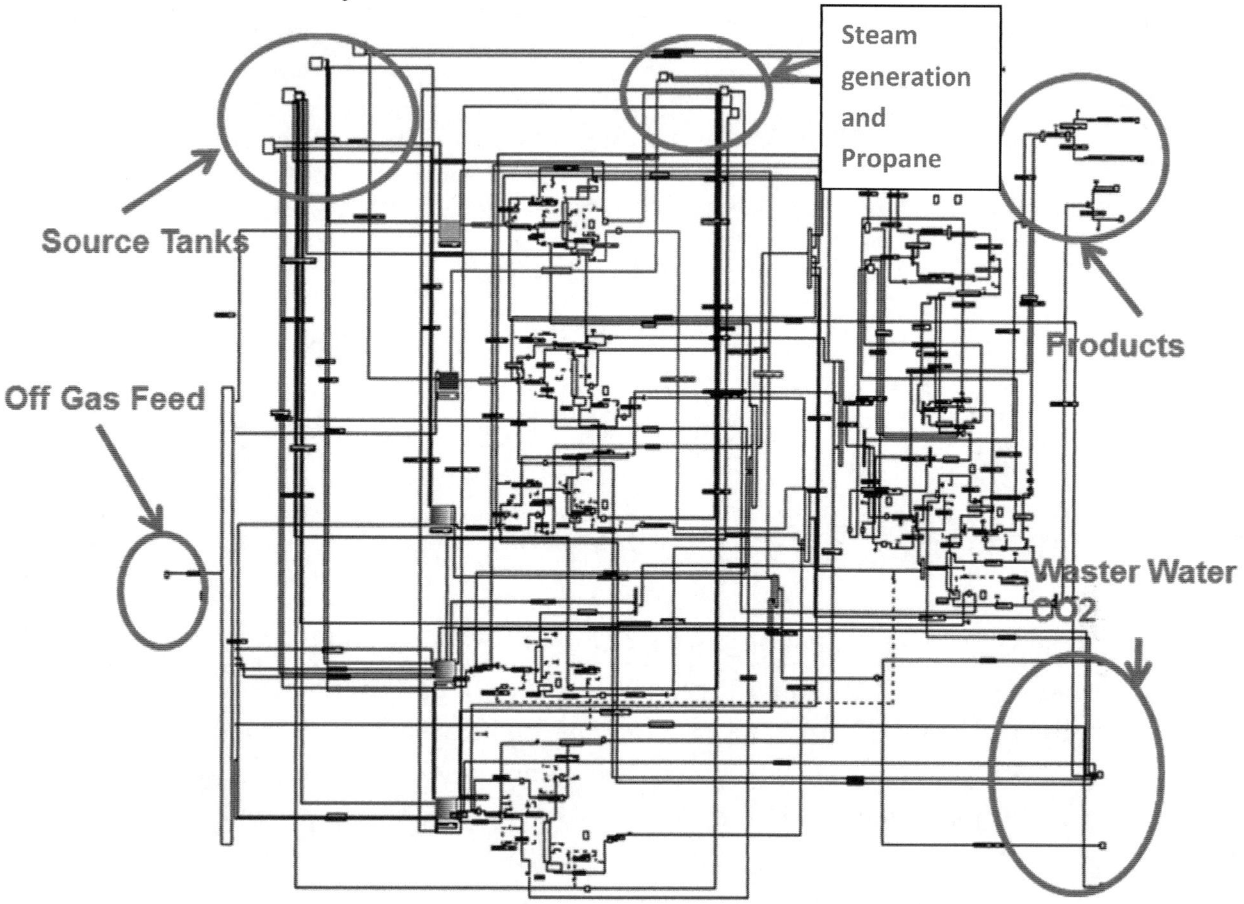

Figure 2

Figure 2 demonstrates the process models with eight compression trains, five high pressure stripping columns, one deethanizer column and a propane refrigeration system. All process units in this project are simulated with rigorous thermodynamic models.

Data Reconciliation

Process model has been integrated with PI historian for data reconciliation purpose. Macro routines with TCL scripts are developed to turn on and turn down major equipment under different operation modes. Data reconciliation results deliver the report for measurement system analysis that indicates major instrumentation issues. Operation, instrumentation engineering and maintenance are able to utilize this report to identify the issue and improve the situation. Furthermore, accuracies of some major KPIs are improved after the improvement of measurement system.

Tags	Unit	Scan	Model	Bias	% ABS(Bias)	Comments
462-FI-310.PV	MMft3/day	114.851	91.047	-23.804	26.14474	
340-Total-Off-Gas.CAL	MMft3/day	393.3	297.296	-96.004	32.2924	
461-FIC-182.PV	MMft3/day	18.471	14.189	-4.282	30.17831	
499-TI-764AI1PV.CV	F	159.65	125.356	-34.294	27.35729	
462-FI-1147.PV	MMft3/day	79.345	61.404	-17.941	29.21797	
334-FI-1005.PV	lb/hr	38297.5	25444.8	-12852.7	50.51209	
461-FIC-12.PV	lb/hr	62396.6	29909.7	-32486.9	108.6166	

Table 2

Table 2 shows an example of data reconciliation results with problematic instrumentations related to the major KPIs. Engineering and operation was able to take immediate actions with the reported warnings and resolved the issue in time.

Online Open Loop Real Time Optimization

The online process model has been up running to provide advisory for operation engineers. The new process set points for high pressure stripping columns such as the feed flow rates and individual column specs are reported on daily basis. A typical adjustment could increase a few MBD of NGL product. When feed composition shift due to variation of oil feed combination and stabilization depth, the NGL recovery can be enhanced and lined out quickly with ROMeo RTO.

Table 3 shows an example of production increase with RTO. The long term benefit is estimated at least $3MM per our study.

Optimization Summary

Variable	UOM	Value	Change
Net Objective	currency/day	5.628E+006	56752.500
Positive Contributors			
+ Products	currency/day	1.119E+007	54915.900
Negative Contributors			
- Steam Feeds	currency/day	5.563E+006	-1836.620

Table 3

ROMeo has also identified a potential increase of 3MBD recovery with simple process modification. As shown in the figure 3, under the optimization mode, ROMeo tends to divert all the overhead gas from deethanizer feed drum to bypass deethanizer overhead condenser. It has been the normal practice for NGL operation since the system was designed. Similar result was concluded from an independent energy optimization study from corporate process optimization group. The online RTO demonstrates capability of evolving process by justifying the bottleneck of current process design.

Other Monitoring System Development

A couple of monitoring indexes have been proposed for the NGL plant and pipeline operations. For NGL process, a fouling index is defined to capture the early stage of fouling. Instead of relying on the steam flow meter, a fouling index is defined with fundamental equations to remove the impact of temperature gradient across the heat exchanger. The fouling index can provide information for unplanned maintenance schedule.

Table 4 shows the examples of fouling indexes for the kettle reboilers. Out temperature at the cold side is assumed to be the bubble point of the hydrocarbon flow. Hot side temperature is assumed to be the dew point of the superheated steam flow. Temperature difference is calculated with the average temperature from both sides. Based on the engineering data, the heat transfer coefficient in red shows the inconsistent dimension which requires further validation during turn around activity.

Fouling Studies

Variable	UOM	Initial	Value
/340_E3_Duty_CU/e3ua	BTU/hr-F	337104.000	317283.000
/340_E3_Duty_CU/e3u	BTU/hr-ft2-F	67.421	63.457
/461_C_101Duty_CU/e103ua	BTU/hr-F	1.900E+006	1.789E+006
/461_C_101Duty_CU/e102u	BTU/hr-ft2-F	383.799	361.470
/462_stripper_duty_cu/e203ua	BTU/hr-F	2.037E+006	1.905E+006
/462_stripper_duty_cu/e203u	BTU/hr-ft2-F	59.916	56.036
/499_E3_Duty_CU/E3UA	BTU/hr-F	1.515E+006	1.515E+006
/499_E3_Duty_CU/E3U	BTU/hr-ft2-F	69.030	69.030

Table 4

Table 5 shows another environmental variable established for carbon dioxide release. The total amount of carbon dioxide can be tracked with online optimization model. With data collection from historian, justification for carbon dioxide capture can be established because supercritical carbon dioxide injection could be the potential technology for depleted oil fields in the Kingdom.

Overview Stream Economics: CO2 Generation

Variable	UOM	Value	Change
Flow	MMft3/day	59.916	0.000

Table 5

Figure 3 shows another important process variable to be monitored is the predicted hydrocarbon liquid condensate. The offgas pressure will drop along the pipeline to the gas plant for C1-C2 separation and sulfur removal. When the initial temperature is higher than the ambient temperature, hydrocarbon liquid condensate is expected in the pipeline. The amount of hydrocarbon liquid condensate can be predicted with online optimization model which has been updated on daily basis. The amount of accumulated liquid can be correlated to the back pressure incidents for future optimization of pipeline scraping frequency and plant design criteria. For instance, the NGL target recovery and frequency of pipeline scrapping shall be determined based on the total cost analysis.

Tracking of hydrocarbon liquid condensate can be achieved with an online model to mitigate back pressure and flaring of valuable products. This new performance index could potentially save the company millions of dollars.

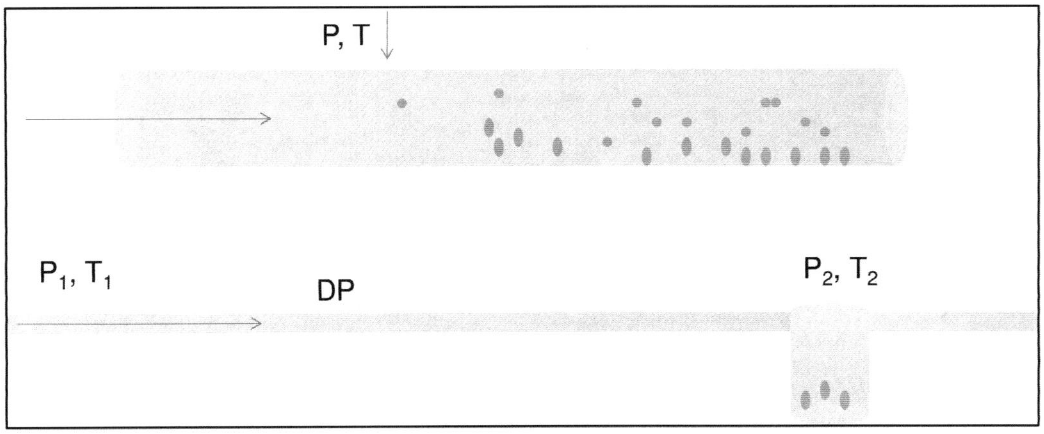

Figure 3

Conclusion

Abqaiq NGL Plant has successfully implemented and tested an in-house developed RTO NGL plant model. A potential increase of $200K to $400K revenue increase for each adjustment has been predicted under the current operating mode and unit operation constraints. In addition to the quantifiable revenue, this presentation also highlights other substantial advantages such as measurement health monitoring, next generation process KPIs, and energy optimization. Future development of this project will focus on supervisory PI Processbook for DCS operators and implementation of engineering details for major equipment.

Acknowledgement

The authors thanks for the supports from the managements of South Area Oil Production Department at Abqaiq and Process simulation and Control System Department at Dhahran under Saudi Aramco Oil Company. This joint technology development has demonstrated a successful story for Saudi Aramco accelerated transformation program.

(29e) Dynamic Simulation of CGC System Under Ethylene Plant Turnarounds

Chuanyu Zhao, Jie Fu and Qiang Xu, Dan F. Smith Department of Chemical Engineering, Lamar University, Beaumont, TX

CGC system (Cracked Gas Compression system), or Multistage compression system, is a vital section during ethylene production. It is located after the cracking furnaces and quench towers, and ahead of the distillation section, which is considered as the "heart" of an ethylene plant because its reliability, stability and sustainability determines the production capacity of that plant. Thus, the CGC system needs to be strictly controlled and operated, especially during plant turnarounds. Generally, plant planned turnarounds include startup and shutdown operations, which have two levels of requirements: i) A feasible and safe startup/shutdown strategy needs to be well developed. Many past turnaround cases failed due to compressor surge, which caused severe equipment damages and unnecessary flaring. ii) An optimal startup/shutdown strategy is better to be pursued if possible in order to minimize startup/shutdown time and reduce flaring emission, raw material usage and energy consumption.

Based on the above consideration, this paper employs rigorous dynamic simulations to help plant turnaround operations, i.e., examine and improve operation strategies of a CGC system for ethylene plant startup and shutdown. For safety consideration, anti-surge process design and control strategies are dynamically evaluated along with alternative turnaround procedures. Dynamic operating trajectories for every stage of CGC are also identified. Assisted by the dynamic simulation, the improvement of plant startup and shutdown performance are demonstrated.

(29f) Dynamic Simulation of Demethanizer and Chilling Train System

Shujing Zhang, Meiqian Wang, Karthik Krishnadevarajan, Qiang Xu and Kuyen Li, Dan F. Smith
Department of Chemical Engineering, Lamar University, Beaumont, TX

The chilling train system is an important part of ethylene plant which is usually combined with demethanizer and a series of flash drums to separate hydrogen and methane from cracked gas, and recover the hydrogen and methane as different products through multi-stream heat transfer with refrigerants. The system of demethanizer and chilling train is the intermediate part of an ethylene plant whose performance can significantly affect the amount of flaring during start-up and thus energy consumption. However, most research efforts are focused on design and steady-state optimization aspects, the dynamic behavior of chilling train is rarely explored. In this study, several control strategies of demethanizer and chilling train system will be simulated and corresponding dynamic responses will be investigated in order to reduce time and flaring during start-up.

General methodology includes the following steps: 1) In the first place, a steady-state model will be built using the sequential modular approach and plant design data; 2) the model will be further validated with normal steady state operating conditions obtained from the DCS (distributed control system) historian; 3) If the steady-state model is satisfied, it will be exported to the dynamic simulation environment. Three types of data will be needed in this transferring, including equipment dimensions, control strategies and P& ID (piping and instrument diagram) parameters, and the heat transfer option; 4) after the dynamic modeling is completed, it will be further validated to match some process upsets in DCS historian; 5) the control strategies will be tested in dynamic simulation to decrease start-up time and energy consumption. A case study is used for the demonstration.

RECENT EXPERIENCES ON MODELLING FOULING OF CRUDE REFINERY PREHEAT TRAINS HANDLING COMPLEX AND HEAVY CRUDE SLATES

* *Edward M. Ishiyama, IHS ESDU, London, UK*
Graham T. Polley, University of Guanajuato, Mexico
Simon J. Pugh, IHS ESDU, London, UK

*Corresponding author: edward.ishiyama@ihs.com; Tel: +44 (0) 20 3159 3300.

Abstract

Fouling of crude oil streams is the major cause of planned and unplanned downtime in oil refineries. As crude slates become heavier and more varied, the refinery operator must decide how to accommodate fouling, by for example selecting which heat exchangers to clean, against increasingly tight budgeting constraints. Often, it has not been recognized that fouling is often unnecessarily severe in poorly designed heat exchangers. In this paper we present recent experience in the application of dynamic fouling modeling using the novel heat exchanger network simulation tools, EXPRESSplus and SmartPM. These experiences illustrate the performance of dynamic fouling models in refinery pre-heat trains and the identification of opportunities for predictable improvements, for example through simple equipment modifications.

Introduction

Data reconciliation and fouling analysis is a vital step in obtaining practical understanding of plant fouling behaviour. With the understanding of the fouling behaviour, and, with the ability to model and predict dynamic fouling in individual heat exchangers [1], it is possible to explore reliable fouling mitigation strategies through simple retrofits and minor changes in operating conditions. This manuscript builds on the case study introduced by Ishiyama *et al.* [1]: opportunities to recover additional process heat through exploiting retrofit opportunities and its effect on the total heat exchanger network behaviour are explored. The studies in this manuscript were obtained using two commercial software EXPRESSplus and SmartPM, available through IHS ESDU.

The case study preheat train from Ishiyama *et al.* [1] is reproduced in Figure 1 and Table 1. The complete geometrical data, stream properties and inlet conditions are given in the original manuscript.

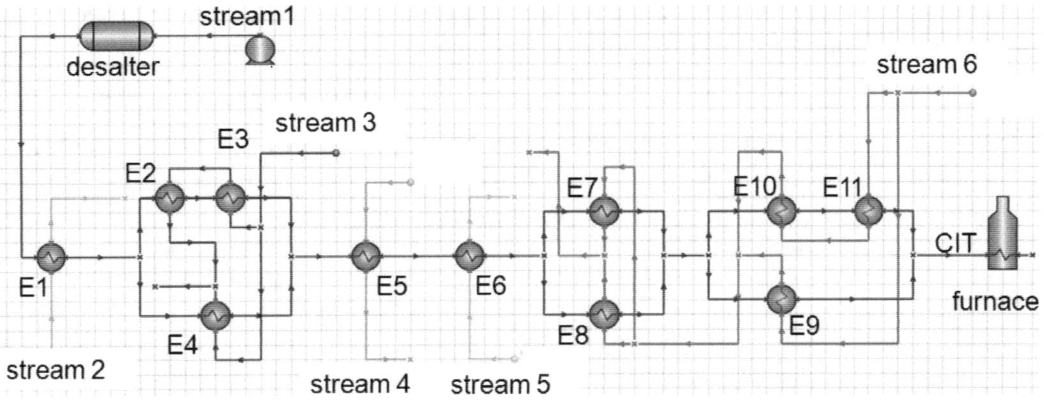

Figure 1: Schematic of a section of the case study preheat train reproduced from Ishiyama *et al.* (2012). The circles represents shell-and-tube exchangers, CIT, denotes the furnace coil inlet temperature.

Table 1: Geometrical details of E9, E10 and E11 (reproduced from Ishiyama *et al.* [1])

Heat exchanger	No. of tube passes	Tube length (m)	Tube internal diameter (m)	Tube external diameter (m)	Total number of tubes	Shell diameter (m)	Tube layout (°)	Number of baffles	Baffle cut
E9	8	6	0.0166	0.0190	1500	1.4	30	13	0.25
E10	6	6.1	0.0148	0.0190	2680	1.4	90	16	0.186
E11	6	6.1	0.0148	0.0190	2680	1.4	90	16	0.186

Exchanger analysis

Table 2 gives a summary of the performances of E9, E10 and E11. Note that the simulation was performed with initial fouling resistances that were obtained through the data reconciliation exercise. Based on the geometric data provided in Ishiyama *et al.* [1], E10 and E11 have the same heat transfer area: still the rate of drop in heat duty for E11 is much faster than E10. E9 has the smallest area but also has the biggest heat duty among the three exchangers (Table 2).

Table 2: Heat exchanger performances at the beginning and at the end of a heat exchanger network simulation during 8 month period. Values extracted from Ishiyama *et al.* [1].

Exchanger No.	Simulated exchanger duty (MW)		Final exchanger duty as a percentage of the starting duty
	Beginning	End (after 8 months)	
E9	7	4.5	64 %
E10	5.5	4.2	76 %
E11	4.4	0.9	20 %

Retrofit analysis

In this study we explore the retrofit analysis associated with E11 in three steps. First, to enhance heat recovery, second, to reduce rate of fouling and, third, to explore effects of the retrofit to the total network.

Ideally, retrofit options have minimum disruptions on plant operations. Such retrofit changes include changes in bundle geometry or the use of tube-inserts; these are changes that could be done off-sight. Here we consider the use of tube-insert for E11. EXPRESSplus was used to obtain an initial analysis on the unit performances. Twisted tapes are a common tube-insert due to its easy fabrication. Here, an example of a tube insert with 360° twist over a length of 100mm is considered. Hitran is also a common tube-insert due to its ability to keep particulates in suspension and commonly used in non-crude streams. For a hot stream flow rate of 150 kg/s, the film transfer coefficient comparison with and without tube-inserts are given in table 3.

Table 3: Comparison of the tube-side film transfer coefficients for E11

Exchanger No.	Plain tube (W m^{-2} K^{-1})	Hitran (W m^{-2} K^{-1}) Loop density = 1	Twisted tape (W m^{-2} K^{-1}) 100mm per 360° turn
E11	800	1100	2200

Use of tube-inserts drastically increases the tube-side pressure drop. It is worth to note that E11 has 6 tube-side passes. The tube-insert results in an increase in pressure drop which is above the allowable limit (Figure 2). Hence, if the tube-insert is to be used, tube-sheets would be required to be modified in order to operate at a reduced number of tube-side passes. It is important to note that these modifications only contribute on enhancing the tube-side heat transfer coefficient and has insignificant changes on the shell-side heat transfer coefficient. Hence for this particular unit the crude is on the shell-side, the effect on fouling with the retrofit is insignificant.

(i)

(ii)

(iii)

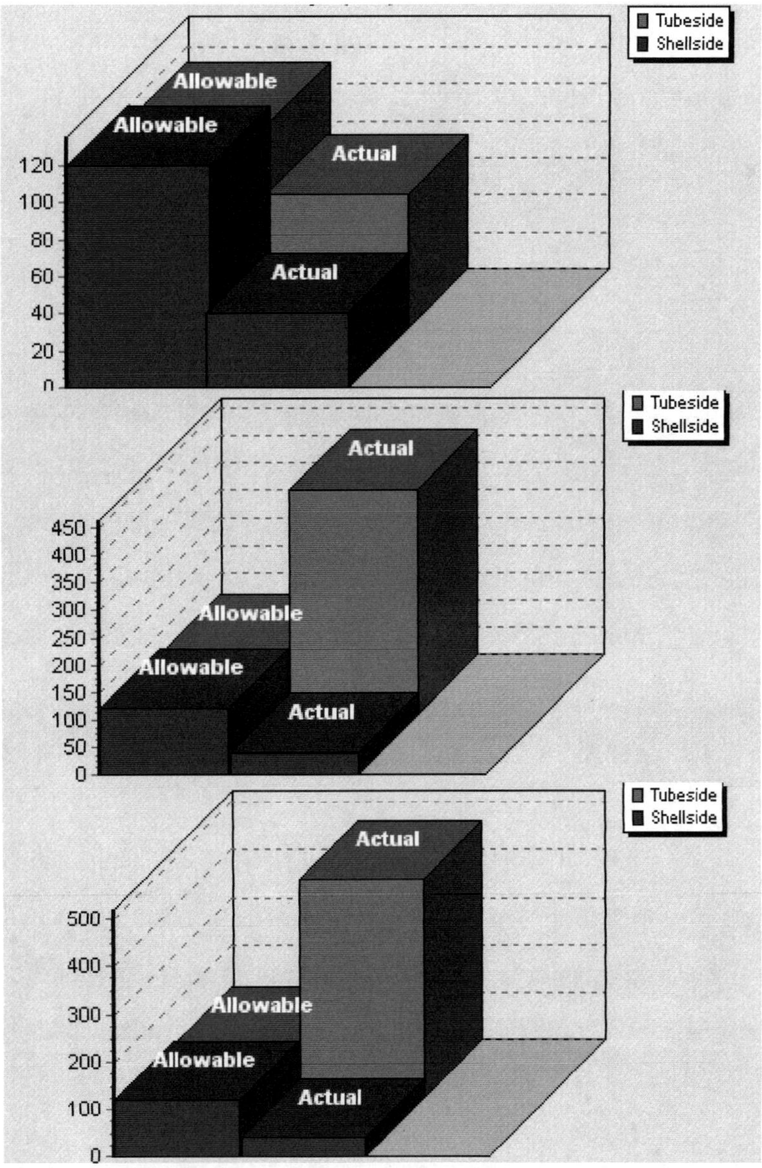

Figure 2: Actual and allowable pressure drops in kPa for, (i) plain tube, (ii) hitran (loop density 1), (iii) twisted-tape (100mm per 360° twist).

It is necessary to point that the crude is on the shell-side. Even, with the possibility to enhance heat recovery, on the tube-side. This has only minor contribution in decreasing the overall rate of fouling. The extent of thermal fouling in a particular exchanger can be represented by the fouling resistance, R_f, given by

$$R_f = \frac{1}{U} - \frac{1}{U_{cl}} \tag{1}$$

Here, U is the overall heat transfer coefficient and subscript cl refers to the clean state at that operating condition. The rate of thermal fouling has been presented in several forms, and also as functions of temperature and flow velocity [2]. Here, we use the fouling rate model described by Polley [3], where the rate of fouling on the crude side, $\frac{dR_f}{dt}$, is be given by

$$\frac{dR_f}{dt} = \frac{A}{h} exp\left(-\frac{E}{RT}\right) \tag{2}$$

A is a pre-exponential constant, h is the film transfer coefficient on the crude side, E is the activation energy and R is the gas constant. Note that equation (2) is a generalized form of the fouling model and can be used for either the tube- or the shell- side, whichever side the crude is present.

It is of interest to perform analysis that involves reduction in the rate of fouling based on equation (2). This could involve major piping changes on-site, such as swapping the crude-stream from the tube-side to the shell-side. The reason for this retrofit is to enhance the film transfer coefficient on the crude-side to reduce the rate of fouling (equation (2)). The retrofit change could be performed when the exchanger is taken off-line for cleaning.

Consider E11 is taken off-line for cleaning and replaced back with the crude on the tube-side (following a major piping change on-site). The heat duty profile is compared before and after the retrofit in Figure 4. First thing to note is the increase in the heat duty as the unit is returned in the clean state. Also the net heat duty contribution is much higher even after considering fouling (i.e. after the retrofit, the heat duty drops from 14 MW to 5 MW in contrast from dropping from 4.5 MW to 0.7 MW in the original case study).

(i)

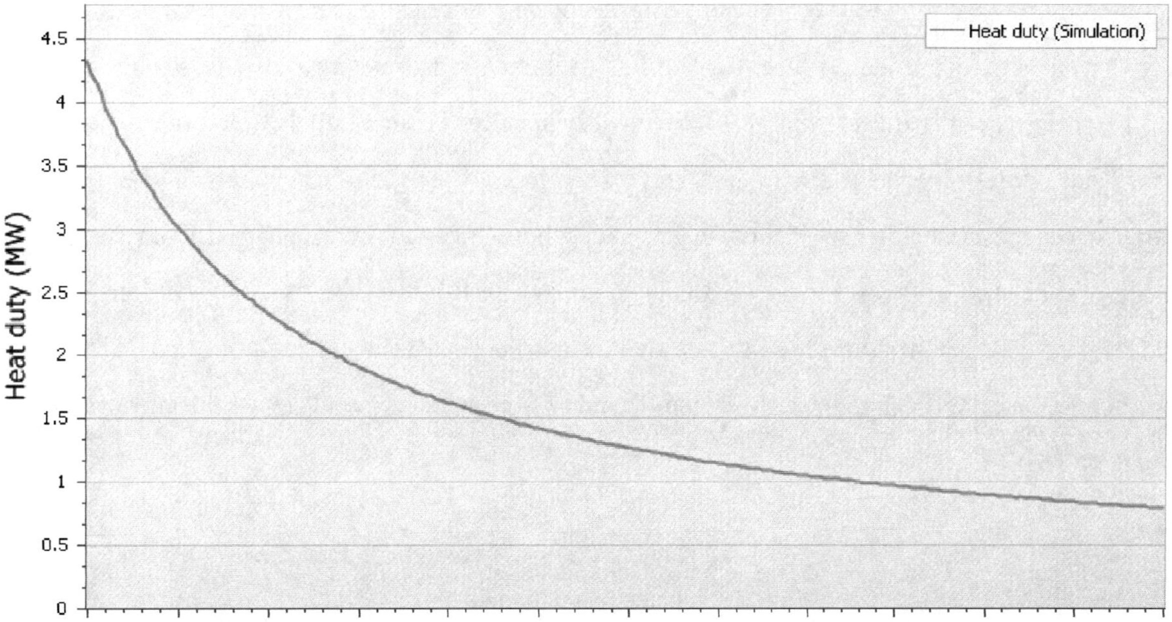

(ii)

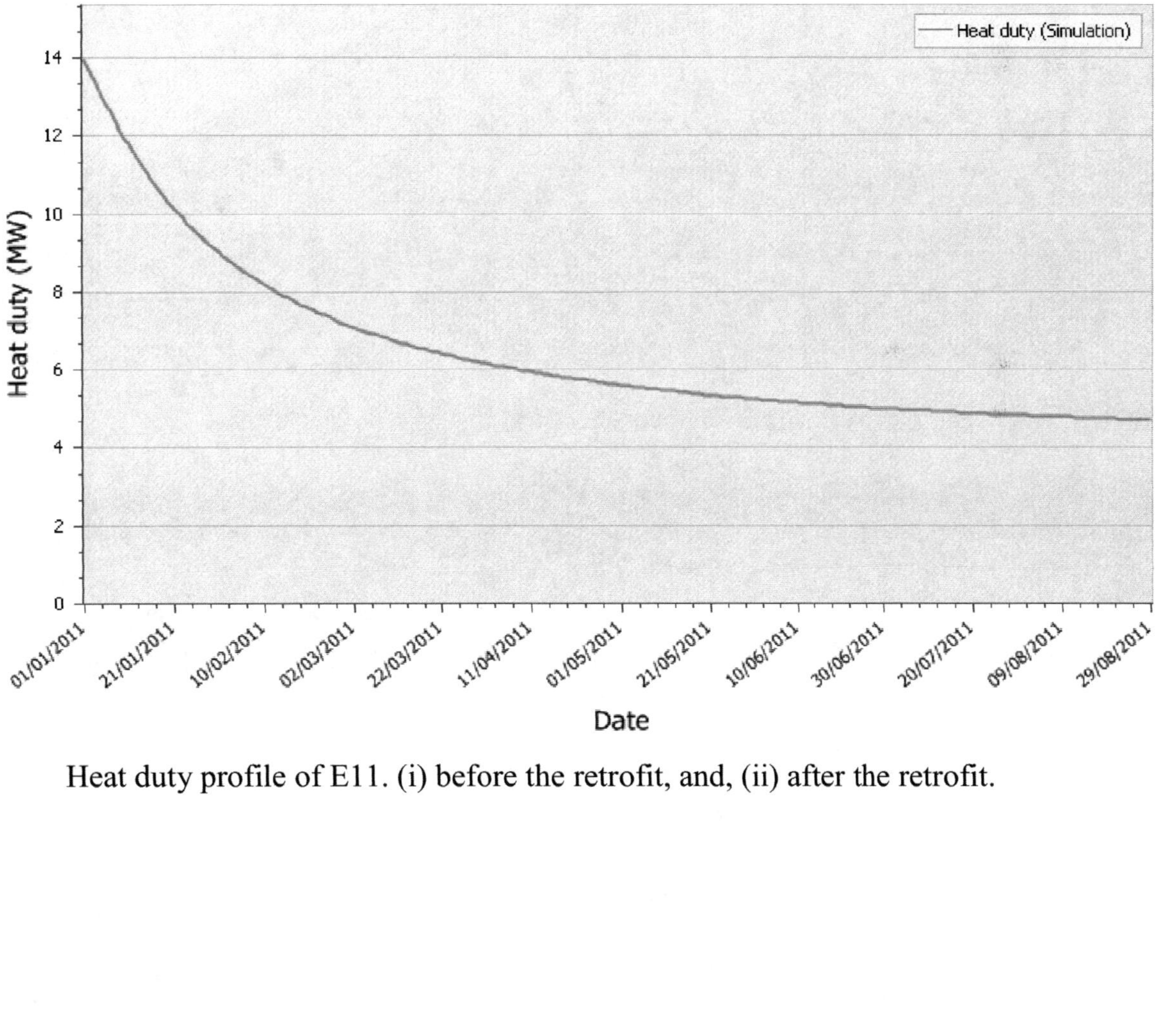

Figure 4: Heat duty profile of E11. (i) before the retrofit, and, (ii) after the retrofit.

Network analyses have an important role in assessing the impact of a retrofit towards the overall plant performance. Figure 5 is the full preheat train showing how the stream across E7, 8, 9, 10 and 11 interact with the exchangers before the desalter. Enhanced heat recovery in E11 would also mean that there will be reduced heat recovery from the exchangers before the desalter. This is a scenario that is described in Polley *et al.* [4], where the heat recovery path play a major role in the whole preheat train performance. In this case study the flow bypass across the three exchangers before the desalter could be manipulated to achieve a desired desalter inlet temperature. Note that the desalter operates at a specified temperature band and its control is vital in maintaining the preheat train operation [5].

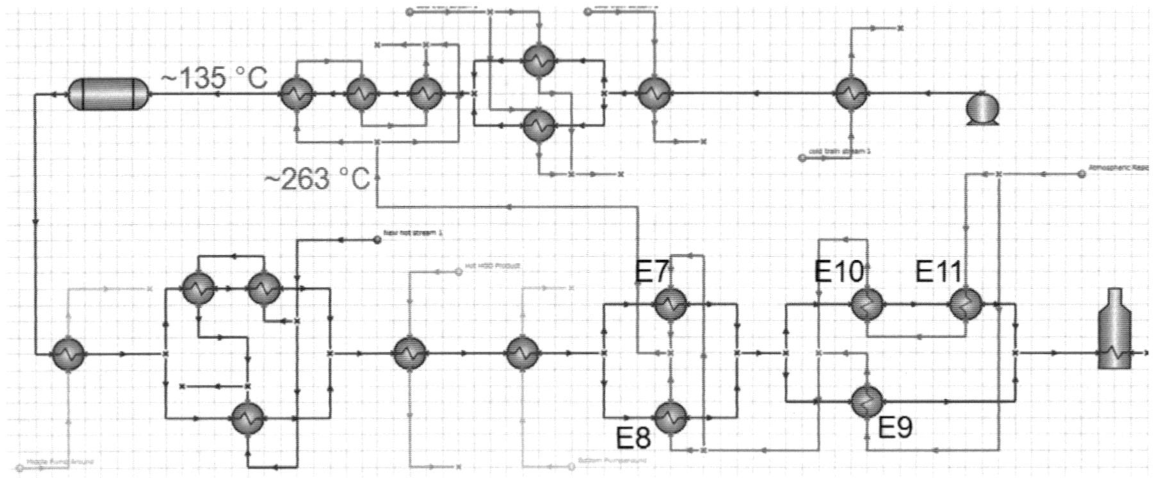

Figure 5: Schematic of the complete preheat train.

Conclusions

A combination of EXPRESS and SmartPM has been successfully used in this study to assess how a retrofit would be able to enhance heat recovery, reduce rate of fouling and also to evaluate the impact of the retrofit on preheat train operation.

Nomenclatures

Roman

A	pre-exponential factor, s^{-1}
h	film transfer coefficient on the crude side, W m^{-2} K^{-1}
R	gas constant, J mol^{-1} K^{-1}
R_f	fouling resistance, m^2K W^{-1}
T	temperature, K
U	overall heat transfer coefficient, W m^{-2} K^{-1}
U_{cl}	clean overall heat transfer coefficient, W m^{-2} K^{-1}

References

[1] Ishiyama E.M., Pugh S.J., Pollety G.T. and Wilson D.I. (2012). Industrial experience in handling cleaning of crude refinery preheat trains, *AIChE spring meeting*, Paper 31e, Huston, Texas.

[2] Knudsen J.G. and Ebert W.A. (1999). The determination of threshold fouling curve for a crude oil. In Bott, T.R. (ed), *Understanding heat exchanger fouling and its mitigation*, Begell house inc., Italy, 265-272.

[3] Polley, G.T. (2010). Review for the development of models for the prediction of fouling rates in exchangers heating crude oil, 11th annual international conference in Petroleum Phase Behaviour and Fouling, Jersey City, June 13-17.

[4] Polley G.T., Tamakloe E., Picon Nunez M., Ishiyama E.M. and Wilson D.I. (2011). Applying thermo-hydraulic simulation and heat exchanger analysis to the retrofit of heat recovery` systems, 12th UK heat transfer conference, Leeds, UK

[5] Ishiyama E.M., Paterson W.R., Wilson D.I., Heins, A.V. and Spinelli L. (2010). Scheduling cleaning in a crude oil preheat train subject to fouling: incorporating desalter control, *Applied Thermal Engineering*, 30(13), 1852-1862.

(31b) Tutorial: The Chemistry of Refinery Fouling

Irv Wiehe, Soluble Solutions, Gladstone, NJ

The chemistry of refinery fouling is well enough understood to mitigate nearly all fouling. The key information required are the location of the foulant, elemental analysis of the foulant, and identification of precursors in the oil flowing through the fouled unit. Most organic fouling is caused by insoluble asphaltenes: on mixing oils, on cooling after conversion, or during conversion (coke). One exception is the polymerization of conjugated olefins after thermal conversion, such as in a fractionator after a coker. Inorganic fouling is most commonly caused by iron sulfide that is the result of corrosion or by sea salts that pass through the desalter. Once the cause of fouling is determined, multiple mitigation solutions can be determined by tracing the foulant precursors to the source.

Design of Compabloc Exchangers to Mitigate Refinery Fouling

E. Tamakloe & G.T.Polley

Dept. of Chemical Engineering, University of Guanajuato

*Email - Gtpolley@aol.com

Welded plate heat exchangers (such as Compabloc^{TM} units offered by Alfa Laval) are finding wide application in the oil and process industries. The flow arrangements used in these exchangers can be complex; for instance, the authors know of one case in which a hot stream flowed through twelve passes whilst being matched against a cold stream that flowed through just 3 passes. It is also common to find that the physical properties of the fluids being handled exhibit quite large changes between inlet and outlet of the exchanger. This means that the individual film heat transfer coefficients are different for each individual pass. In this paper the authors describe how the performance of such exchangers can be analysed.

An important potential application of such exchangers is improved heat recovery at the hot end of crude oil pre-heat trains. Analysis of monitoring data obtained from an operational unit has shown that under conditions of high shear the fouling in these exchangers is very much lower than that occurring in conventional shell-and-tube heat exchangers. However, welded plate exchangers have higher film heat transfer coefficients than conventional units. This can lead to higher wall temperatures. Therefore under some conditions fouling in welded plate units can be greater than that in conventional units. The evaluation of the effects of exchanger geometry on fouling behaviour needs to be an integral part of the design process.

This paper presents two case studies. The first involves an exchanger in which heat from the residues leaving a vacuum distillation unit is used to heat crude oil. The physical properties of the residues are very sensitive to temperature and the heat transfer coefficient on the hot side of the unit is very much lower than that on the crude side.

The second case study involves a heat recovery match between a column pump around stream and crude oil. In this example the heat transfer coefficient on the hot side of the unit is high.

WELDED PLATE HEAT EXCHANGERS

Welded plate heat exchangers are consist of a stack of corrugated heat transfer plates enclosed within stiff rectangular plates bolted to support bars. A gap between these plates and the stack provides the header through which the fluids being heated or cooled are fed into the channels of the stack. A multi-pass flow arrangement is achieved by placing sealing plates within the headers. A typical construction is shown in Figure 1. In this example each of the streams flows across the stack four times. So, the unit has a 4-4 configuration.

The exchangers can be fabricated in a number of configurations. The greater the number of passes used the higher the velocity within the exchanger and the larger the film heat transfer coefficient.

The flow field within the individual channels is complex. However, it approximates a "cross-flow". Therefore, the pass arrangement can have a strong influence on the performance of an exchanger.

Permitted arrangements are those in which the ratio of passes is integer, integer plus one half and integer plus one third. The authors are aware of units having as many as twelve individual passes.

The individual arrangements should be viewed as consisting of "base configuration" repeated a number of times. For instance, the 4-4 configuration should be viewed as a 1-1 arrangement set up as four sections in series. The larger the number of series components the better is the use of temperature driving force.

A general algorithm for analysing the performance of a welded plate exchanger needs to consider the following base configurations: 1-1, 1-2, 1-3, 2-3, 2-5 and 3-4.

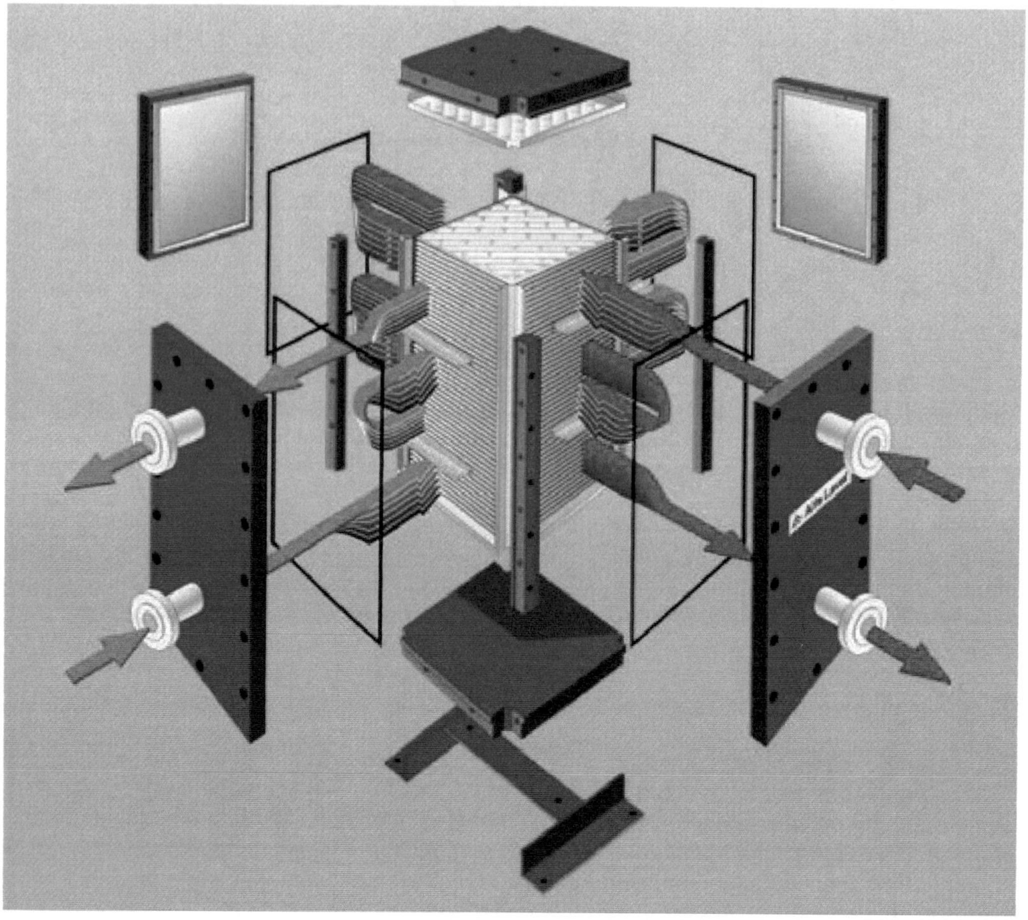

Figure 1. Welded Plate Exchanger Construction (Courtesy of Alfa Laval)

In many applications it is found that the physical properties of one or both of the fluids involved vary significantly between inlet and outlet conditions. This means that the film heat transfer coefficients vary across from pass to pass.

Chemical reaction fouling is an important factor when crude oil is being heated. It is now generally accepted that such fouling increases with increased wall temperature but for fixed wall temperature is reduced by increased fluid shear. Variation of both wall temperature and fluid shear stress (due to variation in fluid viscosity and deposit thickness) across the unit adds complexity to performance analysis.

MODELLING BEHAVIOUR OF A BASE CONFIGURATION

The behaviour of a base configuration can be predicted using individual algorithms for the simulation of a heat exchanger network. For instance, the structure of a 1-2 unit (Figure 2) is equivalent to the exchanger network shown in Figure 3.

The procedure requires knowledge of the overall heat transfer coefficient within each of the sections. These values are dependent upon local physical properties and local fouling resistance (which is time dependent).

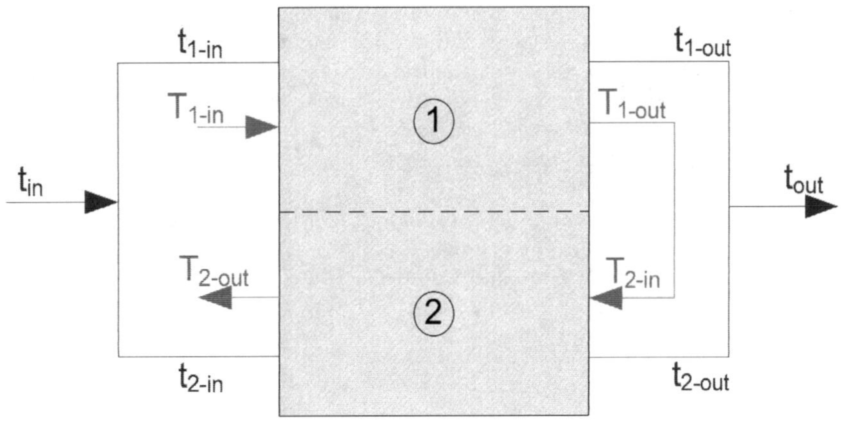

Figure 2. The 1-2 Base Configuration

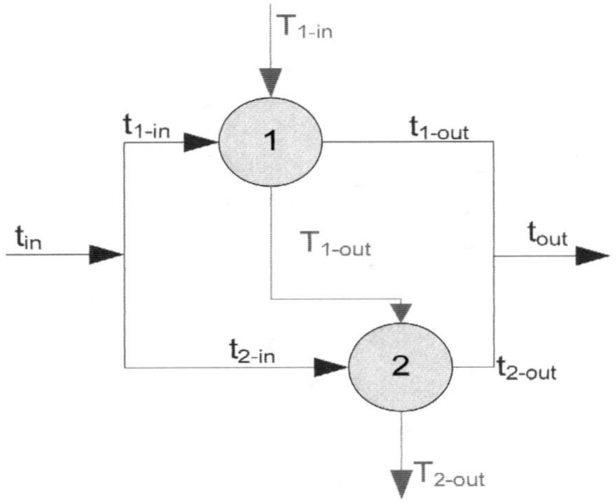

Figure 3. Configuration defined as individual exchangers.

DEALING WITH VARIABLE PROPERTIES

Consider the arrangement illustrated in Figure 3 once more. Since physical properties vary with temperature the heat transfer coefficients in section 1 differ from those in section 2. This variation has an effect upon the temperatures at which the streams leave the individual sections.

At the outset of the design calculation the only know temperatures are t_{in} ($t_{1\text{-in}}$ and $t_{2\text{-in}}$ equate with t_{in}) and $T_{1\text{-in}}$. Initial values for the other temperatures can be determined by assuming constant physical properties (based on the inlet temperatures). The variation of physical properties across each section and the resulting overall heat transfer coefficient for each section can then be determined. The true outlet temperatures can then be determined using an iterative procedure. For instance, for the arrangement shown in Figure 3 the performance of section 1 is determined first. $T_{1\text{-out}}$. is assumed. The resulting value of $t_{1\text{-out}}$ is determined from a heat balance. Performance of the section is determined and the value of $T_{1\text{-out}}$ is calculated. Assumed and calculated values are compared and the assumed value adjusted until the two values equate.

Attention can then be directed at section 2. Here the iteration uses temperature $T_{2\text{-out}}$.

Unfortunately, because the streams flow through each section in a cross-flow manner some approximation has to be introduced into the calculation of the performance of a given section. Three approaches are possible. In the first approach, heat transfer coefficients at the ends of the section are determined and a mean value assumed to apply for the section. The second approach is to calculate physical properties at the end of each section and use mean values in the calculation of the heat transfer coefficients. The third approach is to calculate the "caloric temperatures" for each of the streams flowing through the section and use physical properties calculated at these temperatures in the determination of the heat transfer coefficients.

The determination of caloric temperatures is described by Kern [1950]. They are based on the work of Colburn [1933] who analysed of the performance of a counter-flow exchanger that had a linear variation in the overall heat transfer coefficient over its length. Subsequent studies indicated that the approach could be extended to multi-pass heat exchangers without large loss in accuracy. The caloric temperature approach has been used in this work.

OVERALL ALGORITHM

Consider an exchanger that has two base configurations in series (2-2, 2-4,2-6 etc.) as shown schematically in Figure 4.

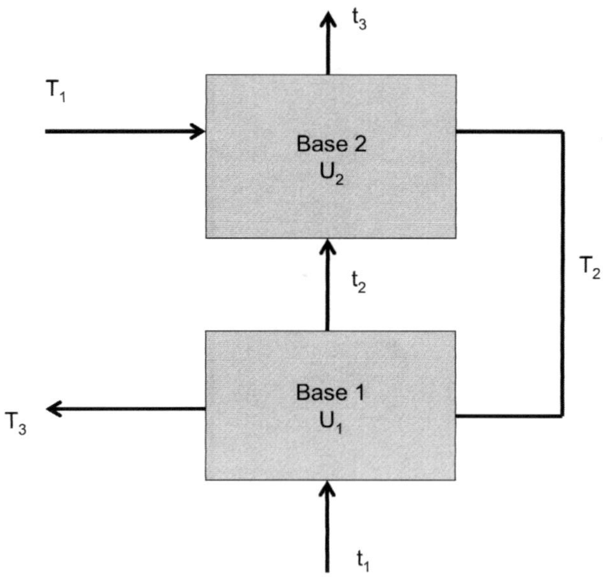

Figure 4. Exchanger with two base configurations in series

The only known temperatures are T_1 and t_1. Again the other temperatures can be determined using an iterative approach. In this case the variable used in the iteration is the hot stream that leaves section 2 (T_2). Knowledge of this temperature allows calculation of the performance of section 1. This yields the cold stream inlet temperature to section 2 thereby allowing the performance of section 2 to be calculated. Temperature T_2 is then adjusted until assumed and calculated values coincide.

The approach is easily extended to cover exchangers involving higher number of base configurations in series.

FOULING IN WELDED PLATE EXCHANGERS

Several models are available for the prediction of fouling rates in exchangers heating crude oil. One major difference in the models is assumptions made regarding the influence of fluid shear stress on fouling rate. Welded plate units can operate at significantly higher shear stresses than achieved in shell-and-tube heat exchangers. Therefore the analysis of data from these units could provide valuable insight into how fouling should be modelled. Unfortunately, the amount of plant monitoring data currently available is sparse.

In Figures 5 and 6 we see how one fouling model (Polley et al, 2011) fits monitoring data from one operational unit. For the operating period shown in Figure 5 the shear stresses within the exchanger were relatively large (around 70 Pa). In the second period the shear stresses were much lower (around 18 Pa).

It can be seen that at high shear stresses the fouling rate is very low. However, at low shear stress the fouling rate is high. In fact at the low shear stress the fouling rate is higher than would be expected in a shell-and-tube heat exchanger operating at this shear stress.

(Note: 18 Pa is about the maximum shear stress that can be achieved in a shell-and-tube unit).

The reason the fouling in the welded plate unit exceeds what could be expected in a shell-and-tube exchanger is that the compact unit operates at a higher wall temperature. In a shell-and-tube unit handling crude oil it is common to find that the heat transfer coefficient for crude flowing through the tubes is significantly higher than that for the hot stream flowing through the tube bundle. With a compact unit the coefficient for the hot stream will be similar, if not even higher, than that of the crude stream with the result that the compact unit operates at higher wall temperature.

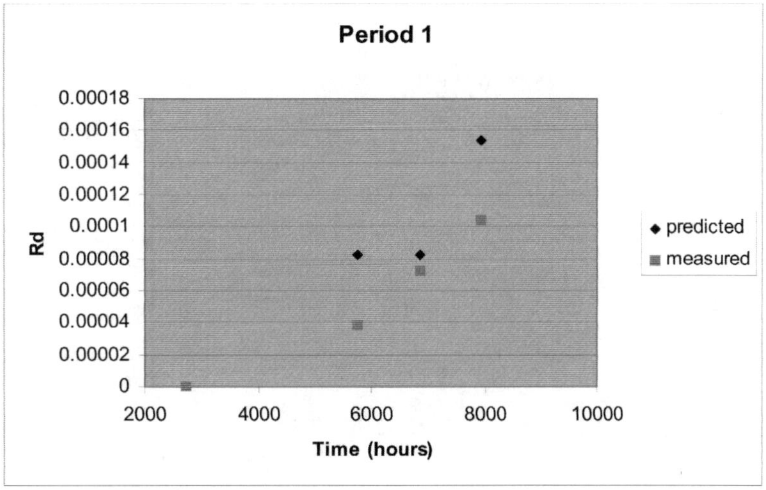

Figure 5. Fouling in welded plate exchanger at high shear stress

Given the algorithms developed in this study it is possible to set pass arrangements that result in better control of wall temperatures and fouling rates.

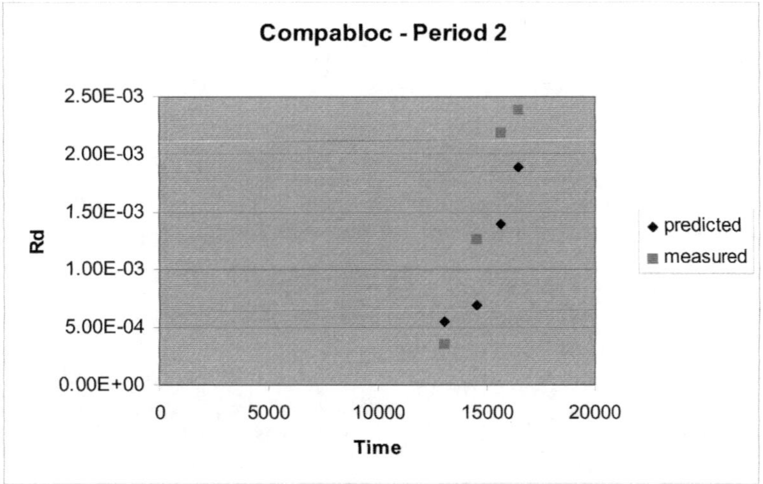

Figure 6. Fouling in welded plate exchanger at low shear stress

EXAMPLE APPLICATION 1

A welded plate heat exchanger is to be designed for use on a split stream positioned at the hottest part of a pre-heat train. The hot stream flowing through the exchanger is residue from a neighbouring vacuum column. The flows and inlet temperatures of the two streams are given in Table 1.

Stream	Inlet Temperature C	Flow kg/s
Crude Oil	235	28
Residue	320	18

Table 1. Example Application: Exchanger Duty

The minimum exchanger outlet temperature that is to be achieved is 270 C. At temperatures below this value it is predicted that the crude will enter the fired heater as a liquid rather than a saturated two-phase mixture. Under these conditions the fired heater will experience high fouling in the convective section leading to loss of fuel efficiency (Polley et al, 2011).

The physical properties of the two streams are given in Table 2.

Property	Crude at 200 C	Crude at 300 C	Residue at 250 C	Residue at 320 C
Density kg/m^3	800	760	891	835
Viscosity cps	0.6	0.5	8	2
Th,Conductivity W/mK	0.07	0.071	0.15	0.14
Heat capacity kJ/kgK	2.62	2.73	2.43	2.7

Table 2. Physical Properties

Thermal Behaviour

It was found that a unit having 240 plates evenly distributed between 6 cold side passes and 6 hot side passes and operating in a clean condition provided a crude outlet temperature of 278.8 C. This is well above the minimum value of 270 C. In the absence of a fouling analysis this design could be accepted.

When the fouling behaviour of the unit is analysed it is found that the rates were high. This is not surprising for the wall shear stress generated on the crude side is just 19.2 Pa (similar to that in period 2 of the monitored unit reported above).

The fouling results in a rapid decline in the performance of the unit such that the crude outlet temperature reaches the minimum allowable value of 270 C after just 80 days of operation. If left un-cleaned the outlet temperature would drop to 263 C after 200 days operation. (Which equates with the behaviour that could be considered to be the best achievable using a conventional shell-and-tube heat exchanger on this duty).

Now let's consider alternative designs. The shear stress on the crude side of the unit can be increased by increasing the number of passes on this side of the unit. Consequently, the next configuration examined had 9 crude side passes and 6 hot side passes. When clean this unit provided a crude outlet temperature of 279.7 C (nearly one degree higher than the original configuration). The shear stress had increased to 41.8 Pa. Consequently, the fouling rate was reduced and it was predicted that the outlet temperature of 270 C was encountered after 200 days operation.

The wall shear stress in this unit is still below what could be achieved. So, a configuration having 12 crude side passes was examined. This unit provided a shear stress of 72.5 Pa (similar to that in period 1 of the monitored unit reported above). This unit provided an outlet temperature of 280.4 C when clean and 275 C after 300 days operation.

The final configuration considered was a 12-12 unit. The advantage of this configuration was that it could be constructed as two 6-6 units operating in series. Such an arrangement would provide efficient operation where two operating scenarios are needed (such as was the case with the monitored unit described above). The outlet temperature on start-up was 282.7 C. After 300 days operation this had fallen to 277 C.

Hydraulic Behaviour

When evaluating the hydraulic behaviour of a unit in which fouling is occurring the effect of fouling on pressure drop has to be evaluated. Pressure drop models have been reviewed by Yeap et al (2004). In this current study we use a smooth core model where it is assumed that at any position along the flow path the deposits are uniformly distributed across the exchanger surface. The thickness of the deposit does vary along the flow path as local temperature and shear conditions change (this calculation is incorporated into the overall model). With a smooth core model the effect of fouling on pressure drop is solely one of flow restriction. The behaviour predicted fro the 12/12 configuration is shown in Figure 7.

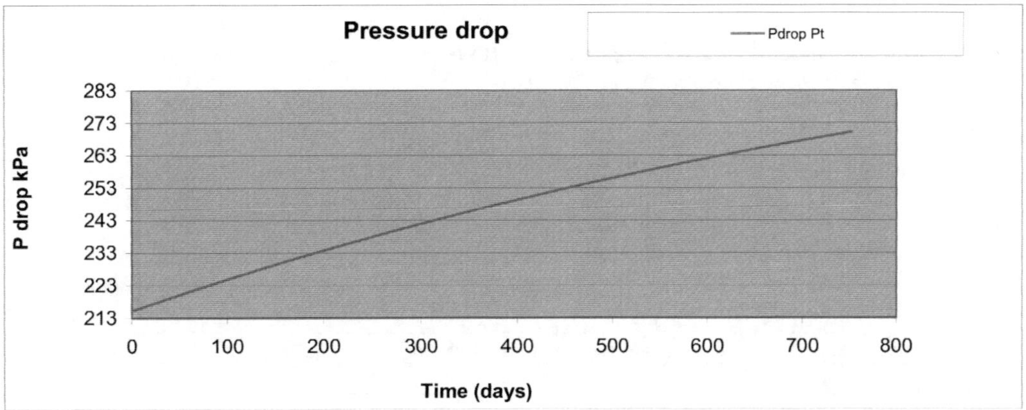

Figure 7. Pressure Drop During Operation the 12/12 Unit

With 12 passes on the crude side of the unit the pressure drop is high even when the exchanger is in a clean condition.

The behaviour predicted for the 9/6 configuration is shown in Figure 8.

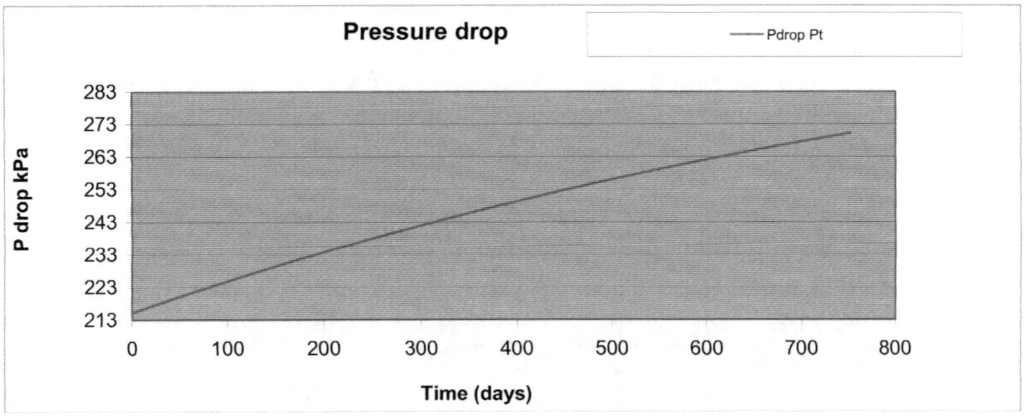

Figure 8. Pressure Drop During Operation the 9/6 Unit

Variation of Behaviour over the Unit

Finally, we examine how the behaviour varies across the whole of the unit for both the 12/12 and the 9/6 configurations. Variations in overall heat transfer coefficient, film temperatures (mean of wall and bulk temperature, used in fouling model), fouling resistance and wall shear stress are listed in Table 3 for the exchangers in a clean state.

Section	Mean OHTC	Mean R_D	Mean Film Temp.	Mean Shear Stress Pa	Mean diameter
1 – 12/12	1135	0	238	72.5	0.01
6 – 12/12	1235	0	257	72.9	0.01
12-12/12	1380	0	293	73.6	0.01
1 - 9/6	840	0	242	41.8	0.01
6 – 9/6	896	0	256	42	0.01
12 – 9/6	1009	0	289	42.4	0.01

Table 3. Clean Condition

A 9-6 configuration is equivalent to 3(3-2) units in series. A 3-2 unit has four individual sections. There are twelve individual sections in the exchanger.

The differences in shear stress are due to changes in physical properties. The overall heat transfer coefficients change by 20% over the length of the exchanger. Increases in film temperature indicate that fouling rates will increase over the length of the exchanger.

Table 4 presents predictions of behaviour following 200 days operation (when the fall off in performance of the 9/6 configuration would require that the unit be cleaned).

Section	Mean OHTC	Mean R_D	Mean Film Temp.	Mean Shear Stress Pa	Mean diameter
1 – 12/12	869	0.00035	240	76	9.88e-3
6 – 12/12	868	0.00039	256	76.4	9.87e-3
12-12/12	770	0.00062	285	80.9	9.76e-3
1 - 9/6	506	0.00091	242	48	9.65e-3
6 – 9/6	480	0.00117	257	50.3	9.56e-3
12 – 9/6	402	0.00162	273	54.2	9.38e-3

Table 4. After 200 days Operation

The 9/6 configuration has a fouling level that is 2.6 times that of the 12/12 unit in both first and last sections.

Comparison of fouled and cleaned behaviour shows that whilst the shear stress in the last section of the 12/12 configuration has increased by 10% over the operating period, that of the 9/6 unit has increased by 28%.

On occasions there is benefit in accepting higher fouling in one location in order to reduce that in another. With a welded plate exchanger this could be achieved by having differing plate counts in the individual sections. For instance, a unit may provide 192 plates distributed through the first 8 sections and just 48 through the last 4 sections of a 9/6 unit. However, in this application the fouling rate in the first section is high and such action would not appear worthwhile.

Conclusion

A 9/6 configuration should be used on this duty. A unit based on clean performance alone (the 6/6 unit) would foul rapidly. A unit using a larger number of passes on the crude side would result in a relatively minor increase in heat recovery at the cost of a significant increase in pressure drop.

EXAMPLE APPLICATION 2

In our first example the hot stream heat transfer coefficient was significantly lower than that for the crude side. This will not be the case in every application.

Where shell-and-tube heat exchangers are used it is often found that the shell-side heat transfer coefficient is lower than the tube-side coefficient. Where crude oil is passing through the tubes a low heat transfer coefficient on the hot stream has advantage of reduced wall temperature. The use of a Compabloc exchanger on the same duty has a disadvantage in that both hot and cold heat transfer coefficients are enhanced. The result can be increased wall temperature that can result in higher fouling rate. It is therefore of particular importance that design is not undertaken on the basis of clean performance alone.

This second example involves a match at the hot end of the pre-heat train between a pump around stream and crude oil. The problem is specified in Table 5.

A design that provides adequate "clean performance" has 200 plates distributed between 4 hot and 4 cold side passes. In a clean state the crude outlet temperature 279.5 C, this is well above the target temperature of 260 C. The overall heat transfer coefficient is 1727 W/m^{2}K whilst the individual film heat transfer coefficients are 6362 W/m^{2}K for the pump around stream and 2833 W/m^{2}K for the crude oil. The hot side coefficient is not only very much higher than could be obtained on the shell-side of a conventional exchanger but is twice as high as that on the crude side. The result is high wall temperatures at both inlet (219 C) and outlet (295.7 C) of the exchanger. The wall shear stress on the crude side is also observed to be quite low at 23 kPa.

The consequence of these high temperatures is very rapid fouling with the crude outlet temperature falling rapidly (Figure 9).

	Crude Side		Hot Side
Flow	36.4	kg/s	64.5
Inlet Temperature	172	C	306
Target Temperature	260	C	
Density	700	kg/m^3	733
Viscosity	0.44	cps	0.12
Thermal Conductivity	0.1	W/mK	0.12
Heat Capacity	2610	J/kg.K	2820

Table 5. Example Application 2

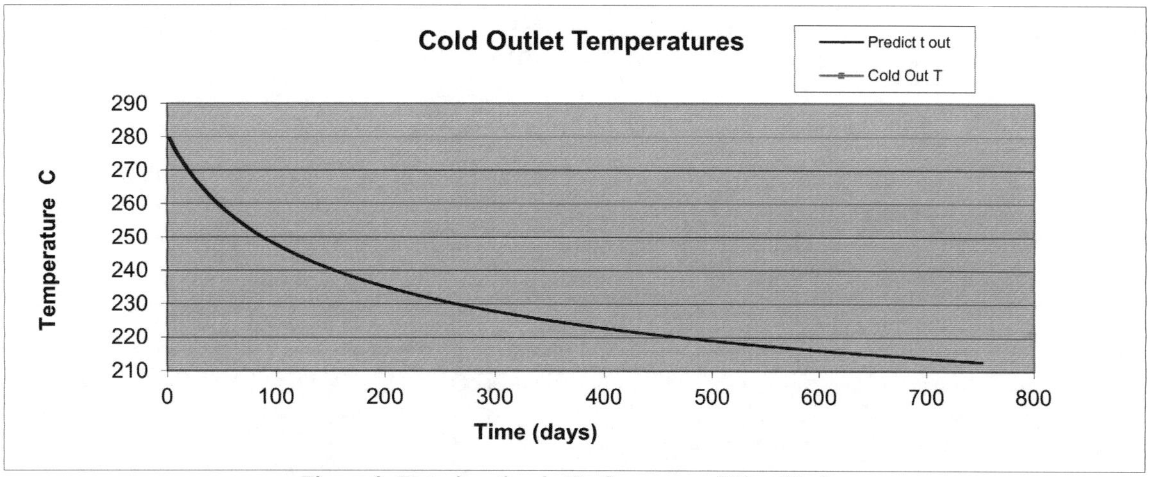

Figure 9. Deterioration in Performance of First Design

To reduce fouling rates we need to reduce wall temperatures. This can be achieved by changing the pass arrangements such that the hot side coefficient falls relative to the cold side coefficient. Rather than use 4 passes on the hot side we can use 2. Rather than use 4 passes on the cold side we can use 8. Such a design would result in a higher overall coefficient (of 1850) but more significantly the balance between hot and cold heat transfer coefficients is changed (now 4100 on hot side and 4400 on cold side). The wall temperatures fall to 205 C at inlet and 291 at outlet. However, the most significant factor is an increase in wall shear stress on the crude side of the exchanger from 23 kPa to 86 kPa. The time related performance of this unit is shown in Figure 10.

The increased wall shear leads, of course, to increased pressure drop. In a clean condition the pressure drop through the 4/4 is 22 kPa. This compares with 169 kPa for the 2/8 unit.

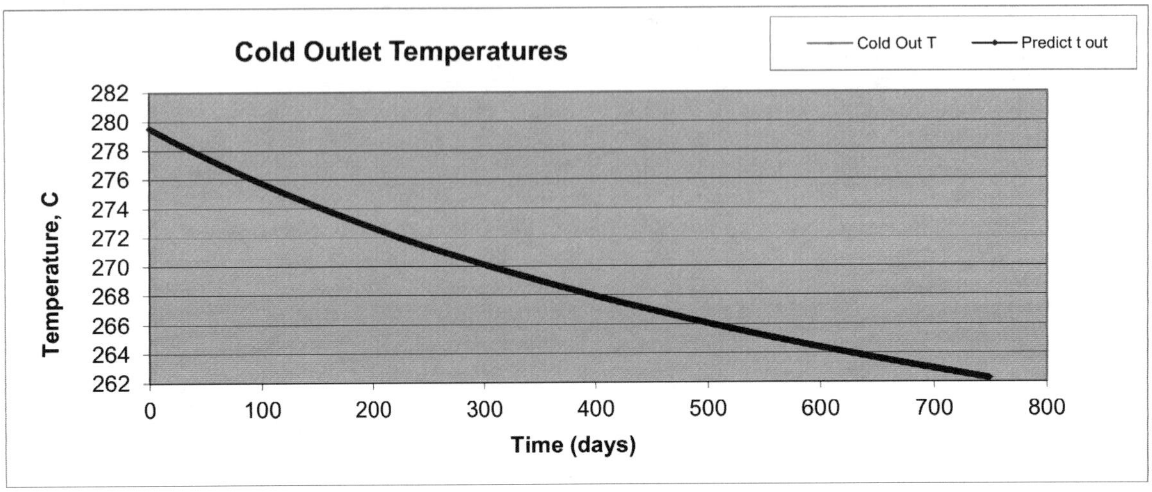

Figure 10. Performance of a 2/8 Design

"RULE OF THUMB"

The above applications and the operating data all indicate that the wall shear stress on crude side of the unit should be at least 70 kPa in order to avoid high fouling rates when welded plate exchangers are being used at the hottest part of a pre-heat train.

CLEANING OF WELDED PLATE EXCHANGERS

Welded plate exchangers cannot be cleaning mechanically. Chemical cleaning has to be used. However, there are specialist companies that can be employed develop chemicals that are specific to the nature of the deposit formed within the exchanger. It has been found that by combining chemical treatment with ultra-sonics these exchangers can be cleaned efficiently (McCartney & Smith, 2010).

CONCLUSIONS

In order to get good performance out of a welded plate heat exchanger it is often necessary to use multiple passes. A range of possible configurations is possible. Some of these configurations are complex. The behaviour of multi-pass units has now been modelled. The modelling has included consideration of the variation of physical properties over the length of the exchanger.

The shear stresses developed in this type of exchanger can be much greater than those possible in shell-and-tube heat exchangers. This has led to interest in the application of this type of exchanger in crude oil pre-heat trains. Examination of monitoring data has indicated that fouling models currently used to predict fouling rates in shell-and-tube heat exchangers may also apply to welded plate exchangers. Consequently, such a model has been incorporated into the modelling of welded plate exchanger behaviour.

Examination of example applications has demonstrated the importance of considering fouling behaviour during the design of this type of heat exchanger. Geometry should not be chosen on the basis of performance under clean conditions even when this appears to be a large over-performance.

The wall shear stress on the crude side of a welded plate exchanger should be at least 70 kPa. Excessive fouling can occur at lower shear stresses.

REFERENCES

Colburn A. P. 1933, "Mean Temperature difference and heat transfer coefficient in liquid heat exchangers", *Ind. Eng. Chem.* Vol **25** 873-877

Kern D. Q. 1950, Process Heat Transfer, McGraw-Hill, New York.

Polley G.T., Tamakloe E. & Picon Nunez M. "Models for chemical reaction fouling", paper presented at AIChE Spring Meeting, Chicago, March, 2011

McCartney T. & Smith S. "Advancements in large scale application of ultrasonic cleaning in removal of refinery fouling", 11th Annual Intnl. Conf. on Petroleum Phase Behaviour and Fouling, June, 2010, Jersey City

Yeap B.L., Wilson D.I., Polley G.T. & Pugh S.J. "Mitigation of crude oil refinery heat exchanger fouling through retrofits based on thermo-hydraulic fouling models", Trans.I.Chem.E.,2004,82A,53-71

(31d) Monitoring of Crude Preheat Train Fouling Using Aspen HYSYS Petroleum Refining

Dhaval Dave, Research and Development, Aspen Technologies Inc, Burlington, MA

Fouling reduces the amount of heat being transferred and hence rest of heat needs to be supplied by furnace and hence it increases processing cost and amount of CO_2 being generated. Furnace can take up the additional load to certain point, but once furnace is bottlenecked feed throughput needs to be reduced to keep the unit running. On mitigation through online cleaning may require expensive chemicals and shutdown of preheat train costs about 45 to 90 k$/ train. The preheat train is present in most of the refinery units such as FCC, Delayed coker, Hydrotreaters and so on. It is estimated that financial losses due to heat exchanger fouling in UK is 450 to 750 M$/year. There is considerable amount of literature material is available on the mitigation techniques and cleaning chemicals. But whatever the mitigation technique, fouling monitoring is indispensible in ensuring that that the cost of fouling is known so that a suitable mitigation technique can be applied and on time.

In this presentation, we describe a framework, where Aspen HYSYS can be used in conjunction with

Aspen Shell & Tube Heat Exchangers and Aspen Simulation Workbook to monitor the preheat train heat exchanger performance.

INDUSTRIAL EXPERIENCE IN HANDLING CLEANING OF CRUDE REFINERY PREHEAT TRAINS

** Edward M. Ishiyama, IHS ESDU, London, UK*
Simon J. Pugh, IHS ESDU, London, UK
Graham T. Polley, University of Guanajuato, Mexico
D. Ian Wilson, University of Cambridge, UK

*Corresponding author: edward.ishiyama@ihs.com; Tel: +44 (0) 20 3159 3300.

Abstract

Oil refineries are a crucial part of any national economy: in the UK for instance, there are nine major refineries, processing over 1.8 million barrels of crude oil per day (Watson and Vandervell, 2006 [1]), consuming energy at the rate of gigawatts (~ 7.9 GW: Marsh-Patrick, 2006 [2]). Heat exchangers (HEXs) play a major role on refineries in saving energy. Distillation is the main consumer of energy on a refinery, and networks of HEXs are connected together in a preheat train to maximize distillation efficiency by limiting the duty on furnaces. These HEXs all suffer from fouling. Based on the world crude oil consumption of ~ 88 MM bbl per day (in 2009 – 2010), around 36 MM tonnes per year of extra CO_2 is estimated to be emitted due to fouling (IHS ESDU, 2011, [3]). In this paper, a case study preheat train is introduced with real plant monitoring data. The case study is successfully used to illustrate the importance of data reconciliation in formulating cleaning schedules with the use of a novel software, SmartPM, available from IHS ESDU.

Introduction

Preheat trains are networks of heat exchangers on a refinery that help to raise the temperature of the crude oil before it enters the atmospheric distillation column for fractional separation. These preheat trains play a major role as they recover up to 70% of the energy that is required to heat the oil to its distillation temperature. The performance of the preheat train reduces over time due to the phenomenon known as fouling, which is the formation of unwanted deposits on heat transfer surfaces. The deposition mechanism can be caused by several species and depends on the local conditions. Handling fouling is a major challenge, and to-date, total elimination of fouling has not been achieved. Hence, cleaning of refinery exchangers is often considered a necessary operating strategy. Since the original work done by Georgiadis and Papageorgiou [4] and Smaïli *et al.* [5], optimizing scheduling in a practical perspective has recieved broader attention [6]. This paper focuses on the operation of crude refinery preheat trains subject to fouling with three aims; firstly, to provide guidance in industrial data reconciliation; secondly, to introduce methodologies in using historical data to predict and improve future performances of preheat trains; and, thirdly, introducing the user friendly software package, SmartPM.

Our discussion starts with data reconciliation, as without a proper understanding of the industrial data, it is impossible to identify the actual degree of fouling in each individual exchanger. In our study, data reconciliation is the process of fitting plant data to a specific model. This is common in many studies, as reported by Ishiyama *et al.* [6] and Bagajewicz [7]. The novelty in the current study lies in incorporating preheat train fouling behaviours in a process plant, and with that knowledge to guide the reconciliation steps. All analysis in this work has been carried out through the novel software package, SmartPM, developed and marketed by IHS ESDU, London, based on research conducted at the University of Cambridge.

A case study preheat train, based on an existing refinery is introduced. The SmartPM package carries out the following steps in sequence:

1. Reconciliation of plant data
2. Use of reconciled data for determining fouling model parameters
3. Use of fouling models to perform future predictions
4. Identification of opportunities for process improvement
5. Resource management through scheduling of cleaning within planned financial targets.

Industrial preheat train

An operating preheat train (PHT) has been selected for this study; the refinery remains anonymous due to confidentiality. The PHT is drawn by SmartPM as Figure 1. For the purposes of this study, only the hot end of the preheat train (downstream of the desalter) is considered. The train processes 100k to 130k bbl/day of crude oil and consist of 11 heat exchangers between the desalter and the furnace. The furnace inlet temperature at normal operation is ~ 275 °C. The preheat train includes tube side and shell side crude heating.

The effect of fouling in the preheat train is reflected by the inlet temperature to the furnace, or 'Coil Inlet Temperature' (CIT, Figure 1). The operating temperature drops by 25 °C over an 8 month period (Figure 2a), over which time the throughput is maintained around 130k bbl/day (Figure 2b), apart from two periods of interruption (with some cleaning) in the 7th month. Geometrical and operational details of the heat exchangers are listed in Table 1, while the thermo-physical properties of the streams are summarized in Table 2.

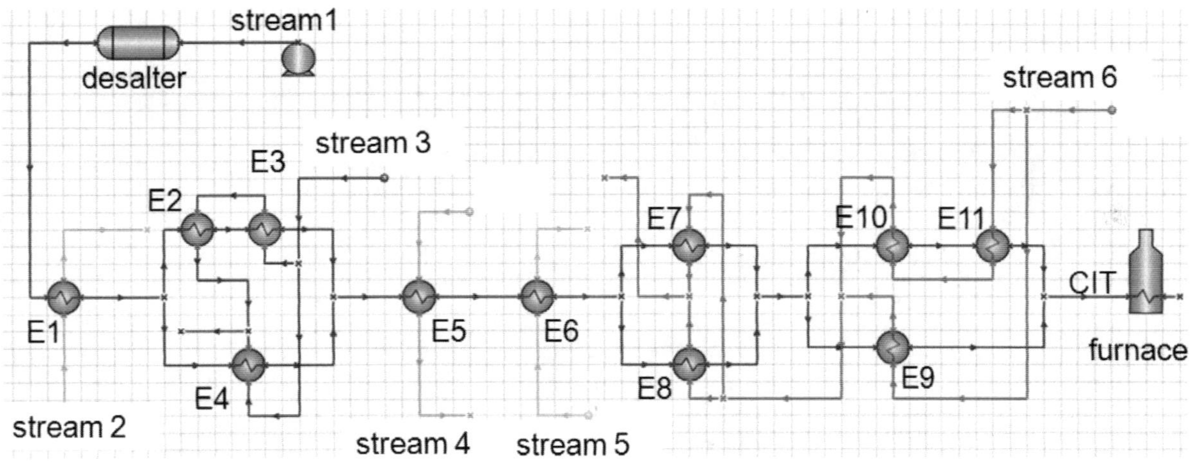

Figure 1: Schematic of a section of the case study preheat train. The circles represents shell-and-tube exchangers, CIT, denotes the furnace coil inlet temperature.

(a)

(b)

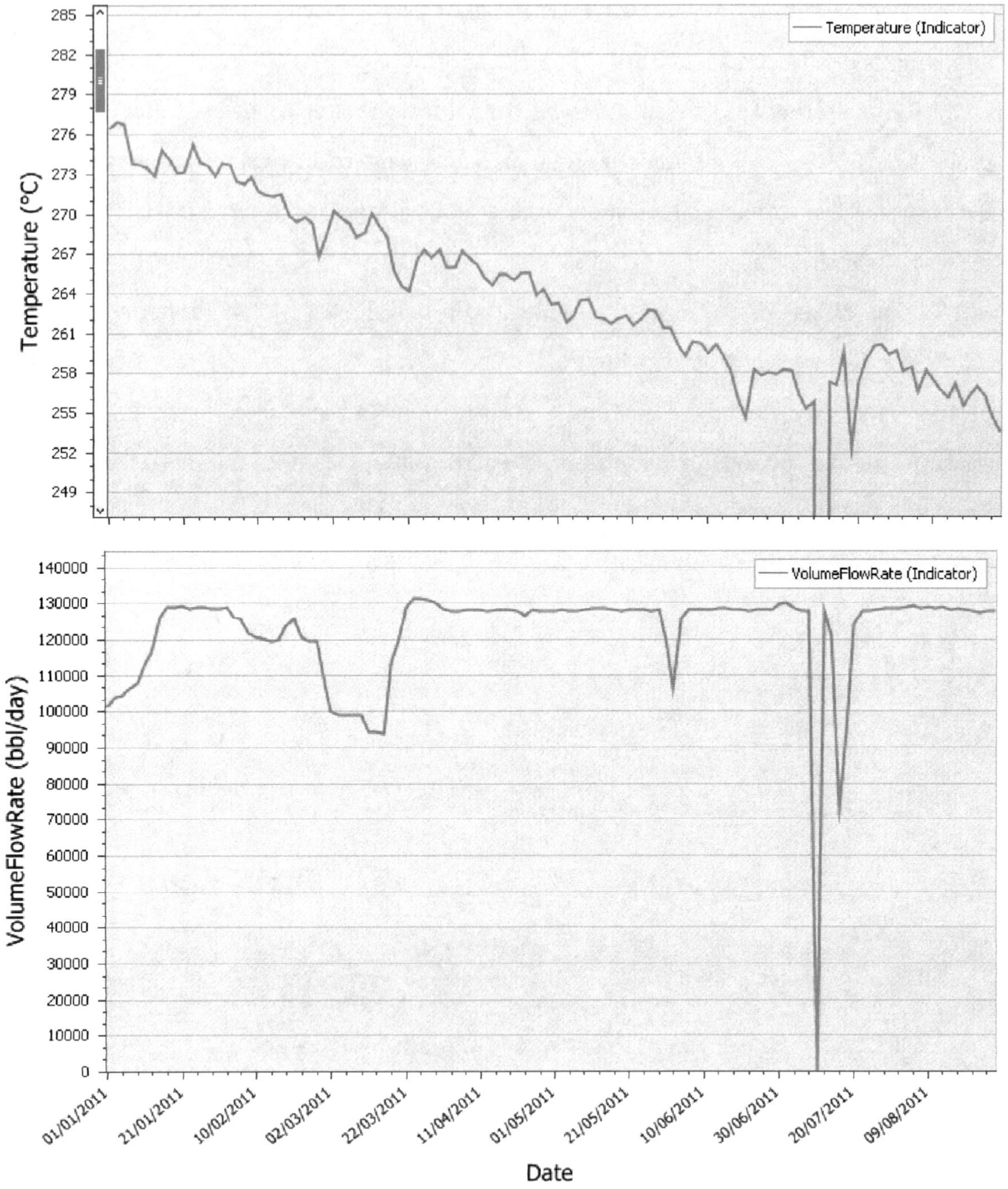

Figure 2: Impact of fouling on PHT operations. (a) Coil inlet temperature, and, (b) crude feed rate.

Table 1: Geometrical details of heat exchangers

Heat exchanger	No. of tube passes	Tube length (m)	Tube internal diameter (m)	Tube external diameter (m)	Total number of tubes	Shell diameter (m)	Tube layout (°)	Number of baffles	Baffle cut
E1	2	7.3	0.0148	0.0190	2000	1.5	90	12	0.2
E2	2	6.1	0.0199	0.0254	1450	1.5	45	8	0.192
E3	2	6.1	0.0199	0.0254	1450	1.5	45	8	0.192
E4	2	6.1	0.0199	0.0254	1450	1.5	45	8	0.192
E5	2	4	0.0199	0.0254	660	1.2	45	14	0.163
E6	2	4.3	0.0199	0.0254	696	1.1	90	10	0.176
E7	8	6	0.0166	0.0190	1500	1.4	30	13	0.25
E8	6	6.1	0.0148	0.0190	2680	1.4	90	20	0.186
E9	8	6	0.0166	0.0190	1500	1.4	30	13	0.25
E10	6	6.1	0.0148	0.0190	2680	1.4	90	16	0.186
E11	6	6.1	0.0148	0.0190	2680	1.4	90	16	0.186

Table 2: Stream thermo-physical properties

	Density, ρ (kg m^{-3})	Thermal conductivity, λ (W m^{-1} K^{-1})	Specific heat capacity, C_p (J kg^{-1} K^{-1})	Viscosity, μ (Pas)
Relations	$\rho = a\ (T\text{-}273) + b$	$\lambda = a\ (T\text{-}273) + b$	$C_p = a\ (T\text{-}273) + b$	$\mu = a\ exp(b\ /T)$
Units	*a* kg m^{-3} K^{-1}	*a* W m^{-1} K^{-2}	*a* J kg^{-1} K^{-2}	*a* Pa s
	b kg m^{-3}	*b* W m^{-1} K^{-1}	*b* J kg^{-1} K^{-1}	*b* K
	a, b	*a, b*	*a, b*	*a, b*
stream 1	-0.444, 874	0, 0.11	7.78, 622.22	7.17E-06, 2608.29
stream 2	-0.655, 976	-0.0012, 0.115	2.76, 2406.9	2.08E-05, 2566.31
stream 3	-1.091, 956	-0.0015, 0.125	5.45, 1518	9.93E-07, 2901.99
stream 4	-0.706, 915	-0.0024, 0.142	5.882, 1205	3.62E-06, 2720.10
stream 5	-1.000, 990	-0.001, 0.11	3.33, 1967	3.98E-06, 2774.99
stream 6	-1.500, 1200	-0.003, 0.213	11.67, -967	1.44E-10, 9097.72

Indicator measurements and their relationship to resource management

It is important to note the availability of indicator measurements throughout the train. Figure 3 is an enlarged reproduction of Figure 1 with indicator measurements (tags) added. The figure is split into two sections to clarify the location of the indicator measurement tags.

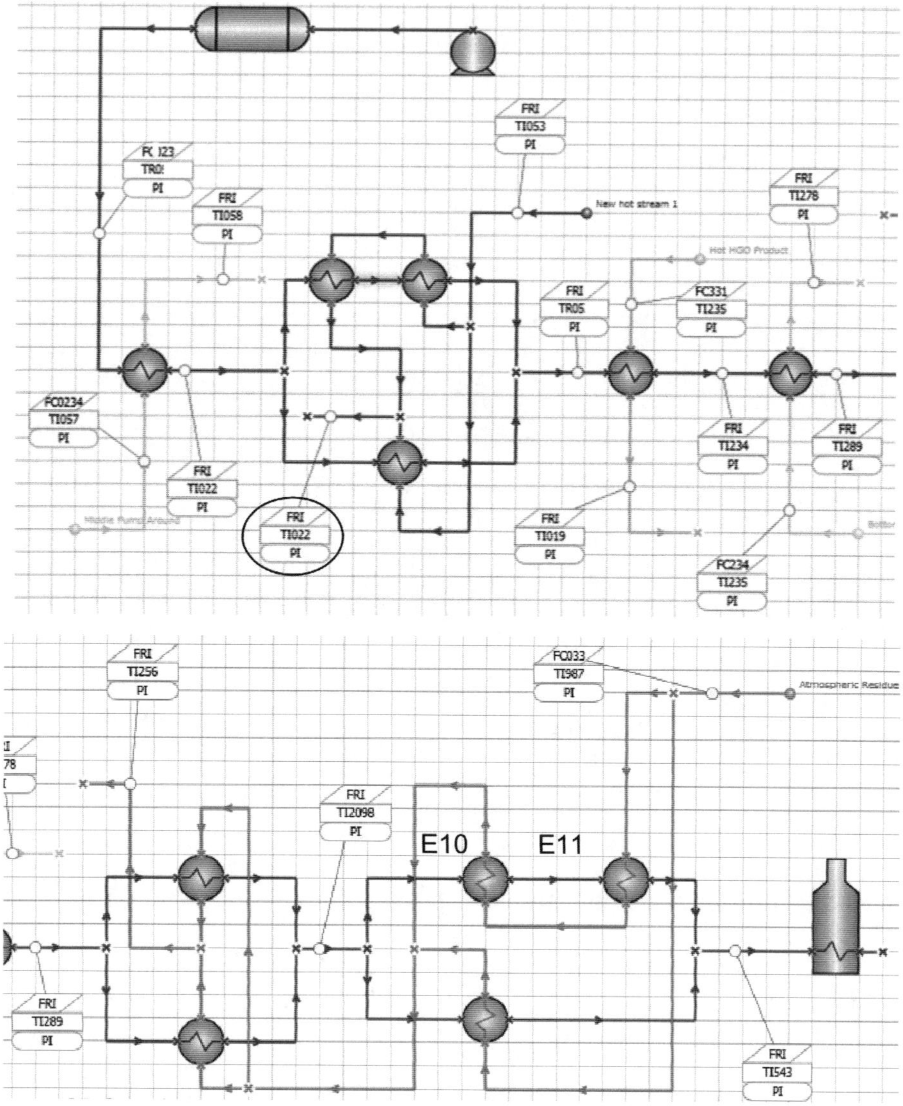

Figure 3: Case study preheat train with locations where temperature/ flow rates are measured. A temperature measurement is presented with TI followed by a numerical label and a flow measurement is presented by FC. PI indicates that gauge pressure measurements could be utilized at these points if available. The circled tag,TI022, represents the indicator discussed in Figure 6(b).

It is evident in Figure 3, as with many industrial set-ups, that not all stream temperatures or flow rates are available. This could be due to inadequate instrumentation or inability to make measurements due to physical constraints (such as difficulty in accessing a particular pipe).

In this paper we are considering an example where the measurement could not be made due to physical constraints – *i.e.* the temperatures between shells in series (*e.g.* E10 and E11 in Figure 3). A common modelling practice is to lump the exchangers together and generate an overall estimate. This is satisfactory in performing an overall thermal analysis of the network but would be a problem when planning cleaning. In particular, there is no means of knowing the degree of fouling in individual exchangers when they are lumped together. It might well be the case that both of these exchangers would need to be taken offline together due to bypass arrangements, but it may equally be true that only one needs to be cleaned. The aim is to guide the operator in advance with decisions on allocating cleaning resources, such as whether to clean all exchangers in a unit or just one.

The extent of thermal fouling in a particular exchanger can be represented by the fouling resistance, R_f, given by

$$R_f = \frac{1}{U} - \frac{1}{U_{cl}} \quad (1)$$

Here, U is the overall heat transfer coefficient and subscript cl refers to the clean state at that operating condition. The rate of thermal fouling has been presented in several forms, and also as functions of temperature and flow velocity [8]. Here, we use the fouling rate model described by Polley [9], where the rate of fouling on the crude side, $\frac{dR_f}{dt}$, is be given by

$$\frac{dR_f}{dt} = \frac{A}{h} exp\left(-\frac{E}{RT}\right) \quad (2)$$

A is a pre-exponential constant, h is the film transfer coefficient on the crude side, E is the activation energy and R is the gas constant. Note that equation (2) is a generalized form of the fouling model and can be used for either the tube- or the shell- side, whichever side the crude is present. In case of E10 and E11, the distribution of fouling between the exchangers is estimated by the ratio of the initial fouling rates (assuming both exchangers are clean) based on equation (2). This methodology yields Figure 4 (a and b) for E10 and E11, respectively. A detailed explanation of the data reconciliation algorithm is given in Ishiyama *et al.* [10].

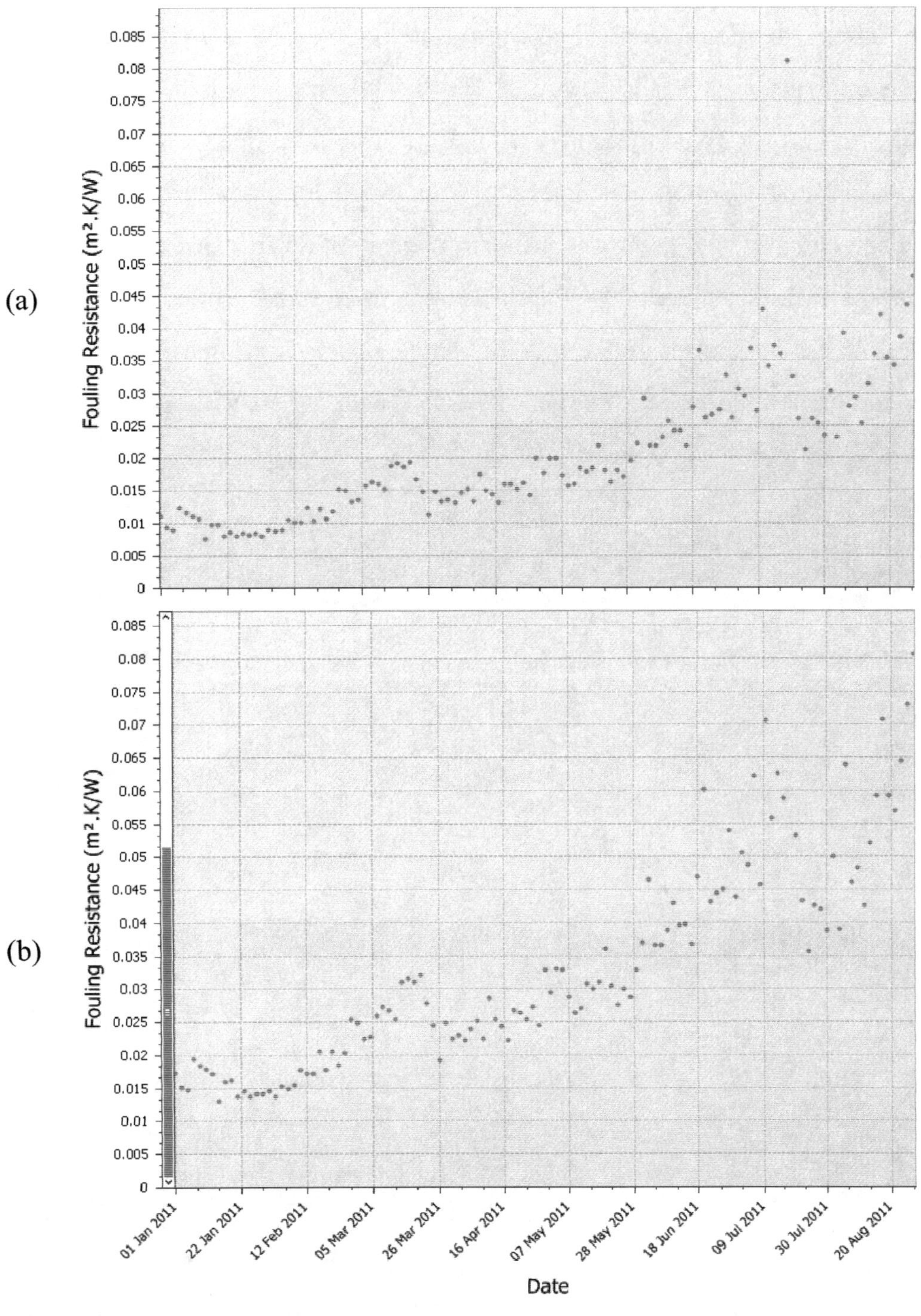

Figure 4: Fouling resistance profiles for exchangers, (a) E10, and (b), E11, respectively, estimated from reconciliation of plant data.

There are two common misunderstandings in handling industrial data. 1) the frequency of data recording and, 2) the condition at which the volumetric flow is made. The frequency of data measurement could vary from every second to every 2 days, *etc*. SmartPM allows the user to enter this information before performing a study so to avoid time scale errors at a later stage (section I in Figure 5). Daily measurements are usually adequate for the purposes of data reconciliation and fouling estimation. Secondly, for a given measured volumetric flow rate, mass flow is strongly dependent on the density, *i.e.* a strong function of temperature. SmartPM highlights this feature through a project management pane (Figure 5), allowing the user to choose either to input volumetric flow at standard conditions (i.e. the flow is converted by the measurement software to the value at 60 °F) or at actual conditions (i.e. values are reported at that specific temperature) - section marked as 'III' in Figure 5. SmartPM also has an option for the user to use a noise cancelling algorithm (section II in Figure 5).

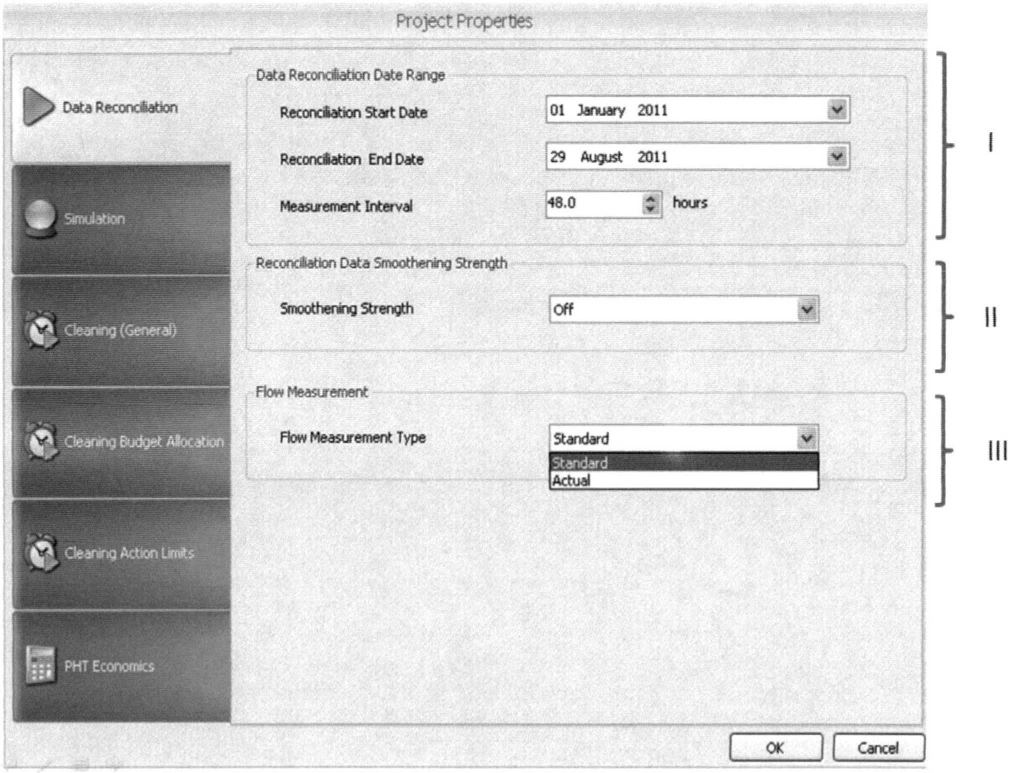

Figure 5: A snap shot of the SmartPM interface that is used to enter information in relation to the length of data reconciliation, frequency of measurement, condition at which the flow is measured.

Where do the uncertainties go?

SmartPM performs data reconciliation though fitting available measurements to a heat balance. It is important to note that there are uncertainties associated with physical properties, measurements (which can also be totally wrong), and in empirical heat transfer correlations. Given these uncertainties, it is unrealistic to expect the hot and cold stream to heat balance perfectly. So, the plant data usually requires some manipulation to close the heat balance, which is necessary for estimating fouling levels in the heat exchangers.

A single exchanger is used in this section to introduce how the uncertainties associated in a heat balance are handled in SmartPM. Consider an exchanger where both inlet and outlet temperatures and flows are measured. The following steps are performed in fitting the data to a heat balance for a single time step:

(1) Estimate R_f using prior behaviour, if available, or a 'best guess'.

(2) Using equation 1, calculate U.

(3) Calculate NTU and effectiveness, based on physical properties, geometry and U.

(4) Calculate outlet temperatures; refine the estimate of R_f until the difference between the measured and calculated cold stream outlet temperatures is below a specified error.

(5) Replace the measured hot stream outlet temperature with the calculated hot stream outlet temperature.

There are shortcomings in weighting all uncertainties on the hot stream outlet temperature: this is simply one way of handling the data. In certain instances, there could be an error in using empirical correlations to calculate film heat transfer coefficients, when specific geometric changes in the heat exchanger are made. In SmartPM equations are available for the film coefficients in a wide range of geometries, including flows through tubes with inserts and through helical baffles. There is also an option to enter heat transfer coefficient scaling factors to re-scale the empirical values to any preferred values, or even apply user-supplied values.

The hot stream outlet temperature (associated with E4, circled in Figure 3 (top section)) and CIT are presented in Figure 6. Note that for this particular stream, the hot stream temperature is varying from 200 °C to 220 °C. The difference between the reconciled and the measured data at maximum is around 5°C for this example. The offset in hot stream outlet temperature is, as discussed, carrying all the uncertainties associated with the heat balance.

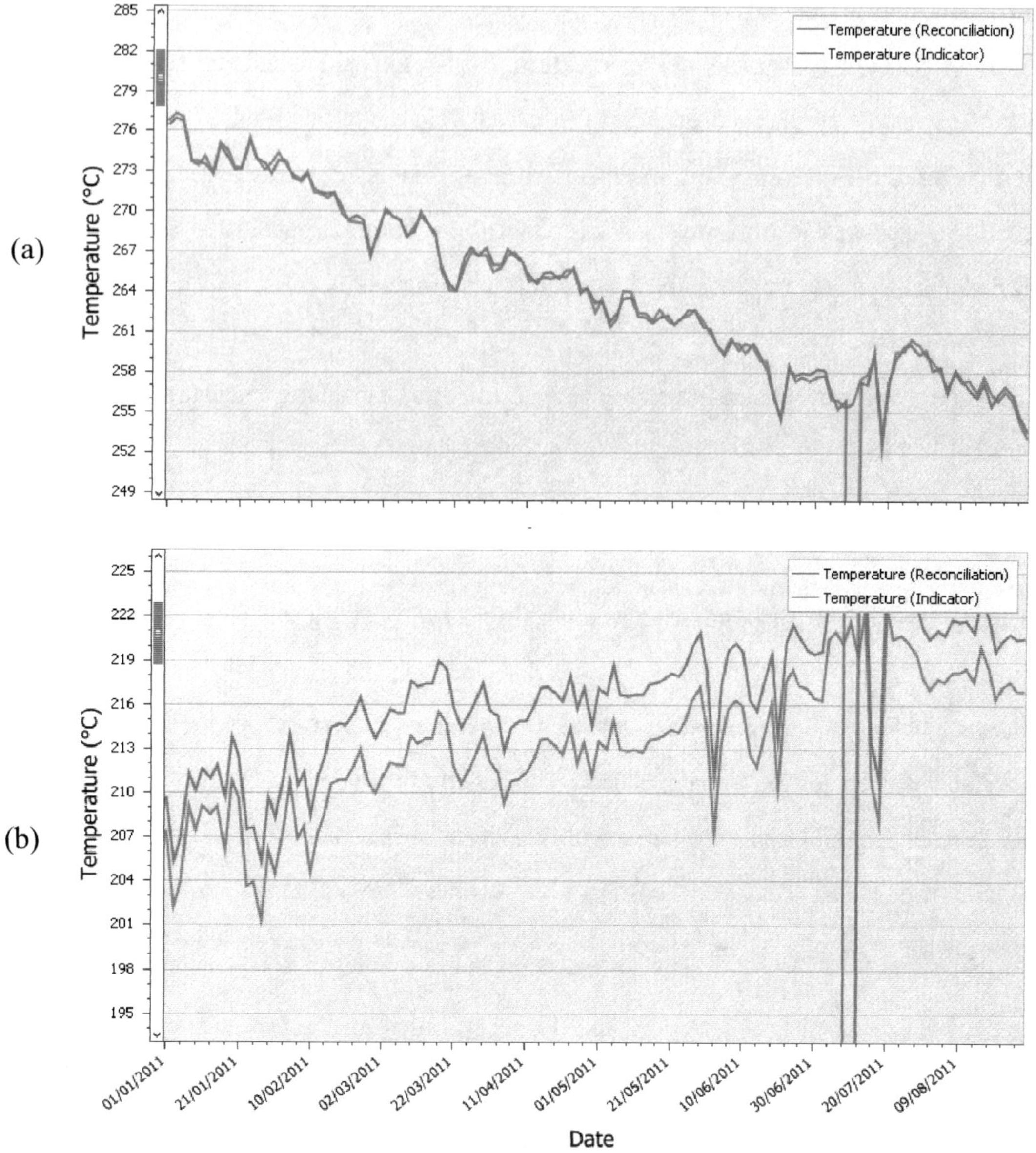

Figure 6: Measured and reconciled temperature over a nine month period, (a) CIT, and, (b) hot stream outlet temperature (based on the mixing of the hot stream outlets of E2 and E4, indicator circled in Figure 3).

Fouling model parameter determination

Once data reconciliation has been performed, SmartPM generates the individual R_f profiles for each exchanger. The slopes of R_f profiles, *i.e.* the fouling rate, are a strong function of operating temperatures and flow. SmartPM allows a very wide range of flow and performance parameters to be plotted; for example, Figure 7(a) shows both fouling resistance and Reynolds number (this is a strong function of flow) plotted together. This plot allows the user to identify regions of relatively stable operation and time-slice the fouling resistance data for each exchanger. This allows SmartPM to generate a matrix of operating periods. The 'Fouling Rate Profile" mode of SmartPM uses the corresponding fouling rates and operating data to evaluate parameters for fouling models such as that described in equation (2).

(a)

(b)

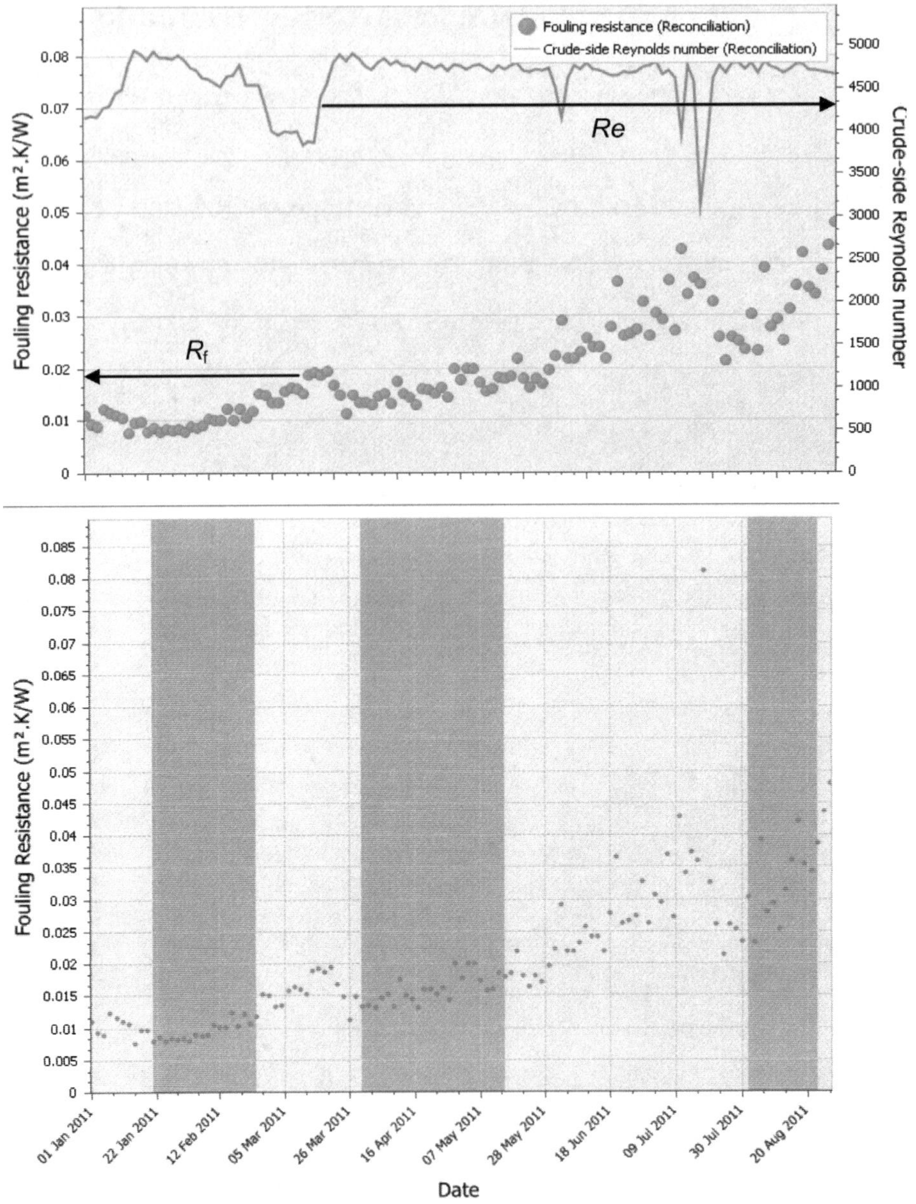

Figure 7: Exchanger E10 performance: (a) R_f (left hand y-axis); *Re* (right hand y-axis) and Reynolds number on the secondary y-axis, and, (b) time sliced fouling resistance profiles.

This time slicing facility also allows the operator to model the fouling performance in periods where particular crude blends are processed, hence allowing the a database of crude slate specific fouling model parameters to be built that can be used in simulation and cleaning calculations.

Use of dynamic fouling models to predict network performance

Dynamic fouling model parameters obtained from data reconciliation are then used in SmartPM to perform prediction analysis. Figure 8 shows a snapshot of the SmartPM interface window where properties for scheduling cleaning are entered. 'IIa' and 'IIb' are a table of data that would be generated once a fouling analysis is carried out as in Figure 7. As discussed under equation (2), a fouling rate model is necessary to perform a network prediction analysis. In Figure 8, III, three fouling models, namely, Ebert-Panchal model, Generalized Ebert-Panchal model and Asphaltene Precipitation model are currently available to be selected by the user. The forms of the models are detailed in Polley [9].

The use of a fouling model (in this case equation (2)) will then predict the exchanger performance, *e.g.* as in the fouling resistance profiles in Figure 9 and the drop in CIT in Figure 10; this example was done to show how close the prediction could be using 'constant' inlet conditions (i.e. constant inlet temperature and flows to the preheat train) to model the known period of operation. Recognizing that all input conditions actually fluctuated over the period, this is a useful rough check on the validity of the fouling model. The time-averaged inlet conditions for the six streams in the simulation studies are summarized in Table 3.

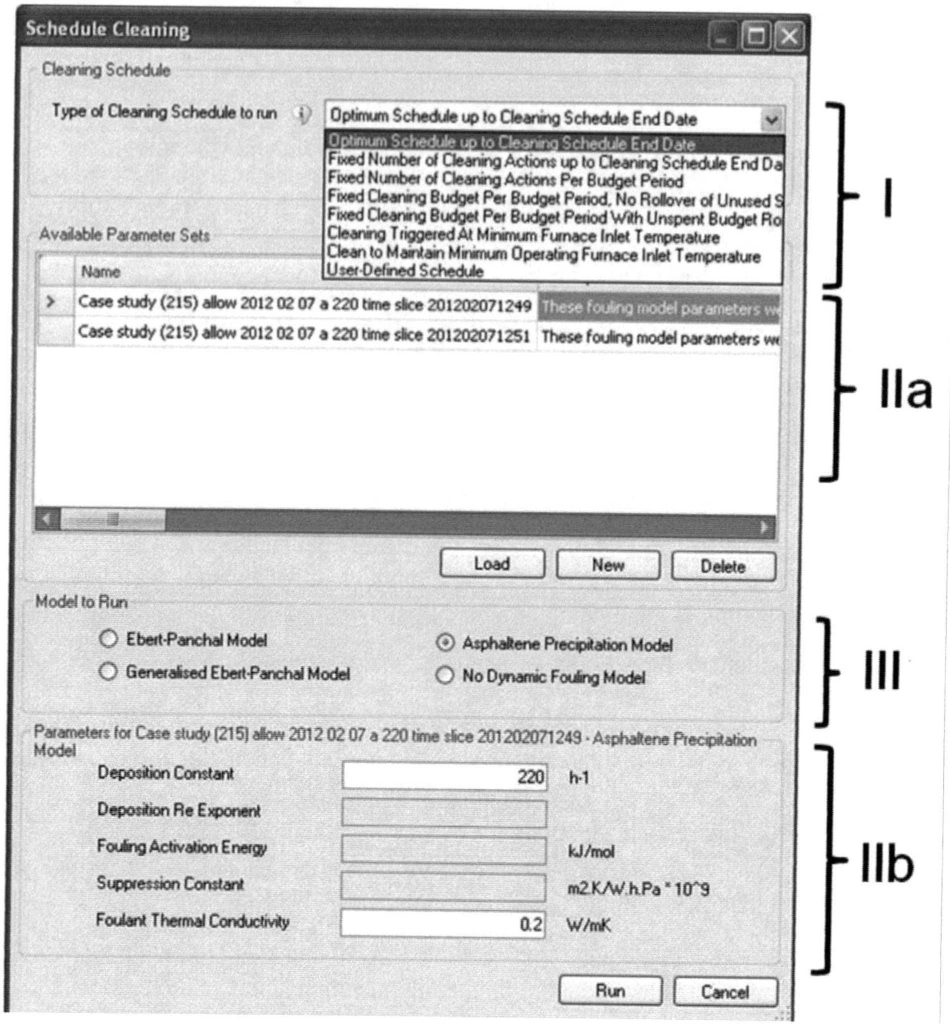

Figure 8: Snapshot of SmartPM scheduling cleaning panel. 'I' is a selection of the methods of cleaning schedules available, 'III' presents the choice of fouling model (Equation 2, is the Asphaltene Precipitation model). 'IIa' and 'IIb' provide parameters for 'III' generated by the data reconciliation process.

Table 3: Values of constant inlet conditions in simulation studies (stream numbers in Figure 1)

	Inlet temperature to the preheat train (°C)	**mass flow rate (kg/s)**
Crude stream 1	138	200
Hot stream 2	250	85
Hot stream 3	240	134.8
Hot stream 4	313	12
Hot stream 5	297	53
Hot stream 6	350	140

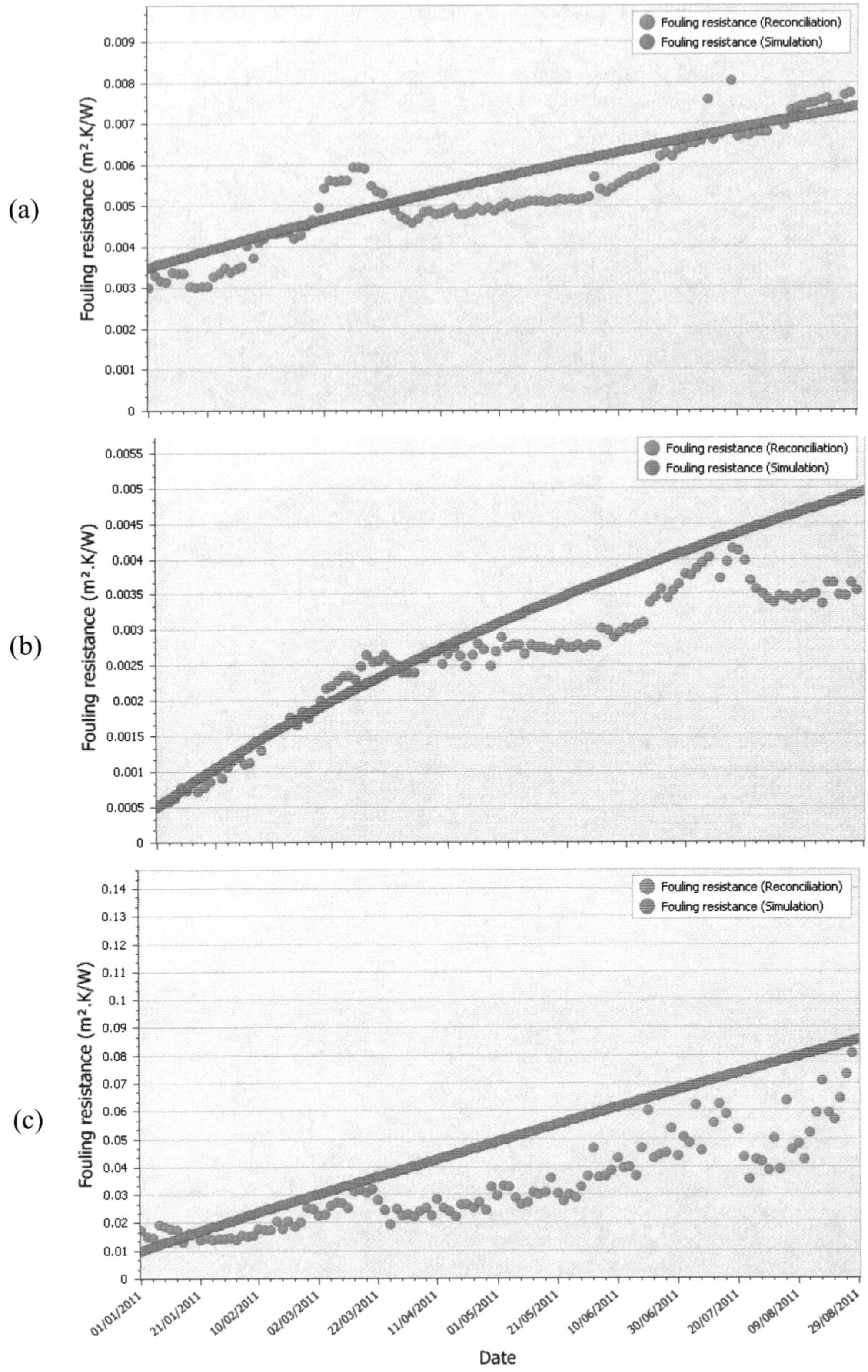

Figure 9: Comparison of predicted and reconciled R_f-time profiles using equation (2) with A of 220 h^{-1}, for exchangers (a) E1, (b) E6, and, (c) E10.

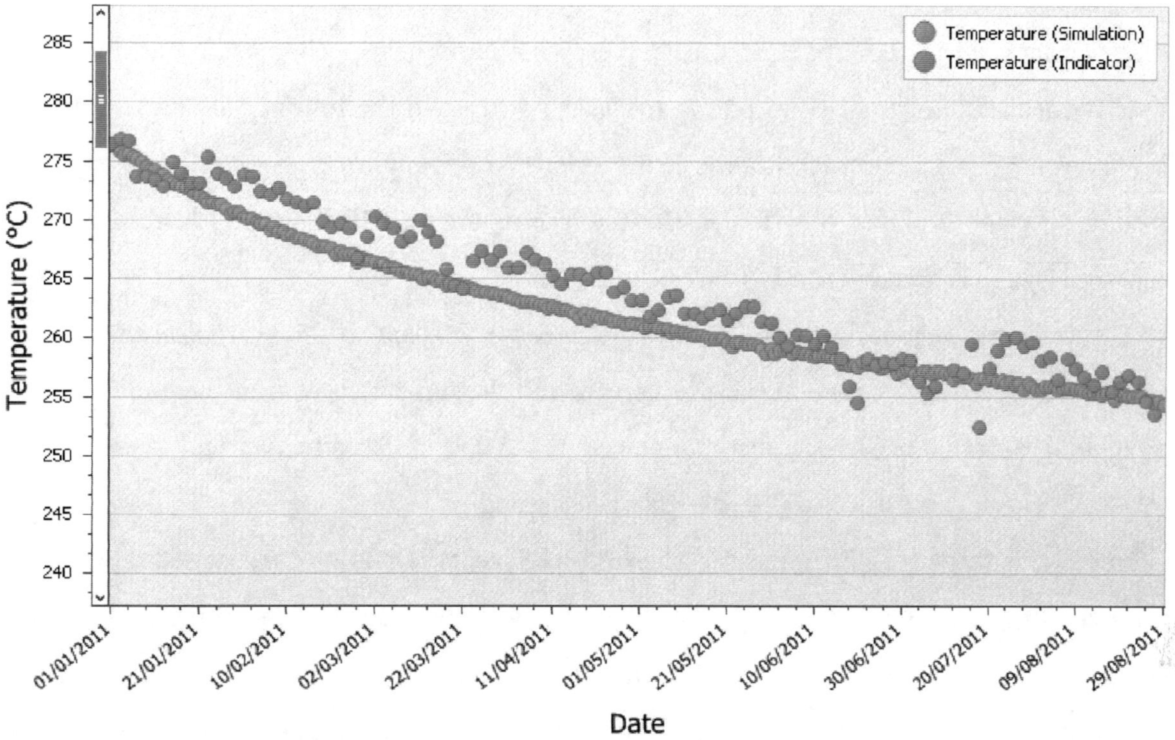

Figure 10: Measured and predicted CIT profile. The prediction is based on a network simulation with inlet conditions as given in Table 3 and with fouling model given by equation (2) with a model parameter, A, of 220 h^{-1}.

Scheduling cleaning actions

It is important to remember that, to decide which and heat exchanger to clean and when, both the capacity (heat duty or total contribution to the network) and the fouling rate have to be considered together. Figure 11 is a 3D representation from SmartPM of how much each exchanger contributes to the network duty and how fast they foul with time. For example, E10 is fouling faster than E9, but heat duty supplied by E9 is greater than E10.

To perform a scheduling calculation, it is necessary to specify the duration that the exchanger spends offline, the cost of cleaning the exchanger, and the expected R_f after cleaning (our case study used a cleaning duration of 7 days, cost of cleaning of 20,000 US$/exchanger and R_f after cleaning of 0 $m^2K\ W^{-1}$). An energy cost of 0.54 US$/kW day is assumed.

Here, schedules are generated for a period of 3 years. Details of the scheduling algorithm are given in Ishiyama *et al.* [11]. The degree of cleaning will vary, depending on the method of cleaning and also on the type of deposit present in the exchanger. The expected fouling resistances after cleaning could be extracted from historical plant data if a proper record of cleaning together with the monitoring data is available. The optimal schedule generated using SmartPM, is shown in Figure 12(a) with the resulting CIT profile (Figure 12 (b)).

All exchangers have been under operation before the cleaning schedule, which is formulated from 1st of January, 2011. Hence, E1 is cleaned at the beginning as it is identified to be beneficial to be initially clean due to its contribution to the network. It is not scheduled to be cleaned at a later stage, mainly as it is the coldest exchanger in the network and the rate of fouling is the least compared to the other exchangers. The most frequently cleaned is E11. This reflects the issue that E11 is one of the hottest exchangers in the network with one of the largest heat transfer areas. Figure 12(a) shows that E10 is cleaned only once, whereas E9 is cleaned 3 times. This is in agreement with the observation in Figure 11 on the heat duty profile; E10 has the lowest contribution to the network even though it is appearing to be fouling fastest. For the optimum schedule (Figure 12a) a total of 12 cleaning actions are performed over 3 years. SmartPM allows the user to consider a range of other options for cleaning scheduling based on maintenance budgets and availability of maintenance staff. Let us consider a case where the plant allows only 8 cleans per operating period due to resource limitations. Figure 13 shows the schedule. For this scenario, E10 is no longer cleaned and E11 is cleaned less often.

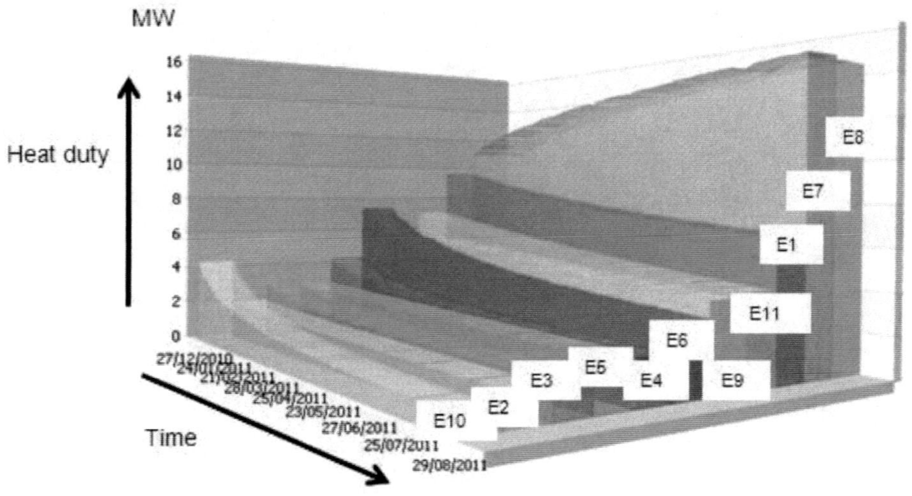

Figure 11: Heat duty profiles for each exchanger, in MW (performance without cleaning).

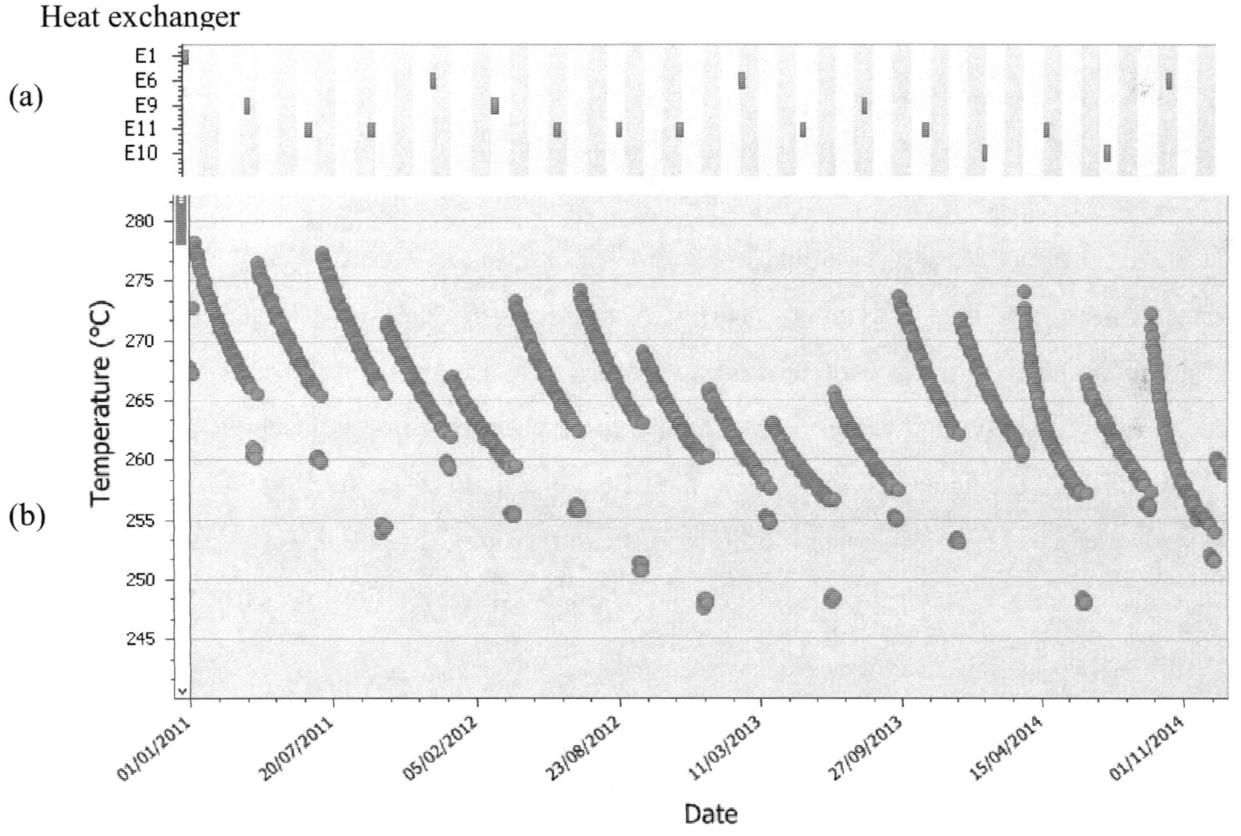

Figure 12: Optimal cleaning schedule for case study PHT: (a) schedule and (b) simulated CIT profile. Boxes in (a) indicate that the exchanger was cleaned. All exchangers are starting with an operating fouling resistance; E1 is identified to be cleaned at the beginning of the schedule.

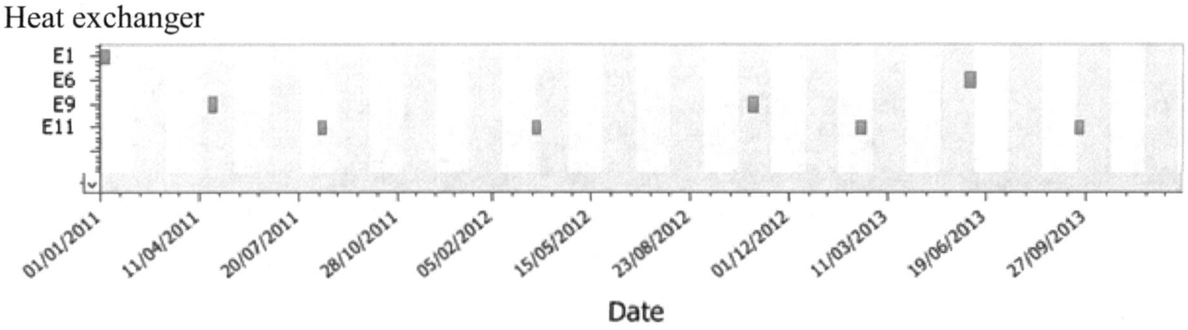

Figure 13: Cleaning schedule when total number of cleaning per operating period is limited to 8.

The shaded area (Area A) in Figure 14 presents the additional furnace cost that is incurred through the furnace. Note that the solid horizontal line is the operation when all exchangers are clean. The economic calculation with scheduling is compared through the difference in Area's A and B in Figure 14. Table 4 is a summary of the process economics under operation of 1100 days for optimum scheduling described in Figure 12.

Network redesign for improved performance

Any of the SmartPM simulation and cleaning modes can use any of the fouling model parameter sets described previously. This can allow the user to investigate the effect on energy, performance and maintenance of processing heavier crudes, for example. These modes can also be used to model energy, performance and maintenance with modified equipment or network structure. This can allow the engineer to investigate modifications that reduce fouling, and then quantifying the savings to justify the project. Such modifications can include the application of tube inserts, switching crude side allocation, adding or (re)moving exchangers, altering flow splits or re-piping within the network.

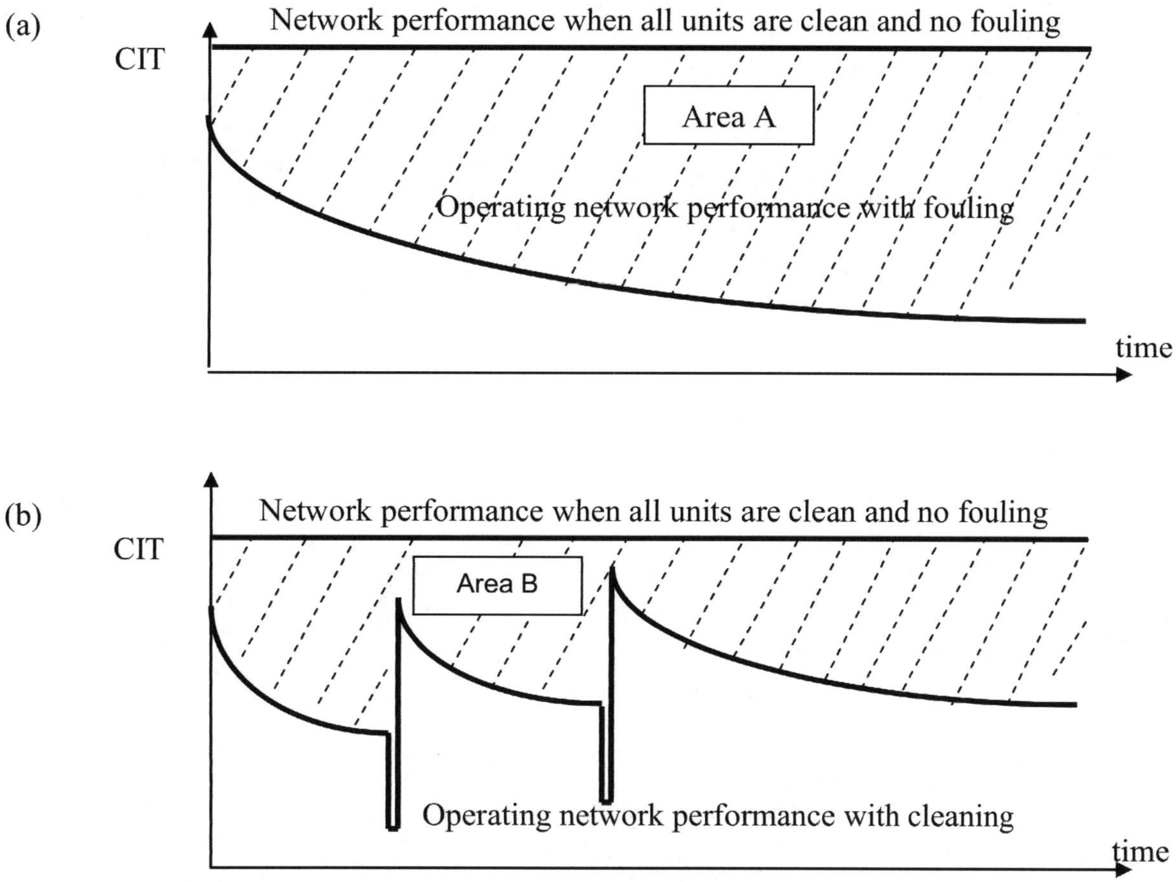

Figure 14: Schematic representation of CIT in a typical refinery PHT. (a) operation without any cleaning action, and, (b) operation with cleaning action. The shaded areas represent the additional energy that is incurred from the furnace. The solid horizontal line represent the CIT that could be achieved when all exchangers are operating clean.

Table 4: Economics for a cleaning schedule extended over 1460 days period (energy costs of

0.54, 1 and 2 US$/kW day). Total saving is given by the difference of energy saving from cleaning actions and the cost of cleaning.

	Number of cleaning actions	Total cleaning cost (US$/year)	Additional CO_2 emission (k tonnes/year)	Additional furnace energy due to fouling (J/year)	Total saving in MM US$/year Energy cost of (US$/kW day)		
					0.54	1	2
Year 1	5	100,000	37.0	6.99E+14	1.07	2.07	4.24
Year 2	4	80,000	41.1	7.78E+14	1.91	3.60	7.28
Year 3	4	80,000	43.7	8.27E+14	2.03	3.83	7.74
Year 4	4	80,000	44.0	8.32E+14	2.24	4.21	8.60

Conclusions

Data reconciliation is highlighted as a vital step in assessing crude refinery preheat train performance before planning cleaning schedules. A novel computer program from IHS, SmartPM, is demonstrated to generate missing stream measurements in order to obtain individual heat exchanger fouling resistance profiles. A case study preheat train with real plant monitoring data was utilized in this work. The fouling resistance profiles generated through SmartPM were used to identify periods of stable operations to formulate parameters for fouling models. With the fouling parameters, the fouling model has been successfully used to simulate the performance of the network as well as to formulate optimum cleaning schedules. The cleaning schedules with industrial constraints, such as limited number of cleaning, is successfully simulated with SmartPM.

Nomenclatures

Roman

A pre-exponential factor, s^{-1}

h film transfer coefficient on the crude side, $W\ m^{-2}\ K^{-1}$

R gas constant, $J\ mol^{-1}\ K^{-1}$

R_f fouling resistance, $m^2K\ W^{-1}$

T	temperature, K
U	overall heat transfer coefficient, W m^{-2} K^{-1}
U_{cl}	clean overall heat transfer coefficient, W m^{-2} K^{-1}

Symbols

λ	thermal conductivity, W m^{-1} K^{-1}
μ	dynamic viscosity, Pas
ρ	density, kg m^{-3}

References

[1] Watson, M. and Vandervell, N. (2006). Meeting our energy needs: The future of UK oil refining. *UKPIA Final report*, 13-24.

[2] Marsh-Patrick, A. (2006). EU emissions trading scheme phase II: review of new entrants' benchmarks – refineries. *Entec UK Limited Final report*. 3-4.

[3] IHS ESDU Energy, Emissions, and Efficiency: How to address heat exchanger fouling to reduce energy and maintenance costs while improving plant performance, *Penn Energy Webcast*, 31st March, 2011.

[4] Georgiadis M.C. and Papageorgiou L.G. (2000). Optimal energy and cleaning management in heat exchanger networks under fouling, *Trans IChemE*, 78, 168-179.

[5] Smaïli, F and Vassiliadis, VS and Wilson, DI (2001) *Mitigation of Fouling in Refinery Heat Exchanger Networks by Optimal Management of Cleaning. Energy & Fuels*, 15 (5). pp. 1028-1056. ISSN 0887-0624.

[6] Ishiyama E.M., Paterson W.R., Wilson D.I., Heins, A.V. and Spinelli L. (2010). Scheduling cleaning in a crude oil preheat train subject to fouling: incorporating desalter control, *Applied Thermal Engineering*, 30(13), 1852-1862.

[7] Bagajewicz M.J. (2000). A brief review of recent developments in data reconciliation and gross error detection, estimation, *Latin American Applied Research*, 30, 335-342.

[8] Knudsen J.G. and Ebert W.A. (1999). The determination of threshold fouling curve for a crude oil. In Bott, T.R. (ed), *Understanding heat exchanger fouling and its mitigation*, Begell house inc., Italy, 265-272.

[9] Polley, G.T. (2010). Review for the development of models for the prediction of fouling rates in exchangers heating crude oil, 11th annual international conference in Petroleum Phase Behaviour and Fouling, Jersey City, June 13-17.

[10] Ishiyama E.M., Pugh S.J., Wilson D.I., Paterson W.R. and Polley G.T. (2011). Importance of data reconciliation on improving performances of crude refinery preheat trains, paper 22d, *AIChE Spring National Meeting*, Chicago.

[11] Ishiyama E.M., Paterson W.R. and Wilson D.I. (2009). Platform for techno-economic analysis of fouling mitigation options in refinery preheat trains, *Energy and Fuels*, 23 (3): 1323-1337.

ExxonMobil
Research and Engineering

Heat Exchanger Monitoring
with
Invensys ROMeo™ Optimization Software

AIChE Spring 2012 National Meeting

Kenneth Teague, Ken Tyner, William Moore
ExxonMobil Research & Engineering

Cleared ExxonMobil Technical Publication – 2011.2043

Abstract

Heat exchanger monitoring for feed preheat exchanger networks incorporates heat exchanger performance tracking and a work process to identify exchangers within the network than can be cleaned for economic benefit. The cleaning studies traditionally have been done with desktop tools using a periodic and manual work process. New applications which automate exchanger performance monitoring and the cleaning studies have been developed in Invensys' ROMeo™ optimization software with many benefits. The heat exchanger viewer is a Microsoft Excel® spreadsheet based supplemental tool that was also developed to address certain modeling challenges, to support user understanding of application decision output and to gain acceptance by heat transfer experts. This presentation is an overview of the applications, the heat exchanger viewer, business impact of current applications, and opportunities created for additional benefit capture.

Exchanger Monitoring Overview

- Starting Point
  - Periodic (Quarterly) studies to identify exchangers to be cleaned
  - Multiple spreadsheet based tools requiring ~60 hours per solution
- Business Objectives
  - Continuously identify heat exchanger cleaning seriatim with minimum effort
  - Automatically historize exchanger performance to drive improvement in cleaning and operational decisions
- Solution Approach
  - Work closely with domain experts
  - Automate calculation sequence
  - Visual reporting and trending of historized performance data
  - Capture additional benefit through platform flexibility and extensibility

Exchanger Monitoring Application

- Data reconciliation
  - ROMeo^{TM} model of the heat exchanger network
  - Reconcile a fouling factor, ***ff***, for every exchanger
  - Determines heat exchanger duties, fouling factors, performance metrics, etc.
- Cost simulations
  - Calculates duty decline when exchanger out-of-service
  - 1 ***per*** "candidate" exchanger, $ff = \infty$
  - Costs incurred for fuel and maintenance
- Benefit simulations
  - Calculates duty improvement when exchanger returned to service
  - 1 ***per*** "candidate" exchanger, $ff = ff_{clean}$
  - Fuel benefit realized over a benefit horizon
  - Coil inlet temperature (CIT) improvements
- Benefit-to-Cost (B/C) Calculations
  - Cleaning decision data (B/C ratios)
- Reporting and Data Export of calculation results

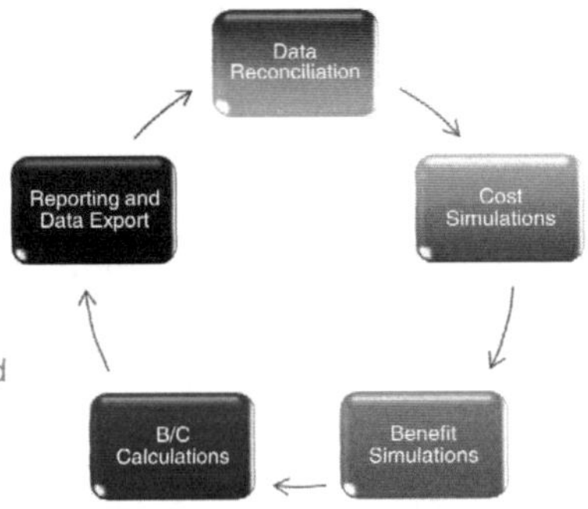

Path to Success

- Success Criteria (Challenges)
  - Automated application must provide equivalent or better representation (model, inputs, constraints, etc.) of the exchanger network relative to current tools
  - User and domain expert must <u>understand</u> solutions
- Strategy
  - Leverage ROMeo^{TM} platform with modeling / automation / reconciliation capabilities
  - Develop a Heat Exchanger Viewer to aid in understanding solutions and to drive rapid model improvement

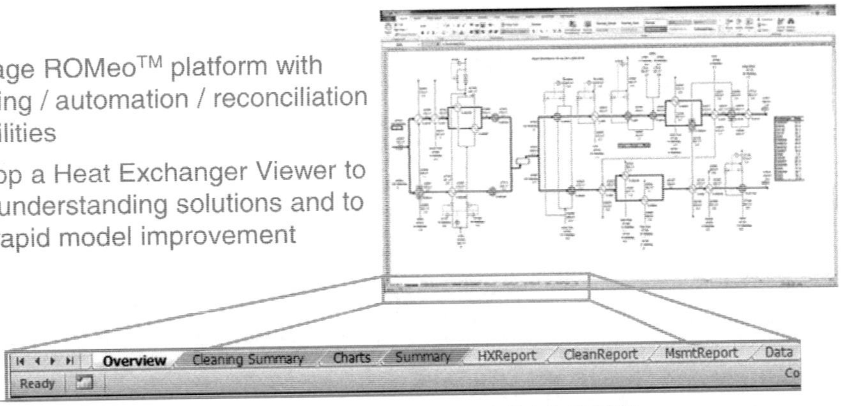

ExxonMobil
Research and Engineering

ROMeo^{TM} Platform Advantage

- Ability to model details (e.g. performance heat exchanger modeling, stream compositions)
- Input data screening
- Model transparency
- Extensibility
- Robust handling of many case studies
- Case failure response
- Monitoring cycle automation (e.g. line-ups)
- Data management (import/export)
- Data sharing with RTO applications (e.g. stream compositions)
- Resource synergy with RTO

ROMeo Flowsheet

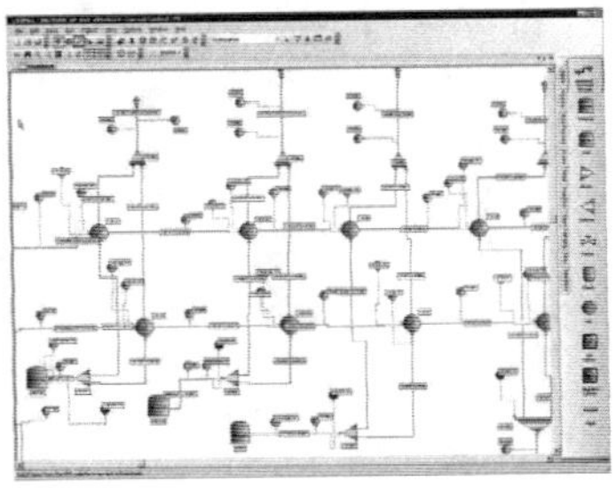

RTS Sequence

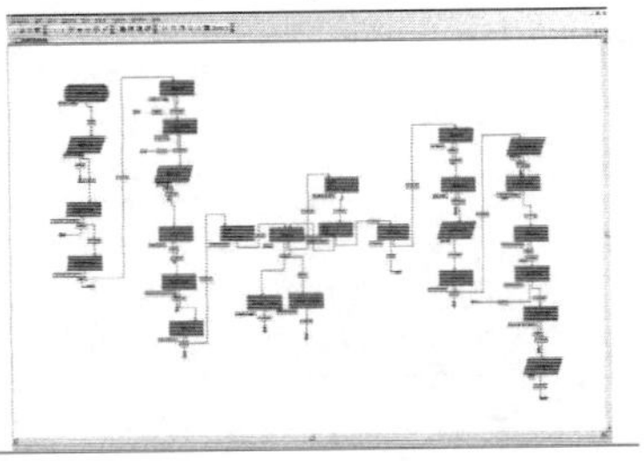

ExxonMobil
Research and Engineering

Model Details Add Value

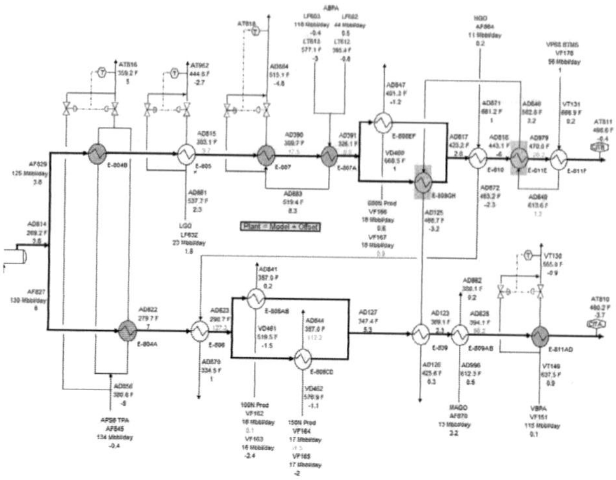

- Model detail must handle exchangers that are…
  - Out of service
  - In alternate service
  - Partially bypassed (not measured)
  - In flashing / two-phase service
- Model incorporates cleaning impacts beyond duty improvements
  - Interplay between exchangers
  - Furnace CIT improvements

ExxonMobil
Research and Engineering

Viewer Supports Understanding and Analysis

- Design Objective: Clearly depict reconciliation case setup and solution (graphical & tabular data)

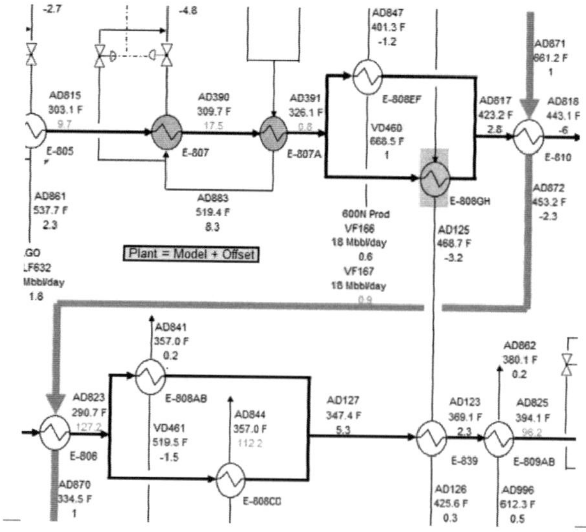

Features

- Summary table of decision output
- Highlight top cleaning candidates
- Color coding for exchangers under duty control
- **Reconciled** and **bias** values displayed (plant value by difference)
- Unweighted biases greyed out
- Highlight large biases

Exchanger	Duty	Cost	Benefit	B/C	CITA	CITB
HX1	48.0	20.0	200.0	10.0	-4.8	14.3
HX2	27.0	20.0	100.0	8.0	-3.5	10.0
HX3	58.1	33.3	200.0	6.0	1.2	0.9
HX4	29.3	37.5	150.0	4.0	-3.5	10.0
HX5	20.9	150.0	300.0	2.0	3.6	0.0

ExxonMobil
Research and Engineering

Viewer Supports Validation

- Example: Identical exchangers with similar service have very different benefit/cost. Overview highlights difference in driving force.

HX	Area ft2	LMTD F	Re (TS)	Re (SS)	# Tubes per Shell	# Tube Passes per Shell	Shell Diameter in	Tube Pitch in
E801A	5864	214.061	34662	38483.77	1120	4	52.25	1.25
E801B	5864	67.276	15792	33602.46	1120	4	52.25	1.25

HX	In-Service Duty MMBTU/hr	CIT A Decrease F	CIT B Decrease F	Qclean (calc) MMBTU/hr	CIT A Increase F	CIT B Increase F
E801A	58.1	-4.6	-3.5	73.4	1.2	0.9
E801B	24.3	-5.1	-3.9	25.5	0.2	0.2

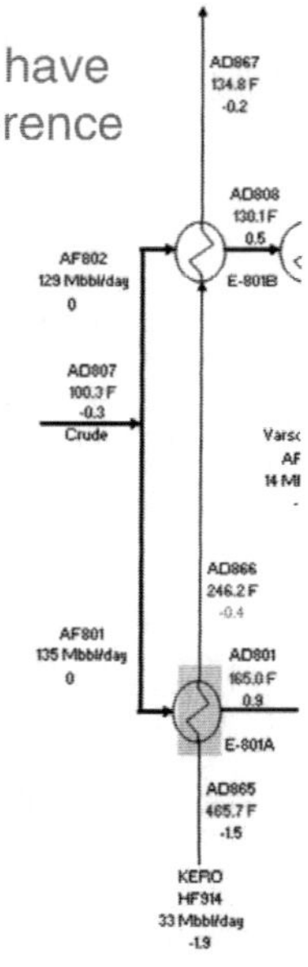

9

ExxonMobil
Research and Engineering

Reporting / Analysis

Previous Strategy
- Infrequent spot analysis
- Data stranded in spreadsheets

Current Strategy
- Frequent trend analysis
- Solution and peripheral data historized

- Trend analysis makes process / measurement variability more apparent and improves confidence in decisions (or highlights uncertain/marginal information)
- Historized data drives:
  - Improved heat exchanger cleaning decisions
  - Supports investigation of operating issues that contribute to fouling
  - Validation of clean duty predictions
  - Predictive model improvements

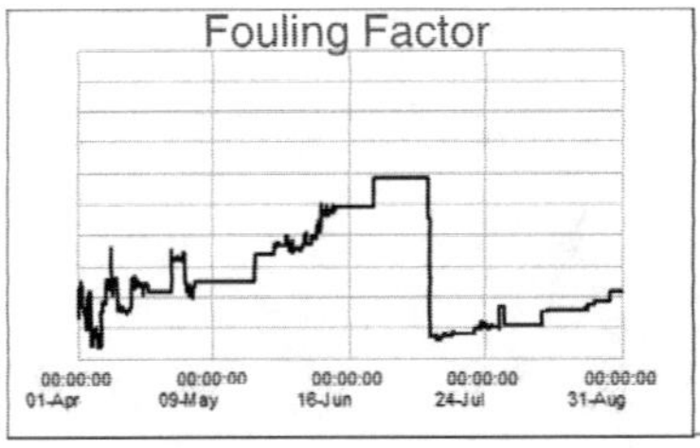

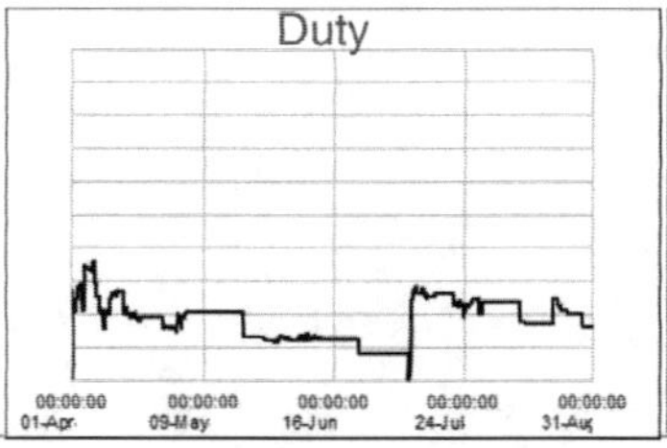

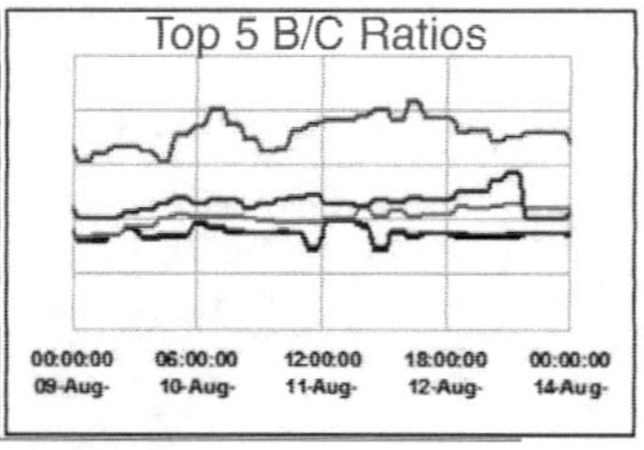

10

ExxonMobil
Research and Engineering

Margin Benefit Opportunities

- Platform and model flexibility support the capture of margin benefits in addition to energy benefits from the "base" application.
- Example: One site's application predicts the incremental throughput expected from cleaning exchangers in a pressure-constrained crude preheat network, utilizing network pressure measurement data. Exchanger monitoring cycle was augmented as follows:
  - Data reconciliation (includes all network temperature, flow and pressure data)
  - Cost/benefit simulations for each candidate HX (energy benefit)
  - Optimization case for each candidate HX (margin benefit)

Crude Rate Improvement Cycle

- Data reconcilation
  - Each exchanger pressure drop modeled as $\Delta P_j = (\Delta P)_{0,j}\left(\frac{F_j}{F_{0,j}}\right)^2$
  - $(\Delta P)_{0,j}, F_{0,j}$ are the reconciled pressure drop and flow for pressure drop j
  - Furnace passes and control valves lumped as single pressure drop and modeled using same equation
- Benefit case is an optimization
  - For candidate exchanger i, let $(\Delta P)_{0,i} = (\Delta P)_{0,i(clean)}$, a value obtained from prior cleaning results; the remaining $(\Delta P)_{0,j}$ and all $F_{0,j}$ retain their reconciliation case values.
  - Assume furnace pass valve positions are constant
  - Objective: Maximize the crude rate against the measured tower inlet pressure. This measured value is the low limit.
  - The flows in any parallel exchanger trains re-balance due to pressure continuity enforcement.

Business Impact

- Annual effort for cleaning study reduced by greater than 75%.
  - Resources now available for analysis and/or other higher value contributions.
- Additional manpower savings and scheduling efficiency:
  - Use reports / Viewer to drive targeted temperature survey, incorporating results into model for future use.
  - Early identification of bad measurements through continuous monitoring – address issues as they arise before they become critical.
- Continuous decision output drives more strategic heat exchanger cleanings
- Monitoring data can be leveraged to evaluate alternate cleaning methods (e.g. chemical versus mechanical)
- Opportunities abound!
  - Modify $\boldsymbol{ff}_{clean}$'s and benefit horizons and determine if cleaning decisions are optimal
  - Add mechanisms for margin improvement benefits

Success!

- Multiple application deployments completed and more planned
- Projects to replace existing tools justified
- Additional value added through platform flexibility
  - Demonstrated by marginal crude rate predictions utilizing wireless pressure data
  - Ability to incorporate additional standard and proprietary models enables related business opportunities to be addressed.

Q&A

All Sulphur Plants are Not Created Equal

Angela F. Slavens, P.E.
Business Development Director, Sulphur Technology
WorleyParsons
Monrovia, CA, USA

EXTENDED ABSTRACT

Sulphur recovery from natural resources is required to meet environmental regulations on sulphur content in final products, as well as SO_x emissions restrictions imposed on processing facilities. While sulphur recovery facilities in oil refineries, gas plants, gasification facilities and various chemical plants may appear to be very similar, there exist key differences in each type of processing facility that create unique, and often critical, design concerns for the sulphur plant. Factors such as feed gas strength (H_2S composition), type and concentration of impurities, feed gas stability, and availability of utilities alter the design and operational factors that must be considered in the corresponding sulphur plant.

This presentation highlights the unique elements that must be taken into consideration when designing or operating a sulphur plant located in a refinery, gas plant or gasification facility. The following parameters are compared for each type of facility:

Acid Gas Quality

- Strength (H_2S concentration)
- Impurities
    - Ammonia
    - Hydrocarbons
    - Organic sulphur compounds
- Stability of flow

Utility Availability

- Hydrogen
- Natural gas
- Oxygen

When acid gas feed quality is poor, robust and reliable sulphur plant operation can be difficult to achieve. However, this presentation Illustrates that when appropriate utilities are available in the facility, acid gas feed quality obstacles can be overcome. On this basis, the features of each type of sulphur plant are compared to provide a perspective on relative ease of design, operation and reliability.

Figures 1-4 provide an overview illustration of each type of facility, with comparative parameters highlighted.

Figure 1. Refinery Sulphur Recovery Unit

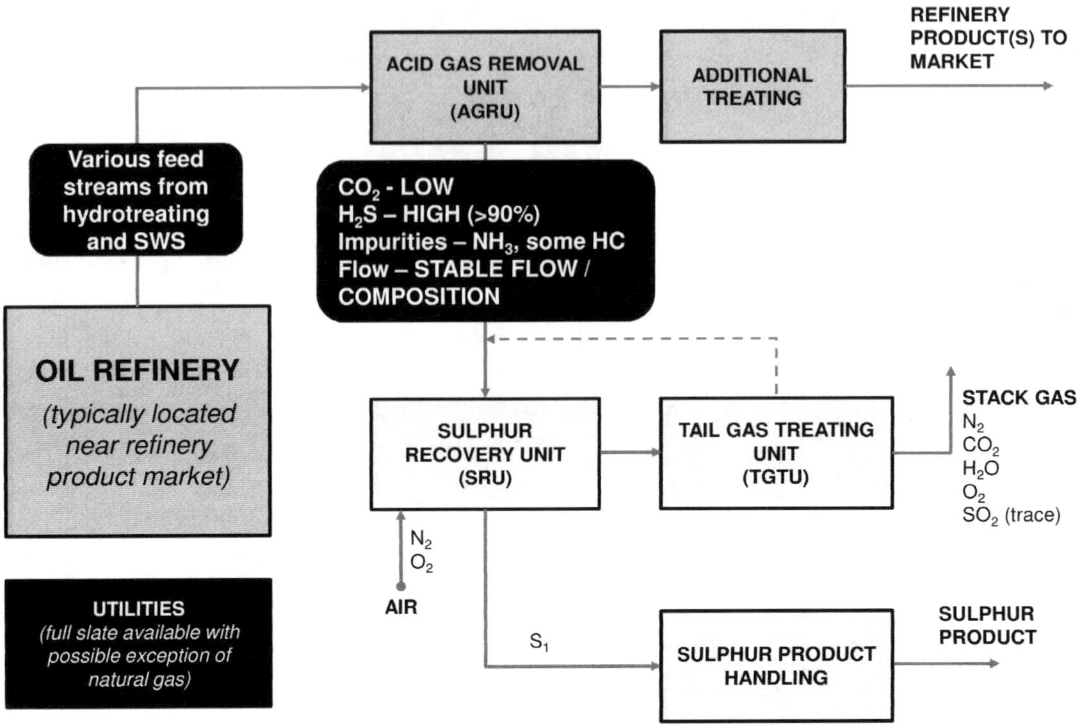

Figure 2. Gas Plant Sulphur Recovery Unit

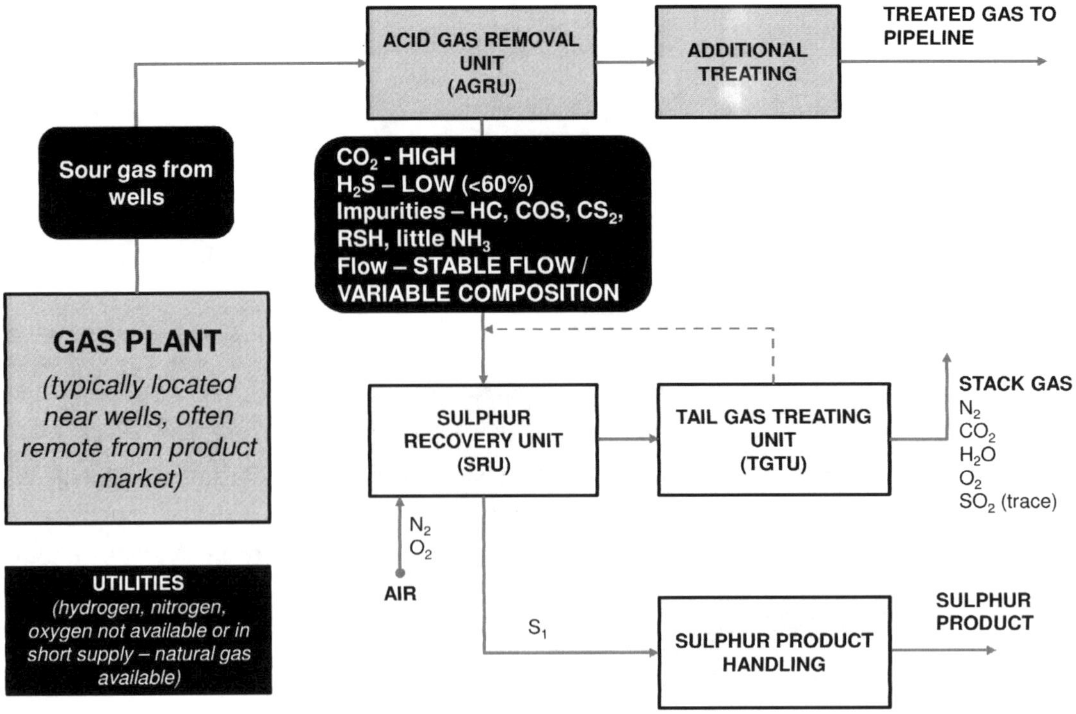

Figure 3. Gasification Facility Sulphur Recovery Unit

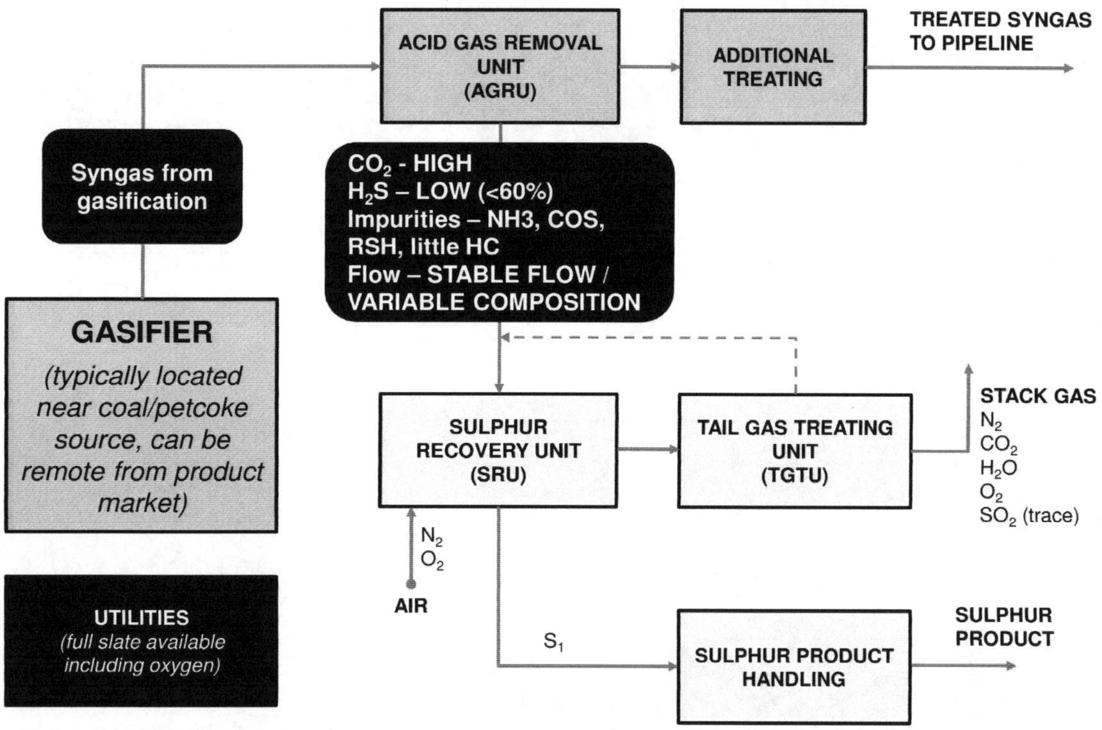

The ease of operation ranking reveals that refinery sulphur plants typically present the fewest obstacles to achieving robust design and reliable operation, due to their high acid gas feed quality, as compared to other types of facilities. Gas plants typically produce the lowest quality acid gas feedstock and often do not have the appropriate utilities available to balance this, thus ranking the lowest in terms of ease of design/operation. Gasification facilities typically produce a somewhat difficult acid gas feedstock but have a full suite of utilities (including oxygen) available to counteract this. Thus, gasification facilities rank in between refineries and gas. The ranking approach employed to achieve these results is illustrated in Figures 4 and 5. Higher rankings indicate more favorable conditions and lower rankings indicate less favorable conditions.

Figure 4. Sulphur Plant Ease of Design/Operation Comparison

FACILITY TYPE	ACID GAS FEED QUALITY					UTILITY AVAILABILITY			
	Acid Gas H_2S vol%	Impurities			Acid Gas Stability	H_2	Nat'l Gas	N_2	O_2
		NH_3	HC	Organic Sulphur					
REFINERY	5	1	4	5	5	5	2	4	3
GAS PLANT	1	4	1	2	2	1	5	1	1
GASIFIER	2	2	3	2	4	5	5	5	5
Weighting	30%	15%	15%	10%	30%	20%	50%	---	30%

Figure 5. Sulphur Plant Ease of Design/Operation Ranking

	ACID GAS FEED QUALITY	UTILITY AVAILABILITY	OVERALL
REFINERY	4.25	2.90	7.15 (MOST FAVORABLE)
GAS PLANT	1.25	3.00	4.25 (LEAST FAVORABLE)
GASIFIER	1.55	5.00	6.55

Although refinery sulphur plants have their challenges (primarily related to ammonia destruction), the ease of design/operation ranking illustrates that refinery SRUs are actually less problematic than SRUs in other types of facilities.

In summary, although many SRUs may look alike, SRUs in some types of facilities create greater challenges than others. Because of this lack of understanding that "all sulphur plants are not created equal," there are some facilities in the world that are not performing well.

Lastly, despite the fact that this presentation provides a generic comparison of the general advantages/disadvantages for different types of sulphur recovery units, it is important to remember that every solution is unique and requires experience and "know-how" to select the most reliable, robust solution with lowest lifecycle cost.

REDUCING CO2 IN THE RECYCLE ACID GAS FROM MDEA-BASED TAIL GAS TREATING UNITS

Ben Spooner, P. Eng

Farsin Derakhshan, P. Eng

INTRODUCTION

The purpose of an amine based tail gas treating unit (TGTU) is to recycle any left-over sulphur components in the tail gas of a Claus sulphur recovery unit back to the front end of the plant, rather than incinerate it. The remaining sulphur components are converted to H_2S and removed from the gas using an amine solution.

CO_2 is also present in tail gas streams and will be partially co-absorbed together with H_2S. This CO_2 co-absorption should be minimised. Any CO_2 removed in the TGTU will be recycled to the SRU front end along with the H_2S, which has negative consequences for the reaction furnace and overall sulphur recovery.

This paper discusses the effect of CO_2 on sulphur plants and how to minimise CO_2 co-absorption or to maximize CO_2 slip through the amine absorber of the TGTU.

Amine Experts believes that a well designed MDEA TGTU should be able achieve a minimum of 85% CO_2 slip.

EFFECTS OF CO_2 ON THE SULPHUR PLANT

The motivation behind maximizing slip though the tail gas treater is to prevent CO_2 from re-entering the sulphur plant. There are several reasons why it is advantageous to minimize CO_2 in the SRU feed, the most important involving the reaction furnace temperature, COS and CS_2 formation and sulphur plant capacity:

1. Reaction furnace temperature: CO_2 does not burn in the reaction furnace, and therefore will lower the furnace temperature. This will have negative effects on ammonia and aromatic destruction, which will then have consequences for the converter beds and the overall sulphur recovery.

   Maintaining reaction furnace temperatures is critical in maintaining smooth and trouble free operation of a sulphur plant. If processing Sour Water Stripper gas, the destruction of ammonia is of primary concern and reaction furnace temperatures of 2250°F+ are required in order to avoid ammonia salt formation and plugging of downstream process equipment. If processing acid gases containing aromatic hydrocarbons, reaction furnace temperatures of 1920°F+ are required to properly combust the benzene, toluene and xylene components. Incomplete combustion of these compounds can lead to poisoning of the downstream catalyst bed. Finally, in order to maintain stable flame operation, a reaction furnace temperature of 1700°F is required.

Though some cracking can occur in the reaction furnace, CO_2 is generally considered an inert compound, and cools the reaction furnace flame. As seen in Figure 1, diluting an acid gas stream with increasing amounts of CO_2 results in a rapid reduction of the reaction furnace temperature.

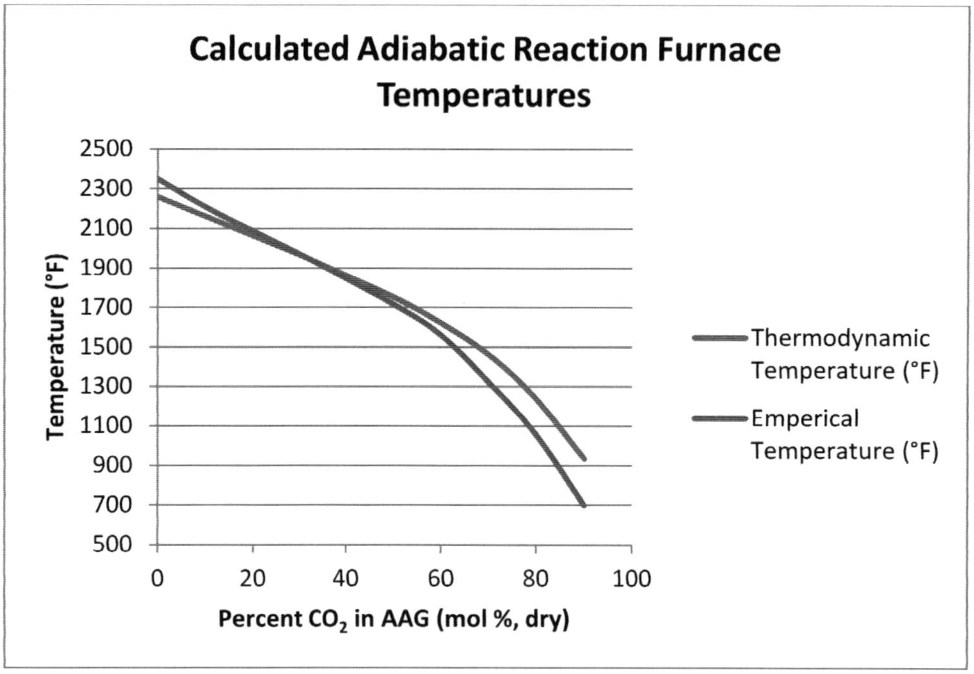

Figure 1 - Calculated Reaction Furnace Temperatures

Figure 1 shows that for a simple acid gas containing varying amounts of H_2S and CO_2 and 1% hydrocarbons, significant operating problems maintaining the required reaction furnace temperature start to occur at very low concentrations of CO_2 in the amine acid gas (AAG). With a CO_2 content of less than 5% (based on this example), it becomes difficult to maintain the required reaction furnace temperature for good ammonia destruction. At a CO_2 content of 30%, BTEX breakthrough could start to occur, and with a CO_2 content of over 50%, problems maintaining a reaction furnace flame become pronounced.

Although not necessarily detrimental to Claus plant operation, higher CO_2 content in the acid gas will require modifications to the operation of the reaction furnace in order to achieve and maintain the required operating temperatures, sometimes at a high capital investment cost or to the detriment of the overall efficiency. These items could include the addition of air and feed gas preheaters, oxygen enrichment, split flow reaction furnace designs, installation of high intensity or other specialty burners, co-firing with fuel gas or acid gas enrichment.

2. COS Formation: Although the chemistry inside a reaction furnace is quite complex and chaotic, one certainty is that CO_2 will partially "crack" in the furnace, resulting in the formation of CO, COS and (indirectly, though a drop in temperature) CS_2. Figure 2 shows the impact that CO_2 has on COS and CS_2 formation rates, using the same simplified acid gas composition discussed previously.

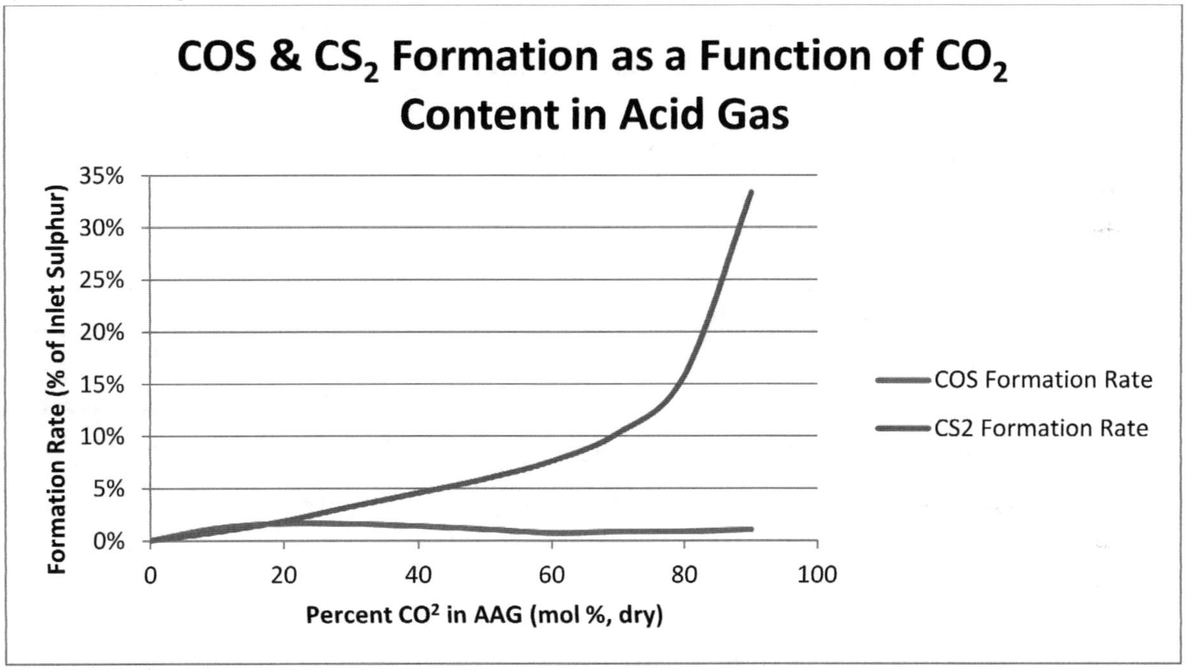

Figure 2 - COS & CS_2 Formation

As can be seen in Figure 2, the increasing CO_2 concentration of the acid gas results in ever increasing formation rates of CS_2. CS_2 formation has been shown to decrease with increasing temperatures, though it is not clear whether simply less CS_2 is formed at higher operating temperatures or whether CS_2 is formed then quickly hydrolyzed(3). However, the addition of CO_2 cools the reaction furnace temperatures which results in increased CS_2 formation. CS_2 is especially unwanted since it binds up two sulphur molecules.

COS and CS_2 formation in the reaction furnace are important, because once formed, COS and CS_2 do not participate in the modified Claus reaction. These compounds must therefore be converted/hydrolyzed back to H_2S downstream of the furnace, either in the first Claus converter bed (operating at high temperatures, typically 600°F to 630°F) and/or utilizing special and often expensive catalyst, or they can be converted back to H_2S in the TGTU hydrogenation reactor. Since amine based TGTU's do not pick up COS and CS_2 in the absorber, any unconverted COS and CS_2 will result in increased sulphur emissions and reduced overall recovery efficiencies.

3. Overall gas capacity: In a world of low sulphur fuels, the demand for increasing the existing processing capacity of refinery sulphur plants is becoming increasingly important. Since CO_2 simply flows straight though the sulphur plant and does not participate in the Claus or modified Claus reactions, it simply takes up space and reduces the amount of sulphur-bearing gases which could otherwise be processed. A sulphur plant already operating at or slightly above its design capacity would quickly run into trouble with an increase in the CO_2 content of the feed gas.

   Apart from the issues related to furnace temperature and increased COS and CS_2 formation, operating a sulphur plant at higher-than-design throughput would see higher operating pressures that could result in:

   - problems with air blowers not being able to deliver sufficient air at the higher operating pressures,
   - heat exchangers not being able to sufficiently cool process gases resulting in more sulphur vapour losses,
   - problems with mass velocities through condenser tubes, resulting in liquid sulphur carryover (both of which would require more reducing gases in the TGTU hydrogenation bed)

   In order to address some of the reduced capacity issues, processing companies would have to consider technologies such as oxygen enrichment or the construction of additional/bigger sulphur plants, which would have a significant impact on the operating or capital budgets of refineries.

EFFECTS OF CO_2 ON THE QUENCH WATER OF THE TGTU

High CO_2 tail gas streams act to lower the pH of the quench water, which can decrease the strength of the protective iron sulphide film. The water will, therefore, be darker in colour and could cause fouling and plugging of the quench system. Ideally, the quench water has a pH of 7-8, however in plants with high CO_2 levels the pH is typically between 6 and 7. Some plants try to correct this by regular caustic addition to the (partly) circulating quench water. As a significant quantity of the quench water is not recycled, caustic injection is, at best, a temporary solution. Regular caustic injection can result in strong pH fluctuations which destabilize the protective FeS film.

CO_2 REMOVAL WITH MDEA

CO_2 does not react directly with the MDEA molecule; instead it dissolves and reacts in the water portion of the solution.

$$\mathbf{CO_2 + H_2O \rightarrow H_2CO_3 \quad (carbonic\ acid)}$$

$$\mathbf{H_2CO_3 \rightarrow H^+ + HCO_3^- \quad (bicarbonate)}$$

$$\mathbf{H^+ + R_1R_2R_3N \rightarrow R_1R_2R_3NH^+}$$

$$\mathbf{CO_2 + H_2O + R_1R_2\ R_3N \rightarrow R_1R_2R_3NH^+ + HCO_3^-}$$

The reaction between CO_2 and water (carbonic acid formation) is a "slow" step; it takes time to occur. Once the carbonic acid is formed, however, the MDEA reacts with it quite quickly and the bond will not be broken again until the amine is regenerated. The removal of CO_2 with a tertiary amine like MDEA is therefore kinetically limited by the reaction rate in the first step.

H_2S on the other hand, reacts directly with the MDEA molecule;

$$\mathbf{H_2S + R_1R_2R_3N \rightarrow R_1R_2R_3NH^+ + HS^-}$$

The reaction between H_2S and the amine is a very fast or instantaneous reaction. This means that H_2S removal is almost always equilibrium limited. Each contact stage or tray in an absorber reaches the H_2S equilibrium between gas and liquid.

The difference in chemistry between the H_2S and CO_2 removal is the key to understanding how H_2S can be removed with a minimum amount of CO_2.

The best strategy to minimize the CO_2 removal with MDEA is to prevent the CO_2 – water reaction from occurring by:

1. Optimizing the amine temperature
2. Optimizing the amine strength
3. Optimizing the amine circulation rate
4. Optimizing the amine feed point of the absorber

5. Choosing a more selective solvent (formulated MDEA or Flexsorb)

1. AMINE TEMPERATURE

As with most chemical reactions, the higher the temperature, the faster the CO_2 reaction takes place. In a TGTU, lower lean amine temperatures will minimize CO_2 pick up because it reduces the reaction between CO_2 and water. It should be noted that high amine temperatures (above 140°F) will also slip amounts of CO_2. Unfortunately, at high temperatures H_2S will also be slipped due to equilibrium limitations. Therefore, operating the absorber at high temperatures cannot be used as an operating strategy.

A low absorber temperature slows down the kinetics of the CO_2 reaction. H_2S removal occurs through a different mechanism which is much less affected by temperature. Therefore, the amine temperature should be kept as low as possible, though typically not lower than the inlet gas temperature. As the gas is coming from the TGTU quench tower, it is saturated with water. An amine temperature lower than the gas temperature would condense water and dilute the amine solvent. It should be noted that the normal temperature guideline for amine systems (maintain 5-10°F temperature difference between gas and amine) is not applicable for amine based TGTU's.

Normal operation is to maintain the lean amine temperature at or within 1-2 degrees of the inlet gas. 90-100°F for both the amine and the gas is ideal.

2. AMINE STRENGTH

Because water is one of the reactants, the more of it in the absorber, the more CO_2 will be absorbed into the amine solution. MDEA can be successfully operated at up to 50% strength (the other 50% being water). Above this strength the viscosity of the solution becomes too high and will negatively affect the mass transfer of H_2S into the amine.

Minimize the CO_2 hydrolysis by minimizing the water content of the MDEA solution. **Maintain solvent strength between 45-50%,** all the while utilizing a proper filtration program to ensure the viscosity of the amine does not get too high. Minimizing SO_2 breakthrough out of the SCOT (or SCOT-type) CoMo reactor will reduce the heat stable amine salt build-up in the system and reduce the chances of viscosity increase due to excess heat stable salt formation.

3. AMINE CIRCULATION RATE

The CO_2 reaction with water (to form bicarbonate) is a kinetically limited reaction. This means that CO_2 builds up at the gas-liquid interface and only reacts with the water and amine as it slowly diffuses to the bulk of the amine solution. In a trayed tower, higher circulation rates increase the height of liquid on each tray, as the weir creates a flow obstruction. The amine "stacks up" against it and provides a larger surface area for CO_2 absorption. In a packed tower, higher amine flows create higher hold-up of amine into each section of packing. The more liquid in the tower, the higher the CO_2 removal will be, as there is more gas/liquid interface for the CO_2 to be absorbed. Over-circulation of the amine is the single largest contributor to poor CO_2 slip in TGTU's.

Furthermore CO_2 increases the loading of the amine which, especially in low pressure TGTU applications, takes up very valuable acid gas holding capacity. Under certain conditions, it is possible that CO_2 will force the amine to release previously absorbed H_2S.

Optimizing (in most cases reducing) the amine circulation rate will always result in a decrease in H_2S and an increase in CO_2 in the treated gas which is exactly the goal.

In Figure 3 below, the effects of circulation rate are shown. As the amine rate is decreased from 200 gallons per minute down to 150, the CO_2 in the treated gas increases by 1,000 ppm, whereas the H_2S decreases from 45 to 40 ppm.

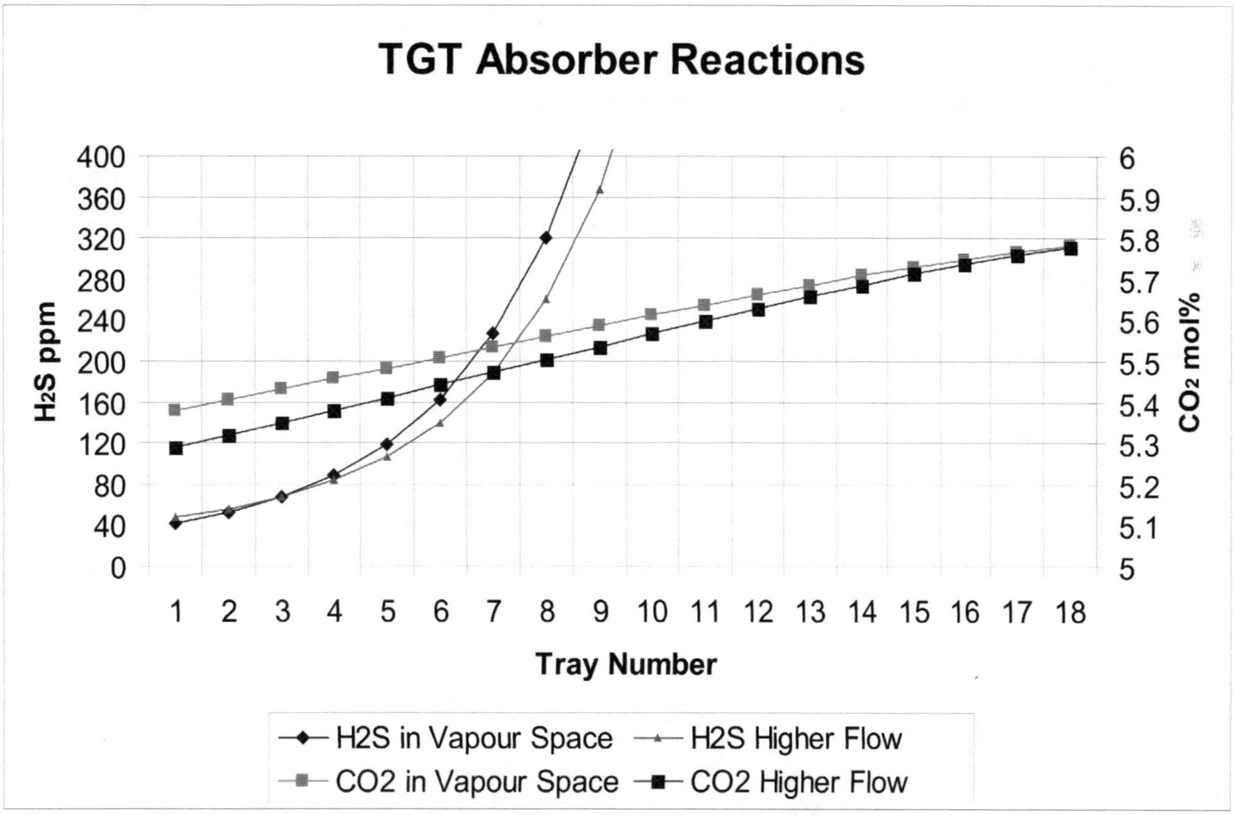

Figure 3: Effect of an increase in circulation rate on H_2S and CO_2 removal

The circulation rate should initially be targeted for a rich H_2S loading of 0.05 mol/mol (once the amine strength has been set, see point 2 above), and slowly decreased with a final target of 0.1 mol/mol. This should be re-enforced by the use of a reliable amine plant simulator.

4. AMINE FEED POINT INTO ABSORBER

TGTU absorbers often have multiple inlet points for the lean amine solution. Several inlet points are normal and can be used to increase or decrease the interfacial contact area between the gas and the amine solution. A shorter contact time (less interfacial contact area) will result in less CO_2 absorption.

The contact time between the CO_2 and amine depends on the internals of the absorber and the height at which the amine is injected. In a trayed absorber, contact time on each tray is determined by the weir height and number of trays. If packing is used, contact time depends on the height and size of the packing. In either case, lowering the injection point of the lean amine will reduce the contact time and interfacial contact area between the gas and amine, thus increasing the amount of CO_2 slip.

Neither the weir height of trays nor total height of packing is a parameter which can be adjusted during normal operation. Changing either of these requires shutting down the system and performing large scale maintenance. This is why most TGTU absorbers have multiple feedpoints built into the design.

It should be noted that there are differences between the selectivity of a packed tower compared to a trayed one. This is due to hydraulics of the internals and the corresponding relationship with mass transfer. A trayed tower has the liquid phase which is highly agitated; packed towers are opposite, whereby the liquid flows over the packing relatively smoothly. The gas flows are turbulent in both. Either type of internals can result in decent CO_2 slip, but the choice between the two must be carefully thought out and researched in the design stage. Because the net liquid hold-up in a packed bed (1-6% of tower volume) is lower than in a trayed tower (8-12% of tower volume) the lower liquid hold-up can result in less interfacial contact area and therefore lower CO_2 absorption. In general, Amine Experts has tested higher CO_2 slip rates on packed towers compared to trays, but this is by no means a universal rule.

A TGTU absorber has typically three lean amine injection points; one at the top tray or top height of packing, and two more at successively lower intervals. The balance is to inject at a high enough point to remove all of the H_2S necessary, but no higher. This procedure is best predicted first on a simulator, as shown in Figure 4 below. Using this type of chart, it is apparent where appropriate injection points into the absorber are, depending on the required H2S specification.

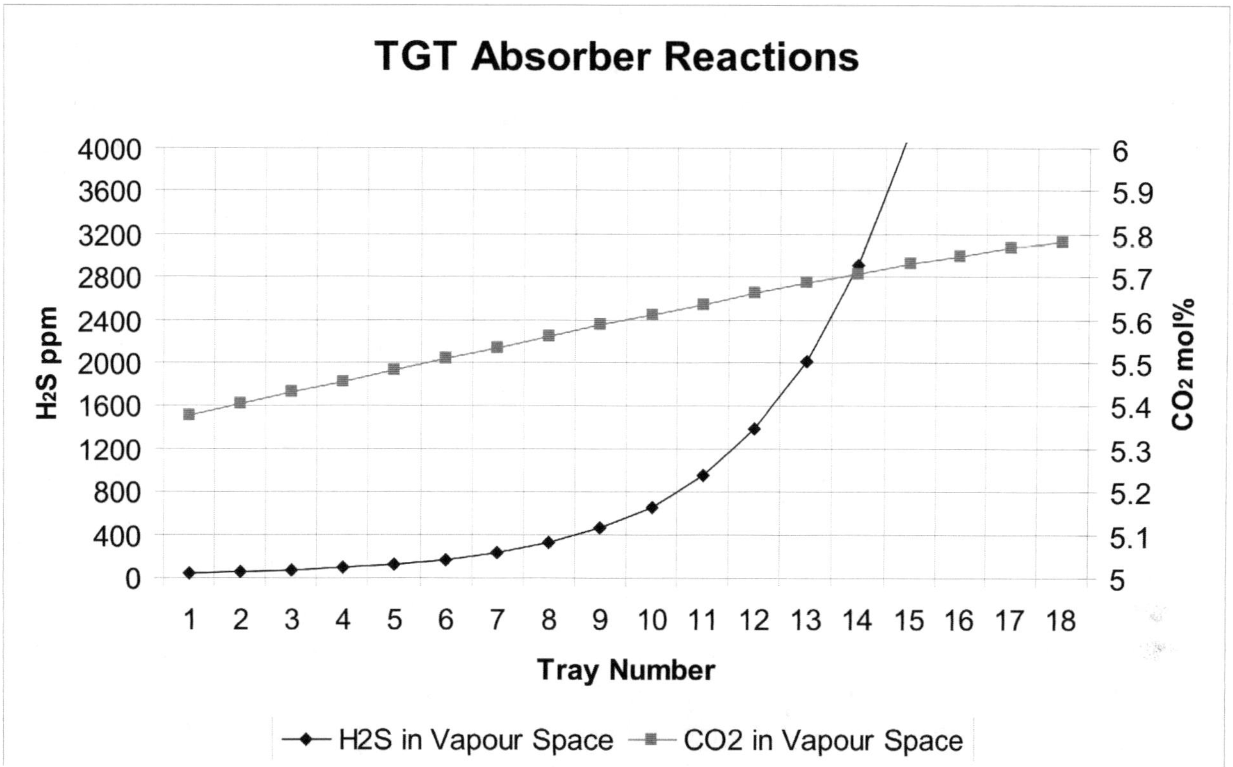

Figure 5: Simulated acid gas removal tray-by-tray in a TGT absorber

If possible, lower the amine injection by one feed point and measure the treated gas H_2S content. If it is acceptable, and simulations agree, continue lowering to another feed point. The main concern is the treated gas H_2S content.

5. CHOOSING A MORE SELECTIVE SOLVENT

Hindered Amine

The most selective solvent for TGTU applications is a hindered amine like the Flexsorb® SE/SE Plus from ExxonMobil Research and Engineering Company. A hindered amine is normally a secondary amine with a bulky group which hinders the direct reaction with CO_2. These molecules combine the low CO_2 reaction rate of a tertiary amine with the base strength of a secondary amine. Particularly the high base strength is very useful at the low pressure of a TGTU absorber because it makes it possible to have a significantly higher rich amine loading than with MDEA. The higher loading allows for a reduction in amine circulation rate, which improves again the selectivity. Flexsorb however is a much more expensive solvent than MDEA, nor is it as readily available. As well, it most likely requires a licensing fee.

Formulated MDEA

Enhanced MDEA formulations for TGTU's have been available for 25 years. These are normally the solvents that contain "pH suppressants" such as phosphoric acid. The reduction in pH allows for easier and deeper regeneration of the amine, especially for H_2S. These formulated solvents are known as: Ucarsol HS-103, GasSpec TG-10 or JeffTreat MS-300. The same effect of improved regeneration is achieved in MDEA solutions containing between 0.5-1 wt% heat stable salts.

By lowering the H_2S lean loading, the H_2S in the treated gas will also drop. This could enable the lean amine to be introduced into a lower feedpoint in the absorber, thus increasing the CO_2 slip as described above. Normally though, the aim of the improved regeneration is to obtain a lower H_2S level in the treated gas. TGTU's are, therefore, also often designed with more trays in the regenerator than normal amine units. This allows for deep H_2S stripping.

A second aspect of these formulated solvents is that they often contain anti-foam. The anti-foam reduces the liquid/gas interfacial contact area because it helps to break the gas bubbles in the solvent faster than normal. In this way it improves the selectivity even when the solvent is not "foaming" (there is always a few inches of foam on the trays of a normal working absorber). This also highlights the importance of minimising foaming problems to reduce the CO_2 pick-up of a solvent. Although TGTU amine systems do not often have foaming problems, it is important to eliminate the foaming as quickly as possible for good selectivity.

REFERENCES

B. Klint, "Hydrocarbon Destruction In the Claus SRU Reaction Furnace"; Sulphur Recovery textbook, Sulphur Experts, 2000

H.G. Paskall, "Reaction Furnace Chemistry and Operational Modes"; Sulphur Recovery textbook, Western Research, 1988

Sames, J.A., Dale, P.R. and Wong, B. (1987): "Evaluation of reaction furnace variables in modified-Claus plants". Laurence Reid Gas Conditioning Conference, Norman, OK.

T.K. Khanamamedov, RH Weiland, "Upgrading Acid Gas Streams"; Sulphur magazine, Sept-Oct 2008

Mahin Rameshni P.E., "How to Improve the Efficiency of Tail Gas Treating in Sulphur Plants"; Worley Parsons

RH Weiland, John C Dingman, "Simulator Provides Guidance for Increasing CO_2 Slip in Gas Treating Applications"; OGT Inc

Selective H_2S removal: Case studies by using a new promoter system in amine units

Gerald Vorberg, Ralf Notz,
OASE® Gas Treatment Excellence
BASF SE, Ludwigshafen/Germany

Kumar Sugavanam,
OASE® Gas Treatment Excellence
BASF Corporation, Houston, TX

Abstract

Applications where selective sulfur removal is required have become an important topic over the last decades. In many applications, e.g. in acid gas enrichment (AGE), in tail gas treatment units (TGTU) as well as in natural gas applications (NG), the H_2S selective treatment including the production of an enriched high quality Claus gas has been one of the favored drivers so far. Meanwhile, the requirement for an adjustable CO_2 slip, the focus on tighter sulfur specifications along with a high degree of flexibility and savings in energy consumption have gained in importance as well.

The principles of selective removal of H_2S with amine-based solvents follow three major routes:

a) various design options and absorber internals, affecting the difference in CO_2 and H_2S mass transfer kinetics

b) severely hindered amines, controlling the selectivity for H_2S removal primarily in the absorber, and

c) generic and promoted tertiary amines, focusing more on the regeneration ability, and thus leading to tighter H_2S specifications.

In particular the combination of a) and c) is widespread and the most common option in industry.

Based on two case studies, this paper will describe the possibilities and leverage of selective designs by using an advanced promoter system for enhanced selective treatment with tertiary amines.

Principles

H_2S selective acid gas removal with amine based solvents is widespread in the industry. Particularly, the use of "generic" amines such as Methyl-di-ethanol-amine "MDEA" is well known in combination with straight forward designs. Although, being not a part of this paper, advanced solutions with sterically hindered amines incorporate useful properties: e.g. fast kinetics and high acid gas load at low acid gas partial pressure and a high H_2S selectivity due to the steric hindrance of the molecule itself. However, in some cases those advantages may be progressively reduced at increasing acid gas partial pressures e.g. most high pressure natural gas applications. Moreover, the relatively high price level of hindered amines is often part of economic considerations between CAPEX and OPEX.

As mentioned above, designs and current operations with generic solutions became very popular due to their "process simplicity" and the broad availability of the amine itself. However, the level of severity has changed making higher demands on existing or even grassroots plants and their subsequent designs. This raises the question, how safe are today's predictions and safety margins?

For instance, a reliable prediction of the "H_2S selectivity" or the CO_2 pick up respectively is not a trivial aspect in the simulation as its impact on sizing and performance can be substantial. In particular, plants being designed and operated at the edge need more attention, but what does this mean in reality? By nature, low partial pressure and/or high temperatures are "natural enemies" of H_2S selective operations and can create headaches for designers and engineers. The achievement of tighter H_2S specifications in low pressure tail gas service or a high degree in operational flexibility of natural gas plants are the new challenges, especially in warmer regions. While tail gas plants usually remain more or less at the same feed gas conditions over their lifetime, the complexity in natural gas plants may strongly increase for the following reasons:

Firstly, a natural gas plant should be suitable for all operational scenarios including summer and winter cases.

Secondly, the feed gas specification of those plants can vary over its lifetime. However, for plants using their treated gas as sales or pipeline gas, the treated gas specification is often fixed to a few ppmv of H_2S and 2-3 vol-% of CO_2. By means, a certain flexibility in selectivity is mandatory.

Thirdly and one should recognize that it is not only the feed gas specification, but also the feed gas rate that may change by the addition or closure of wells and last but not least contractual considerations, requiring different turndown rates. As the dimensions of absorbers and the type of internals are usually fixed, this can trigger substantial fluctuations in terms of H_2S selectivity itself. Usually, good engineering practice allows the addition of safety margins, but to what extent?

From a design perspective, high performance internals are able to cover a wide range of gas liquid ratios while multiple amine feed points provide flexibility in absorber height. However, for an existing plant, operators only have a limited degree of freedom as listed below to adapt the operation to the new situation:

1) Changes in amine feed point (if possible),
2) variations in circulation rate,
3) adaption of the reboiler steam feed rate and possibly a change in lean amine temperature.
4) Further measures, however, with minor impact are changes in the pressure of the flash vessel and the regenerator.

But what about using a promoter or an adaptive promoter technology in the amine systems?

For a long time, the simple acidification of amines was a very "handy" option to boost H_2S selective applications by lowering the H_2S treated gas specification as well as the energy demand based on the following effect.

H_2S is a stronger acid than CO_2, and has a stronger binding energy at the stripper temperature. Consequently, amine regeneration of H_2S down to trace levels becomes progressively more difficult. In particular, designs with very low H_2S specifications in the low ppm-range require a promoter in order to release H_2S more easily from the solution. In this regard, MDEA acidified with phosphoric-, sulfuric- or other acids, becomes regenerated more easily. Even the unintentional "acidification" by formation of heat stable salts (HSS) and their acids - acetates, formates, sulfates, etc. - will also lead to a very efficient H_2S regeneration "boost". However, the level of acidification has its constraints, as there is a strong impact on the equilibrium curve of the acid gas solubility in the solvent.

Today, an alternative for acidified amines is provided by an advanced promoter technology from BASF, which incorporates the advantage of an acidified amine blend, but without the disadvantage of a reduced acid gas capacity. Its mechanism is based on influencing and shifting the pH- and temperature dependent equilibrium load into the regime being required for the individual application.

Case Studies

Two cases studies have been selected to demonstrate the capability of such a promoter technology when working at the edge.

Case 1: Tail Gas Treatment

A tail gas treatment unit (TGTU) has been originally designed for a H_2S treated gas specification of 250 ppmv. Over the years, concepts have been developed to achieve tighter H_2S specifications below 100 ppmv due to upcoming new regulations and a new total sulfur management of the entire plant. Meanwhile, the CO_2 in the feed gas to the TGTU has slightly increased to 51 vol-% over the years. A stepwise acidification of the amine has been applied to reduce the residual amine lean load and the subsequent H_2S specification in the treated gas. However, the reduced acid gas loading capacity required a compensation by increasing the amine circulation rate. As a matter of fact, CO_2 coabsorption increased as well, moving the maximum acid gas capacity closer to the equilibrium. As a consequence, operability of this plant became highly sensitive as H_2S breakthrough could occur suddenly. Meanwhile, a tighter H_2S specification below 50 ppmv has also been considered. By taking current operational data into account, a combined optimization by acidification (H_3PO_4) and circulation rate adjustment at maximum reboiler duty have been considered to get as close as possible to the new requirements. Simulation data have shown a well delimited optimum, however the pinched vapour phase H_2S concentration profile in the absorber clearly demonstrates this as only an academic approach. By contrast, the application of an advanced promoter system from BASF with mentioned properties and its respective performance show a significant higher safety margin combined with less sensitivity on external fluctuations, such as lean amine temperature.

Case 2: Natural Gas Plant (winter/summer scenario)

A natural gas plant has been designed to meet sales gas specification, by means < 5 ppmv H_2S and < 2 vol% CO_2. Feed gas specification has changed over the years due to some new resources. CO_2 concentration increased above 10 vol-% while the H_2S specification remained at the same level. As the absorber height was limited to 25 trays, an increase in circulation rate was the only viable measure to match the CO_2 specification. During the winter season, this scenario worked properly at lean temperatures of around 113 °F (45 °C). However, during the summer season (lean amine temperature up to 132 °F, 55 °C), this kind of operation consequently did not meet specifications. This was mainly that due to limited reboiler and lean rich exchanger duty, a further increase in circulation rate only led to an increase in lean amine loading and a subsequent failure of the H_2S specification. The addition of an acid to the amine (as regeneration promoter) has been considered to lower the H_2S lean amine loading and to match the H_2S specification. But now the problem was shifted to a lowered CO_2 equilibrium load. In other words, the drop in pH value showed a significant impact at elevated temperatures and could not be compensated by the lower overall acid gas lean load. Consequently, plant capacity needed to be reduced by 8 percent.

As circulation rate and reboiler duty have been already at their limits, an alternative design has been provided with the advanced promoter system from BASF. The higher equilibrium load (compared to acidified amine) has provided sufficient capacity to meet the CO_2 specification of 2 vol-%. Furthermore, the improved H_2S regenerability at a slightly reduced reboiler duty gave enough room the meet the H_2S specification even with a certain safety margin.

Summary

Advanced H_2S selective acid gas removal technologies have become increasingly important for existing and new tail gas- and natural gas- treatment applications.

While tighter H_2S specifications are a particular target of tail gas units, the design and technology of world scale natural gas plants focus more on an adjustable CO_2 slip and a high degree in operational flexibility.

Today's plants require a very high degree in flexibility for turndown rates, changing feed gas specifications and conditions. In addition, very tight treated gas H_2S specifications in the low ppm range (e.g <5 ppmv) are becoming mandatory.

As low pressure (e.g. in tail gas units) and elevated amine temperature (e.g. in natural gas plants in summer operation) are "natural enemies" of H_2S selective designs, generic amines and straight forward designs might easily come to its limits and thereby not provide the optimum solution for meeting needed specifications.

MDEA-based technologies are attractive due to competitive solvent prices and generic designs. For OPEX, CAPEX and specification purposes, acidic promoted systems are in use to overcome the relatively high H_2S binding energy and to lower residual H_2S loading in the amine.

However, current acidified state-of-the-art systems also have their limits, especially when operating at the edge as acidification will also trigger an inverse effect on the entire acid gas capacity.

Alternatively, a new advanced promoter technology from BASF has been presented based on two case studies. It provides a substantial increase in regeneration ability while hardly affecting absorption capacity.

(34d) Sour Water Strippers Exposed

Ralph H. Weiland , Optimized Gas Treating, Inc., Clarita, OK and Nathan A. Hatcher, Optimized Gas Treating, Inc., Buda, TX

Sour Water Stripping: The Process Exposed

Ralph H. Weiland, Nathan A. Hatcher

Optimized Gas Treating, Inc.

Sugar Land, TX

ABSTRACT

The sour water system is often considered to be a refinery's sewer. It consists of water used for quenching and scrubbing HDS cold separators, sulphur plant vent gas, blow-down from amine regenerator reflux sections, and waste water from a variety of other operations. Its major components are ammonia and hydrogen sulphide with usually lesser amounts of carbon dioxide and heat stable salts (HSSs are mineral acids), and sometimes amine.

Sour water strippers (SWS) are moderately-large reboiled towers (30 to 60 trays) in which ammonia and other gases are removed from the sour water by steam stripping. Heretofore they have been designed using equilibrium stages. However, tray efficiencies have remained obscure with quoted values anywhere from 15% to 40%, a factor of two range. Consequently, designers have less than complete confidence in the reliability of their final design. The consequence of uncertainty is either overdesign and, therefore, excessive costs or sleepless nights because of an underperforming unit.

Recently a mass transfer rate-based simulation model has become available for designing and troubleshooting sour water strippers. In this paper, we use the model to determine tray efficiencies for ammonia and H_2S stripping, how they vary across the height of a tower, and what operating variables affect them, and how. We also predict quantitatively how the presence of heat stable salts and caustic injection for pH control affect the treat-ability of sour water to specified residual levels of ammonia and H_2S.

(34e) Fate of Ammonia in Refinery Amine Systems

Nathan A. Hatcher, Optimized Gas Treating, Inc., Buda, TX and Ralph H. Weiland, Optimized Gas Treating, Inc., Clarita, OK

Fate of Ammonia in Refinery Amine Systems

Ralph H. Weiland, Nathan A. Hatcher & R. Scott Alvis

Optimized Gas Treating, Inc.

Sugar Land, TX

ABSTRACT

Ammonia ingress and accumulation in refinery and bio-gas amine systems is not a new problem. However, increasing utilization of advantaged crudes with higher nitrogen content may present challenges in today's capital-constrained operating environment. Many refiners have instituted guidelines for purging amine regenerator reflux water for corrosion control. Historically, this has been done empirically based upon periodic lab analysis and adjustment of the purge water rate.

The true amount of ammonia ingress, and its material balance across refinery amine unit is not a topic that has been discussed in great detail, because until now, rate-based mass transfer models have not been available. Using the well-known ProTreat rate-based mass transfer process simulator, this paper addresses the following questions. Where available, comparisons to plant data measurements are provided.

How much ammonia can accumulate based upon choice of regenerator operating conditions?

How much ammonia rejection into the amine acid gas does this correspond to, and is this a significant concern to downstream (i.e., sulfur plant) operations?

Can ammonia build to levels that will cause additional H_2S to be trapped leading to higher lean loadings, reduced treating performance, or even regenerator flooding?

How does the choice of HDS cold separator operating conditions impact the ingress of ammonia to the amine system; just how effective is water washing?

How much ammonia skates through refinery amine treaters?

Ultimately, we attempt to answer whether the ammonia balance on a refinery amine system can be fully characterized based upon the knowledge of a few simple parameters.

ADVANCED CONTROL OF MULTIPLE SULFUR RECOVERY UNITS

Darwin E. Logerot, ProSys, Inc., Baton Rouge, LA

Abstract

Sulfur Recovery in a Refinery entails collecting sour vent gases (containing H_2S) from a number of process sources, including Hydrotreating Units, Hydrocracking Units, Crude Units, and others, then converting the H_2S to a more useable form such as sulfuric acid or elemental sulfur. The feed rates and compositions to the Sulfur Recovery Units (SRUs) are not controllable at the SRU. In addition, most Refineries employ multiple SRUs operating in parallel. This gives rise to a number of process control challenges, including:

Load balancing among the various units
Response to changing loads (total feed load cannot be controlled)
Response to variable compositions
Air/Oxygen load controls
Response to one SRU shutdown (rapid load increase for the remaining SRUs)

This paper will present the ramifications of these control problems along with the methodology used for successful implementation of the advanced control scheme for each.

The Process

Before discussing the advanced control aspects of SRUs, a basic process description is in order. Although the control concepts included in this paper can apply to any sulfur recovery process, the specific type discussed in detail is the Claus process. A Claus unit converts hydrogen sulfide to elemental sulfur.

Figure 1, in the Appendix, shows an overview schematic of the process described herein.

The feed gases to the SRU come from multiple sources in the refinery. Each source will be one of two basic types, either from an amine regeneration unit or from a sour water stripper. The amine regenerators strip H_2S out of an amine stream, after the amine had been used to absorb H_2S from a sour gas stream in one of the refinery units. The sour water strippers remove H_2S and ammonia from refinery waste water.

The feed gas to an SRU is very high in H_2S, with the primary impurities being water vapor, carbon dioxide and trace amounts of amine, hydrogen and light hydrocarbons. The gas from the sour water strippers also contains ammonia. The gases from the two types of sources are usually kept segregated until they arrive at the SRU.

The first step of the process is partial oxidation of the H_2S in the feed gas to form SO_2:

$$H_2S + 1.5O_2 \longrightarrow SO_2 + H_2O \qquad (1)$$

The oxygen fed to the combustion chamber (shown as the partial oxidation burner) is usually supplied from fresh air by an air blower. Many units also supplement the combustion air with pure oxygen to increase residence time and reduce pressure drop.

Two combustion zones are usually employed. The ammonia-rich (sour water) acid gas is fed to the first zone, and the amine acid gas is split between zones. Details of the split are beyond the scope of this paper, but if not done properly, salts tend to form in equipment downstream of the burner causing corrosion and plugging.

The resulting gas mixture is then run over a catalyst bed to accomplish the reaction to sulfur:

$$2H_2S + SO_2 \longrightarrow 3S + 2H_2O \qquad (2)$$

This reaction is usually conducted in several stages, with sulfur condensation and re-heating steps between stages.

The goal of the reaction section is complete, or nearly complete, conversion of the H_2S in the inlet gas. The stoichiometry of reaction (2) dictates that complete conversion is only possible if the oxygen / H_2S ratio at the inlet of the combustion chamber allows exactly 1/3 of the H_2S to be converted via reaction (1). The stoichiometric ratio must be controlled closely; otherwise there will be either excess H_2S or SO_2 in the tail gas.

After the reaction step, there should be only a small amount of sulfur compounds (SO_2 or H_2S) in the tail gas. These compounds must be removed before the tail gas is allowed to vent to atmosphere. Removal of trace sulfur is accomplished by a reduction step, usually using hydrogen to convert any remaining SO_2 to H_2S, followed by H_2S absorption into a circulating amine solution and finally, a tail gas incinerator.

Control Problems

The primary control issues in an SRU arise from two basic areas. Each introduces its own set of challenges for the control system.

1) Process factors:
    - Oxygen to sulfur ratio control at the front end is critical and difficult. The composition of H_2S or SO_2 in the feed gases is usually not known and varies. If the ratio is off in either direction, the result will be excess H_2S or SO_2 to be handled in the downstream equipment. This may overload the reduction, absorption and incineration sections, and may result in excess sulfur emissions.
    - Pure oxygen is expensive, but use of too much air instead of oxygen reduces residence time and increases pressure drop, affecting the oxidation rate and reducing throughput capacity.

    - Each unit has a set of operating and safety constraints, including maximum equipment capacity, MAWP, max temperature, maximum oxygen concentration, and ultimate emissions limits.
    - Proper split of amine acid gas between combustions zones is important to avoid salt formation and corrosion.
2) Most refineries use multiple sulfur recovery units in parallel to handle sour gas from multiple sources, and the source gas rates are dictated by the upstream units. As a result of this factor, the control system must:
    - Respond to uncontrolled load changes. These can be rapid and significant.
    - Ensure that the wild load is properly balanced so that one train is not overloaded while another runs well below its capacity.
    - React rapidly to a sulfur unit trip, raising the loads on the other sulfur units to maintain refinery stability.

In summary, the control problem is a balancing act that must consider all the above factors, keep the units onstream within acceptable operating limits, and maintain sulfur emissions below the permitted maximum.

Control Solutions

The control scheme discussed herein that solves the above problems is based on advanced regulatory control (ARC). ARC basically is an advanced control scheme that does not rely on process models, only on PID control blocks, calculations and selectors. The heart of the control solution is a pressure based load balancing ARC, diagrammed in Figure 2. This scheme uses a header pressure controller as the master. The master pressure controller cascades a setpoint to a total feed gas flow controller. In turn, the total flow controller cascades to a separate feed gas flow controller for each parallel train. As header pressure rises, feed rate targets to each train are increased to handle the extra gas in the system. A key feature of this scheme is the use of adaptive gain in the total flow controller. This means that the proportional gain is adjusted based on the number of Sulfur Recovery units available. A Sulfur Recovery unit is considered unavailable if the unit is down, if the SRU feed flow controller is at a constraint limit, or if the SRU feed flow controller is not in cascade mode. As a result, the overall scheme can react rapidly and appropriately to a sudden change in gas handling capacity, such as a unit trip or a new unit coming on line.

Discussion of the control scheme would not be complete without including the constraint features. Each individual flow controller can be constrained by any one of a number of limits. The constraints can reflect unit capacity, maximum air or oxygen availability, equipment pressure and temperature limits, etc. If any of the pre-defined limits are reached, the train flow controller will no longer accept additional load requests from the total flow controller. In turn, the adaptive gain on the total controller will adjust itself, since a train at a constraint limit is treated as an unavailable unit. So, if one train is limited by a constraint, the control scheme will adjust with the other available trains. Figure 3 is a diagram of the constraint scheme.

Downstream of the header pressure and unit feed flow control, each SRU has its own set of ARCs, designed to keep the unit operation on target. The most critical of these is the oxygen to sulfur ratio feeding the combustion chamber. Control of the oxygen to sulfur ratio is accomplished by an air/O_2 ARC. This ARC first calculates total oxygen requirement based on the feed gas rates. Total oxygen is trimmed by the output of an excess H_2S / SO_2 analyzer and controller on the tail gas. The total oxygen demand is supplied by air first, then by pure oxygen. The output of the oxygen requirement calculation is the setpoint of the air controller. The air controller can be constrained by combustion chamber pressure. If pressure is at its high limit or if the air controller is approaching saturation, any additional load is handled by pure oxygen. Pure oxygen rate is controlled as the difference between the total oxygen demand and the oxygen supplied by the air controller. This scheme assures that the unit will maintain the proper ratio, and that pure oxygen is only used if truly needed. Figure 4 shows the air / oxygen control layout.

Proper acid gas split between combustion zones is also an important consideration. The detailed methodology for the split varies with unit design, and is beyond the scope of this paper. Basically, the control system within each SRU uses its total acid gas set point along with a calculation algorithm to split the total gas into two zone flow controllers.

Benefits

There are a number of benefits resulting from these control schemes. Perhaps the greatest benefits result from greatly improved feed header pressure control. This results in fewer impacts to upstream units, fewer issues when a SRU trip occurs, no need to send sour gas to flare and fewer permit excursions.

Another benefit is better air & O_2 control, which facilitates oxygen savings (pure oxygen is only used when really necessary). Improved control over air to feed gas ratio also increases average sulfur conversion and allows the unit to reduce emissions.

These benefits are realized from a solution that utilizes advanced regulatory control (ARC) without using model-predictive control (MPC). This means that no external software is required, no process models are needed, and the process response to control action need not be linear.

Before vs After

Perhaps the best indication of the success of the control scheme is a reduction in sulfur emissions excursions. Before installation several years ago in a refinery operating three SRUs, the controls did not react appropriately to upstream issues and did not effectively maintain proper oxygen to gas ratios. As a result, the refinery experienced an unacceptable number of operating incidents and had several air permit violations over the course of a year. After installation of this scheme, no violations were experienced in a full year of operation. The cost of the control installation was recovered in a few months from better operation and the cessation of violations. This refinery can now withstand the sudden loss of an SRU without incident.

Acknowledgements

The author would like to acknowledge the contributions of the following individuals, without whom this paper and presentation would not have been possible:

Dustin Beebe, President, ProSys, Inc.

David Solberg, Associate Engineer, ProSys, Inc.

Appendix

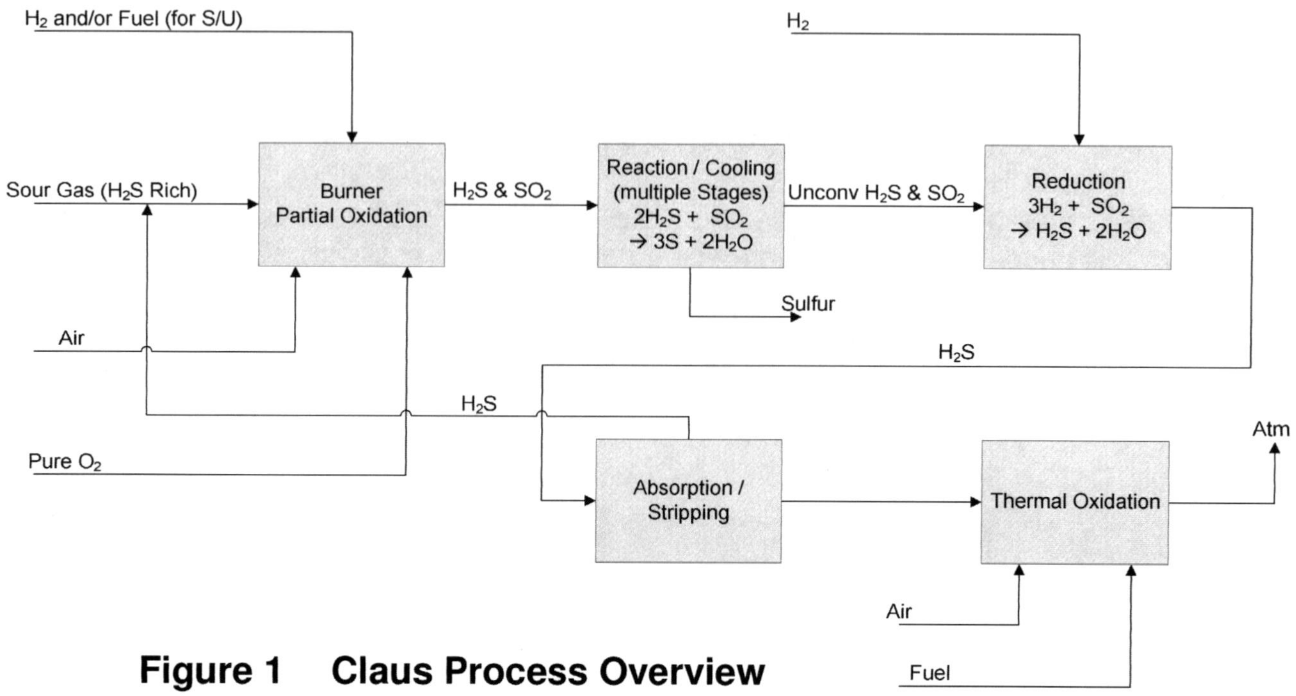

Figure 1 Claus Process Overview

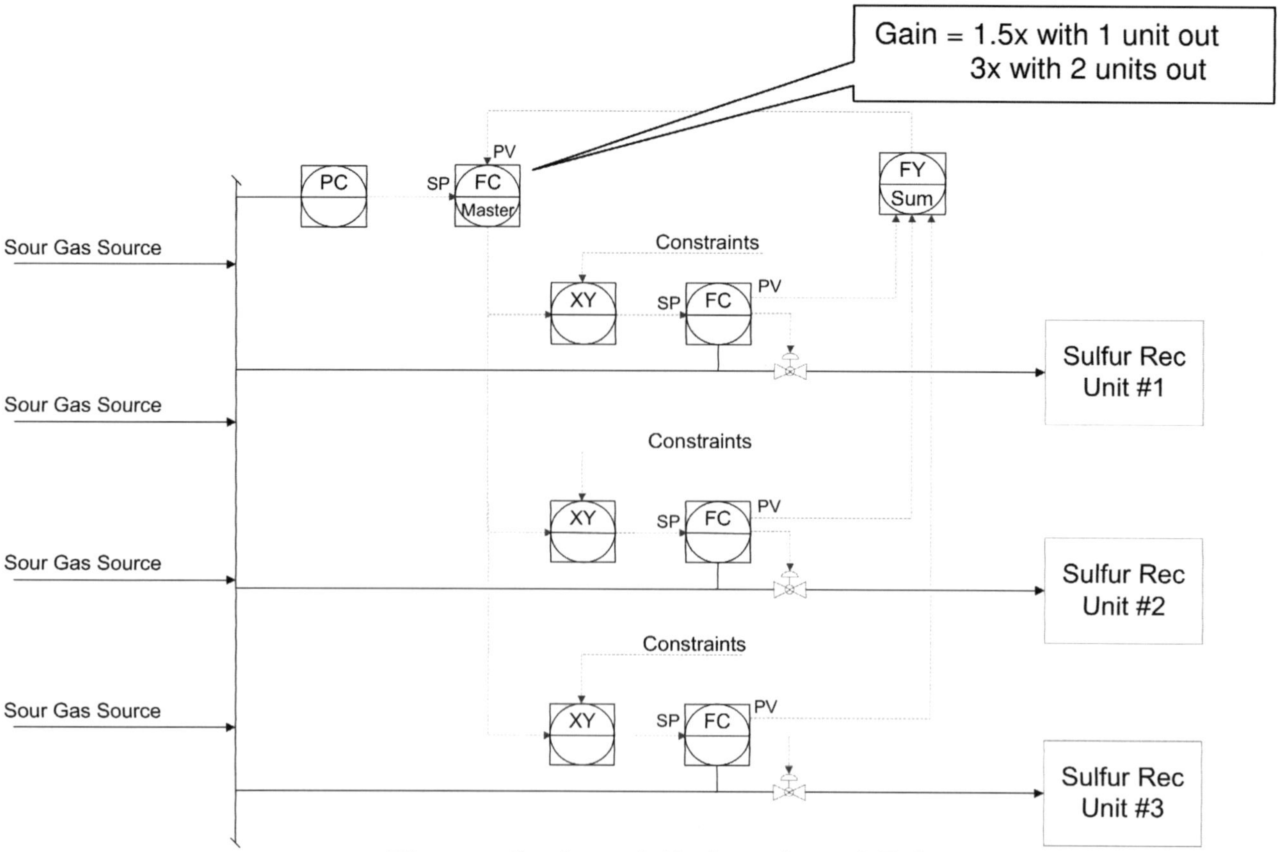

Figure 2 Load Balancing ARC

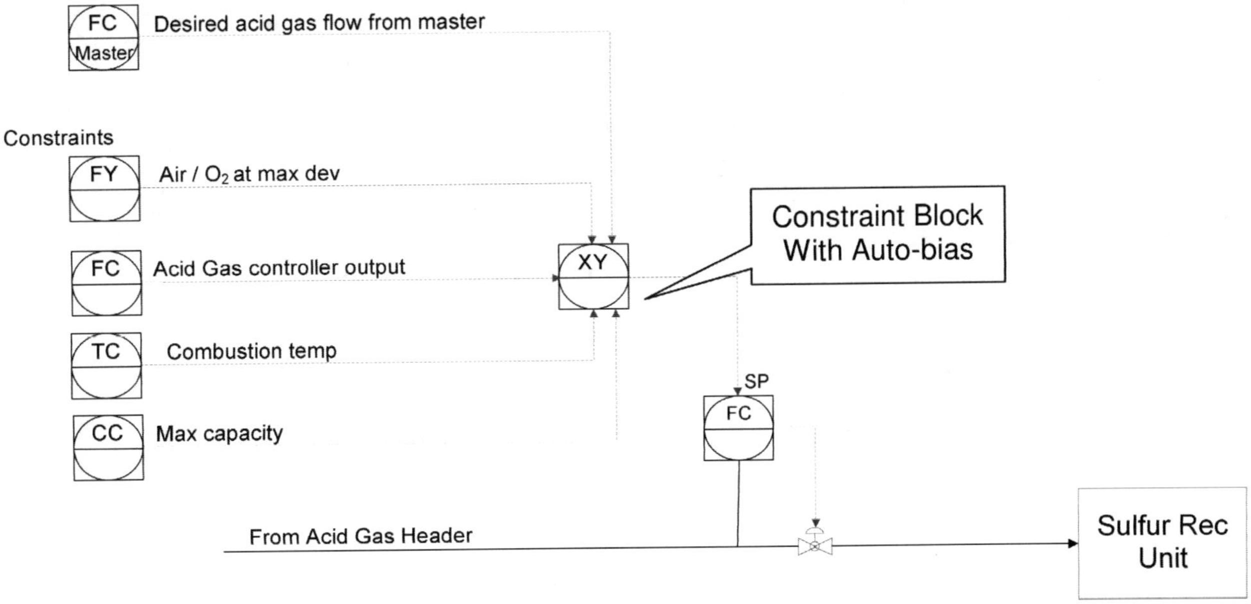

Figure 3 Constraint Configuration

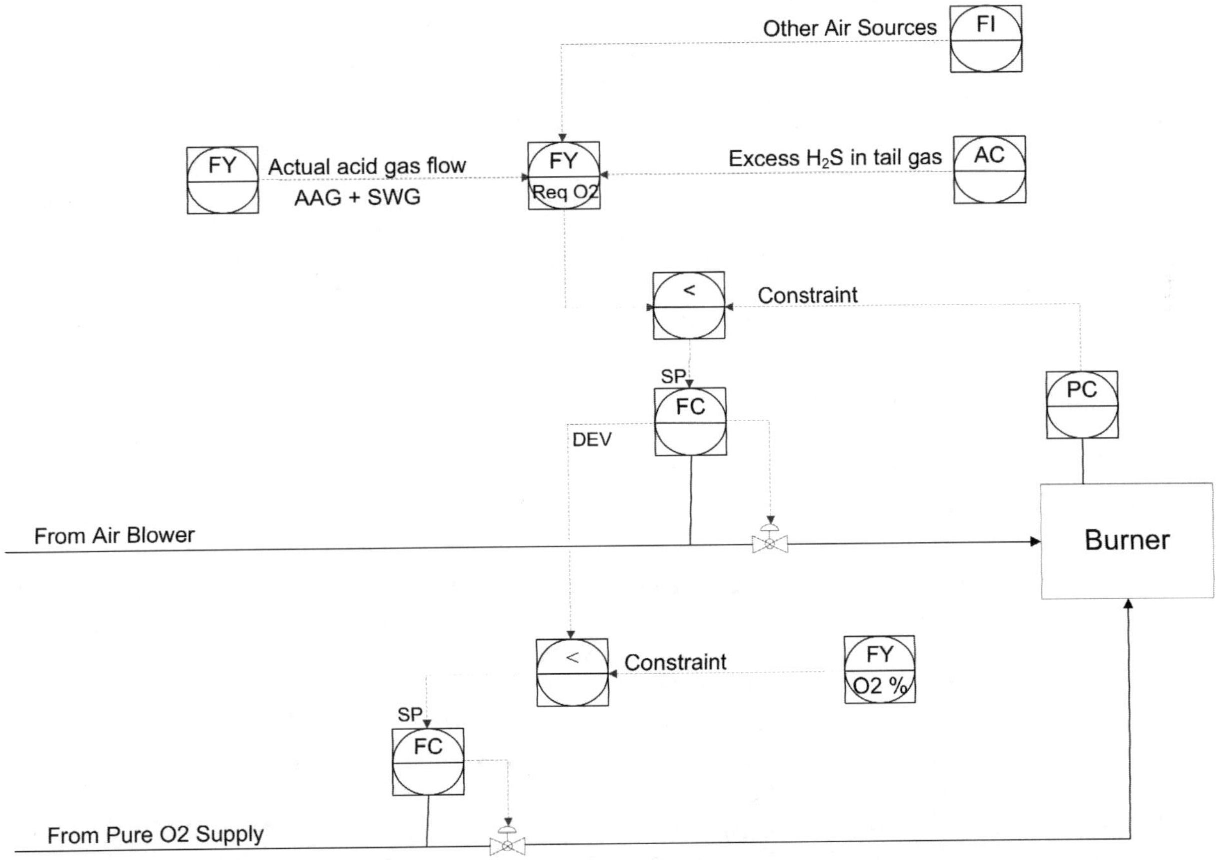

Figure 4 Air / Oxygen Control

Sulfur Recovery and Tail Gas Treating Process Gas Analyzers Stigma, Sins and Salvation

Randy Hauer
AMETEK Process Instruments, USA

Zaheer Juddy
Analytical Instrumentation & Maintenance Systems (AIMS), UAE

John Sames
Sulphur Experts International, Barbados

Presented at the AIChE Spring Meeting
Houston, Texas
April 1-5, 2012

Abstract

Process Engineers have a strong inclination for qualitative and quantitative analysis of process streams. On-Line process gas analyzers are a fundamental requirement for control, optimization and energy efficiency but have a stigma with end users and maintenance managers, sometimes well earned. On-Line process gas analyzers comprise a relatively small share of the capital in a grass roots project but they require detailed attention if they are to be successfully implemented and fully exploited.

The paper discusses practical aspects of the suite of analyzers utilized on a SRU / TGTU complex as well as "best practice" recommendations based mostly on painful experience. The subject is one of the least understood facets of a project. Salvation can be found in the many small details and retaining process analyzer specialists at all phases of a project.

1) Introduction

It would be taken as an article of faith that a chemical engineer and an analyzer engineer would be linked together at every phase of a project. A chemical engineer by definition is most interested and has a large stake in qualitative / quantitative analytical data, the material balance is everything. Sadly this is not usually the case; we communicate through several layers, separated by a caste system.

The objective of the first part of this paper is to give an audience of primarily process design engineers a snapshot picture of process analytics. The authors are grateful for access to two technical papers from a major refining-chemical company who shared their "sins & stigma" experience, this part of the story should only come from an end user.

The second part is a review of analyzer best practices and process analytical innovation for SRUs and TGTUs covering five analyzer applications in some detail. Many of the innovations and enhanced features should be implemented at the FEED, EPC or systems integration (SI) stage but are often not. Better, earlier and direct communication between the process and analyzer disciplines is the solution.

2) Overview of the Process Analyzer Industry

It is worthwhile for a process engineer to have an idea of the breadth and scope of the process analyzer industry. The spend on an individual project is relatively small but the impact far outweighs the cost.

The Big Picture:

- The global process control enterprise is ~USD 136 billion/yr. Process Analytical Instrumentation (PAI) comprises only 6% of this amount, (~USD 8 billion/yr).
- Of this 8 billion the chemical process industries (CPI) accounts for ~70% of all process analyzer applications; further broken down into; USD 2.85 billion on maintenance, USD 1.75 billion on analyzers, USD 560 million on systems integration and USD 450 million on sample handling systems per annum.[1]

The Project Picture:

- According to a major international oil company, integration represents 55% and analyzers 45% of the total cost of an analyzer project. This somewhat contradicts the overall industry figures but can be explained by the fact the CPI spend more on shelters, no doubt influenced by hazardous area (explosion proof) design.[2]
- The 15-year cost of ownership of an analyzer system is equal to the total capital cost of the fully integrated system. Half of this cost is labor, the other half is parts. Of the half invested in parts, ~25% is consumables and ~75% replacement parts.[3]
- Shelter costs for chromatographs are on-par with the cost of the GC and this is when they are ganged eight to a shelter. This is the rationale for locating several analyzers in one shelter, which at times is a source of problem in itself.

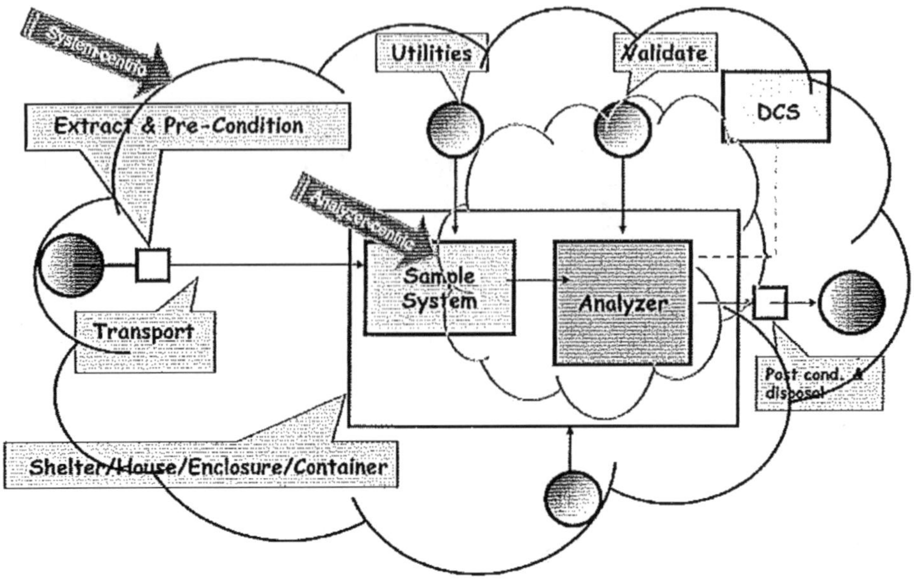

Fig. 1. Analyzer System Scope of Supply (Courtesy of Rob Dubois, "*by-line analytical*")

Trends and Generalizations in the Process Analyzer Industry:
The industry is characterized by widely diverging attributes. It is generally conservative about adopting new technologies, if it works, repeat, repeat, repeat.[1] On the other hand there are significant advances in analytical technology that are game changers in themselves. Here are some general trends;

- There is a revolution in spectroscopy with multi-component measurement capability competing with GCs.[2] Analyzers that are close-coupled to the process requiring very little integration are becoming common. Size and weight matter.
- The price of the shelter and HVAC dominate the price of an analyzer system. It is uneconomical to supply a shelter for 1 or 2 analyzers and large refining / chemical companies are moving away from GCs and analyzer houses for this reason
- One third of all analyzer systems are over-designed; one third are under-designed; perhaps one third are adequate.[4]
- The full capability and features of a process analyzer are rarely utilized. For example; over-range measurement of H_2S/SO_2 in tail gas, COS/CS_2 in tail gas, COS, (CS_2) in TGTU absorber off-gas and Ethernet web-enabled communications.

3) Stigma, Sins, Salvation

The Stigma
Process analyzers suffer a reputation of stigma even from the perspective of dedicated end user analyzer professionals. This is a list of the common complaints.[3]

- High cost (shelters), high maintenance
- Complicated installation, complicated operation
- Low reliability, questionable accuracy

The Sins

Part of the problem is one of experience at the design and EPC level of a project and the amount of time it takes to acquire adequate engineering skills to be addressing the variety of disciplines involved. Most qualified analyzer engineers are employed at the systems integration level and relatively few at the EPC level and practically none at the front end engineering design level. Some examples of how this impacts a project;

- If analyzer engineers had to pick one single problem area that is universal it would be piping design. Piping design done in the absence of a staff analyzer engineer is doomed for problems and rework.
- Instrument data sheets for analyzers that are out of date, 10 years or more, with only minor revisions in between.
- No provision for advances in process analytics, the temptation to define a measurement using a single analytical principle, Gas Chromatography being an example. The process GC is quite simply over applied as a default when a GC vendor is doing the systems integration.
- Critical evaluation of sample system design for specific applications is lacking. Most sample systems are designed based on duplicating previous projects with new features added haphazardly.[3]
- Relative to spending on DCS and discrete devices there are proportionally many more instrument and DCS engineers than analyzer engineers at the EPC level.
- Modern analyzer communications are sadly lacking in an SRU-TGTU project.[4]

The Salvation

The solution lies in retaining an analyzer professional at the EPC and FEED level during a project, any project of any size that has analyzer tags above the "1 to 5" level in terms of complexity (see below). The client must make intervention mandatory. This cannot be someone nominated on an ad hoc basis but rather a recognized full time specialist.

Why is this a problem? No one actually graduates as an analyzer engineer. It is a profession populated by chemical, electrical, instrumentation and mechanical engineers supplemented by physicists, chemists and spectroscopists who migrate to the profession. The skill set can take ten or more years to acquire and demand outstrips supply.

Following are the metrics used by a major oil company based on a three year statistical study of over 10 refinery, oil & gas and sulfur recovery plant complexes.[6] It serves to demonstrate complexity and the nominal maintenance requirements for process analyzers.

Complexity Factor	Type of Analyzer	Estimated Man-hours/month Maintenance
1~5 (Simple)	pH, conductivity, gas detection, O_2	2
6~8 (Physical Property)	Boiling point, flash point, freeze point, RVP, viscosity, etc	3
9 (Environmental)	CEMs SO_2, CO, H_2S, Opacity,	2.5
10~15 (Complex)	Tail gas, GC, Mass Spec, NIR, FTIR	4

Table 1. Grouping of Analyzer Categories for Maintenance Purposes

4) Sulfur Recovery and Tail Gas Treating Analyzer Applications

Following are the four principal on-line analytical measurements in a SRU-TGTU complex as well a new application being driven by safety. Two other applications (sulfur pit vent gas and SO_2 breakthrough to the TGTU) are not specifically covered here

4.1) SRU Acid Gas Feed (H_2S / HC / CO_2 / NH_3)

Feed (acid) gas analyzers are widely applied if not well-understood or fully implemented. Approximately 15% of existing SRUs and slightly more from new design have an H_2S analyzer and now more common, a combined analyzer for H_2S, THC (total hydrocarbon) NH_3, and CO_2. Some analyzer solutions have been over-designed and relatively few have been put into closed loop control.[7] Recently a major SRU engineering designer introduced a complete system including control algorithm and analyzer to achieve the desired result of continuous closed loop control realizing a 0.2% long term improvement of recovery efficiency (15% reduction in emissions) and mitigation of HC upsets to the process.[8]

Solutions have been developed that combine UV and IR detection to measure the H_2S and total hydrocarbon as well as the CO_2 content thereby providing an almost "full analysis". At first glance, a total hydrocarbon (THC) result would have limited value for feed forward control, however, infrared measurement can quantify the carbon count up to C_6. The utility lies in taking the THC result, as total carbon count (methane equivalent) and converting it to "air demand" using the same scale so it can be implemented into the control loop.[11]

Feed gas analysis has possibly the most benefit for the TGTU. It is somewhat counterintuitive, but it is not the sudden appearance of hydrocarbons that makes for problems, but the sudden disappearance. In the absence of feed composition information, the first indication of hydrocarbons in the acid gas is H_2S in the tail gas spikes up, SO_2 goes to zero; the control response is to "add air". Conversely, when the hydrocarbon episode suddenly ceases, SO_2 in the tail gas spikes up and this potentially causes damage to the TGTU amine since it is difficult to reduce the air before SO_2 breaks through the reduction reactor in the TGTU. The ability to anticipate the end of an HC episode at the front end before it transits the SRU to the TGTU is valuable in itself.

Best Practices

- The measurement has to be real time (not a GC). Speciation of the HC is neither required nor possible given the real time constraint.
- Sample system: Make it "hot-wet", it is not desirable to remove the water and in fact it adds an unmeasured bias by reporting the analysis on a dry basis.
- Sample system safety: Buy a solution that incorporates return of the spent gas to process, purging with nitrogen for maintenance purposes. If it frightens the analyzer techs it will fall into disrepair.
- Sample point: As far upstream from the reaction furnace as possible, preferably 5 seconds or more of transit time which approximates the analyzer response. For simplicity and response place the analyzer in close proximity to the acid gas line.
- Sample point safety: Do not place the analyzer in a cabinet and certainly not in a house. The analyzer should be IP-65 for outdoor installation and free ventilation.

- Indicating control is sufficient to realize most of the benefits. A method of simple binary control (e.g. add +5% air for HC upset of x magnitude) can be implemented. Closed loop control with safeguards and customization can be purchased as a package.[8] Operator training is essential so the analyzer is "understood".

4.2) SRU Tail Gas (H_2S / SO_2 / COS / CS_2)

With relatively stable feed compositions, flows and properly operating feedback control of the combustion air flow, control of the excess air to within +/- 0.5% is generally achievable resulting in typical losses of 0.2% for a three stage SRU. Unstable feeds and poor or non-existent control can result in significant losses of 10% or more in the worst cases. The tail gas analyzer, feedback control accounts for up 2% to 4% of recovery efficiency and can contribute more to overall SRU performance than a third converter.[10]

Unlike an acid gas (feed gas) analyzer a case does not need to be made for a tail gas analyzer in the FEED they are considered mandatory. There are two types of tail gas analyzer design; close coupled ("top of the pipe") and sample line type)

Best Practices

- Heat integrity is everything: All too often the sample point has excessive surface area or mass that defeat the electrical heating function of the analyzer (restricted to T-3 norms). Steam jacketing, or ControHeat™ is essential if the nozzle is greater than ~6" in length. MP steam is much preferable to wet LP steam which is better described as a heat sink rather than a heat medium.[11]
- Safety: Do not allow a piping mistake to restrict reasonable access to the sample point. Make this a Haz-Op check list item reviewed by analyzer maintenance.[12]
- Safety: It is unforgiveable not to connect the tail gas analyzer (all SRU analyzers) to the Ethernet analyzer communications system and yet this is very seldom done.
- Vertical tail gas pipe: When a SRU-TGTU complex is designed as a single unit, seldom is consideration given for a proper sample point on the process pipe connecting the final condenser to the inlet of the TGTU. Mounting the tail gas analyzer on a vertical section of pipe is possible but not desirable. Vertical piping sections of 18" or less are totally inadequate for even a sample line based analyzer
- If the SRU is a sub-dewpoint, selective oxidation (Superclaus) type or utilizes TiO_2 catalyst then the tail gas analyzer should be fitted to measure COS / CS_2
- The indicating outputs of [%H_2S] / [%SO_2] should be ranged at least 0-5% so they can be used during upset conditions by the operators and give them confidence the analyzer is operating. This simple feature is never implemented at the EPC stage; we remain slaves to the instrument data sheet first configured circa 1970.[13]
- Moving a tail gas analyzer upstream: It can be done for compelling reasons (turndown, selective oxidation) but be aware of vibration and inline burner issues. Be sure to include the vendor before relocation of the analyzer.

Trends

- The majority of tail gas analyzer installations are top of the pipe (70%) as they save money on shelter and infrastructure. That said, climate, analytical performance and structure constraints often require a sample line based analyzer, in these cases do not make the sample lines longer than 30' if at all possible.

- Hot condensate back flush option for ammonia salts should be selected for all SRUs burning SWS acid gas. It may never be needed but invaluable when it is.
- For safety purposes many some users require "double block" isolation for the tail gas analyzer. This is possible with both types but with some compromise. Do not consider "bleed" with the double block, S_V complicates this and is vented back to process in any case.
- Simple shelter or cabinet: the cost of a single house for one analyzer is prohibitive.

4.3) TGTU (H_2 / H_2S / COS / CS_2)

The most important measured parameter in an amine TGTU is the hydrogen content in the gas exiting the reduction reactor. Good control and measurement of excess H_2 is important to provide stable operation and protect the amine from SO_2 breakthrough. H_2S is a secondary measurement and H_2S can be measured at one of two sample points (or both) depending on the design and requirements. The typical two sample points are the top of the quench tower and top of the absorber column. In addition, the measurement of COS can be made at the top of the absorber.

Up until ~1995 the process GC was the most widely applied analytical method to monitor hydrogen content. Although a GC gives accurate results, it has several disadvantages not the least of which is the technique is non continuous. The TCD detector does require separation to isolate the H_2; in effect it can be treated as a binary mixture. The application was first addressed with discrete TCD and UV analyzers, and later combined the two devices into a single "hot-wet" measurement with an application specific sample system.[14]

Best Practices

- Select a common sample point for the H_2/H_2S, for this combination it does not matter if it is the outlet of the quench or the absorber. If the design has H_2S measurements at both locations consider a redundant H_2 sensor, the cost is minor.
- If there is a need to measure COS/CS_2 the sample point needs to be at the absorber outlet and in fact the additional measurements are only a marginal price addition.
- Locate the analyzer as close to the sample point as possible but also utilize the pipe run to minimize sample line length and locate the analyzer close to grade.
- The sample system has the same requirements as the feed gas analyzer in terms of safety. The H_2S is one (to three) orders of magnitude less vs. acid gas but "frightening" just the same.
- If (environmental) procedures allow for the quench and absorber to be bypassed at start up a temporary sample point at the quench inlet can be designed in for the purpose to stabilize the SRU (H_2) before bringing the quench / absorber on line.[14]
- Amine additives can interfere with the TCD, the solution is to add a Glysorb filter
- As for any SRU analyzer, run Ethernet connection for safety purposes.

Trends

- Of the ~130 analyzers installed by a vendor~110 are at the top of the absorber, the balance at the top of the quench with a few at the inlet to the quench.
- Of the ~110 units installed at the top of the absorber only ~10 have COS, 2 have CS_2 as well, at least one installation is using the H_2S+COS+CS_2 as a pseudo CEM.

- Replacement of process GCs with a combined UV/TCD analyzer is prevalent; it is a mistake to replace a legacy GC with GC.
- SO_2 can be monitored before the quench (in combination with H_2). It requires water removal and has not been widely applied.

4.4) Continuous Emission Monitors (SO_2 / H_2S / Mass)

Depending on the process configuration, SO_2 emission can vary from ~8,000 ppm for a 3-stage Claus SRU; ~1,500 ppm for enhanced Claus; ~100 ppm for standard design TGTU; to single digit ppm SO_2 in some local air quality districts. The measurement standards vary slightly with most SRU/TGTUs in the U.S. measuring SO_2 on a "cold-dry" basis according to the EPA part 60 guidelines. There are alternatives to measuring on a cold-dry basis that are well suited to SRUs and some environmental jurisdictions consider an SRU to be a special case in this regard.

Best Practices

- Consider the CEM to be a specialty item. Consider it to be included in the suite of SRU-TGTU analyzers, not a commodity to be lumped in with all other CEMs
- What sets it apart is incineration of sulfur compounds and how they are measured. Formation of SO_3 is not uncommon and the sample system needs to be designed for this service.
- If the environmental regulations call for SO_2 to be measured in mass terms then the analyzer must use a "hot-wet" technique and not a "dry basis". It is fundamentally wrong to measure flow on a wet basis, SO_2 on a dry basis and use the product of these two parameters to calculate mass flow of SO_2.
- Consider dual ranges (up to 100:1 is possible) if the bypass of the TGTU or enhanced Claus process requires this, it is possible in one analyzer

Measurement of Residual H_2S in the Incinerator Effluent

It is well-known and accepted that equilibrium values of reduced sulfurs exist after incineration with the assumption these are negligible if the incinerator is operated according to design. Proposed US-EPA legislation, [sub part J(a)] was going to require measurement of residual H_2S for refinery SRPs (sulfur recovery plants, defined as SRU +TGTU), but the law was not been promulgated and is currently on hold. In addition, some gas plant SRUs, most notably in Alberta, Canada, have been allowed to increase the amount of reduced sulfur compounds in the interest of saving fuel gas in the incinerator and are required to measure un-combusted H_2S.[15]

There are several ways to make the measurement. Where the H_2S is present in higher concentrations such as the energy optimized incinerators in Alberta, H_2S is (and should be) measured separately from the SO_2. Where the incinerator is optimized for destruction of reduced sulfur compounds after a TGTU, the consensus method has been to oxidize any un-combusted H_2S and other S compounds and residual sulfur and measure as total SO_2 thereby "accounting" for the H_2S. By oxidizing, the analyzer sample system essentially completes 100% of the oxidation job of the incinerator. Some users have reserved the option of segregating H_2S and measuring separately from SO_2 which can be done, but the challenge of resolving 7-8 ppm H_2S in a background of ~150 ppm SO_2 adds complexity. In-situ TDLAS laser techniques have been applied to this application with some success.

4.5) Process Combustion Stoichiometry (O_2 / CO) for Start Up

During start-up (shut-down), as an SRU transitions through fuel gas warm-up to the introduction of acid gas, measurement of O_2 is required. The requirement for most refineries is to start up the entire sulfur recovery plant (SRU + TGTU) in a single sequence with zero tolerance for emission exceedences. Historically, operators have manually taken samples using a portable electrochemical type O_2 analyzer. While giving more or less satisfactory results, the requirement of more stringent operating limits and hazard exposure are reasons to consider a permanent solution.[16]

For this application the most favorable detection principle is tunable diode laser spectroscopy (TDLAS) which measures "actual O_2" and TDLAS has become a favored method of "process" oxygen analysis. Procedures vary as to the amount of free oxygen, recommended during a fuel gas warm-up burn but the strategy should result in no more than 0.25% residual O_2. CO is a very useful parameter for combustion stoichiometry, an absorber in the infrared spectrum and can be integrated into the O_2 analyzer.

The sample handling requirements are similar to a tail gas "ratio" analyzer. While the measurement is not required after acid gas has been introduced and in steady state, there is a period to which this operating mode must be anticipated and the sample system is designed for this duty. A single analyzer can switch between SRU trains or sample points depending on requirements .[16]

5.0) Conclusions, Recommendations

1. As a process design engineer you need to be in direct communication with process analyzer specialists and vendors (with your analyzer engineer sitting next to you).

2. Retain a permanent and dedicated analyzer professional at the EPC / FEED for all projects. The resource for this is www.ADSymposium.org (ISA Analysis Division).

3. Utilize the expertise of specialized vendors at an early stage of FEED design as well as EPC detailed design before the award of a Systems Integration contract. Do not allow good design up front to be subordinated to bad decisions at the end.

4. Be an informed consumer; be wary of (over) applying process GCs on tags where they do not belong. Stay actively involved with SI decisions, SI process and avoid "change order" costs this way.

References:

1) Walton, Stephen., T. McMahon, J. Tatera, "PAI 2009-2012, Analysis & Forecast of Global Markets for Process Analytical Instruments Products & Services", 55th ISA Analysis Division Symposium, New Orleans, Louisiana, April 2010

2) Walton, Stephen., T. McMahon, "Survey of Process Analyzer Systems Integrators", 2010.

3) Podkulski, D. and J. Gunnell." Stop Buying Analyzers". International Forum for Process Analytical Chemistry, January 2011, Baltimore, Maryland

4) Novak, D., "Process Analytics: Are There Dinosaurs Among Us? - The Stigma of Process Analytics". Panel Discussion, ISA Automation Week Houston TX, October, 2010

5) Dubois, R., "Process Analytics: Are There Dinosaurs Among Us? - Myths and Mistakes That May Contribute to Our Extinction". Panel Discussion, ISA Automation Week Houston TX, October, 2010

6) Prakash Rajagopal, Chevron USA. Conversation(s), January-February 2009.

7) Harris K. and R. Hauer. "Measuring Total Hydrocarbon and H_2S in Amine Acid Gas Streams." *Hydrocarbon Processing Magazine*, February 2009.

8) Henning A. and G. Molenaar. "Next Generation SRU Control with ABC+" *Sulphur Magazine*, no.326, (January/February 2010).

9) Myles T, S. Langridge, N Truong "The Advantages of Combined Infrared and Ultraviolet Spectroscopy for Process Measurements." 57th ISA Analysis Division Symposium, Anaheim, California, April 2012

10) Bohme, Gerald and John Sames. "The Seven Deadly Sins of Sulphur Recovery." Paper presented at the Sulphur Conference, Calgary, Alberta, Canada, October 1999.

11) Al-Misfer, A., S.Vedula, R. Hauer, Z. Juddy, "Process Analyzer Best Practices for Sulfur Recovery, Enhanced Claus and Tail Gas Treating Applications". SOGAT Conference, Abu Dhabi UAE, March, 2009

12) Harris, P., "Sample Handling Systems". O'Brien Analytical Users Conference, San Diego CA, February, 2010.

13) Hunt, C., J. Sames, R. Hauer. "Understanding Claus Process Upsets Using Your Tail Gas Analyzer," *Sulphur Magazine*, no.256, (May/June 1998).

14) Harris K., R. Hauer, D. Potter, P. Harris, and B. Lewis. "Field Experience with a Single NDUV + TCD Analyzer on Amine Based Tail Gas Treating Units." Presented at ISA Analysis Division Spring Symposium, April 2007.

15) Klint, B. "Incinerator Optimization/Stack Top Temperature Reduction."

16) AMETEK Process Instruments. 2008. *Continuous Measurement of Process Oxygen for Claus Sulfur Recovery Unit (SRU) Start-ups and Shutdowns.* Pittsburgh, PA.

Practical Process Integration Retrofit
Part 1. Cross-Pinch Method

G.T.Polley

A number of techniques have been developed to aid the process engineer retrofit heat recovery networks. The first of these methods was called the "Cross-Pinch Method" (Tjoe & Linnhoff [1986]) and whilst it is widely practised by specialist consultants it has not, to the author's knowledge, ever been formally described.

The first step in applying the cross-pinch method is the identification of a heat recovery target. This is achieved using an approximate economic analysis that seeks to relate project savings with capital investment.

Economics of Network Retrofit

Economic Criteria for Energy Saving Retrofit

The standard economic criteria suggested for the appraisal of new plants are overall capital cost and return on investment (often determined using discounted cash flow).

Where the purpose of a retrofit project is reduced operating cost different criteria are used. Since, the object is to make a saving in operating cost by making an investment in the plant, rather than employment of 'return on investment' the use of 'payback' is more logical.

'Payback' is a measure of how quickly the money invested is returned to the company. Two forms of payback criteria are in common use. 'Simple payback' is estimated by dividing the investment made by the savings accrued. 'Differential payback' is estimated by dividing incremental investment by incremental savings.

The capital costs of such projects are generally much lower than those associated with the construction of a new plant. However, capital is a limited resource. One means of recognising this is to place a limit on the amount of capital available for the project.

In Figure 1 we show a typical relationship between savings and investment for an energy saving retrofit. As demonstrated, simple payback, differential payback and investment ceiling can all be displayed on this plot. Consequently, once such a relationship is determined for a given project both the criterion which control the project and the amount of energy that should be saved are easily identified.

A project providing the required simple payback is positioned at the intersection between the savings/investment line and a line starting at the origin and having a slope equal to the inverse of the required value.

A project providing the required differential payback is positioned at the point at which a line having a slope that is the inverse of this value is tangential to the savings/investment line.

A project that hits the investment ceiling is also easily identified.

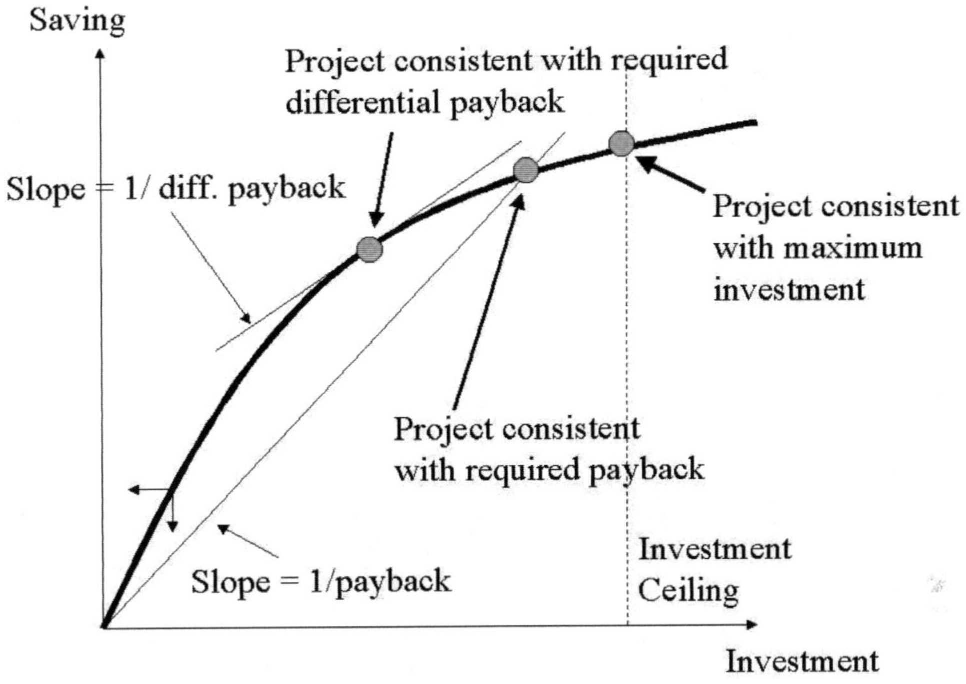

Figure 1. Retrofit Economics

Using Pinch Technology we can generate an approximate savings/investment plot at the very outset of a project. This provides managers with an important decision making tool. Before initiating any detailed design work they can explore project economics. They can examine the likely relationship between investment and payback and thereby set the economic criteria for a project in an objective manner.

Relationship Between Installed Area and Energy Consumption

Pinch Technology is rooted in simple algorithms for the rapid determination of minimum energy and approximate heat recovery area needs as a function of minimum allowable temperature difference [Ahmad & Linnhoff, 1989]. Since, these two important economic factors are both related to temperature approach it is possible to derive a relationship between hot utility consumption and the theoretical amount of heat recovery area needed in the heat recovery network. A typical relationship is displayed in Figure 2.

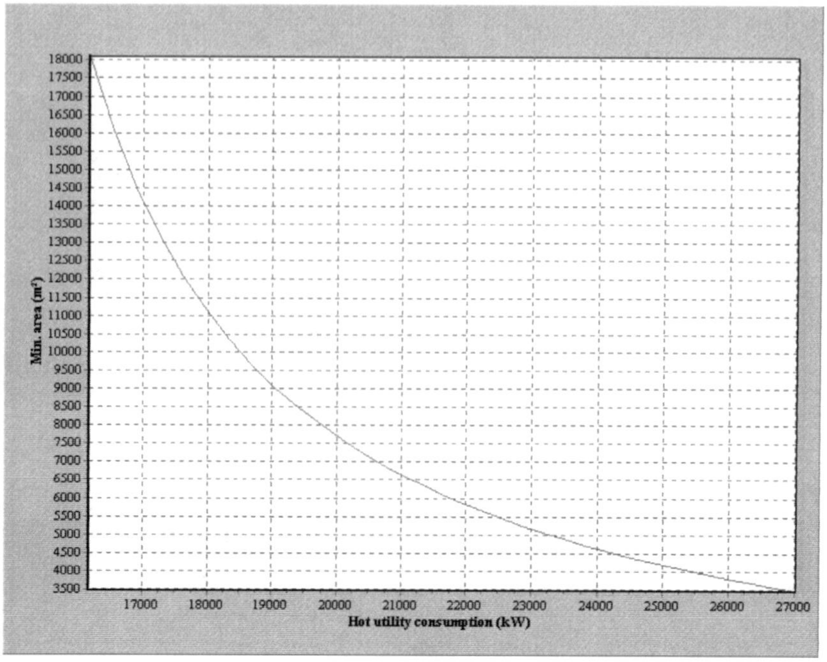

Figure 2 Area-Energy Plot

Relating Energy Saving to Project Economics

The first stage in evaluating the economics of a retrofit is to compare the performance of the existing plant with what is ideally possible. So, the current hot utility consumption and existing heat recovery area provide the co-ordinates for the positioning of the existing plant upon the area-energy plot.

It is common to find that this point (point X) falls above the curve (see Figure 3). This implies that the energy consumption of the plant is high given the amount of heat recovery surface area being used. For instance, in the example shown in Figure 3, the minimum amount of hot utility required by the plant is that associated with point **A** rather than point **X**.

To make an energy saving it is necessary to invest. The investment comes in the form of additional heat recovery surface strategically positioned in order to improve the manner in which the existing surface is used. There is a relationship between the investment made (in the form of additional surface area) and the amount of energy saved. Conceptually, in terms of the area-energy plot, the investment can be considered to follow a relationship similar to that shown in Figure 4.

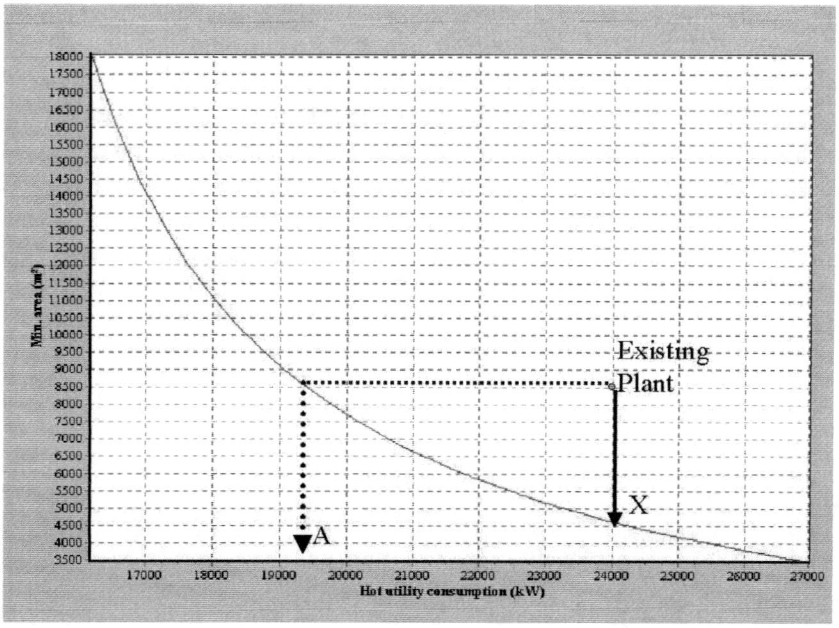

Figure 3. Performance of Existing Plant

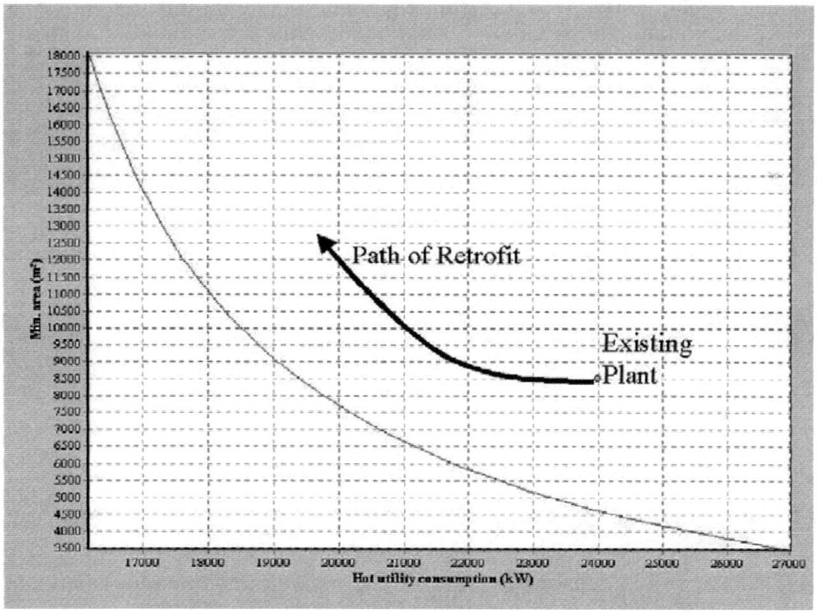

Figure 4 Investment made in Retrofit

If we can determine this 'path of retrofit' we can relate changes in surface area (an 'investment') with changes in hot utility usage ('saving'). We can use this information to derive a savings/investment relationship of the form shown in Figure 1.

In order to determine this 'path of retrofit' we must understand why the performance of the existing plant does not fall on the theoretical line. The 'understanding' underlying the algorithm for the determination of network area is that in a system in which the hot streams have a the same heat transfer coefficients and the cold streams all have the same heat transfer coefficients the network area is minimised if heat recovery matches are made such that hot and cold streams are in pure counter flow. This is equivalent to making heat recovery matches that are 'vertically aligned' between the composite curves

Consider very simple heat recovery system involving a heater, a cooler and a single heat recovery exchanger. Provided the temperature of the hot utility exceeds that of the hot stream entering the exchanger and the temperature of the cold utility is less than that of the cold stream entering the exchanger this system exhibits pure counter-flow. Consequently, if the performance of this system is displayed on the area-energy plot for the system, it falls directly on the theoretical line.

This does not mean that there is no scope for saving energy. If we increase the surface area of the heat recovery unit we increase the recovery and reduce both cooler and heater loads (by the additional amount of heat recovered). The recovery still involves pure counter-flow. So, with this system the 'path of retrofit' lies along the area-energy curve itself.

Now consider the system shown in Figure 5 .

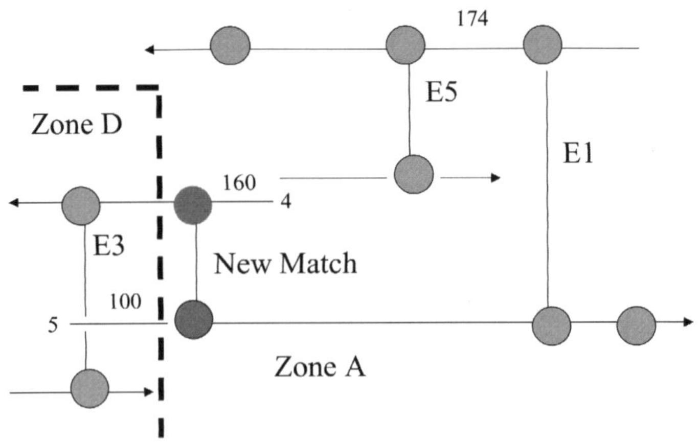

Figure 5 Retrofit Involving New Heat Recovery Match

In this retrofit a new heat recovery match is being made. The match joins two sections (or zones) of the plant. As will become clear when retrofit design is covered, the streams involved in the match and the thermal positioning of the match relative to the others each stream is such that the new exchanger will be close to being vertically positioned between the composite curves. Thus, the area added to the new match will exhibit pure counter-flow.

These observations give us a guide to how a retrofit project will appear on the area-energy plot. The starting point for the 'path of retrofit' may be displaced from the theoretical curve but the path itself will be approximately parallel to the curve. This is a significant finding for it allows us determine 'path of retrofit' from the theoretical area-energy relationship given knowledge of hot utility consumption alone. This, in turn, allows us to consider the introduction of alternative exchanger types (which will have different area-energy relationship) during a retrofit project.

So, we determine the relationship between additional surface area and hot utility saving in the manner illustrated in Figure 6.

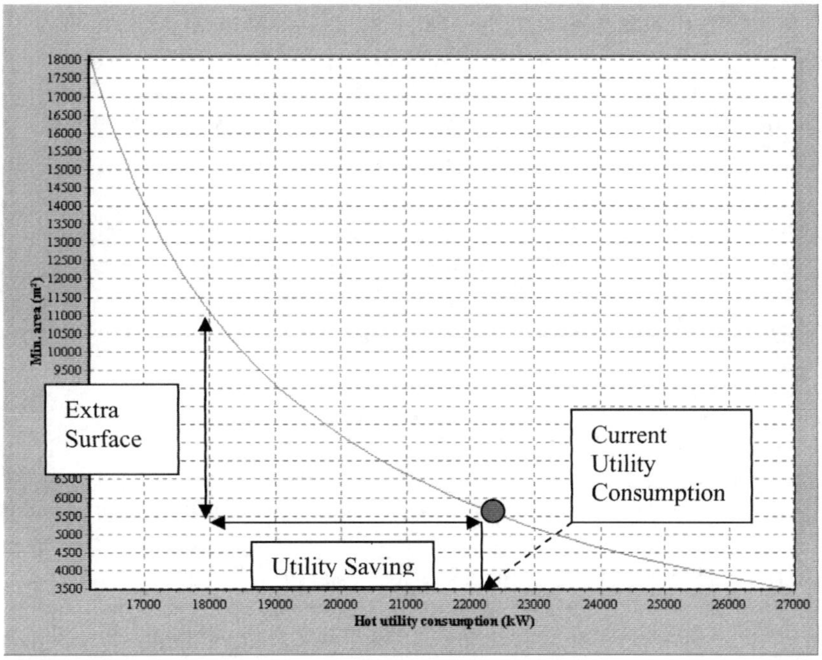

Figure 6 Utility Saving and Added Area

Full Cost Considerations

The economic analysis presented above only yields approximate results. In terms of investment it is centred on the cost of 'heat recovery area'. When identifying design modifications the engineer must be aware of other cost considerations.

The following 'guide-lines' should be adopted.

The cost of plant downtime can swamp other considerations. Therefore, the time required to undertake modifications is an important consideration. The cost of downtime is minimised if the modification can be completed during a scheduled shut down period. The possibility of making modifications in stages that coincide with more than one scheduled shut down should be considered.

The costs of modifications extend beyond the purchase of new exchangers. The cost of installing a new heat recovery match is greater than the cost of installing a new exchanger in an existing location. Linking this with what has been said about downtime, we would introduce the rule: ***changes to plant structure should be minimised.***

The cost of 'exchanger surface area' is not a reliable guide when considering how an existing heat recovery 'match' can be improved. When considering existing shell-and-tube units the designer usually has four options:

1). Enhance the performance of the existing unit using heat transfer enhancement devices.
2). Replace the exchanger bundle with a more efficient one.
3). Replace the exchanger with a totally new unit.
4). Install a new exchanger alongside the existing one.

In the case of the first option the area of the unit is unchanged. In the case of the second option the area that is purchased can be greater than that already present (for we can use extended surfaces or reduce the tube size) but the cost of this surface is significantly less than that associated with the same area in a totally new exchanger. The third option involves the purchase of an exchanger that is larger and more expensive than that involved in the fourth option. However, given that this exchanger can be designed to fit into existing pipe work and onto existing foundations it may be the cheaper option.

Starting Point for Cross-Pinch Method

The Cross-Pinch method ignores the above practicalities on the assumption that they can be addressed during the detailed design stage and the economic analysis is sufficient to provide a 'target' for the energy saving.

So, the economic analysis outlined in Figure 1 is applied. The targeted saving is that meeting whichever of the economic criteria (payback or investment ceiling) limits the project.

The targeted saving is then related to a minimum allowable temperature difference. This, in turn, establishes the location of the Pinch for the revamped plant.

Example Problem

The cross-pinch design method is best described using an example problem. This problem also illustrates the economic analysis described above. The example used is that first presented by Tjoe & Linnhoff [1986]. The structure of the network prior to retrofit is shown in Figure 7.

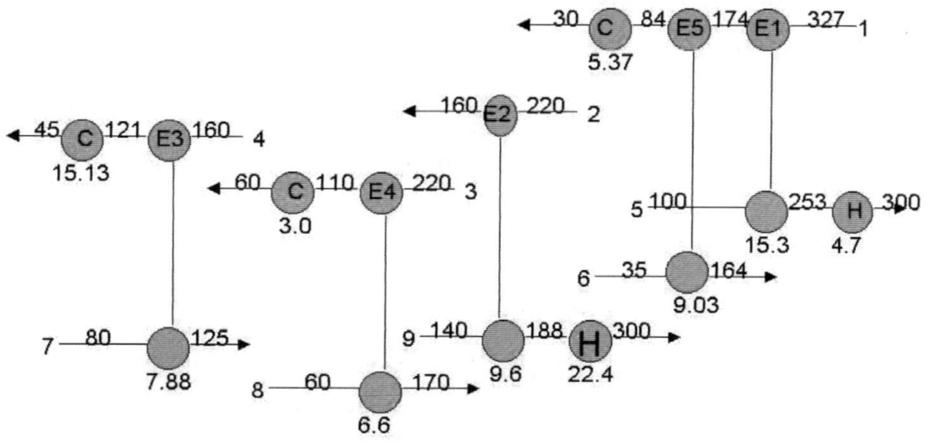

Figure 7. Existing Network

The stream data for the problem are given in Table 1 and details of the existing exchangers are given in Table 2. The required project economics are to achieve a payback of two years or less with a capital investment not exceeding $2M. The cost of energy is $3.2/GJ which, given 8000 hours of operation each year, is equivalent to $110/kWyear. The installed cost of heat recovery area is assumed to be 400 $/sq.m.

The hot utility consumption of the existing plant is 27.1 MW. This is split between heaters on stream 5 (4.7 MW) and stream 9 (22.4 MW).

Looking at the total plant, the relationship between hot utility saving and simple payback is shown in Figure 8. We see that a saving of 5.2 MW is predicted for a two year payback period.

The relationship between investment and payback is shown in Figure 9. We see that the predicted investment is $0.6M, which is well below the specified ceiling of $2M.

We see that the controlling criterion is the payback. Other relevant information coming from the analysis is that <u>the predicted saving of 5.2 MW occurs at a minimum temperature approach of 21 C.</u>

Stream Number	Supply Temp. C	Target Temp. C	Heat Capacity Flowrate MW/C	Heat Transfer Coeff. W/sq.m.C
1	327	30	0.1	800
2	220	160	0.16	500
3	220	60	0.06	2000
4	160	45	0.2	400
5	100	300	0.1	5000
6	35	164	0.07	1000
7	80	125	0.175	500
8	60	170	0.06	200
9	140	300	0.20	800

Table 1. Stream data

Exchanger	Hot Stream	Cold Stream	Area sq.m.
E1	1	5	300
E2	2	9	1220
E3	4	7	935
E4	3	8	726
E5	1	6	830

Total Heat Recovery Area = 4011 sq.m.

Table 2. Exchanger information

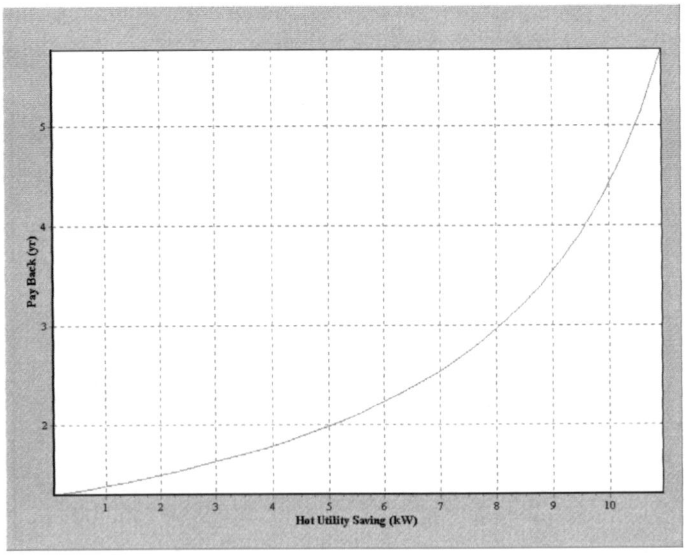

Figure 8. Relationship Between Payback and Hot Utility Consumption

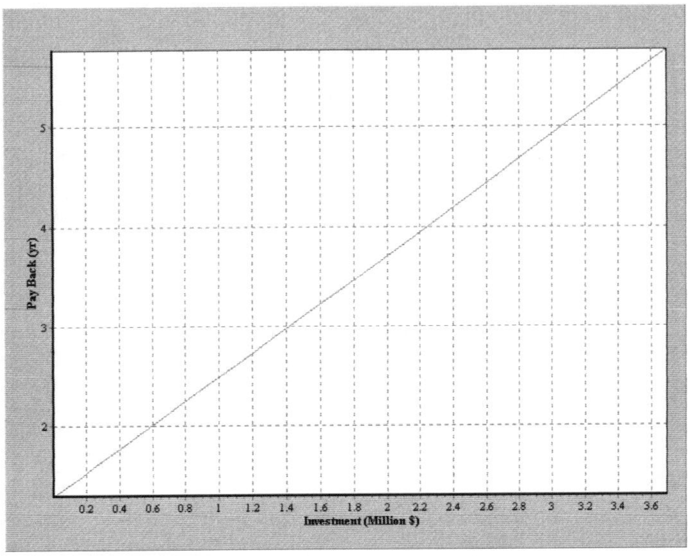

Figure 9. Relationship Between Investment and Payback

Cross-Pinch Design Method

Tjoe [1986] pioneered the use of Pinch Technology in plant retrofit. The philosophy behind his approach was to first identify which elements of the existing network were 'efficiently' placed. In line with the desire to minimise network changes these elements were retained without change. The next stage was to examine those elements that had been identified as not being efficiently placed and modified the them in order to improve their 'placement'. The modifications take the form of adjustments to the temperatures at which the process streams enter these units. It results in the production of packets of unsatisfied 'heat demand' and 'heat supply' (see below). So, the final stage of the design to make heat recovery matches and heat exchanger enhancements that match these supplies and demands.

The concept is illustrated as follows. We stress that exchangers are not moved in physical space but in 'thermal space'. Consider the exchanger illustrated in Figure 10. This unit is in a fixed physical location. Let's assume that we want to change the temperature at which the hot stream enters the exchanger. Since, the exchanger is of fixed size, the temperature at which the cold stream leaves the exchanger will fall (lower temperature driving force) and that at which the hot stream leaves the exchanger will rise. Thermally we have the situation shown in Figure 11.

How can we turn this thermal situation into physical reality. We may want to maintain the duty associated with the match. In this case we enhance the performance of the existing exchanger (e.g. we replace the bundle with one having low finned tubes). This takes up the heat available in the hot stream present after the existing exchanger and uses it to satisfy the demand made by the cold stream. The heat available from the hot stream prior to the match is taken up in a new heat recovery match. So, the network arrangement illustrated in Figure 12 shows the heat exchanger "hardware" needed to provide the identified "shift" shown in Figure 11.

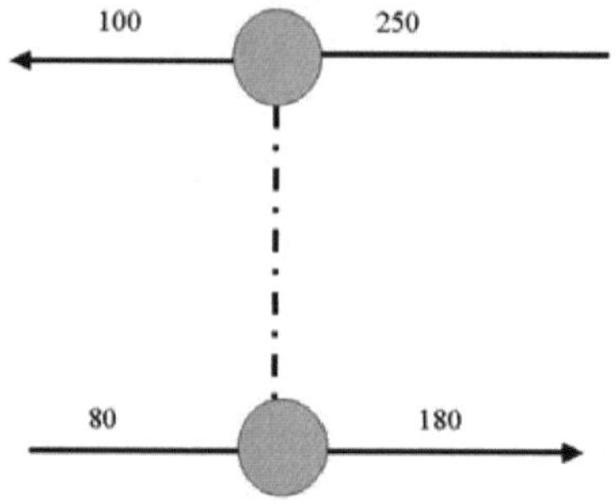

Figure 10. Exchanger to be 'improved'

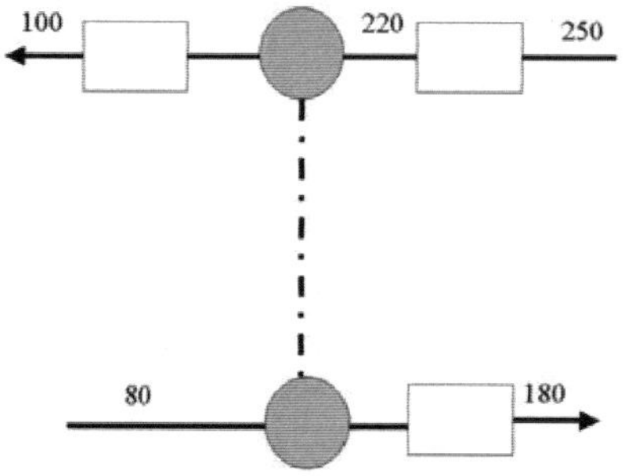

Figure 11. Exchanger Shifted in Thermal Space

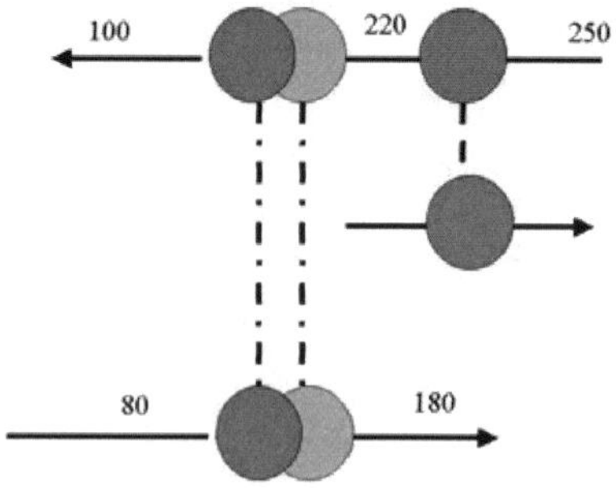

Figure 12. Physical Realisation of Shift

When the Tjoe Method is applied it is usually found that once the poorly place exchangers have been shifted the residual 'heat loads' are located in the 'Pinch' region in a manner similar to that illustrated in Figure 13.

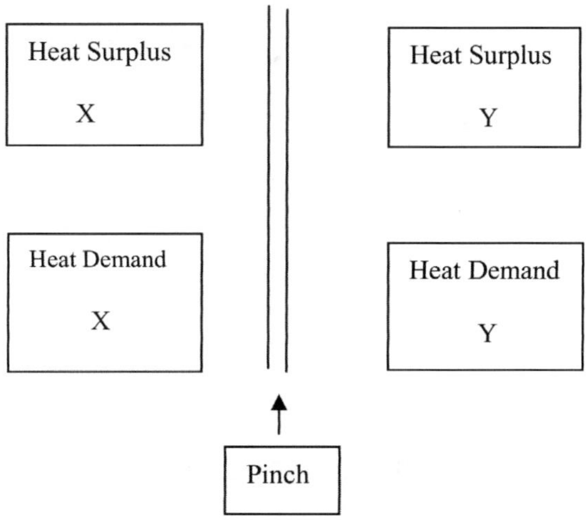

Figure 13. Residual Heat Loads : After Exchanger 'Shifting'

This result suggests that rather than going through the three stage procedure the engineer should concentrate on the matches that transfer heat across the Pinch [2]. These matches should be identified and either shifted or broken. Then the resulting 'heat pockets' re-matched to yield the modified design.

Using the example problem to demonstrate the method, the economic analysis predicted a utility saving of 5.2 MW with an temperature approach of 21 C. The Pinch location is hot stream temperature of 161 C, cold stream temperature of 140 C.

Given this information we can located the exchangers that are transferring heat across this Pinch (Figure 14). We observe that exchangers E1 and E4 transfer heat from above to below the Pinch. Exchangers E5 and E2 transfer a quantity of heat from below to above the Pinch (for it operates at a closer temperature approach than 21 C).

We now quantify these heat flows. This is conveniently done using the "Pinch Table" which can be generated once the location of the Pinch is identified.

In this table the quantities of heat demanded/supplied by individual streams on each side of the Pinch are listed. Cross-pinch heat flows are given by the differences between demand/supply on either side of the Pinch.

The Pinch Table for the example problem is given in Table 3.

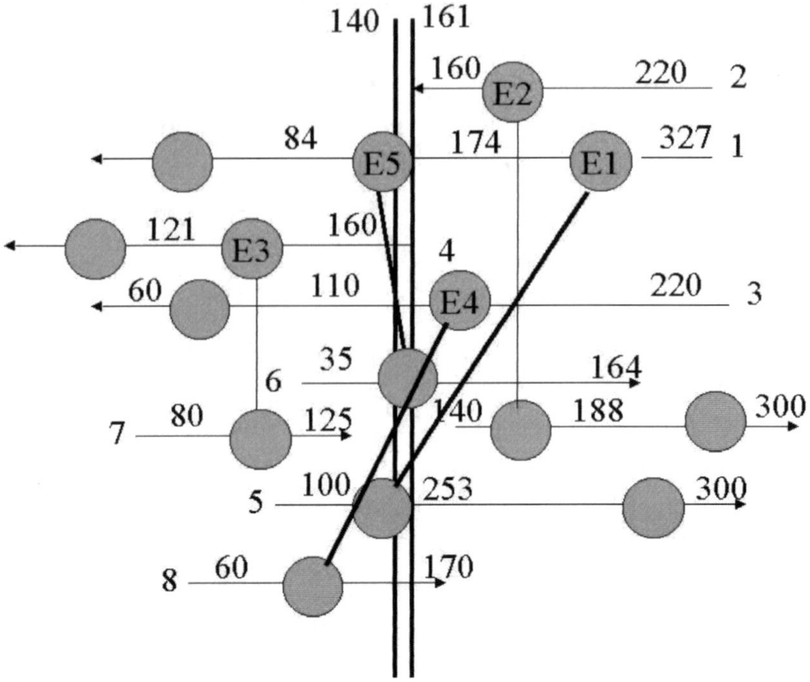

Figure 14. Cross-Pinch Exchangers

		Below	Pinch		Above	Pinch	
Stream	Cp	Low Temp.	High Temp.	Load	Low Temp.	High Temp.	Load
E1 (hot)	0.1				174	327	15.3
E1 (cold)	0.1	100	140	4.0	140	253	11.3
E2 (hot)	0.16	160	161	0.16	161	220	9.44
E2 (cold)	0.2				140	188	9.6
E3 (hot)	0.2	121	160	7.8			
E3 (cold)	0.175	80	125	7.9			
E4 (hot)	0.06	110	161	3.06	161	220	3.54
E4 (cold)	0.06	60	140	4.8	140	170	1.80
E5 (hot)	0.1	84	161	7.7	161	174	1.3
E5 (cold)	0.07	35	140	7.35	140	164	1.68

Table 3. Pinch Table

From this table we see that exchanger E1 transfers 4.0 MW of heat from above to below the Pinch (the hot and cold loads on each side of the pinch differ by 4.0).

Exchanger E2 transfers 0.16 MW from below to above the Pinch (it operates at a temperature approach of 20 C rather than the target value of 21 C).

Exchanger E4 transfers 1.74 MW of heat from above to below the Pinch.

Exchanger E5 transfers 0.35 Mw from below to above the Pinch.

If we do not modify exchangers E2 and E5 (those transfer heat from below to above the Pinch) we see that we have a potential saving of 5.74 MW (rather than the target value of 5.2 MW).

In terms of network modifications, the highest benefit comes from an adjustment to the operation of exchanger E1. The cold stream passing through the unit has a high demand of heat above the pinch (16 MW). The hot stream passing through the unit provides a lot of heat above the pinch (16.6 MW). A match between the two streams is fully justified. The problem is clearly the positioning of that match in thermal space. The pinch analysis shows that the match must be fully positioned above the Pinch. This can be achieved by introducing a new heat recovery match that satisfies the below pinch heat demand (Figure 15).

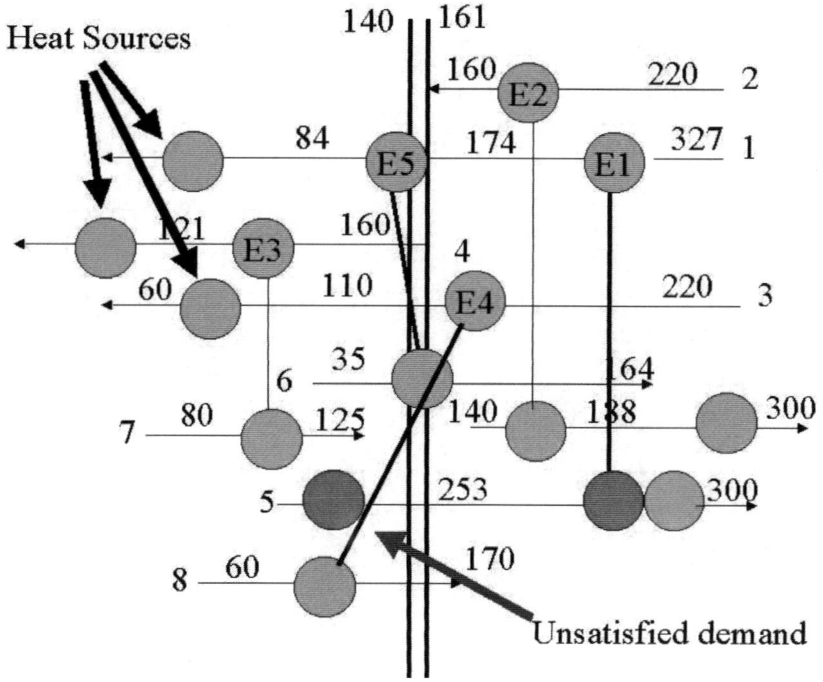

Figure 15. Repositioning of E1

The unsatisfied heat demand for cold stream 5 can be satisfied using heat currently being rejected through coolers. If the modification is to be restricted to a single change only one option is available. That is a match involving hot stream 4. The modification to the network structure, which results in a reduction in 4 MW in the heat provided to the heater on stream 5, is shown in Figure 16.

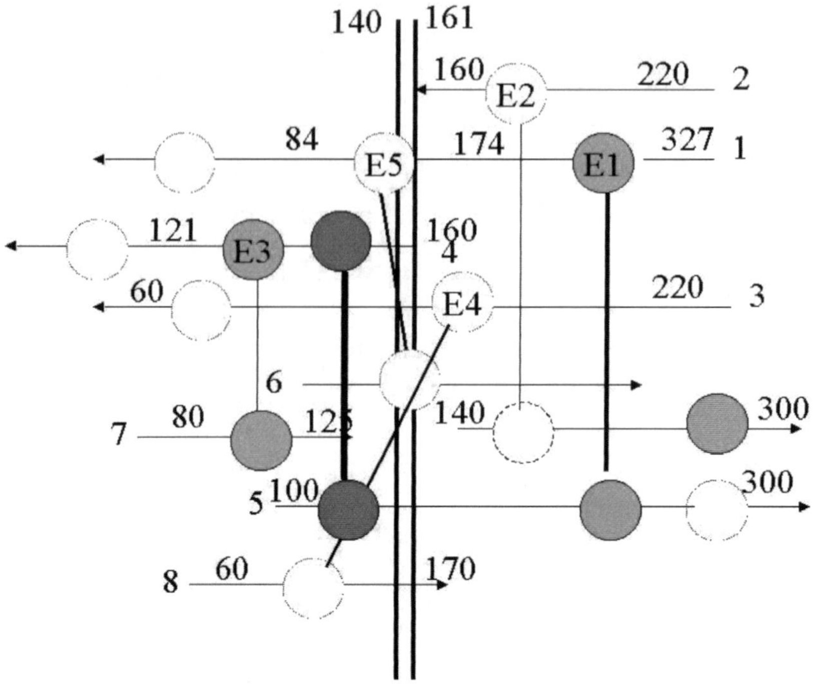

Figure 16. Modification 1.

With this modification enhancement of the performance of exchangers E3 and E1 will be required. Evaluation of the modification can be made by systematically varying the load on the new match and evaluating the size of the new exchanger <u>and</u> the modifications that have to be made to exchangers E3 and E1.

Turning our attention to exchanger E4, we see that this transfer across the pinch can be rectified by increasing the amount of heat extracted from the hot stream positioned below the pinch (Figure 17).

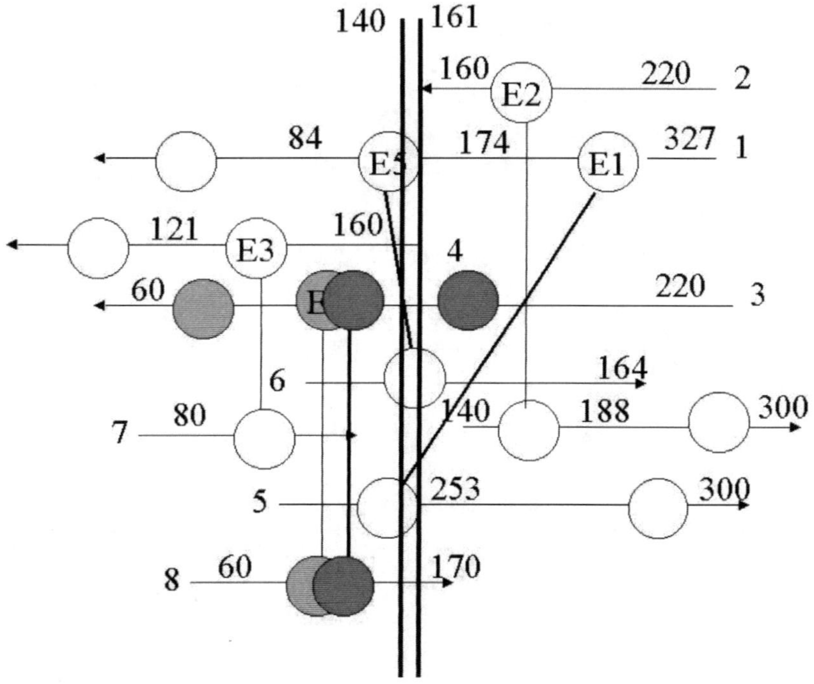

Figure 17. Improving performance of E4

This improvement in the use of heat from stream 3 provides a 'pocket' of heat positioned above the pinch. this can be used to provide part of the heat demanded by stream 9. The likely modification would be a new match between streams 3 and 9 placed in parallel with exchanger E2.

This modification would be evaluated by systematically varying the load on the new match between streams 3 and 9 and evaluating size of the exchanger involved and the enhancements required to exchangers E2 and E4.

So, the second (and final) modification results in a reduction of 1.74 MW in the heat supplied through the heater on stream 9, is shown in Figure 18.

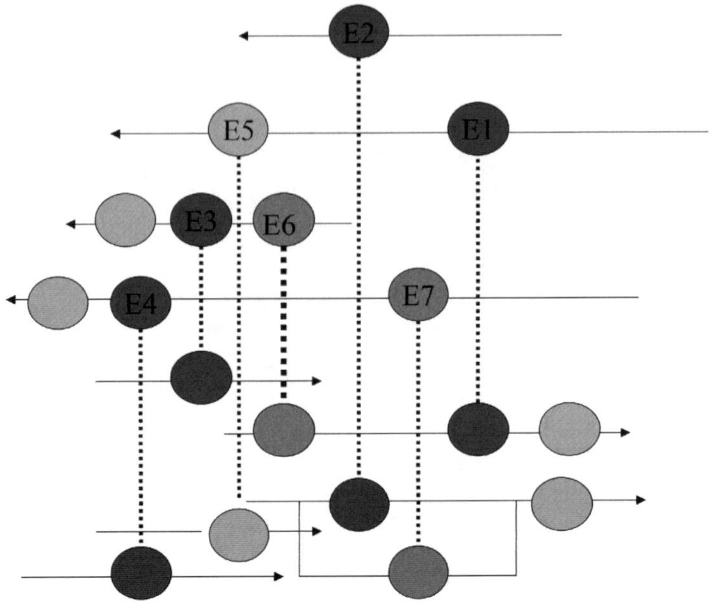

Figure 18. Final Project

The modifications are summarised in Table 4.

Modification 1	
New Exchanger E6	New Match. Additional heat recovery from Stream 4, reducing load on the heater on Stream 5.
Modification of E5	Enhancement of performance to counter-act the reduction in temperature driving force. Need to maintain duty on E5.
Modification of E3	Enhancement of performance to counter-act reduced temperature driving force. Unit has fixed duty.
Modification 2	
New Exchanger E7	New Match. Additional heat recovery from Stream 3, reducing load on the heater on Stream 9.
Modification of E4	Enhancement of performance to counter-act reduced temperature driving force. Unit has fixed duty.
Modification of E2	Enhancement of performance to counter-act reduction in flow of cold stream. Unit has fixed duty.

It consists of two new heat recovery matches and the enhancement in performance of four of the existing exchangers.

The modification of existing matches is dependant upon the geometry of the existing units. In rising order of complexity, it can take the form of:

- Use of heat transfer enhancing tube inserts
- Replacement of existing tube bundle (with or without heat transfer enhancement, e.g. low fin tubes)
- Replacement of existing exchanger(s)
- Placement of new exchanger(s) alongside the existing ones

The need for modification is also dependent upon how stringently existing duties must be met. If, under the revised conditions, the existing exchanger only effects the outlet temperature of a stream by a few degrees this may be deemed acceptable and the unit left unchanged.

Discussion

The Cross-Pinch method for retrofit design has been demonstrated and found to be practical.

The utility savings achieved in the design are, for the case examined, higher than those identified in the economic analysis used to provide a starting point for the design.

Where there are a number of flows from above to below the Pinch, some of which are relatively small, engineering judgement needs to be applied

References

Tjoe T.N. & Linnhoff B. "Using pinch technology for process retrofit", Chem.Engng., 1986 April 28, pp47-60

Ahmad S. & Linnhoff B. "Supertargeting: Different process structures for different economics", ASME J.En.Res.Tech., 1989, 111(3), 131-136

Tjoe T.N. Ph.D. Thesis 1986, UMIST, UK

Practical Process Integration Retrofit

Part 2. Remaining Problem Analysis & Driving Force Plot

G.T.Polley

The "Cross-Pinch Method" is not the only "tool" derived during the initial stages of the application of "Pinch Technology" to retrofit projects. Other significant developments were "Remaining Problem Analysis" and the "Driving Force Plot".

References

Tjoe T.N. & Linnhoff B. "Using pinch technology for process retrofit", Chem.Engng., 1986 April 28, pp47-60

Ahmad S. & Linnhoff B. "Supertargeting: Different process structures for different economics", ASME J.En.Res.Tech., 1989, 111(3), 131-136

Tjoe T.N. Ph.D. Thesis 1986, UMIST, UK

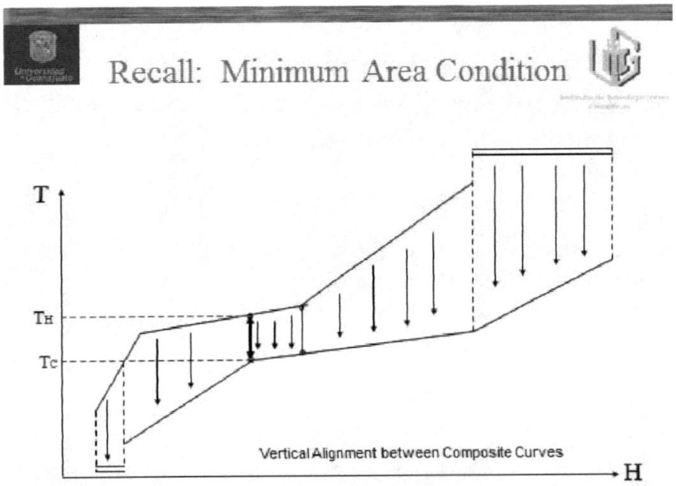

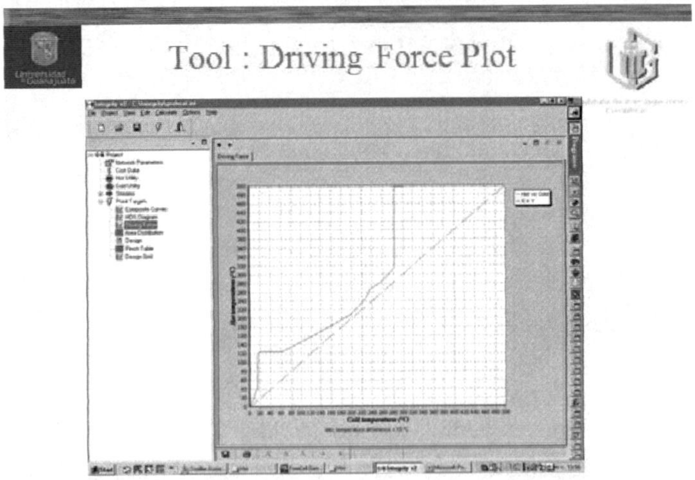

Changing Slope

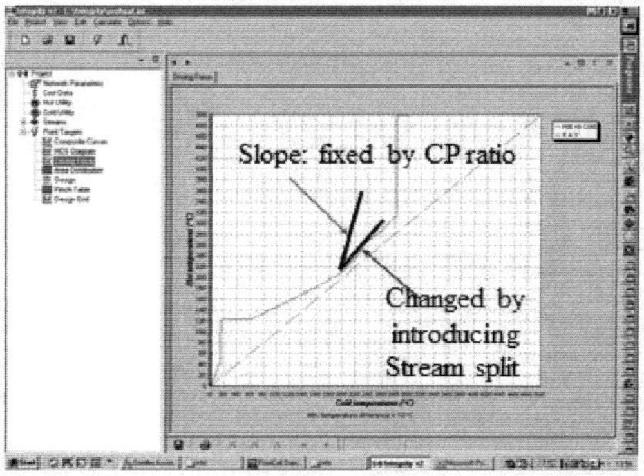

Use in New Design

- Identify Match
- Check against DFP (quantitative)
- Evaluate (RPA)

Use in Retrofit

- Identify which exchangers are badly positioned
- Identify adjustments that can be made

Changing terminal temperatures

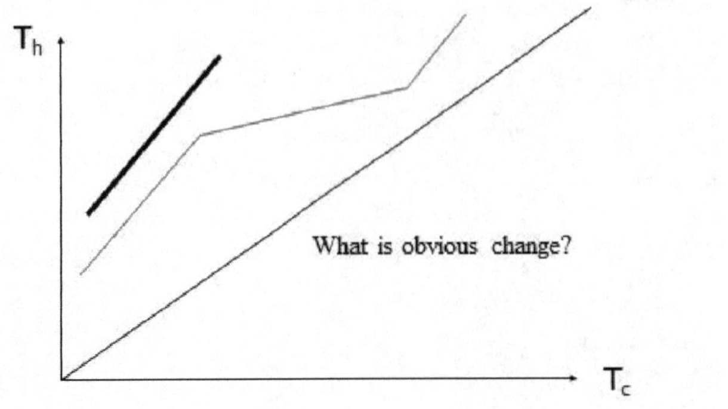

Changing terminal temperatures

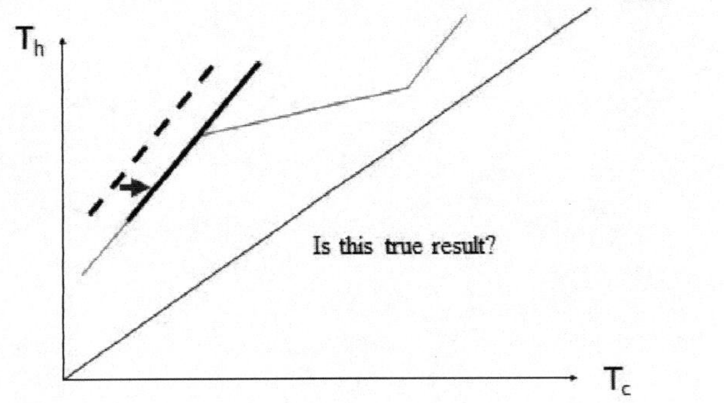

Relate to stream supply and target temperatures

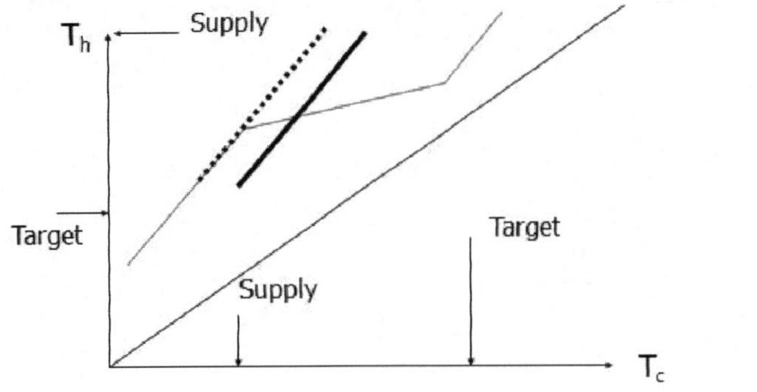

Relate to stream supply and target temperatures

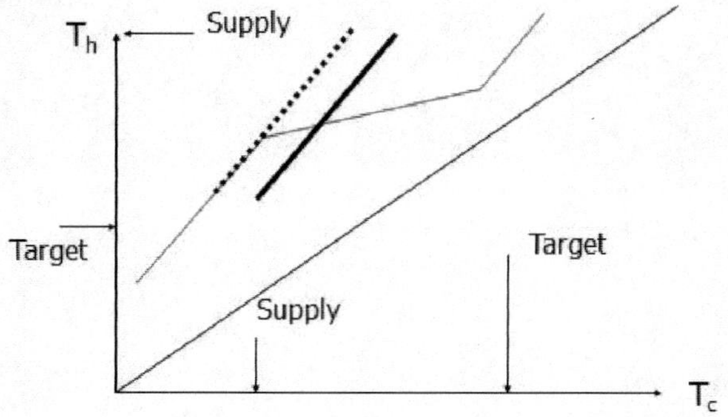

Relate to stream supply and target temperatures

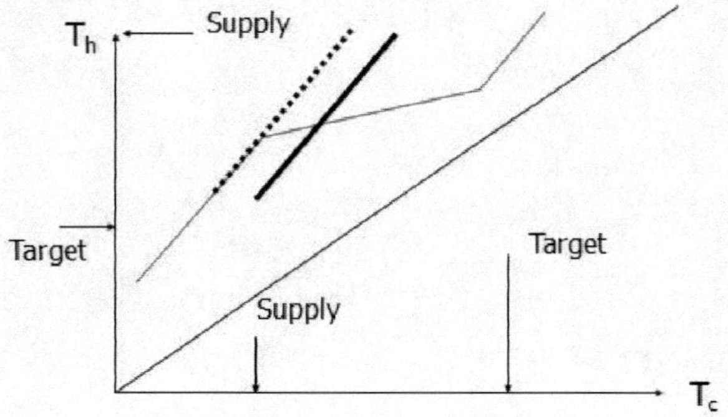

Driving Force Plot

- Identifies how well actual profile fits ideal
- Guidance on how fit can be improved
  - stream splits
  - changes to inlet temperatures

Qualitative Measure

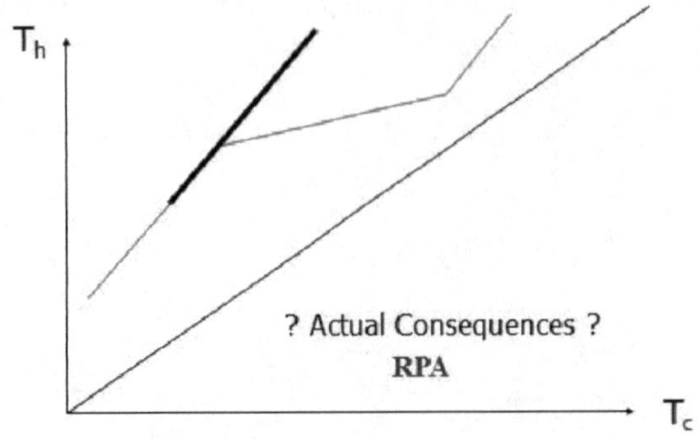

Remaining Problem Analysis

IN THE DESIGN OF NEW NETWORKS

"Remaining Problem Analysis"

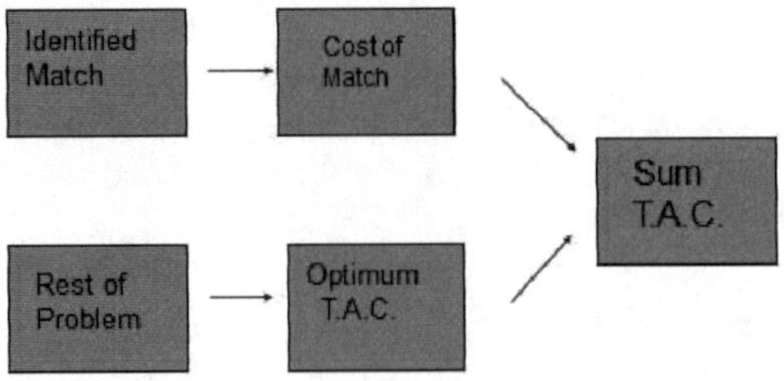

Rest of Problem = Stream Data – Elements involved in match

Evaluation of Match

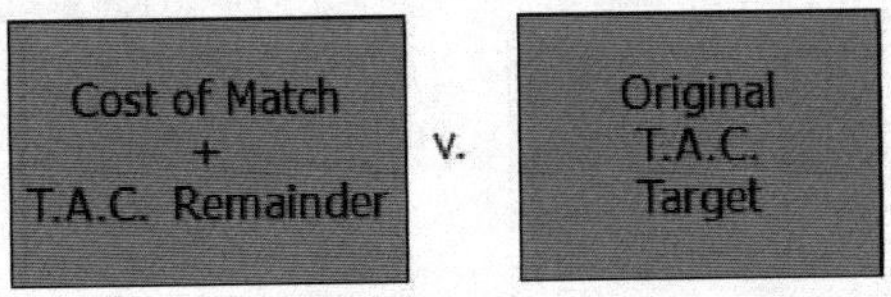

Every Step

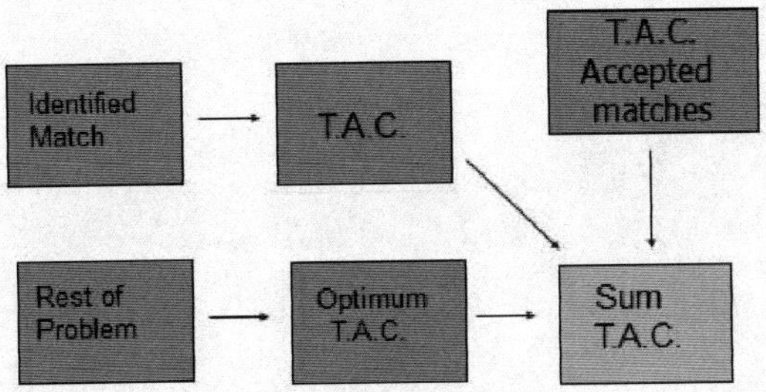

IN RETROFIT OF EXISTING NETWORKS

Standard Targeting

- Determine target E
- Equivalent dTmin
- Area to be added

"Remaining Problem Analysis"

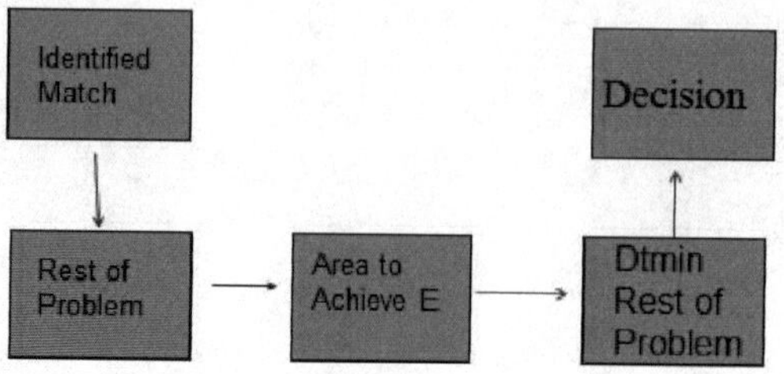

Rest of Problem = Stream Data – Elements involved in match

Evaluation of Match

Area for match plus rest of problem compared with "area target" – acceptable fraction ?

Dtmin for remaining problem compared with target value – if close then solution of rest of problem is not compromised

Decision

If no area penalty expected and rest of problem not compromised - accept match – remove stream elements from problem "set"

Otherwise: look at re-positioning match in thermal space using the DFP

If position of match cannot be improved – reject match
Final stage of design is formation of new matches (Usually in Pinch Region).

Next Step

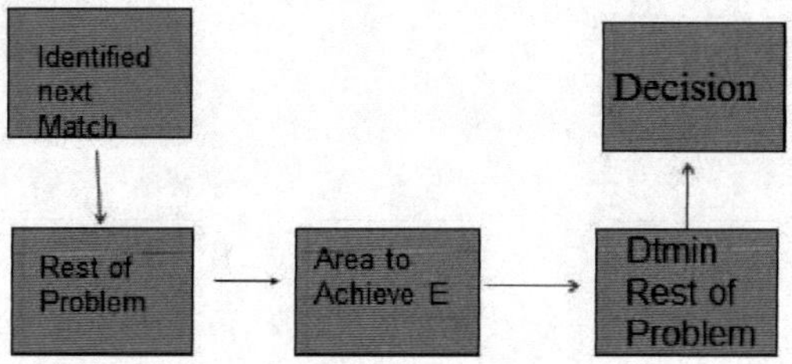

Rest of Problem = Stream Data – Elements involved in match – elements already accepted

Warning

Order in which existing matches are evaluated affects the "numbers" generated.

So, can influence outcome of the analysis.

Practical Process Integration Retrofit

Part 3. Path Analysis

G.T.Polley

The concept of transferring heat along paths was first suggested by Dhole [1]. It was subsequently investigated by Adiarso [2] and Van Reisen [3,4]. It forms the basis of a technique known as “Network Pinch” introduced by Zhu & Asante [5].

When using path analysis the engineer concentrates on minimising the number of changes that are made to the existing network structure for such changes generally involve quite large capital investment.

The identification of existing heat recovery paths and possible new paths is usually quite straight forward and does not require the use of special software.

The technique should not be used in isolation. It is worth undertaking a cross-pinch analysis to establish if structural change is justified. For instance when software employing the network pinch analysis was applied to a large problem Al Riyami et al [6], it failed to identify simple energy saving opportunities associated with re-boilers positioned well below the pinch [7].

Procedure

There are three requisites to the realisation of extra energy in a heat recovery networks. These are the presence of a source of heat that can be recovered, the presence of a heater currently providing heat to the network and a “path” between the heater and the heat source. This is illustrated in Figure 1 which shows one element of a larger heat recovery network. Here heat is currently being rejected to the environment through a cooler on stream H1 whilst heat is being applied to the network through heater H1. We observe that there is a “path” between this cooler and the heater on stream C2. This path involves streams C1, C2 and H2 and exchangers E1, E2 and E3.

Heat currently being rejected through the cooler can be recovered by improving the performance of E1.

This action will result in an increase in the temperature at which C1 enters E2. The result will in reduced heat transfer in this unit and an increase in the temperature at which H2 enters E3. This will then result in increased performance of this unit giving an increase in the heat recovery on stream C2 and reduction in the heater load on this stream.

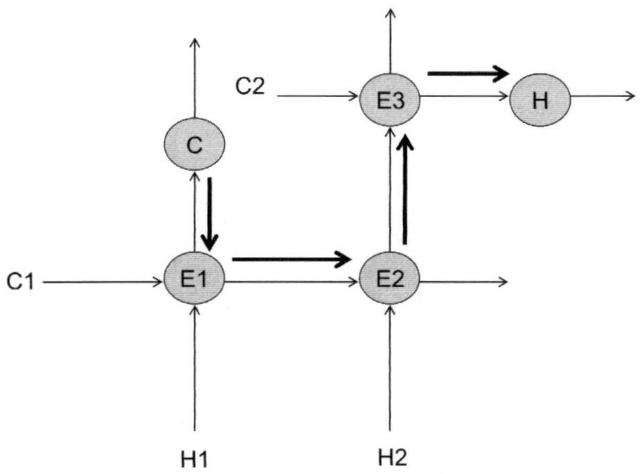

Figure 1. Element of Existing Network Having Path Between Cooler and Heater

The reduction in the performance of exchanger E3 leads to an increase in the temperature at which stream H2 is returned to the process.

Simulation of the transfer of heat along the "path" is necessary for the determination of how much energy can be saved. This needs to be coupled with evaluation of the options for improving the performance of the individual heat exchangers associated with the improvement in heat recovery. These options include using heat transfer enhancement, replacing the tube bundle with one providing higher performance and replacement of the unit with one having higher area. Total replacement of an exchanger is favoured over the installation of a second exchanger alongside an existing one for that requires additional supporting structures and piping modifications.

The procedure recommended here is to undertake a series of simulations where step changes are made to the recovery achieved in exchanger E1. For each of these steps the following information is generated:

1. Temperature field around E1
2. Temperature field around E3
3. Heat saving at the heater.

Heat exchanger analysis software is then used in order to determine the options for achieving the saving on E1, and (if found necessary) how the operation of E3 can be restored.

Commercial tools are available for the implementation of this procedure. These include the SmartPM™ simulation program and the EXPRESS™ exchanger analysis program offered by IHS ESDU [8].

Path Analysis is particularly useful in the revamping of crude oil pre-heat trains.

Example: Increased Heat Recovery in Pre-Heat Train

The typical pre-heat train structure is illustrated in Figure 2. The residues leaving the distillation column provide an important source of heat for pre-heating the incoming crude oil. These residues often have the highest supply temperature of any of the streams matched in the pre-heat train. It is therefore common to find a heat recovery match involving this stream close to the fired heater feeding the column itself.

Heat has to be extracted from the residues to a temperature that is below that of the de-salter installed within the train. Consequently, they are often involved in a heat recovery match positioned below the de-salter.

In this situation the residues stream provides an effective path between coolers positioned below the de-salter and the fired heater itself. The path for our example problem is shown in Figure 3.

The major advantage of such a path is that it "bypasses" many of the existing heat recovery units and therefore the number of exchangers that may require modification is reduced.

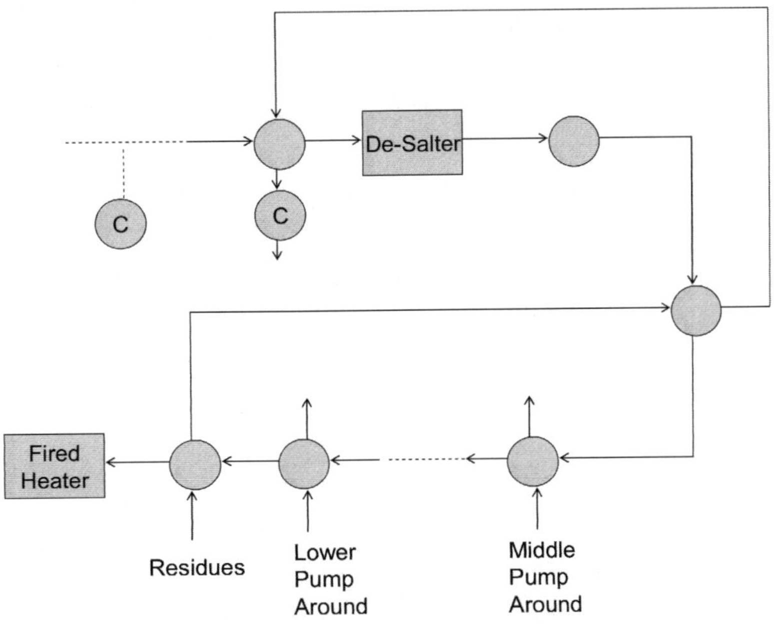

Figure 2. Pre-Heat Train Structure

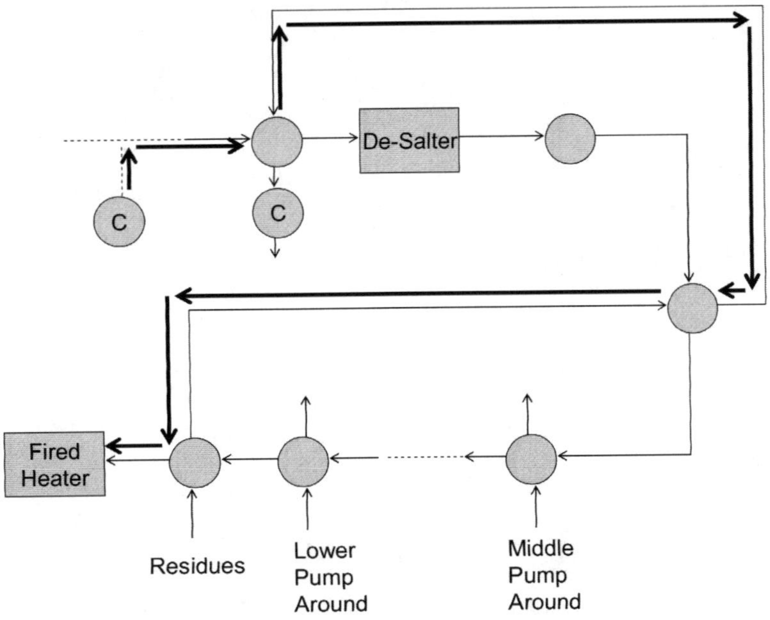

Figure 3. Heat Recovery Path

In this structure increased recovery at the hot end of the train will result in reduced performance of the second unit used for recovery from the residues (positioned part way between de-salter and fired heater) and in the unit positioned below the de-salter.

An effective retrofit would therefore involve the following actions (shown in Figure 4):

1. Increase heat extraction from residues at hot end of train
2. restore performance of unit handling residues part way up the train, in order to eliminate disturbances to exchangers positioned above that unit
3. Improve heat recovery below the de-salter in order to compensate for reduced heat recovery from residues in the match made below the de-salter and thereby maintain de-salter temperature

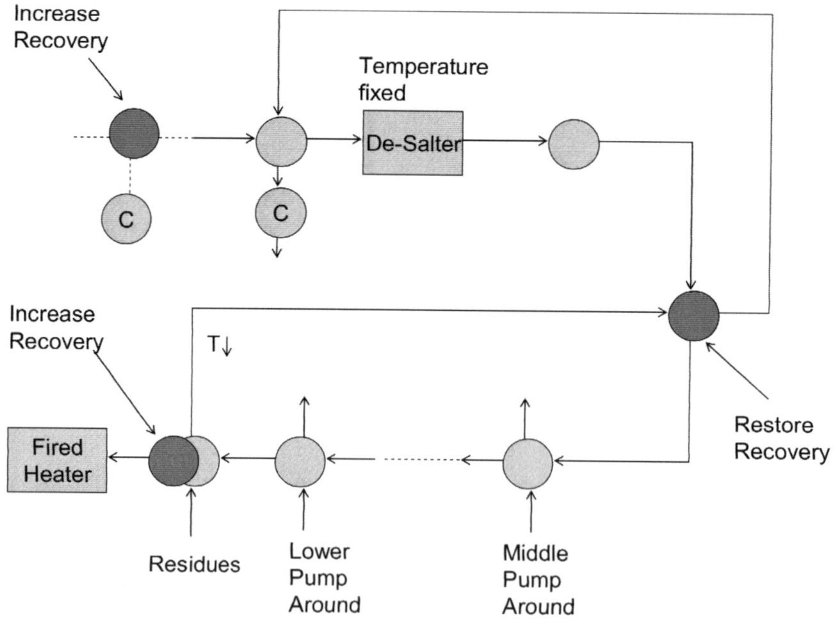

Figure 4. Retrofit of pre-heat Train

Features to be Observed in Path Analysis

There are three features of a network that need to be given careful consideration during a retrofit project. These are the need for "bridges", the presence of "blocks" (which give rise to the *Network Pinch*) and process operations having a fixed heat load.

Bridges

Network behaviour cannot be divorced from process requirements. Process operations can place barriers on heat recovery paths. For instance, in a pre-heat the temperature at which a de-salter operates may fall between narrow limits. Improvement in heat recovery below the de-salter may result in infringement of the upper limit. In this case, improvement in heat recovery can only be achieved if one or more streams involved in the heat recovery train is matched on both sides of the de-salter. That is, a stream forms a "bridge" across the process operation limiting the heat recovery.

In the example outlined above, the residues provided the required bridge.

Blocks

Network structures can place restrictions on the amount of heat that can be transferred along a path. For instance, consider the structure shown in Figure 5.

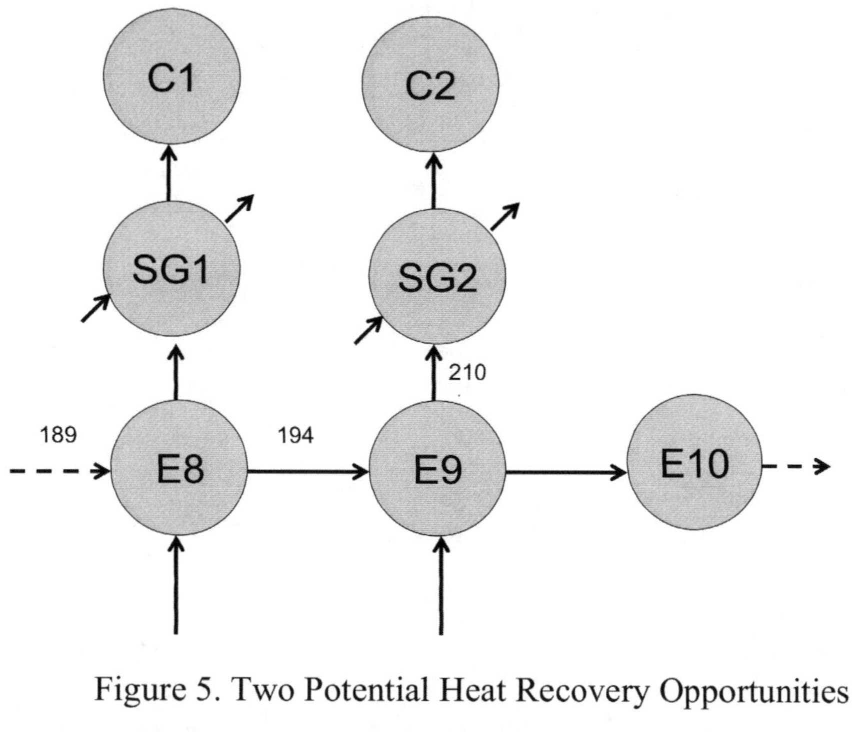

Figure 5. Two Potential Heat Recovery Opportunities

Two opportunities for saving heat can be observed. Heat from either cooler C1 or steam generator SG1 can be transferred to the cold stream by improving the performance of exchanger E8. Similarly, heta from either cooler C2 or steam generator SG2 can be transferred to the cold stream by improving the performance of exchanger E9.

Examination of the local temperatures indicate that the temperature at which the cold stream leaves E8 (194 C) is close to that at which the hot stream leaves E9 (210 C). Therefore, this temperature cannot be raised significantly before there is a decline in the heat recovery from E9. So, E9 is providing a "block" on the path from C1.

The block can be removed by using a stream split on the cold stream (Figure 6). Then E9 does not block the path from C1. The temperature at which the hot stream enters E9 is reduced and its performance may improve.

The validity of the modification needs to be established using proper heat exchanger analysis.

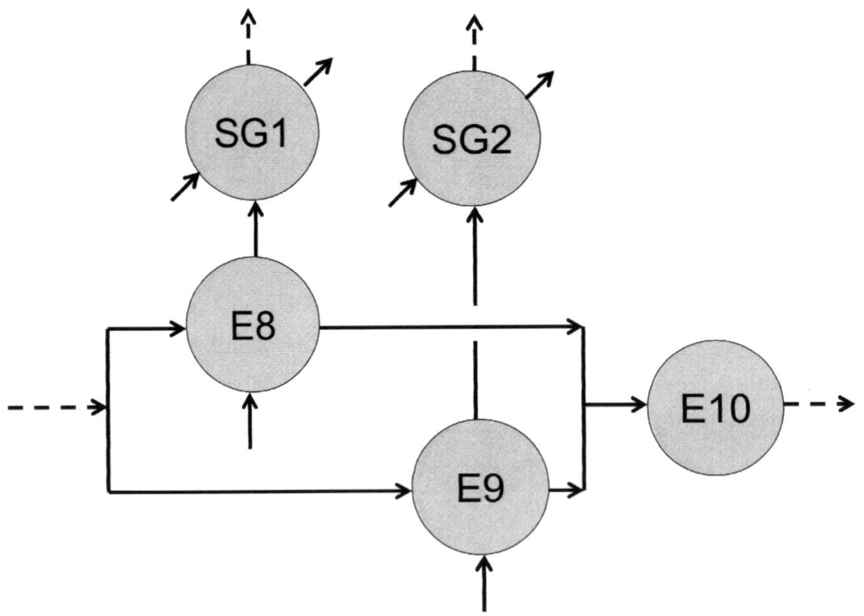

Figure 6. Removal of Block by Installing Stream Split

Fixed Loads

Care needs to be taken regarding exchangers that have a fixed heat load. In the case of pre-heat trains this involves pump around streams.

Paths that bypass such exchangers (such as that exploited in example 1) are to be preferred.

Where units cannot be bypassed then heat exchanger analysis programs need to be used in order to determine how the performance of these exchangers is best restored (and how sensitive the exchangers subsequently are to disturbances to changes to inlet condition and time related factors– such as fouling).

References

[1] V.R.Dhole, Private Communication, UMIST, 1990

[2] Adiarso, "Application of Problem Decomposition in Process Integration Retrofit", M.Sc. Dissertation, UMIST, 1993

[3] Van Reisen J.L.B., Grievink J., Polley G.T. & Verheijen P.J.T. "The placement of two-stream and multi-stream heat-exchangers in an existing network through path analysis", Comp.& Chem. Engng., 1995, 19,S143-S148

[4] Van Reisen J.L.B., Polley G.T. & Verheijen P.J.T. "Structural Targeting for Heat Integration Retrofit", Appl. Thermal Engng. 1998, 18(5), 283-294

[5] Asante N.D.K. & Zhu X.X. "An automated approach for heat exchanger network retrofit featuring minimal topology modifications", Comp.& Chem. Engng., 1996, 20,S7-S12

[6] Al Riyami B.A., Klemes J. & Perry S., "Heat integration retrofit analysis of a heat exchanger network of a fluid catalytic cracking plant", Appl.Thermal Engng. 2001, 21, 1449-1487

[7] Polley G.T. & Kumana J.D. "Energy Saving Retrofit of FCC Plant", NPRA Annual Meeting, March 2007, San Antonio, Texas

Practical Process Integration Retrofit

Part 4. Dealing with Large Projects

G.T.Polley

The "Cross-Pinch Method" is suitable for use in identify retrofits when the number of streams is small (the example used in the demonstration of this method had seven streams, Polley [2011]). However, "Pinch Technology" is based on the assumptions that capital cost expenditure is controlled by the cost of heat exchangers and factors such as network operability can initially be ignored. With small projects operability considerations are easily undertaken by an experienced process engineer as the revamp is developed. With large projects this is difficult. The evaluation of the capital cost of a revamp can often be dominated by the cost of piping a modification to "civils". With small sized projects an experienced process engineer can incorporate a consideration of these factors as a network structure is developed. With large projects this is not possible.

The solution to this problem is to decompose large networks into a number of smaller self contained networks.

The scope for transferring heat between these zones can be determined using the procedure developed by Tamakloe & Polley (AIChE Spring Meeting, 2011).

Energy Saving Retrofit of FCC Plant

G.T.Polley[1] & J.D.Kumana[2]

1. Instituto de Investigaciones Cientificas, Universidad de Guanajuato, (gtpolley@aol.com, tel +44-1229-585-330)
2. Consulting Services Department, Saudi Arabian Oil Company, (jimmy.kumana@aramco.com, tel +9663-874-6157)

The use of process integration techniques to optimize the energy efficiency of process plants is now a widely-accepted practice. However, the results obtained vary considerably as there is no "standard" way of doing things, and despite having a sound theoretical basis for the method, actual application remains an art.

In 2005, Saudi Aramco arranged for three different groups to develop retrofit designs for a given problem. Each groups used a different methodology. The aim was to assess which approach gave the best results. The problem chosen for the study came from the literature (Al-Riyami et al). None of the groups used the methodology used in the original study. Consequently, four different approaches could be assessed. It had previously been observed that the problem definition contained a number of errors (reflecting real conditions in the field). The company was interested to see if these would be picked up by the contestants, and how they would be resolved.

The study highlighted the importance of considering practical design and operational issues such as:

1. the handling of inconsistencies in plant performance data
2. determination of which process streams should be included in the analysis
3. how the heat load-temperature profiles for vaporisation duties should be set
4. what heat transfer coefficients should be used
5. identifying errors in the basic flowsheet

6. consistency between utility costs
7. how differences in basic exchanger cost can be handled
8. how fouling of new exchangers can be mitigated
9. the importance of considering piping costs

Problem Definition

The FCC plant used in the exercise was that described by Al-Riyami et al (2001). The flowsheet is presented below.

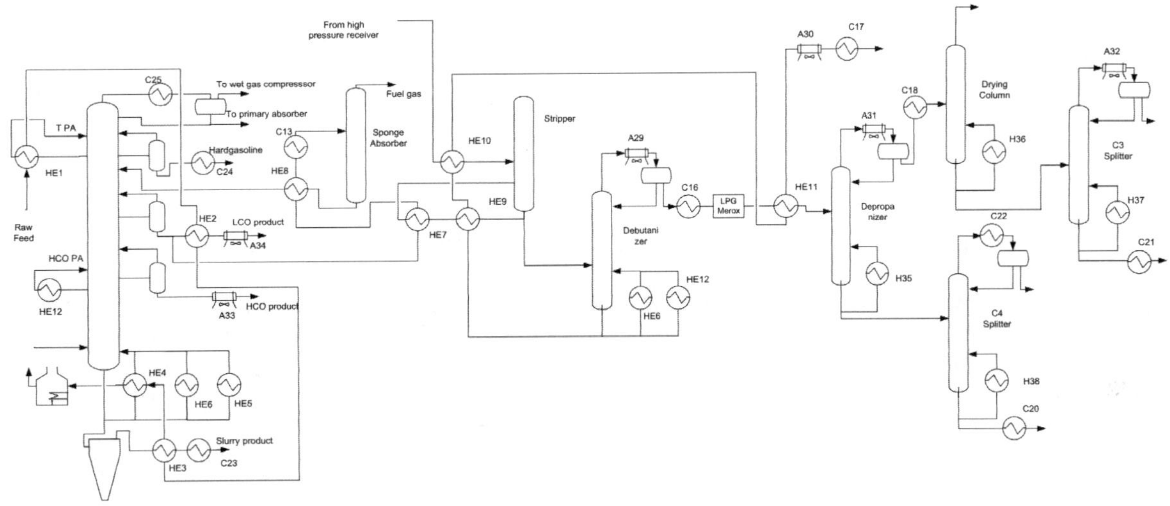

Figure 1. FCC Plant Flowsheet.

The process streams are listed in Table 1.

Stream No.	Process Flow	Stream No.	Process Flow
H1	Bottom p/a - main col.	C1	Raw feed
H2	Slurry product	C2	Steam generation
H3	HCO p/a – main col.	C3	Sponge abs. return
H4	HCO product	C4	Stripper feed
H5	LCO to sponge abs.	C5	Depropaniser feed
H6	LCO product	C6	Debutaniser reboiler
H7	Top p/a – main col.	C7	Stripper reboiler
H8	Hard gasoline	C8	Depropaniser reboiler
H9	Main col. condenser	C9	Drying col. Reboiler
H10	Compressor intercool.	C10	HC3 splitter reboiler
H11	HP outlet	C11	HC4 splitter reboiler
H12	Top p/a prim. abs.		
H13	Bottom p/a prim. abs.		
H14	Debutaniser condenser		
H15	ME ROX feed		
H16	Debut. gasoline		
H17	Depropaniser cond.		
H18	Drying col. feed		
H19	HC3 splitter condenser		
H20	HC3 spl. bottom prod.		
H21	HC4 splitter condenser		
H22	Light HC4 prod.		
H23	Heavy HC4 prod.		

Table 1. Stream Designations

The existing heat recovery network is outlined in Figure 2.

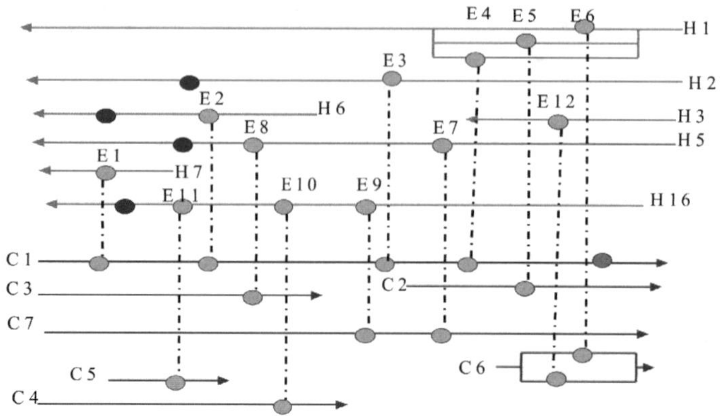

Figure 2. Existing Network

The stream data extracted by Al-Riyami et al have been related to individual exchangers in Table 2. This consists of exchanger inlet and outlet temperatures, heat loads and stream heat transfer coefficients.

The plant uses a range of hot and cold utilities. The following unit costs were specified: cooling water 2.66 \$/MWhr, air cooling 1.61 \$/MWhr, fuel gas 9 \$/MWhr, LP steam 10.2 \$/MWhr and MP steam 13 \$/MWhr.

The material of construction to be used in the heat exchangers is dependant upon design temperature. If this exceeds 200 C stainless steel is required. Otherwise, carbon steel can be used.

Al Riyami et al assumed that the installed capital cost of a unit was given by the equation:

$$C_E = a + bA^{0.81}$$

Where the base cost (a) is 33422 and the material factor (b) is 1872 for carbon steel and 4103 for stainless steel.

In reality the capital cost of a heat exchanger is a function of shell diameter and tube length with shell diameter being the dominant factor. Consequently, a power law equation based is only likely to apply if all exchangers are fabricated to a fixed length.

In 'targeting' it is advisable to determine capital cost on the basis of a cost per unit area. This divorces the cost estimate from influence of the number of new units required and the size distribution across these units.

Unit area cost is determined by dividing the cost of a 'standard' sized unit (e.g. 300 sq.m.) by the assumed size. On this basis the unit area cost of a carbon tube unit is 745 \$/m^2 and that of a stainless steel unit is 1500 \$/m^2.

Stream	Supply Temp. C	Target temp. C	Heat Load MW	Heat transfer coefficient W/sq.m.K
E4-H1	366	218	4.283	150
E4-C1	144	184	4.283	150
H-C1	184	385	21.521	150
E5-H1	366	232	3.698	150
E5-C2	100	195	3.698	500
E6-H1	366	224	41.48	150
E6-C6	184.5	193	41.48	200
E12-H3	293	238	8.787	150
E12-C6	184.5	193	8.787	200
E3-H2	366	160	1.726	150
E3-C1	132	144	1.726	150
C-H2	160	60	0.838	150
E2-H6	224	132	1.912	120
E2-C1	102	132	1.912	150
C-H6	132	50	1.704	120
E1-H7	158	108	5.669	120
E1-C1	60	102	5.669	150
E7-H5	238	226	1.211	120
E7-C7	95.4	101.7	1.211	200
E8-H5	226	93	10.2	120
E8-C3	58.3	192.6	10.2	200
C-H5	93	38	4.066	120
E9-H16	193	117	3.261	250
E9-C7	78.6	95.4	3.261	250
E10-H16	117	87.7	1.414	250
E10-C4	38	55	1.414	120
E11-H16	87.7	62.1	0.872	250
E11-C5	38	50	0.872	120
C-H16	62.1	38	0.8	250
H-C8	118	124	16.85	250
H-C11	95.4	102	5.87	250
H-C10	51.2	54.3	0.76	250
H-C9	48.2	49.7	1.055	250
C-H8	104	40	0.294	120
C-H9	97.2	38	16.26	200
C-H10	68.1	38	4.376	200
C-H11	63.1	38	4.606	120
C-H12	52.5	38	0.245	120
C-H14	96.4	46	50.27	150
C-H15	46	38	0.36	150
C-H17	53	46	14.125	250
C-H21	67.2	60	7.336	200
C-H23	102	38	1.816	120
C-H4	293	52	1.633	150
C-H13	51.5	38	0.074	200
C-H18	46	38	0.196	150
C-H19	46.4	38	0.856	150
C-H20	53	38	0.032	200
C-H22	60	38	0.206	150

Table 2. Stream data

Finally, the economic criteria for the retrofit are that $5M is available for investment but a payback period of 3 years or less must be achieved. The plant was assumed to operate for 7 920 hours/annum.

Appraisal of Problem Data

It is the authors' opinion that once project data has been collated it should be critically evaluated. Such evaluations should cover: check of flowsheet consistency; assessment of which streams should be included in the integration analysis; critical evaluation of temperatures and loads; and critical evaluation of assumed (or, measured) exchanger performance.

With the exception of the original problem authors, the various groups undertook such appraisal to a greater or lesser extent.

1). Flowsheet Consistency

Examination of the flowsheet only yielded one concern. The boil-up to the Stripper Column C7 is shown to be provided by two reboilers (units E7 and E9) operating in series. Such an arrangement is unusual. If the units are vertical thermosiphons or kettle reboilers the arrangement would be impractical. In the case of a horizontal thermosiphon it would be possible but ill advised. In each case the expected reboiler arrangement would be units operating in parallel.

2). Assessment of Streams to be Used for Integration

The stream data is observed to include a compressor inter-cooler flow. The purpose of an inter-cooler is reduction of compressor power consumption through the reduction of volumetric flow-rate (cooling of the gas reduces density and hence volumetric flow at fixed mass flow). Pressure drop plays a significant part in the design of inter-coolers. The designer will generally look for as low a pressure loss as possible (for losses must be made up by within the next compression stage). This, in turn, means that the engineer should seek a match involving as cold a stream as possible (which is usually the cold utility). Integration of this stream was considered ill advised and the stream should be removed from consideration. Two of the groups made this observation.

3). Evaluation of Temperatures and Loads

Examination of the stream temperatures and individual exchanger loads indicated a number of inconsistencies.

The raw feed to the main fractionator (C1) passes through exchangers E1, E2, E3 and E4. The heat capacity flow-rates of this stream have been computed at each of these locations using the stated exchanger heat loads and temperature increases. The results are given in Table 3.

Location	Heat Capacity Flow-rate
E1	135
E2	64
E3	144
E4	107

Table 3. Consistency of Data for Stream C1

It is clear that the data is inconsistent. This is not unusual in 'real' projects. The policy adopted by the authors is to request clarification and confirmation of the data, but continue with analysis. The sensitivity of the project results to errors in stream data need to be assessed and reported if clarification is not forthcoming.

The data for the de-butanized gasoline (stream H16) was also found to be inconsistent. This stream passes through exchangers E9, E10, E11 and a cooler. The heat capacity flow-rates for this stream are given in Table 4.

Location	Heat Capacity Flow-rate
E9	42.9
E10	48.2
E11	34.1
Cooler	93.2

Table 4. Consistency of Data for Stream H16

Turning attention to individual re-boilers, the temperature rise reported for the stripper column re-boilers appears very high. The inlet temperature is given as 78.6 C. The outlet temperature is 101.7 C (a temperature rise across the re-boiler of over 23 C). Most re-boilers operate with re-circulation of liquid feed. The outlet stream is a two-phase mixture which usually has a mass quality of between 0.05 and 0.15. The quoted temperature change is what would be expected if the liquid was being totally evaporated (where vaporization starts at the bubble point temperature and finishes at the dew point of the mixture at which point all of the liquid has been evaporated). The suspicion is that the data has come from a simulation involving total vaporization rather than from plant measurement or from equipment simulation.

4. Evaluation of Heat Transfer Coefficients

The stream heat transfer coefficients listed in Table 2 are very low. If these have been extracted from plant measurements then they indicate poorly sized equipment and severe fouling.

The magnitudes of film coefficient and fouling resistance that can be expected for various process duties are listed in Table 5.

Operation	Film h.t.c. W/m^2K	Fouling resistance m^2K/W	Duty h.t.c. W/m^2K
Single Phase			
Gas @ 1 bar	100	0.0002	100
Gas @ 20 bar	600	0.0002	540
Gas @ 50 bar	800	0.0002	690
Gas @ 100 bar	1000	0.0002	835
'Light' organic	1600	0.0005	890
'Medium' organic (hot)	800	0.0005	570
'Med.' organic (cold)	300	0.0005	260
'Heavy' organic (hot)	200	0.0008	175
'Heavy' organic (cold)	100	0.0008	90
Aqueous	5000	0.0001	3300
Condensation			
Light, no inerts, 0.1 bar	1300	0.0002	1050
Light, no inerts, 0.1-1 bar	2000	0.0002	1430
Light, no inerts, > 1 bar	4000	0.0002	2220
Medium, no inerts	1500	0.0002	1150
Heavy, no inerts	800	0.0002	690
Light with inerts	300	0.0002	285
Steam	7000	0.0001	4120
Boiling			
Light organic	1500	0.0003	1035
Medium organic	1000	0.0003	770
Heavy organic	800	0.0005	570
Aqueous	3000	0.0003	1580

Table 5. Magnitude of Heat Transfer Coefficients Expected on Various Process Duties

This table is based on the recommendations made by Saunders (1984).

Energy can be saved by both improved integration and by improvement of exchanger performance. The two elements should be separated. 'Typical' coefficients should be used in the integration analysis (such as those listed in Table 5).

Improvement in the performance of existing exchangers can be brought about by monitoring performance, determining the cause for poor performance, modelling fouling and the identification of fouling mitigation strategy.

The heat transfer coefficients provided above are typical of what is observed when exchangers have been grossly over-sized. Under such circumstances the operator often needs to use bypasses in order to obtain acceptable plant performances. These bypasses are closed as the exchangers foul.

The selection of proper heat transfer coefficients is important for it affects many, if not all, of the economic calculations undertaken during the integration analysis. Only one group based their analysis on expected rather than reported heat transfer coefficients.

The Techniques Compared in the Study

The group that originally presented the Case Study used a mathematically based procedure named 'Network Pinch Analysis'. This procedure, originally developed by Asante & Zhu (1997) uses specialist software.

Specialist consultant #1 used their own proprietary 'automated retrofit design' software program to identify plant improvements. Details of the techniques used in this software are not available.

Specialist consultant #2 used the long established 'cross-pinch' analysis.

Finally, the study used techniques being recommended to Saudi Aramco engineers. These techniques involved a combination of 'problem decomposition' and 'cross-pinch' analysis with a strong emphasis on the simultaneous consideration of heat exchanger technology.

Results of Network Pinch Analysis

These workers identified four potential retrofits. These had some common features. For instance, each used the inter-cooler stream (H10) to drive the drying column reboiler (C9). Each used the HCO product stream (H4) to drive the de-propaniser reboiler (C8).

Their preferred design involved further integration of the de-propaniser reboiler (C8). A second re-boiler was placed in series with the new H4-C8 match. This reboiler was driven by the LCO stream feeding the sponge absorber (H5).

The use of the inter-cooler stream was discussed above. It is worth noting that the compressor needing the inter-cooler is in a separate part of the plant to the drying column re-boiler. The inclusion of this stream in the integration problem has resulted in the 'network pinch analysis' suggesting an impractical match.

The use of re-boilers in series has also been discussed above. Given the temperatures of the streams involved it could be argued that these reboilers could easily be placed in parallel. However, the temperature of the hot streams involved in these matches also provides a problem. The HCO product has an inlet temperature of 293 C. The LCO stream has a supply temperature of 238 C. The reboiler is required to operate at a temperature of 118 C. The temperature differences involved in these matches are very high (120 – 175 C). It is likely that a reboiler operating under such large temperature differences will experience premature dryout. Severe fouling and operability problems may be expected. It is likely that the heat transfer specialists would reject such a proposal.

The project recommended by this group was reported to have a potential saving of 8.349 MW. However, 1.055 MW came from integration of the inter-cooler, 1.742 MW from the use of HCO product to drive the de-propaniser reboiler and 1.107 MW from the use of the LCO stream to drive that reboiler. So, of this suggested saving it is likely that at least 3.9 MW would be considered impractical following detailed design review.

Results from Automated Retrofit Design

The group that used automated retrofit pointed out that the stated stream heat transfer coefficients were much lower than would be expected and that if these values were used it would not be possible to obtain a project that had a payback of less than eleven years.

They did not re-place the coefficients with more reasonable values. They highlighted the problem and then sought to demonstrate the capabilities of their approach by developing a retrofit for the limiting investment rather than required payback.

They included the compressor inter-cooler stream in the analysis. As a consequence their computer program suggested that this stream be used to drive the reboiler on the HC3 Splitter. The reported saving was 0.76 MW.

Their automated design program also came up with an infeasible design for re-boiler on the Stripper (C7). Comment has already been made that the use of two re-boilers in series on this duty was questionable and that the arrangement needed to be checked. The automated design program placed a new re-boiler (driven by the overheads from the de-butanizer column seeking to save 1.75 MW) in front of one of the existing units (E7) and a second new re-boiler (this one driven by heat from the LCO Product seeking to save 0.74 MW) positioned between that unit and the second existing re-boiler (E9). Rather than have two re-boilers in series there are four re-boilers in series. Such a modification would not be consistent with the hydraulics of thermo-siphon systems.

Three new exchangers provided additional heat recovery to the Raw Feed (C1).

The overall saving was reported to be 4.72 MW.

Our judgement would be that the savings associated with the use of the heat from the inter-stage cooler and the integration of the stripper boil-up are unrealistic. Only the savings from heat recovery to the raw feed should be considered.

The procedure failed to recognise that the Drying Column boil-up was being undertaken well below the pinch. The duty covers a temperature span of 48.2 to 49.7 C and is currently undertaken using LP Steam. The potential saving is around 1 MW.

Results from Cross-Pinch Analysis

The third group examined the basic data and observed that the smallest temperature approach encountered on the existing plant was 24 C. They then examined the economics of under-taking a retrofit for a temperature approach between the composite curves of 20 C. However, since they assumed used the heat transfer coefficients specified in the data they found that a large surface area demand resulted in payback in excess of the required 3 years and relaxed the temperature approach to 30 C.

From a comparison between the Grand Composite Curve derived at a temperature approach of 30 C and the existing utility system they identified that there was scope to reduce LP steam consumption and to switch between fuel use and MP steam generation. However, they then observed that there was no economic incentive for such a switch given the small differences in tariff. Consequently, they concentrated on reducing LP steam consumption.
The group then conducted a 'cross-pinch' analysis at the proposed temperature approach. This analysis indicated a potential saving of 6.83 MW. However, since there were a significant number of existing exchangers transferring heat across the 'pinch' they concluded that a lot of modifications would be necessary to make the full saving.

They subsequently proposed a re-design that involved five changes and made an estimated saving of 5.35 MW.

The proposed modifications were:

1. The heat load on existing exchanger E1 to be reduced by 3.54 MW so that the 'top pump-around' could be used to drive a new re-boiler on the HC4 Splitter. This re-boiler should be placed in parallel with the existing unit.
2. The resultant shortfall in pre-heat of the column feed was to be made up through two new heat recovery matches. The first involved the installation of a new exchanger matching the feed with main column overheads.
3. A new match between the column feed and HCO product (prior to entry into the air cooler) was also necessary.
4. The HC3 Splitter Column was to be driven by an integrated re-boiler. The proposed hot stream was the overheads from the de-butaniser column. This saves 0.76 MW.
5. The Drying Column was also to be driven by an integrated re-boiler. Again, the overheads from the de-butanizer column were to be used. This saves 1.054 MW.

The existing re-boiler on the HC4 splitter currently consumes 5.87 MW. After the proposed modification its duty would be reduced to 2.33 MW. This is a substantial change. Consequently, the performance of both the existing re-boiler under reduced load and the control needs of the column have to be established before this modification can be accepted.

The top pump-around driving the new reboiler would enter the unit at a temperature of 158 C. The saturation temperature of the column return is 102 C. This is quite a large difference and the behaviour of reboiler operating under these conditions also needs to be established.

The de-butanizer overheads are available at a temperature of 96.4 C. The HC3 splitter boil-up is provided at a temperature of 53 C and the drying column boil-up is at a temperature of 50 C. Again the temperature driving forces are around 50C and checks on design suitability have to be made.

The load of the current de-butanizer condenser is very large (50.3 MW) so the proposed integration is not likely to impose control or operability problems. However, the piping costs associated with such a modification are likely to be high (vapour streams required large diameter pipes).

Results from Combined Problem Decomposition and Cross-Pinch Analysis Supplemented by Consideration of Exchanger Technology

In this study the stated heat transfer coefficients were not accepted and the values given in Table 5 were used.

Using 'problem decomposition' analysis the problem was divided into two sub-problems. The first of these consisted of the existing heat recovery network used in the fractionation area <u>plus</u> the demand made by the de-propaniser reboiler (C8). The second consisted of the heat recovery network used in the gas processing area <u>plus</u> the other un-integrated reboilers (drying column reboiler C9, HC3 splitter reboiler C10 and HC4 splitter reboiler C11).

With regard to the fractionation area the inefficiencies identified using cross-pinch analysis were transfer of 3.4 MW across the pinch from poor use of heat from the top pump around (H7); 1.02 MW from poor use of heat from the LCO product (H6); and, 1.12 MW from failure to integrate the HCO product stream (H4).

Correction of the first two inefficiencies involved introducing new heat recovery matches to heat the raw feed (C1) below the pinch. This could be achieved by making a heat recovery match between the raw feed and the overheads from the main column (H9). However, the maximum load on this match was estimated to be just 3.67 MW. The shortfall of 0.75 MW could be made up through new matches involving Slurry Product (H2) and the LCO Product (H6) below pinch temperature. The cost of installing two small additional matches may not be economically justified and the recommendation was to solely correct the inefficiency in the use of the top pump around (a saving of 3.4 MW). So, only the match between the raw feed (C1) and the column overheads (H9) was recommended. The heat made available in the top pump around was used to drive a new reboiler on the de-propaniser column.

Heat from the HCO product (H4) was also used to drive a new re-boiler on the de-propaniser column.

Given the high supply temperatures involved in these matches the use of indirect integration was recommended. The stream was used to heat water which was then pumped across the plant to drive the re-boiler. The use of plate-and-frame exchangers was suggested for the use of such a unit would significantly reduce the cost of the recovery system.

The use of indirect integration was supported by a consideration of piping costs (piping three separate streams across the plant to three separate re-boilers would be excessively expensive).

The overall energy saving in the fractionation area was predicted to be 4.6 MW.

In the gas processing area they recommended that heat currently rejected through the cooler on the de-butanised gasoline stream (H5) to drive the re-boilers on both the drying column and the HC3 Splitter. This stream is a liquid so the piping costs will be relatively low and will allow reasonable control through the application of a bypass.

They observed that this stream (H5) and the LCO Return stream (H16) were both being used inefficiently and recommended the following actions:

1. install new match between the de-butaniser overheads (H14) and the Sponge Absorber Return (C3) prior its match with the de-butanised gasoline (E8). This release heat from the de-butanised gasoline (H5) at a high temperature;
2. re-pipe (or replace) the exchanger heating the Stripper Feed (E10) such that it is driven by heat from the Heavy C4 Product (H23) rather by heat from the LCO Return (H16);
3. use the heat made available in the LCO Return (H16) and de-butanised gasoline (H5) to provide boil-up on the HC4 Splitter (C11). The load on the existing re-boiler would be reduced from 5.87 MW to 2.47 MW. Consequently, this unit may need to be replaced.

The overall energy saving in the gas process area was also 4.6 MW. So, the total identified saving for the plant was 9.2 MW.

Observations

1. The nature and cost implications associated with individual streams (e.g. the compressor inter-cooler) must be considered at the outset of a project. This is particularly the case when procedures highly dependant on numerical analysis is used. Failure to do so is likely to result in impractical solutions.
2. Limitations imposed by heat exchanger technology need to be recognised. Existing numerical procedures appear weak in this area when re-boiler duties are considered. The result is that these procedures again produce impractical solutions.
3. Retrofit projects present opportunities to correct errors in the design of existing plant. Errors in exchanger design and application are common. Many plants have grossly over-sized exchangers that have been allowed to foul in order to obtain reasonable operability. The direct use of heat transfer measurements from plants is to be discouraged. Analysis should be based on expected heat transfer coefficients. Failure to do so leads to poor economic decisions.
4. The nature of streams should be considered in identifying heat recovery matches. Why pipe a vapour across a plant when a liquid can be used?
5. Effects of improved integration on the behaviour of existing utility units must be made. The argument that since the heat load is being reduced they can be ignored is untrue. For instance, re-boilers have a limited turndown.
6. The effects of temperature difference on plant performance needs to be considered. The existing plant was reported to exhibited high fouling levels. If the reported performance is correct the values found in some of the exchangers will be due to the use of excessive temperature driving forces leading to very high fouling rates.

Final Word

The fairness of the competition may be questioned. All parties had the same information but the time each party allocated to the task was at their own discretion. Since, consultants work on day-rate the time they allocated to the work was probably short. These groups certainly have the capability of including consideration of exchanger technology in their work (provided they don't rely on automated numerical procedures) but will only do so if requested by the customer. The customer needs to appreciate the importance of such a consideration. However, the consultant needs to appreciate that proposal of impractical modifications is not in the interest of any party.

References

Al-Riyami B.A., Klemes J & Perry S. Appl.Termal Engng. 2001, 21, 1449-1487

Asante N.D. & Zhu X.X. Trans.I.Chem.E. 1997,75[a]

Saunders E.A.D. Heat Exchangers, 1984

(52e) Practical Process Integration Retrofit Part 5. Software Needs

G. T. Polley, Chemical Engineering, University of Guanajuato, Guanajuato, Mexico

Practical Process Integration Retrofit

Part 5. Software Needs

G.T.Polley

In this final paper in our series on process integration retrofit we look at what is needed in terms of computer software in order to undertake retrofit projects. Three types of program are identified:

Integration analysis

Exchanger design and analysis

Network simulation

Optimizing bitumen upgrading scheme – modeling and simulation approach

Jinwen Chen and Mugurel Munteanu

CanmetENERGY, Natural Resources Canada
One Oil Patch Drive, Devon, AB, T9G 1A8, Canada

Abstract

The present study focuses on modeling and simulation of the bitumen upgrading and refining schemes using the HYSYS modeling software in conjunction with pilot plant experimental data obtained at CanmetENERGY, as well as any available commercial operation data. Two existing upgrading schemes were investigated: coking-based and hydroconversion-based, which are commonly used by oil sands companies. The major upgrading units, such as atmospheric and vacuum distillation columns, coker/hydroconverter, and hydrotreaters, were depicted in detail and the commercial operating conditions for each of them were identified.

The coking-based upgrading scheme was simulated under various operating scenarios. Using existing data, mass balance was performed and a synthetic crude oil yield close to 82wt% was achieved based on the initial diluted bitumen feedstock. The hydroconversion-based upgrading scheme was also studied in this work by replacing the coker with a hydroconverter (an ebullated bed reactor) and by utilizing commercially available data under different operating conditions.

Introduction

About 1.6 million barrels of combined mining and in-situ based bitumen is currently produced. Over 50% of this production is upgraded to synthetic crude oil (SCO). To date, virtually all of the mining based bitumen is upgraded. Much of the synthetic crude is processed in Canadian refineries today, but increasingly large volumes will be marketed in the northern tier US states as the industry expands its output[1].

Production of transportation fuels from Canadian bitumen feedstocks requires either new integrated upgrading and refining facilities, or converting existing refineries that use conventional crudes to allow higher input of bitumen feedstocks. In either case, it is important and useful to optimize the entire upgrading and refining scheme under different process configurations and product scenarios to minimize the process related energy intensity/ consumption, at the same time to achieve the best economic benefits. Such optimization can provide guidelines to either existing upgrading and refining operations or process design for new upgraders and refineries.

There are two major commercial primary bitumen upgrading processes: coking and hydroconversion. Historically coking has predominated as the choice for primary upgrading. As the first step to produce a bottomless (zero residue) SCO, it handles the higher solids and water content in mining based bitumen more easily. The by-product coke helps to trap solids, and concentrate and remove metals as well as some of the sulphur and nitrogen. However, the total liquid product yield is relatively low due to the formation of coke. In comparison, ebullated bed catalytic hydroconversion process has much higher total liquid product yield due to the hydrogen addition. In some commercial operations the conversion of bitumen is not 100% in the hydroconversion unit, generating a small portion of residue that is further processed with a coking unit.

Both of the two primary processes produce liquid products with boiling ranges similar to conventional crudes. However they have high concentrations of impurities, such as sulphur and nitrogen. The secondary hydrotreating processes remove these impurities to produce sweet blending stocks for the SCO without changing much the boiling range of the liquid. The SCO boiling range is essentially controlled by the primary upgrading step. In reviewing some of the major challenges that the oil sands industry is facing, bitumen upgraders need to capitalize on, or address, the following:

(a) Take advantage of some relatively minor upgrading at the recovery stage
(b) Take advantage of the necessity to move to alternative energy and hydrogen sources, particularly internally generated residues, which is a trend with very large impact on main upgrading process selection
(c) Address major environmental concerns in an integrated way
(d) Meet future crude quality requirement with existing facilities

The oil sands industry by its very size is in a position to influence technology development for their relatively unique needs. It is important to identify the possible avenues for better upgrading technology for current and future projects. The most important technology developments for existing commercial processes are: coking, ebullated bed hydroconversion, moderate primary upgrading, hydrotreating, hydrocracking, and catalyst development.

The objective of this study is to identify the technologies and different operating conditions that directly meet the future trends in current upgrader performances. Improvement in process efficiency and bitumen utilization will benefit from advanced modeling techniques. Advanced modeling can also help to achieve better process control. Advanced simulation results also provide valuable information for understanding bitumen processabilities and increasing marketability, reducing GHG emissions and other related environmental impacts in bitumen upgrading and refining.

Modeling and Simulation Bases

Figure 1 presents a simplified generic flow diagram of current upgrading process. The diluted bitumen from the extraction and froth treatment plant is distilled in the diluent recovery unit (DRU), or atmospheric distillation unit (ADU). The diluent is recovered and recycled to the extraction plant. The bitumen constitutes the feed for the vacuum distillation unit (VDU). The distilled products from the VDU are routed directly to naphtha, light gas oil and heavy gas oil hydrotreaters, and the vacuum-topped bitumen is either thermally cracked in the coker or catalytically hydroconverted in the ebullated bed hydroconverter. The total liquid product (TLP) from the coker or hydroconverter is fractionated into naphtha, light gas oil and heavy gas oil, which are combined with the corresponding fractions from the VDU. The combined naphtha, light gas oil and heavy gas oil are further hydrotreated in three individual hydrotreaters. The liquid products from the hydrotreaters are blended to form the final upgrading product SCO, which is shipped to refineries in Canada and the US by pipeline for further refining[2,3].

Table 1 summarizes the properties of typical Athabasca bitumen and its distilled products. The bitumen contains about 4.3wt% of sulfur and the API gravity is about 10.6.

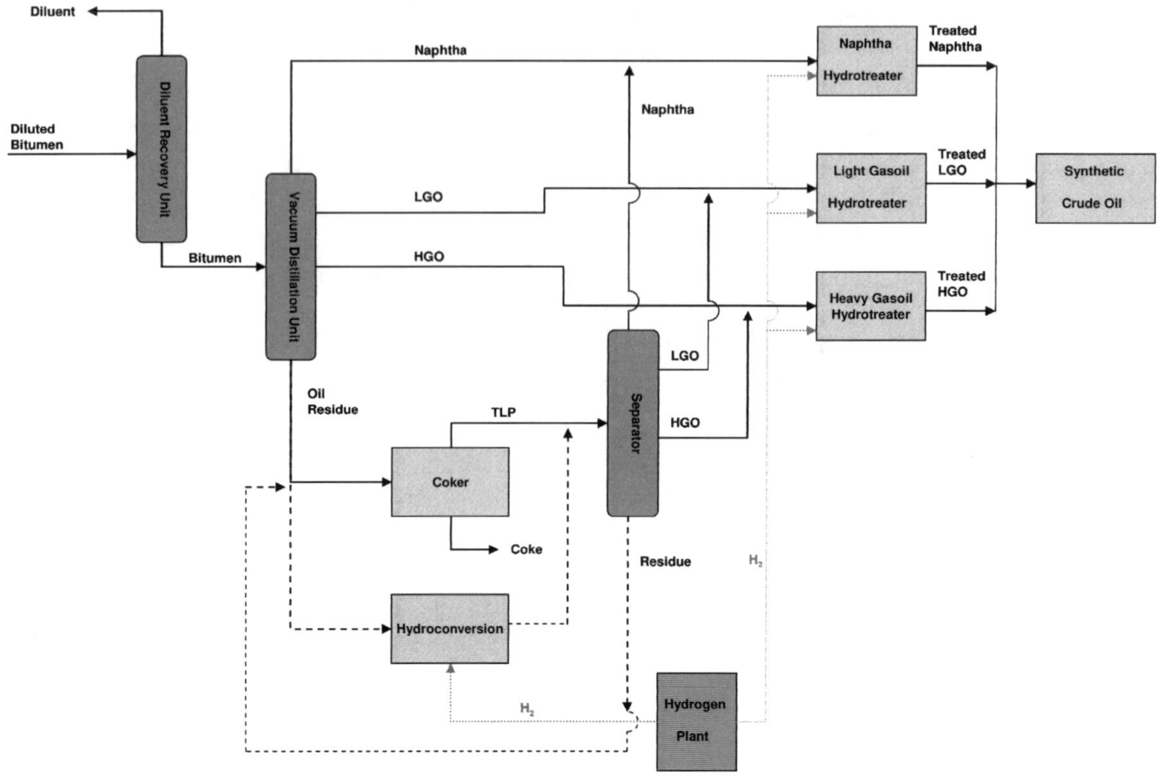

Figure 1. Simplified generic schematic diagram of upgrading process

Table 1. Properties of typical Athabasca bitumen and its distilled products

Property	Unit	Bitumen	Bitumen atmospheric bottoms	Vacuum bottoms	Naphtha	Light gas oil	Heavy gas oil	Synthetic crude oil
Density	g/cm^3	0.9960	0.9957	1.0670	0.7820	0.9410	1.0020	0.8650
API gravity	°API	10.6	10.6	1.1	49.4	18.9	9.8	32.0
Sulfur	wt%	4.3	4.3	6.0	1.7	3.6	4.3	0.15
Nitrogen	wppm	5423	5421	10590	237	1600	3780	800
Hydrogen	wt%	10.5	12.3	8.5	12.4	10.9	10.2	12.6
Carbon	wt%	82.82	85.3	83.7	85.1	84.8	84.2	86.6
H/C atomic ratio	-	1.53	1.8	1.4	1.7	1.5	1.5	1.8

The present work focuses on modeling and simulation of the bitumen upgrading and refining schemes using the HYSYS modeling software in conjunction with pilot plant experimental data obtained at CanmetENERGY, as well as any available commercial operation data and published literature data[4-12]. Two existing upgrading schemes are investigated: coking-based and hydroconversion-based. The major upgrading units, atmospheric and vacuum distillation columns, coker/hydroconverter, and hydrotreaters, are depicted in detail and the commercial operating conditions for each of them were identified.

Results and Discussions

Simulations are performed for both schemes, coking-based and hydroconversion-based with the same bitumen-to-diluent ratio as feedstock (75:25 vol/vol). The processing capacities are assumed to be the same at 100K barrel/day diluted bitumen. The properties of the bitumen feed, intermediate streams and the final SCO for both simulated schemes are presented in Tables 2 to 4 (coking-based scheme), Tables 5 to 7 (hydroconversion-based scheme), and Table 8 (synthetic crude oil), respectively. The upgrading units are modeled and calibrated adequately in order to simulate commercial operating conditions and to generate data of industrial interest. Since there are no available models for coking and ebullated bed hydroconversion in HYSYS, pilot plant experimental data obtained at CanmetENERGY and published in literature are used to develop the coker and hydroconverter models. Simulation results show that a SCO yield of about 81wt% can be achieved based on the initial diluted bitumen feedstock for the coking-based scheme. In contrast, the hydroconversion-based upgrading scheme generates a SCO yield of about 92wt%. The difference in the SCO yield between the two schemes is due to the fact that the ebullated bed hydroconversion unit completely converts the vacuum bottoms into gaseous and liquid products without generating any solid residual material. As seen in Table 8, the SCO from hydroconversion-based scheme has slightly lower concentrations of sulfur, nitrogen and aromatics than that from coking-based scheme. This results from the hydrodesulphurization, hydrodenitrogenation and hydrogenation, in addition to the hydrocracking of heavy molecules, occurring in the ebullated bed hydroconverter.

The major upgrading units are investigated in detail and calibrated accordingly to commercial operating conditions. In this paper, the authors present only one set of operating conditions and parameters for the two above mentioned upgrading schemes as shown in Table 10. It is noted that the atmospheric distillation column, the vacuum distillation column and the hydrotreaters are operated under the same conditions for both schemes. The only difference is that one uses coking and the other one uses hydroconversion. The operating conditions are summarized in Table 9.

As seen in Tables 11 and 12, even if the hydrotreaters are operated under the same conditions, the hydrogen consumptions for diesel and gas oil in the hydroconversion-based upgrading are significantly higher than those in the coking-based upgrading. This is because feed flow rates to these two units in the hydroconversion-based scheme are much higher than those in the coking-based scheme (Tables 4 and 7). Consequently the hydroconversion-based upgrading scheme has a higher SCO yield at the cost of higher hydrogen consumption.

Table 2. Coking-based scheme streams properties

		Bitumen	Diluent	DilBit	Diluent return	Atmospheric bottoms
Mass flow	kg/h	494800	126600	620800	126600	494300
Nitrogen content	wppm	5423.0	0.0	4318.0	10.8	5420.0
Sulfur content	Wt%	4.3	0.0	3.4	0.15	4.3
Aromatics content	vol%	61.5	0.5	46.2	3.4	60.4
Density	g/cm^3	0.9960	0.7643	0.9381	0.7670	0.9949
API gravity	°API	10.6	53.6	19.3	53.0	10.7
Conradson carbon residue	wt%	9.5	0.0	7.6	0.0	9.5

Table 3. Coking-based scheme streams properties

		Vacuum distillation				Coker				
		Naphtha	Light vacuum gas oil	Heavy vacuum gas oil	Vacuum bottoms	Coker lights ends	Coker naphtha	Coker diesel	Coker gas oil	Coke
Mass flow	kg/h	6672	75720	188000	223900	19140	33590	26870	78370	60460
Nitrogen content	wppm	14.68	202.8	1552.00	10590.0	0.0	88.0	1132.0	3100.0	18370.0
Sulfur content	wt%	0.20	1.5	3.5	6.0	0.0	0.9	3.8	3.9	7.9
Aromatics content	vol%	4.3	26.1	54.6	82.1	0.0	11.0	50.0	57.0	89.0
Density	g/cm^3	0.7802	0.8836	0.9753	1.0670	0.6500	0.7535	0.8814	0.9798	1.0690
API gravity	°API	49.9	28.6	13.6	1.1	86.2	56.3	29.1	12.9	0.9
Conradson carbon residue	wt%	0.0	0.0	1.4	19.9	0.0	0.0	0.1	1.5	20.4

Table 4. Coking-based scheme streams properties

		Naphtha hydrotreater		Diesel hydrotreater		Gas oil hydrotreater	
		Mixed naphtha	Hydrotreated naphtha	Mixed diesel	Hydrotreated diesel	Mixed gas oil	Hydrotreated gas Oil
Mass flow	kg/h	40260	37290	102600	102000	266400	261200
Nitrogen content	wppm	75.9	22.0	446.2	432.2	2007.0	319.2
Sulfur content	wt%	0.8	0.08	2.1	0.14	3.6	0.1
Aromatics content	vol%	9.9	6.0	32.4	26.4	55.3	50.3
Density	g/cm^3	0.7578	0.7591	0.8830	0.8516	0.9766	0.8811
API gravity	°API	55.2	54.9	28.7	34.7	13.4	29.1
Conradson carbon residue	wt%	0.00	0.00	0.03	0.00	1.44	0.00

Table 5. Hydroconversion-based scheme streams properties

		Bitumen	Diluent	DilBit	Diluent return	Atmospheric bottoms
Mass flow	kg/h	494800	126600	620800	126600	494300
Nitrogen content	wppm	5423.0	0.0	4318.0	9.9	5421.0
Sulfur content	wt%	4.3	0.0	3.4	0.14	4.3
Aromatics content	vol%	61.5	0.5	46.2	3.2	60.5
Density	g/cm^3	0.9960	0.7643	0.9381	0.7653	0.9957
API gravity	°API	10.6	53.6	19.3	53.4	10.6
Conradson carbon residue	wt%	9.5	0.0	7.6	0.0	9.5

Table 6. Hydroconversion-based scheme streams properties

		Vacuum distillation				Hydroconverter					
		Naphtha	Light vacuum gas oil	Heavy vacuum gas oil	Vacuum bottoms	Residue (vacuum bottoms + recycle)	Lights ends	naphtha	diesel	gas oil	Residue (recycle)
Mass flow	kg/h	1781	80990	187500	224000	348400	16230	17150	76370	86880	124600
Nitrogen content	wppm	12.	194.0	1553.0	10590.0	9202.0	0.0	0.0	0.0	0.0	6645.0
Sulfur content	wt%	0.2	1.5	3.5	6.0	4.2	0.0	0.0	0.0	0.2	0.9
Aromatics content	vol%	3.8	25.1	54.6	82.1	69.2	0.0	0.0	0.0	10.0	44.0
Density	g/cm^3	0.7723	0.8805	0.9754	1.0670	1.0780	0.6495	0.7423	0.8783	0.9767	1.1000
API gravity	°API	51.7	29.2	13.6	1.1	-0.3	86.4	59.1	29.6	13.4	-2.8
Conradson carbon residue	wt%	0.0	0.0	1.4	19.9	17.7	0.0	0.0	0.0	0.0	14.6

Table 7. Hydroconversion-based scheme streams properties

		Naphtha hydrotreater		Diesel hydrotreater		Gas oil hydrotreater	
		Mixed naphtha	Hydrotreated naphtha	Mixed diesel	Hydrotreated diesel	Mixed gas Oil	Hydrotreated gas oil
Mass flow	kg/h	18930	16690	157400	159900	274400	277500
Nitrogen content	wppm	1.2	0.4	99.9	21.2	1061.0	126.9
Sulfur content	wt%	0.02	0.00	0.8	0.16	2.5	0.07
Aromatics content	vol%	0.4	0.2	12.9	6.4	40.5	29.9
Density	g/cm^3	0.7450	0.7465	0.8794	0.8516	0.9758	0.8664
API gravity	°API	58.4	58.0	29.4	34.7	13.5	31.8
Conradson carbon residue	wt%	0.00	0.00	0.00	0.00	0.96	0.00

Table 8. Synthetic crude oil properties

		SCO Coking-based scheme	SCO Hydroconversion-based scheme
Mass flow	kg/h	400500	454100
Nitrogen content	wppm	320.3	85.0
Sulfur content	wt%	0.12	0.10
Aromatics content	vol%	39.5	20.5
Density	g/cm^3	0.8607	0.8561
API gravity	°API	32.9	33.8
Conradson carbon residue	wt%	0.0	0.0
Yield	vol%	93.7	106.8
Yield	wt%	81.0	91.9

Table 9. Diluent recovery and Vacuum distillation column operating conditions

Diluent Recovery Unit		Vacuum Distillation Column	
Number of stages	24	Number of stages	24
Feed stage	24	Feed stage	8
Diluent return stage	Overhead liquid outlet	Naphtha withdraw stage	Overhead liquid outlet
Atmospheric bottoms stage	Bottom liquid outlet	Light vacuum gas oil Withdraw stage	2
Pressure (top)	120 kPa	Heavy vacuum gas oil Withdraw stage	7
Pressure (bottom)	140 kPa	Vacuum bottoms Withdraw stage	Bottom liquid outlet
Diluent D86 90% Temperature	206.1°C	Pressure top	2 kPa
		Pressure bottom	5 kPa
		Naphtha D86 90% temperature	221°C
		Light vacuum gas oil D86 90% temperature	345°C
		Light vacuum gas oil D86 90% temperature	524°C
		Naphtha yield	1.4wt%
		Light vacuum gas oil yield	15.3wt%
		Heavy vacuum gas oil yield	38.0wt%
		Vacuum bottoms yield	45.3wt%

Table 10. Coker (coking-based scheme) and Hydrocracker (hydroconversion-based scheme) operating conditions

Operating conditions		Coker	Hydroconverter
Reactor catalyst density	kg/m^3	N/A	1500
LHSV	h^{-1}	N/A	0.48
Reactor pressure	MPa	Atmospheric	16
Reactor temperature	°C	482.2	430.0
Sulphur conversion	wt%	N/A	71.5
Conradson carbon residue conversion	wt%	N/A	48.5
Nitrogen conversion	wt%	N/A	51.6
Aromatics conversion	wt%	N/A	39.1
H_2 consumption	SCF/bbl	N/A	1512
Total liquid product yield	wt%	70.6	92.2
NH_3 yield	wt%	0.45	4.8
H_2S yield	wt%	2.0	3.0
Light ends initial boiling point	°C	-60.0	-60.0

Table 10 - continued

naphtha initial boiling point	°C	37.0	37.0
diesel initial boiling point	°C	242.7	204.0
gas oil initial boiling point	°C	402.2	371.0
Residue initial boiling point	°C	N/A	565.0
Light ends yield	wt%	8.6	5.1
Naphtha yield	wt%	15.0	5.3
Diesel yield	wt%	12.0	23.8
Gas oil yield	wt%	35.0	27.0
Coke/Residue yield	wt%	27.0	38.8
Residue conversion	wt%	100.0	61.2

Table 11. Operating conditions for hydrotreaters in coking-based scheme

		Naphtha hydrotreater	Diesel hydrotreater	Gas oil hydrotreater
Number of reactors		2	2	2
Number of beds/ reactor		2	2	2
Reactor catalyst density	kg/m^3	1500	1500	1500
Reactor ID	m	2.5 m	2.5	2.5 m
Catalyst loading	kg/bed	18,000	18,000	18,000
Bed voidage		0.37	0.37	0.37
Reactor pressure	kPa	5,000	8,000	13,260
Reactor temperature	°C	285.0	351.2	365.6
Gas/oil ratio	STD m^3/m^3	300.0	300.0	584.2
Sulphur conversion	wt%	90.0	93.3	96.8
Conradson carbon residue conversion	wt%	0.0	100.0	100.0
Nitrogen conversion	wt%	71.1	3.1	84.1
Aromatics conversion	wt%	39.7	18.6	9.1
H_2 consumption	SCF/bbl	170.0	581.3	912.6
Product yield	wt%	99.9	99.6	99.4

Table 12. Operating conditions for hydrotreaters in hydroconversion-based scheme

		Naphtha hydrotreater	Diesel hydrotreater	Gas oil hydrotreater
Number of reactors		2	2	2
Number of beds/ reactor		2	2	2
Reactor catalyst density	kg/m^3	1500	1500	1500
Reactor ID	m	2.5	2.5	2.5
Catalyst loading	kg/bed	18,000	18,000	18,000
Bed voidage		0.37	0.37	0.37
Reactor pressure	kPa	5,000	8,000	13,260
Reactor temperature	°C	285.0	351.2	365.6
Gas/oil ratio	STD m^3/m^3	300.0	300.0	584.2
Sulphur conversion	wt%	88.8	78.5	97.0
Conradson carbon residue conversion	wt%	0.0	100.0	100.0
Nitrogen conversion	wt%	67.6	78.8	88.0
Aromatics conversion	wt%	33.2	50.4	26.1
H_2 consumption	SCF/bbl	170.0	892.4	1628.0
Product yield	wt%	99.8	99.5	99.6

Summary

In this work, two major commercially used bitumen upgrading schemes are simulated with HYSYS. The properties and flow rates of the intermediate and final products in each scheme are estimated and compared. Models for the atmospheric distillation column, the vacuum distillation column, the coker, the hydroconverter, and the hydrotreaters are calibrated according to the pilot plant and industrial data. It is shown that the hydroconversion-based upgrading scheme has a SCO yield of about 92wt% which the coking-based upgrading scheme has a SCO yield of about 81%, which are quite close to commercial operation values. In addition, the SCO generated from the hydroconversion-based scheme has higher quality (lower sulphur, nitrogen and aromatics contents) than that generated from coking-based scheme. The simulation is versatile and multiple options can be considered for different scenarios, and therefore can be considered as an important tool to guide bitumen upgrading process design and operation.

Acknowledgments

Partial funding for this study was provided by the Canadian Interdepartmental Program of Energy Research and Development (PERD 1.1.3).

Literature

1. Bitumen & Very Heavy Crude Upgrading – Long Term R&D Opportunities - LENEF Consulting Limited, 1994
2. Sok, Y., *Journal of Japan Petroleum Institute*, 51, 1-13, 2008
3. Murali, C., Voolapalli, R. K., Ravichander, N., Gokak, D. T., Choudary, N. V., *Fuel*, 86, 1176-1184, 2007
4. Verruschi, E., Freitez, J., Gonzalez, Y., Dassori, C. G., *Journal of Computational Methods in Science and Engineering*, 9, 175-195, 2009
5. Jarullah, A. T., Mujtaba, I. M., Wood, A. S., *Chemical Engineering Science*, 66, 859-871, 2011
6. Ordorica-Garcia, G., Croiset, E., Douglas, P., Elkamel, A., Gupta, M., *Energy and Fuels*, 21, 2098-2111, 2007
7. Martinez, J., Sanchez, J. L., Ancheyta, J., Ruiz, R. S., *Catalysis Reviews*, 52, 60-105, 2010
8. Morawski, I., Mosiewski, J. M., *Fuel Processing Technology*, 87, 659-669, 2006
9. Farahani, H. F., Shahhosseini, S., *Chemical Product and Process Modeling*, 6, 1-21, 2011
10. Ding, F., Ng, S. H., Xu, C., Sok, Y., *Fuel Processing Technology*, 88, 833-845, 2007
11. Cotta, R. M., Wolf Maciel, M. R., Maciel Filho, R., *Computers and Chemical Engineering Supplement*, 791-794, 1999
12. Soderbergh, B., Robelius, F., Aleklett, K., *Energy Policy*, 35, 1931-1947, 2007

(60b) Refining & Petrochemicals - Technology Solutions Through Integrated Optimization

Clayton Sadler, Research and Development, UOP LLC, Des Plaines, IL

Interest in integrating refining and petrochemical complexes continues to grow due to synergies between these two groups of technologies. The advantages of integration include upgrading of otherwise low value byproducts, improved energy efficiency, and greater vertical integration. There are many challenges when selecting the proper technologies for the integrated complex, including product slate and mix, feedstock sourcing, and local regulatory requirements that require a thorough understanding how to optimize the intermediate streams to their highest value potential. This presentation will step through an integration example showing how different technology selections can take a maximum fuels refinery all the way to a maximum petrochemical facility. It will also include technology optimization examples that highlight the importance of proper technology selection for different integrated complex configurations and objectives.

(60c) Creating Opportunities From Challenges: Maximizing Propylene Yields From Your FCC

Bob Mehlberg, UOP, Des Plaines, IL

Fluid Catalytic Cracking (FCC) technology has been the refining industry's conversion mainstay since WWII for gasoline production. Over the decades, it has proven to be the cheapest form of carbon rejection in the conversion of heavy gasoil feedstocks into lighter, more valuable fuel products. However, emphasis on energy independence and climate change is leading to significant changes in the motor gasoline market as ethanol addition and vehicle efficiency reduce the demand for petroleum-derived blend components, thus creating a challenge for FCC-based refineries. At the same time, petrochemical producers are looking for propylene yields exceeding 20 wt% and BTX yields approaching 10 wt% of gas oil feed. This demand for petrochemicals offers a potential answer to this challenge by re-purposing the FCC unit as a means to increase propylene yields.

This presentation will summarize UOP's response to customer requirements for a high propylene-yielding FCC process and through case studies. It will also highlight key features of and learnings from the development of UOP's 2-stage RxPro FCC process.

AIChE

EXTENDED ABSTRACT
Energy Efficiency And Cost Saving Economy(Pre and Post War) In Libyan Oil Refineries

Govindaswamy Balachandran[1]; M.J.Prince[2]

Department of Petroleum and Offshore Engineering, AMET university

India

ABSTRACT

Libya is the second largest producer of oil in Africa next to Niger and is the major supplier of petroleum crude to Europe (EU). According to the Oil and Gas Journal (OGJ), Libya holds close to 44 billion barrels of oil reserves, the largest in Africa. Energy Information Agency (EIA) data indicate that during the year of 2008 total oil production (crude plus liquids) was approximately 1.88 million barrels per day (bbl/d).

The Petroleum refining industry in the Libyan Arabic Jamariyah is developing well and is attractive due to low sulphur content crude and high API gravity. Because of low sulphur content it needs much less refining processes as compared to high sulphur content crude oils and yields more straight-run gasoline and middle distillates.The Refining Industry operates five Refineries (Brega, Ras Lanuf, Sarir, Zawia, and Sirtie) in Libya with a capacity of 348,000 bbl/d. National Oil Corporation (NOC) owns these refineries and operates them through its subsidiaries and employs approximately 8000 employees. The Libyan Refining industry produces petro-products of value exceeding US $ 30 billion (Oil Export, Natural Gas and Refining). Refineries spend little more than 50 percent on operating costs (Excluding capital cost and Depreciation) and on Energy which is the main cost factor. Energy efficiency is vital for cost reduction in Refining industries.

Production of Energy is the main source of emissions in the refining industry and needs consideration for reduction in emissions as well as in operating costs in the Libyan oil Refineries. This paper emphasizes on the major energy efficiency methods, cost savings and opportunities for LIBYAN PETROLEUM REFINERIES in the international oil and gas market. The trends, planning, structure and production, energy used in refining and the conversion processes have also been discussed.Libyan oil industry needs the investments of american and international oil companies after the fallout of Gadaffis regime..

Key words: Petroleum Refining, Low Sulphur, Straight-run Gasoline, Middle Distillates, Energy Efficiency, and Cost Reduction

1 – Associate Professor (Dr.), Department of Petroleum and Offshore Engineering, Amet university,India. (Author)

2 – Assistant Professor, Department of Petroleum and Offshore Engineering, India

E-mail: govindaswamy.balachandran@gmail.com

EXTENDED ABSTRACT

1.Introduction

According to EIA and Citac Libya has five domestic refineries, with a combined nameplate capacity of approximately 380,000 bbl/d.

significantly higher than the volume of domestic oil consumption (227,000 bbl/d; the rest is exported). Libya's refineries include: 1) the Ras Lanuf export refinery, completed in 1984 and located on the Gulf of Sirte, with a crude oil refining capacity of 220,000 bbl/d; 2) the Az Zawiya refinery, completed in 1974 and located in northwestern Libya, with crude processing capacity of 120,000 bbl/d; 3) the Tobruk refinery, with crude capacity of 20,000 bbl/d; 4) Brega, the oldest refinery in Libya, located near Tobruk with crude capacity of 10,000 bbl/d; and 5) Sarir, a topping facility with 10,000 bbl/d of capacity.

The plants are unsophisticated and consume a lot of energy. They are not capable of producing products that correspond to the specifications required by the consuming countries such as unleaded petrol, reformulated gasoline and low sulphur gas oil. The refineries suffered very badly from the technological and equipment limitations imposed by sanctions. The steam cracker at Ras Lanuf needs repair and the inability to purchase

new parts caused disruptions at Az Zawiya.

In addition to the domestic refineries, Libya has operations in Europe. Oilinvest the overseas arm of NOC, has a network of refineries with a 300,000 bpd capacity. Libya is a direct producer and distributor of refined products in Italy, Germany, Switzerland, and (since early 1998) Egypt. In Italy, Tamoil Italia, based in Milan, controls about 7.5% of the country's retail market for oil products and lubricants, which are distributed through around 2,200 Tamoil service stations.

1.1.In 1995, new units came on line at the Ras Lanuf site. These included a 3,300 bpd catalytic reforming unit and a 5,600 naphtha hydro-treatment unit. Libya is seeking a comprehensive upgrade to its entire refining system with a particular aim of increasing output of gasoline and other light products. It plans to build a new 20,000 bpd refinery in Sebha to process crude from Murzuk. The anticipated output from Sebha will be 268,000 tonnes/year of gas oil, 184,000 tonnes/year of gasoline, 84,000 tonnes/year of jet fuel and 274,000 tonnes/year of fuel oil. The last will be used in a new power station that is also planned for this area.

An export refinery with a capacity of 200,000 bpd is planned for Misurata

1.2.NOC owns these refineries and operates them through its subsidiaries. In the late 1980s, it had plans to raise the total capacity to more than 620,000 b/d by the mid-1990s. One project then under consideration was a 200,000 b/d export refinery at Misurata. But the planners have since changed their minds.

For years NOC has been reviewing options to build a new refinery in the south-west of the country. The project is part of a larger plan aimed at bringing on board JV partners to support NOC's efforts to upgrade and expand its refineries.

1.3 In the late 1980s, it had plans to raise the total capacity to more than 620,000 b/d by the mid-1990s. One project then under consideration was a 200,000 b/d export refinery at Misurata.

But the planners have since changed their minds. Now there is talk of a new refinery at Mellita which is estimated to cost about $8 bn.

For years NOC has been reviewing options to build a new refinery in the south-west of the country. The project is part of a larger plan aimed at bringing on board JV partners to support *Noc.*

NOC vacated its head offices in Tripoli in late 2001 and moved to Ras Lanuf, a downstream industrial centre which is an eight-hour drive from the Libyan capital. The relocation took several months to complete and

1.4.The main industrial zone in Libya, Ras Lanuf is the site of the biggest petrochemical and refining plants in the country. A first-phase complex consisting of three units went on stream in April 1987 (more than six months behind schedule, with NOC's subsidiary Rasco facing problems which ranged from poor performance by some contractors to delays in Libyan funding). Brown and Root of the US was the main technical consultant and a variety of foreign contractors were involved in engineering and construction work. Rasco contracted Hemijska Industria Pancevo (HIP) of Yugoslavia to run the complex. The capacity of each unit was as follows: 330,000 t/y of ethylene, with naphtha

Ras Lanuf Oil & Gas Processing Company (Rasco) has confirmed that it has terminated the second phase development contract on the Ras Lanuf petrochemical complex, awarded to a consortium of companies from the former Yugoslav republics. The consortium, led by Bosnia-based Energoinvest, was awarded the estimated $130 million contract in 1990 to build four units at the site. Rasco says that it is now in the process of retendering the work (MEED 7:12:90).

The original Energoinvest consortium included Croatia's INA Project, Serbia's HP Engineering & Development Company and an overseas office of the US-based Brown & Root. The last company pulled out of the group in late ...

Summary: The council has confirmed a partnership project with Emirates Union star for development of Ras Lanuf refinery in bulk of investment comes to 2 billion dollars at participation percentage comes to 50% between the two parties

The council has confirmed a partnership project with Emirates Union star for development of Ras Lanuf refinery in bulk of investment comes to 2 ...

National Oil Corporation
Official selling prices of the Libyan crude Oil 2009

Crude	Benchmark	Aug.	Sep.	Oct.
Es Sider	Avg. Brent Dated	+0.05	+0.05	-0.05
Brega	Avg. Brent Dated	+0.40	+0.40	+0.30
Sharara	Avg. Brent Dated	+1.00	+0.80	+0.70
Zueitina	Avg. Brent Dated	+0.15	+0.15	+0.15
Sirtica	Avg. Brent Dated	-0.30	-0.30	-0.30
Amna	Avg. Brent Dated	-0.65	-0.65	-0.65
Sarir	Avg. Brent Dated	-1.10	-1.10	-1.05
Abu Attifel	Avg. Brent Dated	+0.30	+0.30	+0.30
Mellitah	Avg. Brent Dated	+0.90	+0.70	+0.60
Bouri	Avg. Urals (Med)	-1.90	-1.90	-1.90
Al-Jurf	Avg. Urals (Med)	0.00	0.00	0.00

Figure 1 Libyan Sedimentary Basins

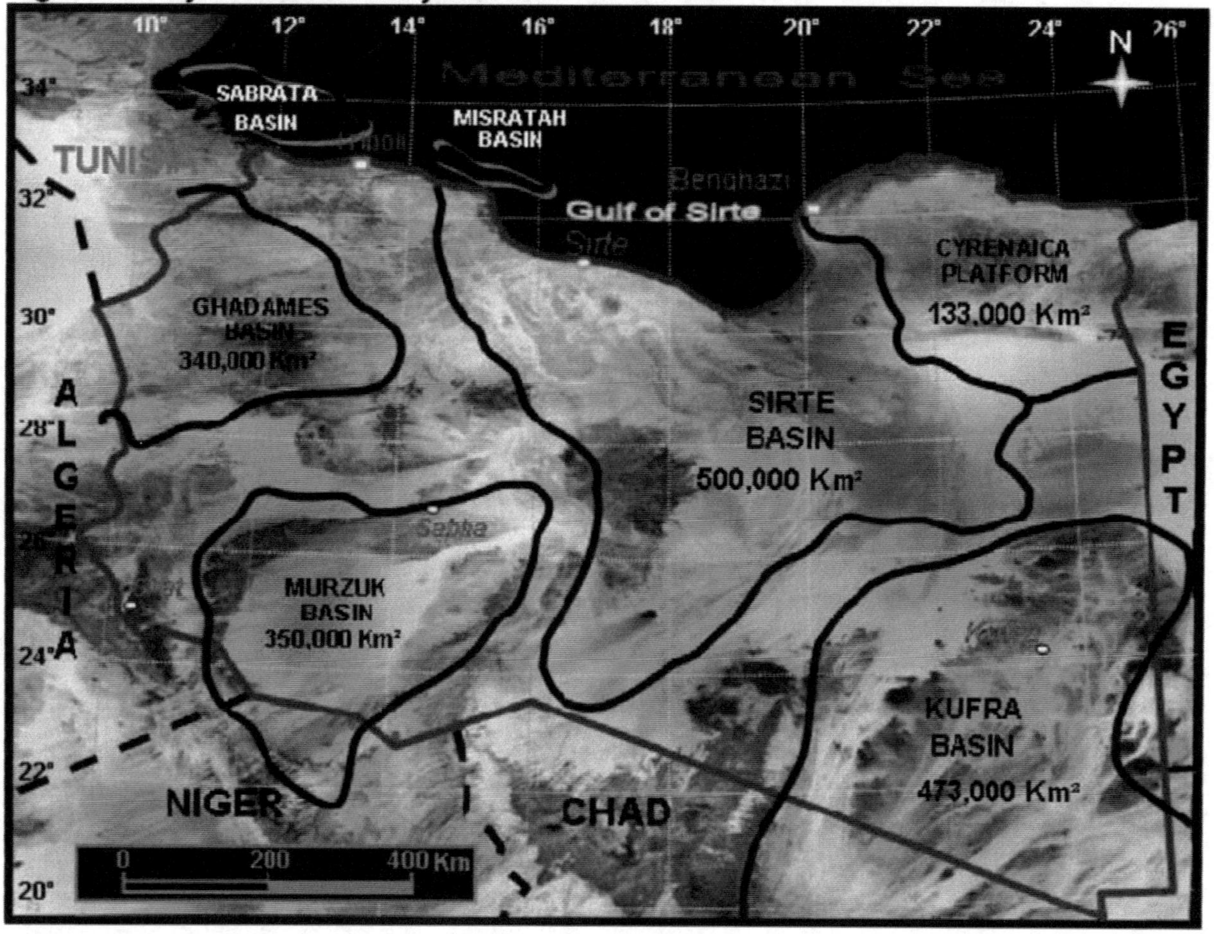

<u>**2.Properties of Major Libyan crude oils-Table:1**</u>

Property	Essider	Sharara	Amna	Abuttifel	Sarir	Brega	Sirtica	Bouri	Aljurf	Zuetina
A P I °	37	42.1	37.1	43.3	37.5	42.4	41.1	25.9	30	40.8
Total Sulfur wt%	0.39	0.07	0.17	0.06	0.17	0.22	0.42	1.86	2.01	0.35
Total N_2 wt%	0.17	-	0.07	0.01	0.1	0.09	0.12	0.17	-	0.1
Pour Point °C	9	-46	21	39	21	-9	-9	9	-6	12
Viscosity@ 37.8°C cSt	4.897	2.83	9.7	25	10.73	2.696	2.935	18.62	8.7	3.682
Vanadium ppm wt.	2	0.7	2.7	0.4	<1	0.8	1.5	14	33	0.4
Nickel ppm wt.	7	0.1	< 0.1	0.5	1	0.5	3.5	12.6	16	1.5
Con. Carbon wt%	2.6	1	2.98	0.4	4.6	1.5	2.1	2.3	-	2
Asphaltenes wt%	0.53	0.3	0.25	0.09	0.3	0.16	0.46	3.3	0.27	0.19
R V P Kpa	36	1.75	29.25	5	12	54	54	23	10.25	35
Acidity mg KOH/g	0.2	-	-	0.08	0.085	0.09	0.1	0.07	0.13	0.09
(K) Factor	12.1	12.1	12.4	>12.5	12.5	12.2	12.1	11.7	-	12.2

Cut Vol%	Light Naphtha	Heavy Naphtha	Kerosine	Atm. Gas Oil	Atm. Residue	Vac. Gas Oil	Vac. Residue
Essider	6.2	19.3	9.7	21	40.7	27.8	12.9
Sharara	7	23.9	16.8	22.8	29.3	24	5.3
Amna	4.8	16.5	8.3	21	47.3	30	17.3
Abuttifel	4.1	14.3	9.4	23.7	48	34.4	13.6
Sarir	5.4	15.2	7.9	19.3	49.3	31.6	17.7
Brega	8.1	24.3	10.8	21.6	31.8	23.5	8.3
Sirtica	8.6	23.9	10.1	20.3	31.9	21.6	10.3
Bouri	3.1	11.6	7.6	21	56.4	34.7	21.7
Aljurf	2.8	12.7	11.25	22.11	51	30	21
Zuetina	6.5	24.5	11.4	21.5	34.3	23.8	10.5

Libya is an oil-producing country for four decades and expected to continue production for years to come.

Although Libyan crude oil contribution is only 2% of the current global crude oil production, Libyan crude oils are preferentially demanded in the international markets due to their superior quality.

The high API gravity and low sulphur content of most of the Libyan crude oils make them very suitable for blending into the feed charge of both hydroskiming and conversion refineries. Their inclusion in the feed charge lowers the processing cost and improves the quality and yield of the desirable light products.

Many straight cuts derived from Libyan crude oils can be directly blended into the fuel pools.

The atmospheric residues from most of the Libyan crude oils have low sulphur and heavy metals content and hence are suitable for processing in RFCC units or for low sulphur fuel oil production.

Anode grade coke and high quality bitumen can be produced from selected vacuum residues with favorable qualities.

2.1.Energy Efficiency and cost saving Economy in Libyan oil Refineries.POST-WAR.

1.Energy efficient Gear Motors-Use of QUANTIS GOLD,a new C- face energy efficient gear motor.

Introduction of this product combines Bolder/Dodge QUANTIS line helical and Right-angle helical bevel C-face gear motors with the companys Reliance Super-E premium Efficient motors to create what is said to be the Motor energy efficient gear motor package available in the Refining Industry.

2.2. **Minimising power consumption in slurry transport in Libyan refineries.**

Transporting slurries through pipeline is common in the solids handling,mineral and hydrocarbon industries.,because of use of a pipeline ensures dust free environment ,requires substantially less space

Makes full automation possible and require less operating staff.on the other hand it needs higher

Operating pressure and considerable demands for high-quality pumping equipment and control system.

2.3.Cost effective heat transfer media:Flexisaddle low-pressure drop(LPD) 1-in.heat transfer media is

Available for regenerative thermal oxidizers(RTO)S.The medias high-open area and aerodynamic designs

Claimed to reset in over 20% lower pressure drop than standard saddle media designs for the same

Heat-transfer efficiency.

The unique three rib-design improves heat transfer through more efficient use of medias surface area,.

The strong ceramic three-rib structure limits resting and withstands rugged chamber loading.

FLEXXISADDLE LPD therby minimizes dead areas that collect particulate.As a result particulate buildup is

Reduced by 40%.

2.4.Improving energy efficiency for distillate Hydro treaters in Libyan Refineries:Use of Case study-Distillate hydrotreaters

Are large energy consumers.As more stringent sulfur specifications are introduced,refiners must increase usage for hydrotreating purposes.as much as 10% of refinerys total energy consumption is

Attributed to hydrotreating products.Furthermore,these units can be quite energy efficient.this case study examines the economisc of retrofitting distillate hydrotreaters options that can improve energy efficiency:1.increaising the surface area of feed preheat exchangers 2.Installing a hot separator.

Distillate hydrotreaters are one of the main energy consumers within a Refinery.the energy efficiency

Of a process unit or of a complete refinery can be compared against a selected reference value.Various

Methods are used-statistical,historical performance and on engineering basis.

The Best technology(BT) benchmark,whereby energy efficiency of an existing unit is compared with

The best design target.The ratio of the actual and the targeted consumption yields a BT index for the

Process unit. Fig.1 shows a conventionally designed diesel hydrotreater.In this case,the BT index was estimated at 545%.Part of this inefficiency is attributed to the units electric power consumption,which is

Generated at suboptimal total efficiency(24%).the feed is preheated to 313 C,although the reactor effluent is available at 385 C.thus the size of exchangers E-1 and E-3 are increased,more heat could be

Transferred between the hot strem and cold stream.

The other major inefficiency is the absence of a hot separator.the effluent is condensed in the air cooler

Upstream of the high-pressure and low-temperature(HPLT) gas/liquid separator.It is proposed that **a hot separator,Fig.2 combined with enhanced heat transfer between the the feed and effluent,can lower unit energy consumption to practically a BT value.**

The proposed enhanced heat transfer is to be accomplished by using plate-type heat exchangers.such solution is feasible, and a number of units have been designed,built and operated in this way.

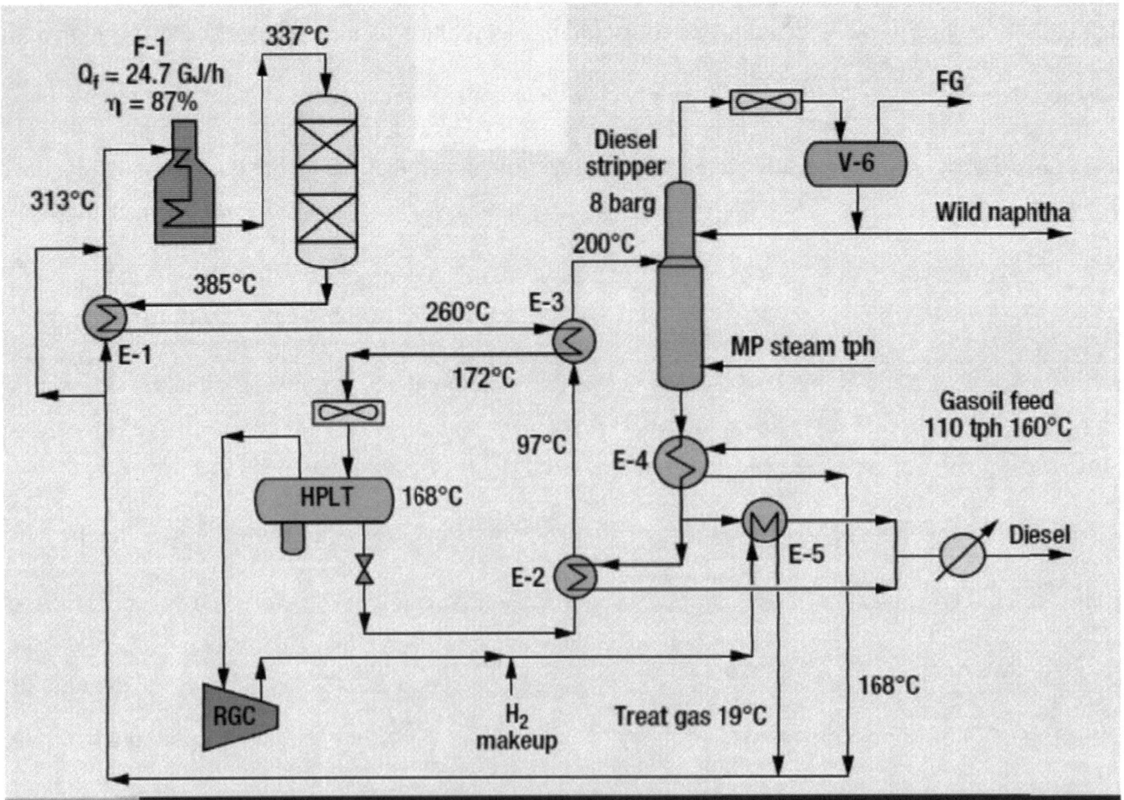

FIG. 1 Flow diagram of a conventional distillate hydrotreating unit.

TABLE 1. Retrofit options to increase energy efficiency in diesel hydrotreaters

	Project	Saving GJ/h, process	Value, thousand €/yr	Investment, € million	Payback, yr	Furnace duty, GJ/h
	Base case	0	0	0	0	24.7
A	Install twisted tubes in E-1 and E-3	5.2	380	1.7	4.5	18.8
B	33% area increase (1 new shell in each exchanger)	5.2	335	1.6	4.8	18.8
C	67% area increase (2 new shells)	7.9	490	2.3	4.6	16.3
D	100% area increase (3 new shells)	9.2	537	2.9	5.5	15.5

Note: The fuel cost used to calculate the savings is €6/GJ. The CO_2 reduction benefit is added to the fuel-saving benefit at CO2 price of 20 €/ton, making the total fuel cost 7.3 €/GJ.

TABLE 2. Hot separator project economics

Project	Saving GJ/h, process	Value, thousand €/yr	Investment, thousand €	Payback, yr	Furnace duty, GJ/h
Base case	0	0	0	0	24.6
Hot separator	8.8	640	2,710	4.2	14.2
Separator + add area to E-3	13.0	885	3,310	3.7	9.6

Note: The separator investment cost (2.71 million €) is calculated as follows: hot drum: 1.6 million € (D = 2.1 m L = 9.6 m, 55 bar stainless); piping: 200,000 €; modifying and re-rating E-3: 910,000 €.

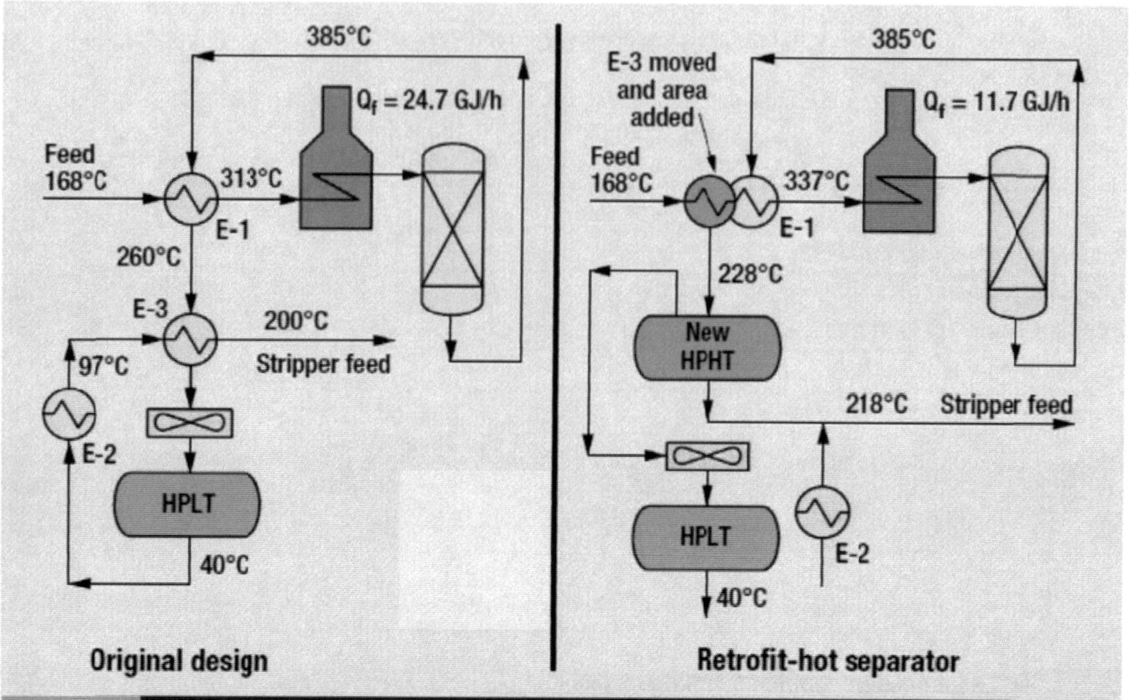

FIG. 2 Flow diagram of the original and retrofit for distillate hydrotreating unit.

After the fallout of Libyas,Gadaffi regime in October-2011,the NTC(National transition council) new government must consider the new methods in modernizing the outdated Libyan Refineries.

References:

1.Handbook of Petroleum Refining-Nelson.

2.Hydrocarbon processing-July-2011

3.Libyan Petroleum industry handbook-2004 Edition,Libya.

4.NOC-National oil corporation-Energy datas,Tripoli,Libya.

Platformer Debutanizer Revamp & Optimization

Rakan S. Bilaus, Sajeesh Padmanabhan, Abdulrahman H. Al-Harbi, Neelay Bhattacharya

Abstract:

Saudi Aramco Yanbu' Refinery & Process & Control System Department engineers worked jointly to troubleshoot and resolve the high reformate RVP that is imposing blending constraints, resulting in reduced iC5 and isomerate blending and RON giveaway. The joint study concluded with Debutanizer revamp that was implemented in early 2011. As a result, the reformate RVP saw a drop from 4.5 to 3.8 psia. However, during summer season, several performance issues were raised. Optimization and overcoming cooling limitations have resulted in performance improvements and further drop in RVP from 3.9 to 3.7 psia. The reduction of platformate RVP allows for an addition of 1 MB of i-C5 to the gasoline pool, which is equivalent to $5.17MM/year.

Background:

As Part of Saudi Aramco Yanbu Refinery Excellence Program, Yanbu Refinery worked jointly with Process & Control Systems Department to improve the reformate RVP. The high reformate RVP causes problems with gasoline blending leading to high RVP in summer months, resulting in reduced iC5 and Isomerate blending. Consequently, the gasoline production is reduced with an RON giveaway. In order to increase gasoline production and eliminate RON giveaway, reformate RVP needed to be reduced from 4.5 psia to 3.8 psia. Accordingly, a joint study was initiated to look into the available options to upgrade Yanbu Refinery Platforming Debutanizer Column to achieve the targeted RVP at full column capacity.

Process Background:

The Platforming unit has the purpose of upgrading the low octane heavy naphtha from Naphtha Hydrotreating Unit to a high-octane material for motor fuel blending. The principal product is the high octane reformate (sent to Saturated Gas Concentration for vapor pressure adjustment), whereas a hydrogen rich gas is produced as a by-product and sent to Naphtha Hydrotreating Unit and Light Straight Run Naphtha Hydrotreating Unit. The Unit consists of the following sections:

- Reaction section, composed mainly by three heaters and three reactors, in which the Reforming reaction takes place.
- Compressor section, composed by three centrifugal compressors for the recycle gas service and by two alternative compressors for the hydrogen make-up gas sent to Hydrotreating units
- Product Separation section, in which reformate is recovered from the hydrogen rich gas by means of a chilling unit.

The Saturate Gas Concentration processes the Unsterilized Platformate produced in the Platforming Unit and the treated LPG coming from the LPG Merox Unit. Products are the Stabilized Platformate, Propane, Butane and/or mixed LPG. Components lighter than Propane are sent to the Refinery Fuel Gas Network.

The Unit consists of the following sections:

- Debutanizer Tower reboiled by means of the fired heater that separates the Unstabilized Platformate in LPG as overhead product and Stabilized Platformate as bottoms product.
- Deethanizer Tower reboiled with MP steam that removes components lighter than C3 from LPG.
- Depropanizer Tower reboiled with MP steam that splits LPG in Propane as overhead product and Butane as bottom product.

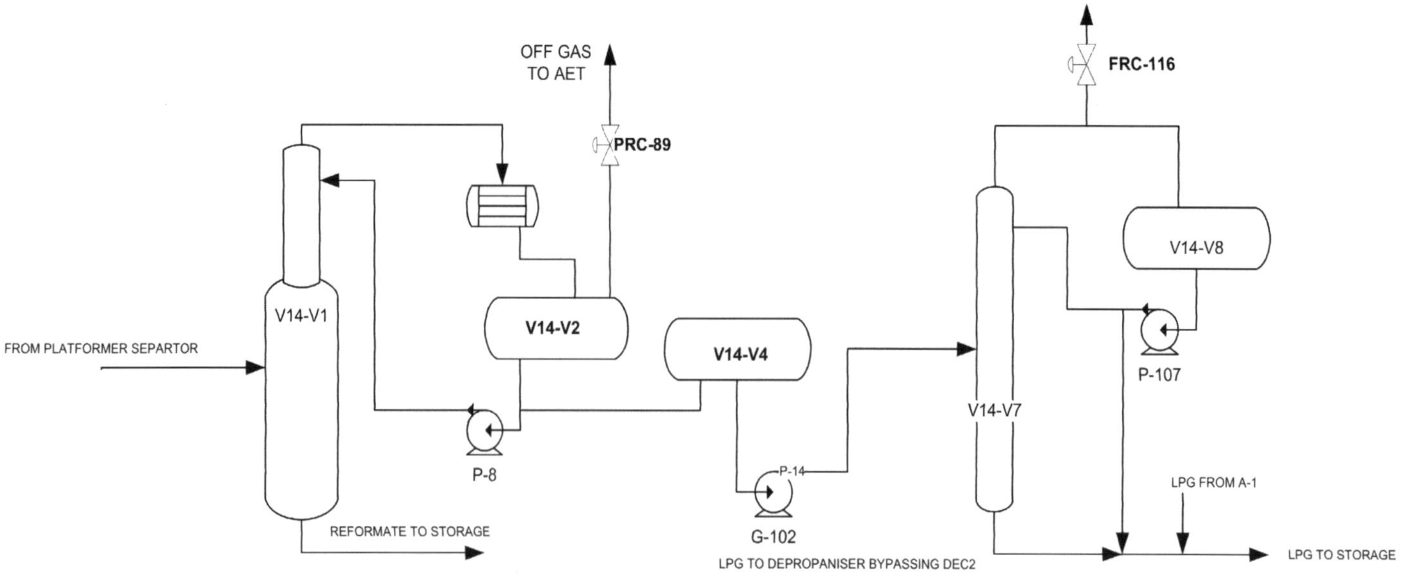

Figure 1: Saturated Gas Concentration Unit Process Flow Diagram (DeC4 & DeC2)

Unit Operation:

The Debutanizer Column was designed by Chiyoda in 1977 and the column internals were provided by Koch Glitsch. When the platformer unit was converted to CCR in 2006, the column was deemed suitable for the new process flows. The only modifications implemented were new Feed-Bottoms exchangers and new Overhead trim coolers. The man function of the column is to stabilize the reformate and produce LPG as the overhead product. The main specs were <1.0mol% C4 in reformate and <1.0mol% C5 in LPG. The reformate was estimated to have an RVP of 3.8 psia. However, the reformate RVP has been consistently at 4.5 psia at full unit capacity. It is possible to reduce the reformate RVP by increasing the LPG draw to decrease the C4 components in the reformate. However, this will result in higher C5 content in the LPG causing it to be off-specification. Furthermore, the reflux temperature was having considerable impact on column operations. The design temperature was 38°C, but over time, the reflux temperature has gradually shifted higher. During summer peaks, the temperature had been reported as high as 50°C.

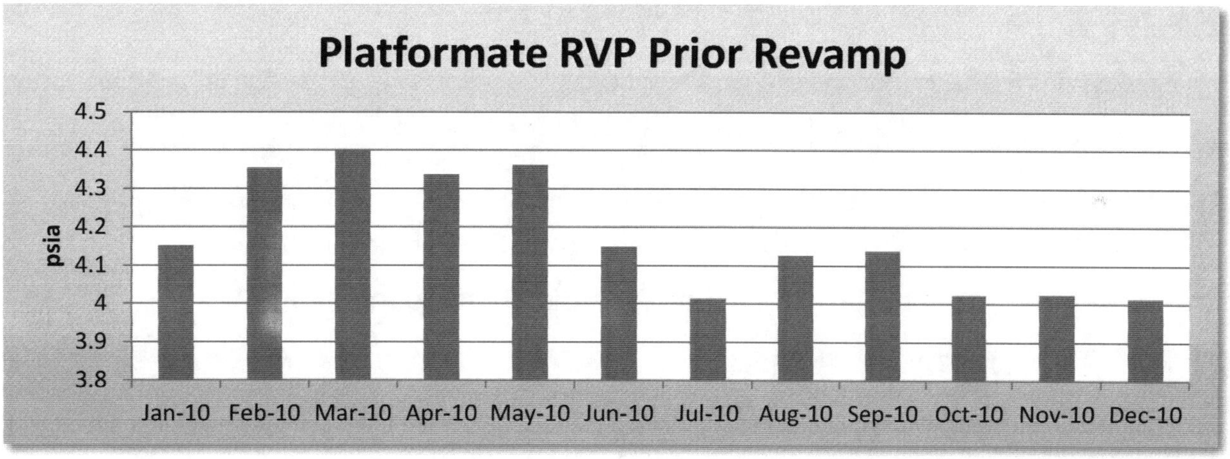

Study Basis & Scope:

PFDs, PI Data System and Chiyoda mechanical drawings of the Saturated Gas Concentration unit were used as a basis of the study. The study would provide the following outcome:

1. Provide detailed simulation of the unit with tray by tray rating calculations
2. Identify the hydraulics limitations contributing to low tray efficiency
3. Provide sensitivity analysis for reflux temperature
4. Povide changes to internals proposed by Glitch to overcome limitations

Outcome of Study:

- **Design Check Through Simulation**

  Engineering contractor evaluated the Debutanizer column for the new process conditions when the fixed bed Platformer was converted to a CCR. As per their calculations, the column internals were adequate to produce a reformate RVP of 3.8 psia. In order to validate the process design, the debutanizer column was simulated with full capacity with the design conditions. The results clearly indicated that for the design process conditions a 30 tray Debutanizer column is adequate for producing reformate having RVP of 3.8 psia.

- **Hydraulics Limitation Assessment**

  Once the process calculations were verified, the next step was to perform the hydraulic check for the rated dimensions to check for any limitations. A detailed tray by tray analysis showed high flooding % (80-89%) in the rectification section. To verify the results, Glitch was contacted to perform a detailed hydraulic check for the column. Their comments were as below:

  1. Trays# 1-15 are working close to their limits (Jet Flooding 83%, Limit is 85%)
  2. High exit velocity of tray# 5 (1.58 m/s)
  3. Velocities and head losses under downcomers of Tray#16-30 are above max
  4. Vapor tunnel on off-center downcomer of tray# 30 was not provided
  5. Inlet momentum through the existing feed distributor is too high.

- **Estimation of Operating Tray Efficiency**

  Though the Debutanizer column has sufficient trays to provide a reformate quality of 3.8 psia, the poor performance can be attributed to smaller diameter of the rectification section for sieve tray application, improper vapour and liquid disengagmement at the feed tray, higher than recommended downcomer exit velocity and improper distribution of the reboiler return vapour. The rectification efficiency was found to be about 60% and the stripping section efficiency was about 40%.

- **Reflux Temperature Sensitivity Analysis**

  Since the reflux temperature had been gradually increasing above the design temperature of 38˚C, it was important to understand its effect on the column performance. It was found that at reduced capacity, the reflux temperature has negligible effect on reformate and LPG quality. However, quality of reformate and LPG is very sensitive to reflux temperature at full capacity.

Study Recommendations:

Based on the outcome above, the recommendations to improve reformate RVP to 3.8 psia, were as the following:

1. Replace the existing trays with Minivalve trays at Tray#1-15
2. Replace the existing distributor with a new Flashing Feed Pipe
3. Replace existing lower downcomer panels to increase clearance
4. Replace 2/3 of existing off-center downcomer panels of tray#30 with vapour tunnel

Implementation:

The study was completed with recommendation on August 2010. Yanbu Refinery managed to implement all the recommendations during the total refinery shutdown in February 2011.

Results:

Implementation of the study recommendation has resulted in a reduction in reformate RVP from 4.5 psia to 4.1 psia after shutdown. The debutanizer operating parameters after the start up were not optimized to meet the targeted RVP.

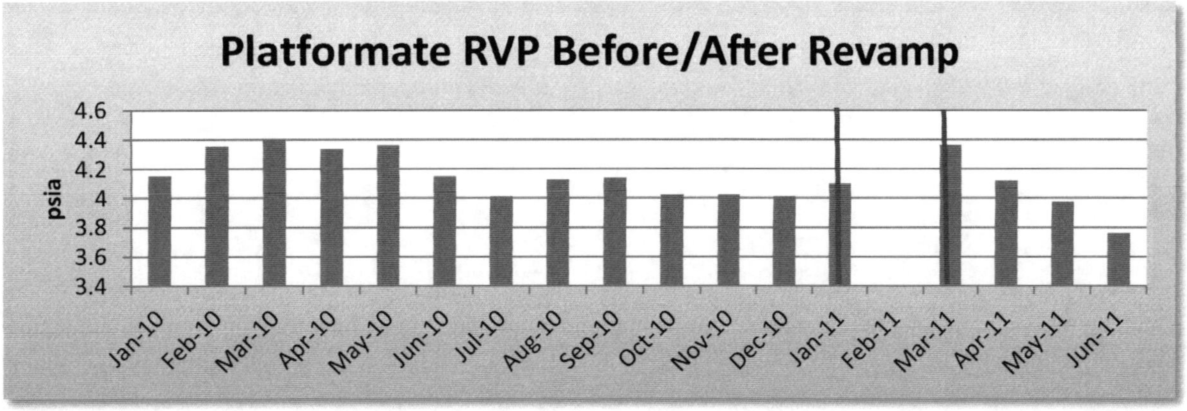

Operating Parameters Optimization:

Debutanizer column started operating with reboiler outlet temperature of 236°C, Tray-8 temperature of 100°C and a reflux ratio of 4.2. However, these operating parameters were not achieving the target RVP. As such, the operating parameters were adjusted to a reboiler outlet temperature of 240°C, Tray-8 temperature of 105°C and a reflux ration of 5.0. As a result, the reformate average RVP dropped to 3.8 psia. However, during summer peak, the reformate RVP increased to 4.0 psia.

Overhead Cooling Limitations:

During summer season, the reflux temperature was reported high in the range of 40-50°C which was affecting column operations. Performance assessment on the debutanizer overhead trim coolers V14-E-103A-D indicated that they are operating at a performance of 35-40%. The trim coolers were designed to use 500 m3/hr of sea water for cooling, while the total sea water flow was around 150 m3/hr, which clearly pinpoints a problem in the sea water side.

Overcoming Cooling Limitations:

To overcome the cooling limitations, Backwash practice was conducted for the debutanizer overhead trim coolers V14-E-103A-D, where the flow of sea water is reversed. As a result, the reversed flow of sea water removes a large amount of plugging material from the system. Consequently, the reflux temperature dropped from 45°C to 33°C allowing for more cooling. The realized benefits of the Backwash were tremendous as the overhead cooling capacity was greatly improved. Consequently, the platformate RVP has dropped from an average of 3.9 psia to an average of 3.5 - 3.7 psia.

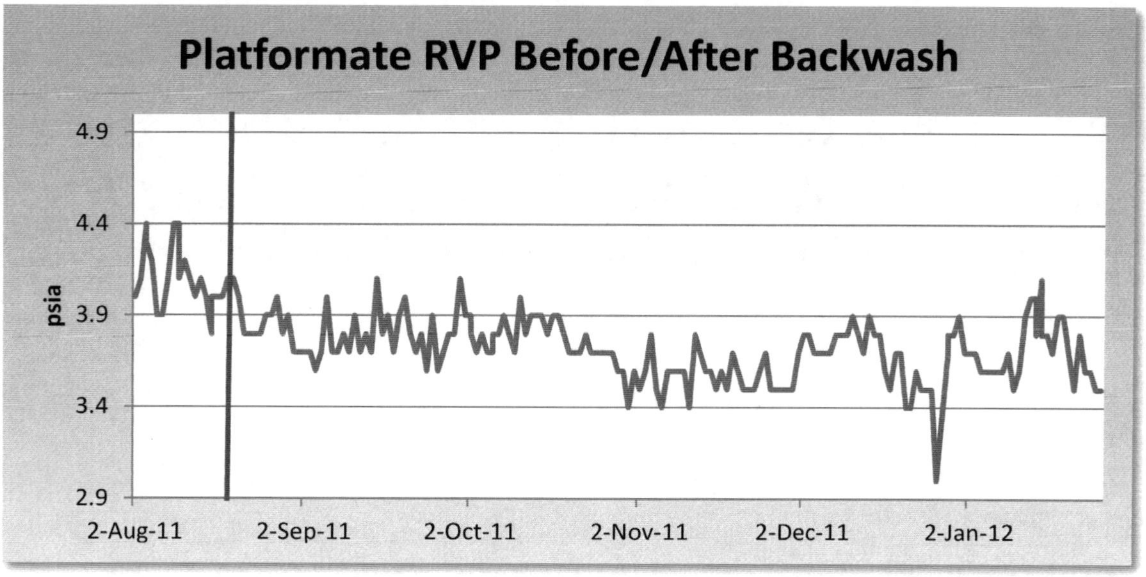

Benefits of Debutanizer Optimization & Cooling Improvements:

The reduction of platformate RVP allows for an addition of 1 MB of i-C5 to the gasoline pool. The price difference between i-C5 and gasoline is $26.5 per barrel for the month of September, which is equivalent to $14370 per day (i.e. $5.17 MM/yr). Furthermore, as a result of backwashing, the debutanizer off gases were reduced by more than 1000 m3/hr resulting in an increase in LPG production by 300 BPD.

Backwashing the debutanizer overhead trim coolers V14-E-103A-D resulted in additional benefits including:

1. Increased sea water flow by more than 100 m3/hr
2. Increased LPG production by 300 BPD
3. Decreased reflux temperature by more than 10˚C
4. Improved debutanizer overhead cooling system significantly
5. Relaxed operation of the debutanizer column

(60f) Variable Neighborhood Search with Active Specification Switching Repair Strategy Applied to Crude Distillation Energy Integration Problem

Gregory McKinne Robertson, Chemical Eng., Louisiana State University, Baton Rouge, LA and Jose A. Romagnoli, Chemical Engineering, Louisiana State Univeristy, Baton Rouge, LA

Variable Neighborhood Search with Active Specification Switching Repair Strategy Applied to Crude Distillation Energy Integration Problem

G. Robertson, J.A. Romagnoli

Louisiana State University, Baton Rouge, LA, U.S.A

Supply chain management (SCM) is the oversight of materials, information, and finances as they move in a process from suppliers to manufacturers to wholesalers to retailers and then ultimately to consumers. Global solutions for supply chain problems can be computationally burdensome due to the size of the problem and the number of alternatives. Therefore, both within and among companies, objectives are naturally separated horizontally along the supply chain, with entities along the supply chain having their own local objective. The decision making can also be separated vertically into layers by the time horizons they consider.

The longer the time horizon a decision considers, the less frequent the decision is made and the greater its ramifications. Here, we have broken down decision making into four layers (strategic, tactical, production, and operational.) Strategic level problems are long term, around 5 years, and can include building or expanding a facility and contract negotiations. Within the time period of the strategic layer problem, many medium term tactical layer decisions are made such as resource allocation and transportation problems. Production layer problems include the scheduling of production units and blending problems. Operational layer problems are online problems including plant diagnosis, fault detection, and process unit optimization. Challenges to the integration of problems in different layers are reconciling their different implementation frequencies and the magnitudes of order of their cost. Decomposition of problems both vertically and horizontally leads to a feasible but not necessarily an optimal solution.

Process industry supply chains are striving to improve efficiency and profitability (Shah, 2005). Integrating different levels within a supply chain can improve profitability. Many planning problems have been addressed using managerial judgments where complex interactions between different decision-making levels were disregarded. Recent developments in mixed integer process optimization provide new tools to help solve more complex problems of a company's hierarchy (Puigjaner and Heyen, 2006.) The decisions made by planning, scheduling, and control functions have a substantial economic impact on process industry operations - estimated to be as high as US $10 increased margin per ton of feed for many plants (Puigjaner and Heyen, 2006.) Hence decision support tools for scheduling and planning can have a profound effect on the profitability of an enterprise.

Modern petroleum refining has become an extremely competitive business due to the deteriorating quality of crude oil coupled with tighter product specifications and more stringent environmental regulations. Furthermore, refineries today receive shipments of crude from a variety of sources. These crude oils are of different quality and composition and usually blending can improve the economics of the refinery. Therefore, refineries deal with a dynamic schedule of incoming crude which cause them to frequently change unit's operating conditions to reduce operating expenses including environmental impact. Final products in a petroleum refinery are created by combining feed stocks emerging from distillation units, reformers, and catalytic cracker. The input streams which have varying chemical compositions and physical properties, are sent into common tanks or pools to be mixed into final products.

Lee et al. (1996) and Shah (1996) first addressed the CSP by developing an MILP model for short-term refinery scheduling using discrete time formulation. Jia et al. (2003) approached the same problem using unit and event based continuous time formulation. Reddy et al. (2004) designed an approximation algorithm in order to address the extensive time requirement to solve the CSP in order to reach a feasible point.

Robertson et al. demonstrated the importance of integrating the problems on the different decision making layers by incorporating the CSP with the heat integration of the distillation units and demonstrated the method integrating the different decision making layers. Sylvia et al. proposed a global optimization approach for integrating the CSP and the final product blending pooling problems with a lagrange decomposition approach. This can be troublesome for larger scale problems. Here we illustrate an algorithm possessing a more robust operational model capable of operating underneath a scheduling algorithm.

Metaheuristics are solution methods of optimization problems that orchestrate an interaction between local improvement procedures and higher level strategies to create a process of escaping local optima and performing robust searches of solution spaces. Hybrid metaheuristics take advantage of the strengths of their individual metaheuristic components to better explore solution spaces. On the front of applications, metaheuristics are used to find high-quality solutions to an ever-growing number of complex, ill-defined real world problems.

An optimization problem is typically modeled as an objective function along with variable relationships and constraints. In an optimization problem, a set of variables, decision variables (D.V.), are manipulated to optimize (minimize or maximize) an objective function while other variables, parameters, are held constant. Among the constraints are physically limiting constraints as well as relationship constraints violated by a certain combination of decision variables which can be more difficult to define in a mathematical model. In process system engineering groups, large models are constructed to mimic complex processes in software such as Aspen tech. A process simulation can be utilized as the model for an optimization problem.

When a simulation is used as a model, the optimizer is a separate entity which treats the model as a black box. The optimizer sends a set of D.V.'s to the model, and the model returns values necessary to solve the objective function. If the set of D.V.'s are outside the solutions s pace, the model will not converge. A converged solution is one where all equations in the model are satisfied.

The model can also not converge if the optimizer makes large jumps in the solution space. For a nonlinear set of equations to be solved simultaneously, modeling equations are typically solved through an iterative procedure. Despite the advances of algorithms such as the inside-out distillation algorithm, the convergence is a function of the initial estimate given to the algorithm. The modeling solver's initial estimate contains the values of the previous point solved. Therefore, increasing the solution space can help find global optima; however, taking larger steps increases the probability of a non-converging solution by worsening the model solver's initial condition.

In order to handle complex optimization models with solution spaces made of ill-defined constraints, we demonstrate a hybrid metaheuristic where Variable Neighborhood Search is aided by a Threshold Acceptance search. This hybrid operates separately from the model and includes a repair strategy for a non-converging result of a D.V. set. The method is applied to a heat integration problem of a distillation train to have robust optimization under a higher level scheduler and demonstrates competitive improvement and superior handling of non-convergent models when compared to SQP, Fletcher-Reeves, and Newton-Rhapson optimization techniques with no repair strategy.

Keywords: Variable neighborhood Search, Refinery Optimization, Heat Integration Distillation, Crude Oil Scheduling

(65a) Simultaneous Biofuel and Electricity Generation Using CO2-Neutral Microbial Fuel Cell

Marci Kerls, Chenical Engineering, Texas Tech University, Lubbock, TX, Alim Dewan, Chemical Engineering, Texas Tech University, Lubbock, TX and M. Nazmul Karim, Chemical Engineering Department, Texas Tech University, Lubbock, TX

Limited availability and the environmental impacts from fossil fuels require that we find alternative energy sources that are renewable and safe. The microbial fuel cells have gained renewed attention for the production bioelectricity and biofuel from renewable sources. We have developed a CO_2-neutral microbial fuel cell for producing bioethanol, algae oil, and bioelectricity simultaneously. In our system, beer yeast was fermented in the anodic compartment of the fuel cell. The CO_2generated in the fermentation was introduced to the cathode as carbon source for microalgae,*Chlorella vulgaris.*The power generation was characterized using potentiodyanmic polarization, and intermitted energy harvesting scheme that uses cyclic charging and discharging of the capacitor. The power generation was correlated with the growth kinetics of the yeast and the algae. The algae oil was quantified using Soxhlet extractor with hexane as the solvent. The bioethanol production in the anodic compartment was measured using HPLC. The power generation was ~ 30 $mW/m^2$of the electrode surface. We found that 40 g/L of ethanol could be produced in the anodic compartment. The algae oil production was 10% of the dry weight of the algae.

NOVEL METHODS TO PRODUCE BIOFUELS FROM BIO-OIL DERIVED FROM FAST PYROLYSIS OF BIOMASS

Sanjeev.K.Gajjela, Sathish.K.Tanneru, and Philip.H.Steele*

Department of Forest Products
Mississippi State University
Box 9820
Mississippi State, MS 39762
*Corresponding author: *sgajjela@cfr.msstate.edu*

Biography

Dr. Sanjeev Gajjela graduated with a B.S in Industrial Chemistry and M.S in Organic Chemistry from India. In 2010 he received Ph.D. in Forest Products from Mississippi State University (MSU). Currently he is working as an Assistant Research Professor in the Department of Forest Products at MSU for almost a year with research and development duties. In this role he coordinates the research activities of MSU faculty members and graduate students focused on production of second generation biofuels and chemicals from renewable energy feedstock. Dr. Gajjela was runner-up for outstanding graduate research assistant for his bioenergy work at MSU for the year 2010. Dr. Gajjela has filed 2 U.S, 1 PCT and 3 provisional patents. Each of these patents has the potential to be licensed in the future. He has published several peer reviewed articles. He is also very well recognized in scientific communities such as American Chemical Society and American Institute of Chemical Engineers.

Introduction

Dwindling petroleum resources combined with concerns over global climatic change and economic and socio-political issues associated with energy security and foreign oil imports from volatile nations have highlighted the need to develop the renewable and sustainable fuels and chemicals from the existing infrastructure. To cope with this issue the Energy Independence and Security Act of 2007(EISA) estimates production of 36 billion gallons of renewable fuels by 2022 that includes 21 billion gallons of advance biofuels. Lignocellulosic based biofuels must represent at least 16 billion gallons of these advanced biofuels[1, 2]. Lignocellulosic biomass is the most abundant and inexpensive source of carbon that can be utilized for renewable fuels/chemical purposes. According to DOE/USDA 2005 billion ton study, it is estimated that United States alone can sustainably produce a billion tons of biomass per year without interrupting or encouraging the deforestation and this biomass can be used for the advanced biofuels production[3].

Bio-oil is produced by fast pyrolysis of biomass at 400 to 550°C in the absence of oxygen. The yield of bio-oil is relatively 60-70% dry-weight basis or higher depending on the production process. Bio-oil is a dark brown colored liquid with pungent phenolic odor and bio-oil chemical properties vary with feedstock type and applied pyrolysis condition[4-6]. As a fuel raw bio-oil has environmental advantages when compared to fossil fuels because on combustion, bio-oil produces half the NO_x, negligible quantities SO_x emissions and it is CO_2 neutral when compared to petroleum fuels[7, 8]. Raw bio-oils demonstrate some negative properties such as significant water content, high acidity, immiscibility with petroleum products, viscosity increase over time and a distinctive odor that have prevented its commercial use to date for other than pilot and demonstration projects[9].

Raw bio-oils have been tested as engine fuels but these upgraded fuels invariably cause engine damage regardless of the engine type tested. It is universally agreed that they will need to be significantly upgraded to allow their use for internal combustion engines[8].

Upgrading of raw bio-oil

Presently, bio-oil upgrading techniques include hydrodeoxygenation,[9-13] catalytic cracking,[14-16] and steam reforming[17, 18] mainly to reduce the oxygen content present in the bio-oil. But these techniques require extensive capital cost, complicated equipment, high hydrogen consumption and are usually limited by catalyst coking and reactor clogging. Apart from these methods, there are also other economic methods in which total oxygen removal is not required. These methods utilize oxygen present in bio-oil to produce products which are combustible and stable oxygenated compounds such as esters and alcohols, in which an esterification process is involved. Scheme 1 represents a general esterification pathway, in which acids react with alcohols in the presence of a solid acid catalyst produce esters and water.

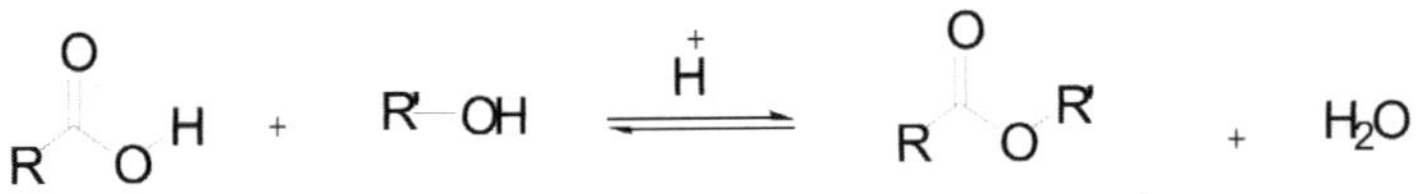

Scheme 1. Esterification reaction pathway.

Bio-oil is very complex in composition, and contains large amounts of oxygenates such as carboxylic acids, ketones, phenolic derivatives, aldehydes etc. The presence of high proportions of acids in bio-oil makes it highly corrosive. It has been demonstrated that esterification via methanol and ethanol will reduce the acidity of the bio-oil, thereby improving bio-oil quality and stability to some extent. To date researchers have used alkaline, resin acid, super critical fluid technology and solid acid catalyst for esterification of bio-oil[19-23]. Although much research has been dedicated to the upgrading of bio-oil via esterification, most of efforts focused on the catalysts and with little elucidation of the increase in energy density and overall length of the carbon chain.

The carboxylic acids present in bio-oil typically include formic acid, acetic acid, propanoic acid and low molecular unsaturated organic acids. Among them formic and acetic acids account for 80% of the overall total acid content. In normal esterification reactions carboxylic acids present in bio-oil are only converted to esters but there are an abundance of other compounds such as aldehydes and ketones left behind. It is well known that aldehydes greatly affect the properties of the bio-oil. Aldehydes can readily undergo spontaneous oligomerization reactions to form compounds of higher molecular weight with phenols during storage. Until now very little research has been performed converting these compounds to esters[24].

Esterification of raw bio-oil produces esters with maximum heating value of 27 MJ/kg and also the liquid yield is suppressed by the low molecular weight carbon chain[25]. The present upgrading technologies such as hydrodeoxygenation consume very high quantities of hydrogen for the fuel production as well as early catalyst deactivation during the process. To solve these issues, we employed a patented catalytic olefination process to raw bio-oil to produce a boiler fuel with improved fuel quality and better physical/chemical properties.

Experimental

Lignocellulosic biomass-based bio-oil was produced with the MSU auger reactor from pine wood particles of 1-3 mm diameter by fast pyrolysis with 65% yield. Physical properties of both raw and olefinated bio-oil, such as pH, water content, acid value, density, viscosity and higher heating value (HHV) were measured. Percent water was determined by ASTM Method D-1744. Viscosity was determined with a Stony Brook Model PDVa-100 viscometer at 50°C. Acid value was obtained by dissolving one gram of bio-oil in an isopropanol/water mixture and titrating to a pH of 8.5 with 0.1 N KOH. The pH was determined indirectly by adding one gram of bio-oil to 50 ml of water, stirring, and measuring the pH with an Orion Model EA920 pH meter. The density of bio-oil was determined at 20°C. The calorific value was measured as HHV by an Ika-5000 bomb calorimeter. Elemental analyses of raw bio-oil were determined by CE-440 elemental analyzer. All the samples were analyzed by GC-MS.

Results and Discussion

Olefin and alcohols selected for these reactions range from C1 to C10. Both gaseous and liquid olefins are also considered to decrease the process expenditure. For this example, we have selected 1-octene as the olefin, 1-butanol as alcohol, and nickel based catalyst. All these reactants were added to a high pressure autoclave and with reaction temperature at 250°C for 2 hr with continuous mixing (1000 rpm).

After the reaction, the reactor was cooled to room temperature and vented to atmospheric pressure. The reactor content was collected and the two liquid layers were separated by centrifugation for 1 hr at 4000 rpm. Two phases were separated by separating funnel and yields were measured accurately. All phases and products were weighed to determine the mass balances and are reported in Fig 1. The byproducts produced in this process included water and off-gas. Water contains some dissolved organics and esters and off-gas are mostly O_2, CO and CO_2.

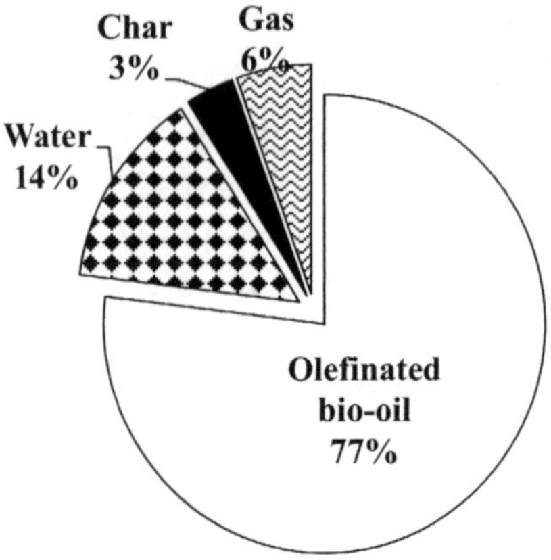

Figure 1. Mass balance of the olefination reaction.

From Figure 1, it is observed that olefination reaction yields 77% of the upgraded olefinated product i.e. cellulosic biodiesel, 14% of water, 3% of char and 6% gas. The physical and chemical properties of the cellulosic biodiesel are given in Table 1. After distillation of the product, the total product yield was 60 wt%.

From the olefination process 14 wt% of clear low-acidity water was formed. This process water can be purified to environmental release standards by filtering through activated carbon. The olefinated bio-oil produced from the patented olefination process of the raw bio-oil was water free and with lower acid content.

Table 1 compares the properties of our cellulosic biodiesel properties, following guaiacol separation, with those of the ASTM D 6751 standard for biodiesel. The cellulosic biodiesel properties are completely within compliance for the properties of sulfur content and kinematic viscosity. For density the fuel is slightly out of compliance but this issue is easily adjusted by increasing the distillation range distribution. Increasing the density by modifying the boiling range distribution will be performed such that the increase the HHV will be brought into specification. The oxygen content is 6.2 percentage points above standard requirement.

Table 1. Comparison of olefinated bio-oil properties with ASTM D6751 standard biodiesel.

Properties	**Test method**	**ASTM D 6751 biodiesel**	**Olefinated bio-oil**
Density, g/mL	D 7371	≥ 0.88	0.85
LHV, MJ/kg	D240	~36	~34
C, H, N, O, wt%	D 5291	77, 12, <0.05,11	70,12.7,<0.1, 17.16
Sulfur, % mass	D 5453	0.05 max	<0.01
Total acid number, mg KOH/g	D 664	0.50 max	20
Water content, wt%	D 2709	0.050 max	0.01
Kinematic viscosity, 40°C, cSt	D 445	1.9-6.0	2.5

GC/MS spectral comparison

Figure 2 gives GC/MS characterization of raw bio-oil and olefinated bio-oil. From Figure 2 it is observed that there is clear variation between all the spectra. Raw bio-oil spectrum contains mostly oxygenated compounds in the form of alcohols, ester, carboxylic acids, anhydrosugars, phenolic compounds and its derivatives. Presence of all these compounds is attributed to the negative properties of the raw bio-oil.

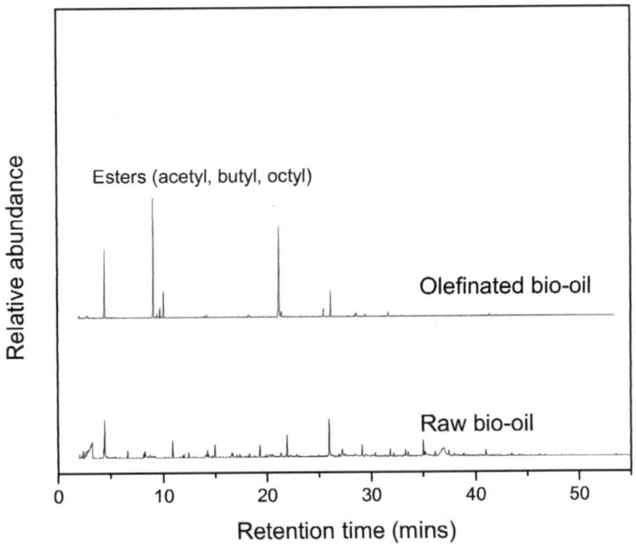

Figure 2. GC/MS spectral comparison of raw bio-oil and olefinated bio-oil.

Olefinated bio-oil contains sharp peaks of esters ranging from C4 to C15 are observed. These esters include methyl, acetyl, butyl and octyl and their derivatives. Due to the presence of higher molecular weight esters the final product has approximately double the high heating value of raw bio-oil and also better fuel properties in terms of viscosity, acid value and water content.

Conclusions

Olefination of bio-oil with solid acid heterogeneous catalyst produced a 60% yield of a fuel suitable as a high energy boiler fuel. The fuel contained zero water content and exhibited a two-fold increase in HHV. Viscosity was reduced by 90% compared to raw bio-oil. Similarly, there was a 60% decrease in both oxygen content and acid values when compared to raw bio-oil. The process produced the boiler fuel with insignificant catalyst coking. The fuel produced is a clear liquid with a slight yellow coloration. The odor is pleasant.

Future research will be directed at optimizing the reaction pathways and purification of guaiacol via extraction from the distillation residue to provide a marketable high-value co-product.

Acknowledgements

This research is based upon work funded through the Sustainable Energy Research Center at Mississippi State University and is supported by the Department of Energy under Award Number DE-FG3606GO86025.

References

1. U.S. Department of Energy. Biomass Multi-Year Program Plan, O.o.t.B.P., Energy Efficiency and Renewable Energy, 2009; available at. Available from: http://www1.eere.energy.gov/biomass/pdfs/biomass_program_mypp.pdf.
2. Beaudry-Losique, J.G.A.R.B.E.V.C.F., August 21–22, 2007; available at. Available from: http://www1.eere.energy.gov/commercialization/pdfs/biomass.pdf.
3. *http://feedstockreview.ornl.gov/pdf/billion_ton_vision.pdf.*
4. Mohan, D., C.U. Pittman, and P.H. Steele, *Pyrolysis of Wood/Biomass for Bio-oil: A Critical Review.* Energy Fuels, 2006. **20**(3): p. 848-889.
5. Huber, G.W., S. Iborra, and A. Corma, *Synthesis of Transportation Fuels from Biomass: Chemistry, Catalysts, and Engineering.* Chem. Rev., 2006. **106**(9): p. 4044-4098.
6. Bridgwater, A.V. and G.V.C. Peacocke, *Fast pyrolysis processes for biomass.* Renewable and Sustainable Energy Reviews, 2000. **4**(1): p. 1-73.
7. Bridgwater, A.V., *Review of fast pyrolysis of biomass and product upgrading.* Biomass and Bioenergy, 2011. **In Press, Corrected Proof**.
8. Czernik, S. and A.V. Bridgwater, *Overview of Applications of Biomass Fast Pyrolysis Oil.* Energy Fuels, 2004. **18**(2): p. 590-598.
9. Elliott, D.C., *Historical Developments in Hydroprocessing Bio-oils.* Energy Fuels, 2007. **21**(3): p. 1792-1815.
10. Wildschut, J., et al., *Hydrotreatment of Fast Pyrolysis Oil Using Heterogeneous Noble-Metal Catalysts.* Industrial & Engineering Chemistry Research, 2009. **48**(23): p. 10324-10334.
11. Senol, O.I., T.R. Viljava, and A.O.I. Krause, *Hydrodeoxygenation of methyl esters on sulphided NiMo/[gamma]-Al2O3 and CoMo/[gamma]-Al2O3 catalysts.* Catalysis Today, 2005. **100**(3-4): p. 331-335.
12. de Miguel Mercader, F., et al., *Production of advanced biofuels: Co-processing of upgraded pyrolysis oil in standard refinery units.* Applied Catalysis B: Environmental, 2010. **96**(1-2): p. 57-66.
13. Elliott, D.C. and N. G.G., *Liquid fuels by low-severity hydrotreating of biocude*, in *Developments in Thermochemical Biomass Conversion*, A.V. Bridgwater and D.G. Boocock, Editors. 1996, Blackie Academic & Professional: London. p. 611-621.
14. Bulushev, D.A. and J.R.H. Ross, *Catalysis for conversion of biomass to fuels via pyrolysis and gasification: A review.* Catalysis Today, 2011. **171**(1): p. 1-13.
15. Zhao, Y., et al., *Aromatics Production via Catalytic Pyrolysis of Pyrolytic Lignins from Bio-Oil.* Energy & Fuels, 2010: p. null-null.
16. Karimi, E., et al., *Red Mud as a Catalyst for the Upgrading of Hemp-Seed Pyrolysis Bio-oil.* Energy & Fuels, 2010: p. null-null.

17. Putun, E., F. Ates, and A.E. Putun, *Catalytic pyrolysis of biomass in inert and steam atmospheres.* fuel. **In Press, Corrected Proof**.
18. Iliopoulou, E.F., et al., *Catalytic conversion of biomass pyrolysis products by mesoporous materials: Effect of steam stability and acidity of Al-MCM-41 catalysts.* Chemical Engineering Journal, 2007. **134**(1-3): p. 51-57.
19. Tang, Y., et al., *One-Step Hydrogenation–Esterification of Aldehyde and Acid to Ester over Bifunctional Pt Catalysts: A Model Reaction as Novel Route for Catalytic Upgrading of Fast Pyrolysis Bio-Oil.* Energy & Fuels, 2008. **22**(5): p. 3484-3488.
20. Xu Ying, C.J., Zhang Qi, Wang Tiejun, Wang Chenguang, *Upgrading of Bio-Oil by Esterification over Solid Base Catalyst.* Petrochemical Technology, 2010. **35**(7): p. 615-618.
21. ZHANG Qi, C.J., WANG Tie-jun, WU Chuang-zhi, Xu Ying, ZHU Xi-feng, *Study on upgrading of bio-oil catalyzed over solid acids.* Journal of Fuel Chemistry and Technology, 2006. **34**(6): p. 680-684.
22. XIONG Wanming, F.Y., LAI Daming, GUO Qingxiang, *Upgrading of Bio-oil via Esterification Catalysted with Acidic Ion-exchange Resin.* Chemical Journal of Chinese Universities, 2009. **30**(9): p. 1754-1758.
23. Tang, Z., et al., *One Step Bio-Oil Upgrading through Hydrotreatment, Esterification, and Cracking.* Industrial & Engineering Chemistry Research, 2009. **48**(15): p. 6923-6929.
24. Xu, J., et al., *Bio-Oil Upgrading by Means of Ozone Oxidation and Esterification to Remove Water and to Improve Fuel Characteristics.* Energy & Fuels, 2011. **25**(4): p. 1798-1801.
25. Mahfud, F.H., *Exploratory Studies on Fast Pyrolysis Oil Upgrading* in *Department of Chemical Engineering*2007, University of Groningen: Groningen, Netherlands. p. 158.

(65c) Opportunities and Challenges for Advanced Biofuel Industry

Ashish Guptaand Ravindra Agrawal, Process, KBR, Houston, TX

Biofuels have been criticized for leading to rise in food costs. Advanced biofuels steer clear of this controversy as they are produced mostly from non-food crops. The US government renewable fuel standard mandates 21 billion gallons/yr of advanced biofuels by 2022 while current production levels are significantly low. In order to rise to this tremendous opportunity, several companies are taking different technological pathways for the commercial production of advanced biofuels. This presentation will discuss some recent projects at KBR for advanced biofuel production. Some of the hurdles in commercialization of advanced biofuels and future industry projections are discussed.

Green Diesel Production – An Overview

Stuart S. Shih
Gainesville, Virginia, USA
Email: ssshih@live.com

Abstract

Currently, there are two types of renewable diesel fuels available: biodiesel and green diesel. The biodiesel or FAME is the transesterification of vegetable oil or animal fat with methanol. The green diesel, which uses the same or similar feed stocks as biodiesel, is produced by a hydrotreating process that is widely employed in the modern petroleum refineries. Hydrogen is added to oil or fat resulting in significantly improved quality of the renewable diesel fuel. It can be produced in a standing-alone unit or in a co-processing unit with conventional petroleum feedstock for the petroleum diesel fuels production.

The paper will discuss the feedstock selection and its impacts on the product quality. The process chemistry, catalyst selection, and heat balance will be also presented. Some characteristics of the green diesel process, such as high hydrogen consumption, high heat release, and unique issues related to the carbon dioxide and carbon monoxide as the process byproducts will be included.

Introduction

To protect environment and to reduce green house gas emission, many countries are mandating the usage of renewable fuels. For renewable diesel fuels, FAME (fatty acid methyl ester) is the current focus. However, amount of FAME that can be blended with petroleum diesel is limited, about 7%. Further increasing the renewable diesel content will require production of other types of renewable diesels, such as green diesel by hydrotreating of fatty acids or Fischer-Topsch diesel derived from biomass. We will focus on the hydrotreating route to produce the renewable diesels (or green diesels).

Feed Stocks

The green diesels can be produced either in a stand-alone unit or co-processing fatty acid or FAME in a conventional petroleum hydrotreating process. Both approaches are very similar, using the hydrotreating technology developed for the petroleum refining industry.

There are more than 300^+ oil-bearing seed plants that can provide the fatty acids for the green diesel production. Of course, FAME and animal fats also can be the feed stock for the green diesel production. The feed stock selection is a complex issue that is beyond the scope of this paper. However, some unique properties of the fatty acid can significantly impact on process operation, process cost, and green diesel property. Key parameters are double bond content (expressed as iodine number), fatty acid chain length, free fatty acid content. In addition, trace metals in feed stocks such as phosphorous, calcium, iron, and dirt can impact on catalyst stability and selection. Higher olefin content will consume more hydrogen and release more reaction heat. High conjugated olefin can promote polymerization at high temperature to foul the catalyst and equipment. Too high free acid content such as in animal fats and trap greases can cause corrosion that may require equipment material upgrade. The fatty acid chain length will determine the cold flow property of the green diesel, which generally measured as cloud point. Consequently, green diesels produced from fatty acids that have a high percentage of C_{20}^{+} carbon chain length might require a post upgrading process (such as isomerization) to lower the cloud point.

Reaction Pathways

Many reaction pathways have been proposed. There are two major reactions: double bond saturation and oxygen removal. The double saturation reaction is straight forward and highly exothermic as shown in Fig. 1.

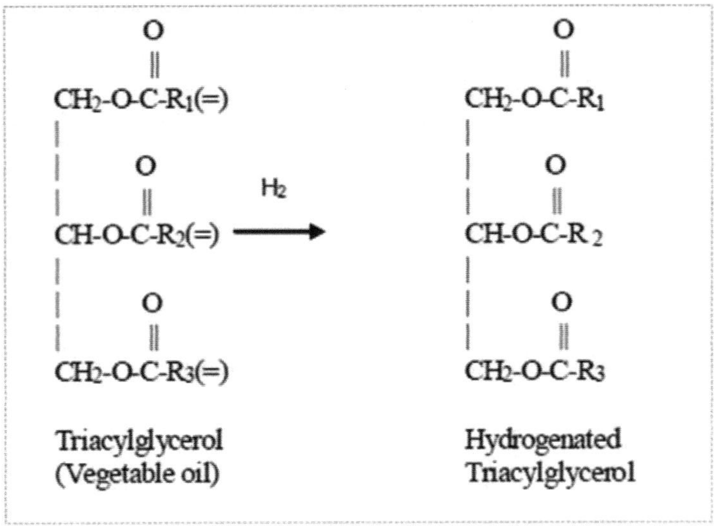

Fig. 1: Double Bond Saturation Reaction

The oxygen removal reaction is more complex, involving some intermediates. In general, it is believed the ester bonds to be broken before the oxygen is removed. Consequently, the formation of propane (if the feed is vegetable oil) or methane (if the feed is FAME) is formed quite before oxygen is removed to fatty acids as reaction intermediates. There are three pathways to remove the oxygen: decarboxylation, decarbonylation, and hydrodeoxygenation. Decarboxylation and decarbonylation consume less hydrogen and are endothermic. The hydrodeoxygenation will consume 3 moles of hydrogen per carboxyl acid group and is highly exothermic. However, it preserves the carboxic carbon in the final green product. Consequently, the green diesel yield can be 7-5% higher. As discussed later, catalyst can affect the reaction path or selectivity among these reaction reactions. The selectivity of the hydrodeoxygenation is easily monitored by GC to measure the even carbon number in the product. The distribution of C_{18} and C_{17} in the product is an effective way to measure the degree of hydrodexygenation reaction.

The separate of decarboxylation and decarbornylation reactions are complicated by the fact that CO and CO_2 are generally reaching equilibrium through the water shift reaction.

- Decarboxylation
  - R-CH2-COOH → RCH3 + CO2 (1)
- Decarbonylation
  - R-CH2-COOH + H2 → RCH3 + CO + H_2O (2)
- Hydrodeoxygenation
  - RCH2COOH + 3H2→ RCH2CH3 + $2H_2O$ (3)

Depending on the hydrogen cost, one might prefer the decarboylation (and decarbonylation) reaction pathway over the hydrodeoxygenation. Catalyst is the tool manipulating the reaction pathway.

Actually, the reaction pathways could be more complex. Besides the carboxyl acids, alcohols also have been detected. Fig. 2 shows the reaction pathways including the alcohols[1].

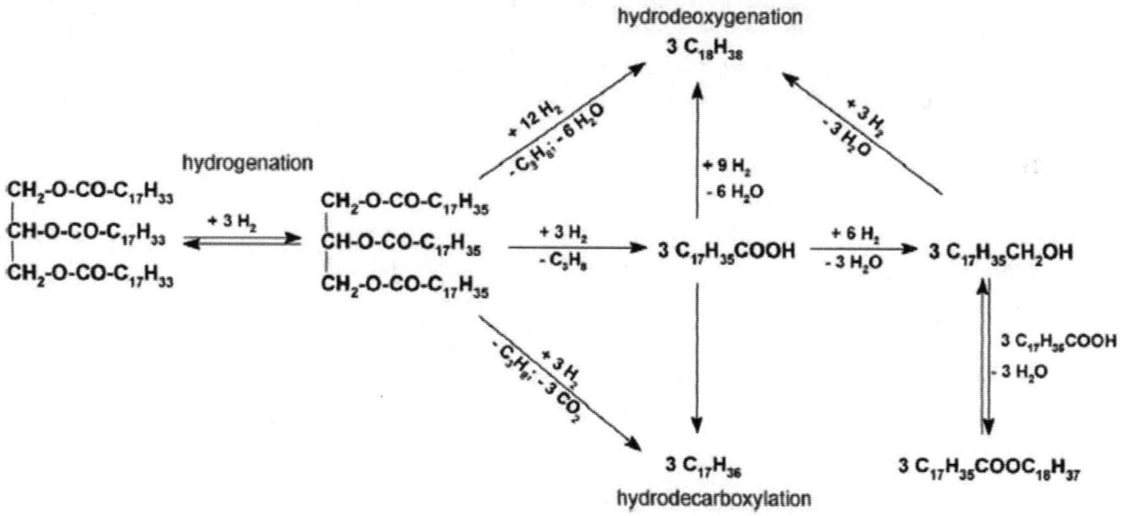

Fig. 2: Reaction Pathways

<u>Catalyst</u>

As discussed, catalyst can affect the reaction pathway or selectivity. Nickel was found to be dominating the decarboxylation and decarbonylation reaction, while molybdenum is highly selective for the hydrodeoxygenation reaction. As expected, the selectivity of the NiMo catalyst falls between these two catalysts.

The reaction pathways also are affected by the operating conditions. For NiMo catalysts, it has been found that low temperature and high hydrogen partial pressure favor the hydrodeoxygenation reaction[1].

Process

As discussed, the green diesel can be produced either in a dedicated process or in a co-processing unit with conventional petroleum feedstock. For a dedicated green diesel process, the catalyst selection could be straight forward, depending on the hydrogen cost. For the co-processing case, the catalyst selection is more limited since it still needs to preserve the desulfurization activity (or other hydrotreating functions) for treating the petroleum component.

For the co-processing, a wide range of feedstocks and operating conditions have been studied. In general, high pressure units are preferred to ensure a high degree of the oxygen removal of the renewable components.

The renewable component in the co-processing unit can affect the original catalyst desulfurization activity. It is found that NiMo is less affected by the renewable component than the CoMo catalyst. In general, oxygen removal is faster than the sulfur removal.

Heat release is a major issue for the green diesel production. To mitigate the excess heat release, one can either recycle the product back to the reactor or conduct multiple injections of the renewable feedstock into the reactor.

Production of CO and CO_2 from the renewable feestocks also causes some challenges for design and operation of the green diesel units. CO_2 can be removed by conventional amine system, but not for CO. Consequently, CO can accumulate in the recycle gas stream to a high concentration, which will depress the hydrogen partial pressure and will increase the CO's poisonous effects on the catalyst. The CO concentration can be reduced either by increasing purge gas rate or by installation of a methanation reactor to remove the CO[2].

Similarly, the accumulation of methane or propane in the recycle gas stream needs to be managed accordingly.

CO_2 can dissolve in water to increase equipment corrosion. Consequently, equipment that contacts with CO_2-saturated process water might require a material upgrade.

Summary

The production of green diesels is needed to meet high renewable diesel contents in the diesel pool mandated by the government policies. Conventional hydrotreating technologies developed for the petroleum refining industry can be applied to the green diesel production.

References

(1) D. Dubica and L. Kaluza, "Deoxygenation of vegetable oils over sulfide Ni, Mo, and NiMo catalysts", Applied Catalysis A: General 372 (2010) 199-208.
(2) R. Egeberg, K. Knudsen, and S. Nystrom, E. L. Grennfelt, and K. Efraimsson, "Industrial-scale Production of Renewable Diesel", PTQ Q3 (2011) 59-65.

() Musings on Management

Frank van Lier, The Lubrizol Corporation, Wickliffe, OH

It really is all about the people!

Hiring good people.

Retaining them.

Adjusting for different personal styles, cultures, encouraging diversity, embracing different perspectives.

Motivating people to work together in successful global teams.

Achieving top notch results that are better than any individual can achieve alone.

Maintaining the focus.

Getting the results.

Management is all about the excitement managers gets from having their people and teams grow and succeed.

(84a) Eliminating Difficult Start Ups with State Based Controls

Dustin Beebe, ProSys, Inc., Baton Rouge, LA and Jonathan Huggins, Performance Monomers, Dow Chemical, Deer Park, TX

Some processes are just difficult to start up. Every year, significant production is lost due to start up delays, product quality issues after startup, mistakes or gaps in operator expertise and other reasons. In fact, the U.S. Chemical Safety Board has estimated that the total loss due to operator error is $10B per year. In addition, because 70% of process incidents occur during start-up and shutdowns, safety can also be positively impacted by automating the process start-up.

Process start-ups can be completely automated and places the operator in supervisory mode approving the continuation of the process at specific intervals. The automation of start-up allows every start up to be exactly the same, no matter which operator is at the controls. The automation eliminates potential errors and overshoots which preserves the integrity of the startup and eliminates most recycle, purges, safety trips and lost production.

Using state based controls has cut average start-up time by a substantial amount, some as much as 50% or more. This is accomplished by streamlining the process and using the logic tools for automation. Automated start-up improves profitability and enhances flexibility of the process. Project ROI estimates show that the project typically pays for itself in about 3 months.

ADVANCED PROCESS CONTROL - PROVE THE VALUE

ROBERT B. HAWKINS
LYONDELL CHEMICAL COMPANY
CHANNELVIEW, TX

JAMES E. HACKNEY
EQUISTAR CHEMICALS, LP
CHANNELVIEW, TX

ABSTRACT

In the early days of the development of advanced process control (APC) techniques, there was much discussion regarding the potential economic benefits of this new approach. Each project had to be carefully justified, generally by using an estimation methodology unique to the project. As the field matured, standard approaches to benefit estimation appeared and were commonly used. Projects were often executed without a formal economic analysis because the benefits of APC were generally accepted, especially in plants with several successful implementations. However, the economic benefits of APC were recently called into question because of an increased focus on real-time performance measurements across the plant. It is no longer sufficient to simply state that the benefits of APC are large and obvious. Real-time measurement of the application's performance is quickly emerging as a requirement as companies place more emphasis on identifying improvement opportunities and properly apportioning limited resources. This presentation will outline the problem and suggest potential approaches to providing real-time measurements of the economic performance of an APC application.

BACKGROUND

The desire for a quantitative measure of the value delivered by advanced process control (APC) applications developed soon after the process industry began using computers to make real-time adjustments to their processes. The approach proposed was based on reducing the variability of the process and moving the average operating point closer to the limiting specification (Stout & Kline, 1976) (Latour, Sharpe, & Delaney, 1986). This approach was used effectively to prioritize APC application development efforts to maximize benefits capture and justify expenditures on new projects.

Today, the situation seems to have changed. The sophistication of the tools available for building APC applications has improved considerably since the first applications were developed. The methods for estimating the benefit of these applications have not kept pace (Bauer & Craig, 2008). Analysis of the estimation methods (White, 2003) (Muske, 2003), on-line estimators (Bailey, Pedersen, & Hawkins, 1992), detailed analysis procedures (Marlin, Barton, Brisk, & Perkins, 1991), and new performance metrics (Kern, 2005) (Latour, 1996) have been suggested to provide additional information regarding the value of the applications. However, the general adoption of sophisticated estimation approaches has been limited.

APC technology has matured and most of the largest applications, at least in some industries, have been implemented. Focus is shifting from the value that might be derived from the next project to the value that is being delivered by the existing applications (Canney, 2003). Although some applications run for years without any problem, most require maintenance to keep them generating full benefits. Proper allocation of resources to application maintenance can only be sustained if credible estimates of the value being generated by the applications are available.

Credibility of APC Benefit Estimates

Skepticism regarding the real value delivered by APC is increasing (Friedman, 2005) (Friedman, 2006) (Wang, 2011) (Kern, 2008). Some APC applications have a high on-stream factor and yet are clamped to the point that there is little improvement in operation when the application is running. If these applications are claimed to be valuable just because they are running (i.e., benefit = fraction time online × stated nominal benefit), then those who see how the applications actually perform will be skeptical about their benefits – as they should be.

They may be skeptical for several reasons. Often the stated benefits seem too good to be true. There may be a limited understanding of how the benefits were determined, which leads to the belief that the benefits are attributable to causes other than the APC application.

It is possible for the assumptions of the offline benefit estimate to be invalid when the controller has been running for a while. That can mean the benefits *are* much reduced or that the benefits are delivered in ways other than those assumed in the original estimate. If there is no online indication of this change, then people might conclude that the original benefits estimate was overstated and the controller has little value.

Skepticism can lead to lack of support for existing applications because the resources will be assigned to other projects that are expected to deliver benefits. Surveys show that only a fraction of installed APC applications are delivering the level of benefits that are possible (Brisk, 2004), largely because inadequate attention is given to the application and its infrastructure. The reduced support for APC applications will cause their performance to degrade further and thus reduces the benefits, confirming the belief that there was no value being delivered.

The perception that APC applications are not valuable can linger for long periods of time and make justification of new applications much more difficult. Note that this does not imply that there is no real benefit delivered from APC applications, nor does it imply that there is no value in making offline benefits estimates.

Benefits of Advanced Process Control

A challenge in estimating the benefits of APC is agreeing on the definition of what should be included in the determination of the benefits. APC applications involve more than the modeling of the process, the control algorithm, and the communication with the process. Upgrading instrumentation may improve understanding of the process. Discoveries about the process itself, such as identification of the true constraints or the real economic drivers, may allow for better identification of the optimal operating point of the process. Regulatory control configuration changes can yield substantial benefits from more stable operation.

The benefits of an APC project generally fall into three categories:

1. Changes in the operation of the process resulting from the project activities
2. Changes in process stability due to regulatory control improvements
3. Functionality added by the APC system itself

The decision of what changes to include in defining APC benefits can have a dramatic impact on the numbers calculated. Some may want to know the total benefit of the entire project, including any non-APC changes made. Others may want to know the contribution of just the APC application itself. Basically, the meaning of any stated benefit must clear to be credible.

Complete Project Benefits

A successful APC project involves much more than just the design and commissioning of the APC application itself. The early stages of the project include a close scrutiny of the process operating conditions, the economic objectives, the typical operating environment, and the effectiveness of the basic regulatory control systems. Significant changes are made in each of these areas to enable the APC application. These changes can produce substantial improvement in the economic performance of the process even before the actual APC application is implemented.

One viewpoint is that the benefit of the APC *project* is the sum of *all* of the activities, not just the change in process operation made by the APC application. Thus, the benefit needs to be estimated by comparing of the state of the system before and after the project is completed.

The project benefits includes the APC application itself and improvements made possible by changes to the regulatory controls. Additionally, it includes benefits from new instrumentation, personnel changes, and "institutional" learning (e.g., the APC application discovered a more profitable way to operate the process that is adopted even when the application is turned off).

Process Control Benefits

The period of time that an APC project is being executed often includes changes to the economic and process environment that are unrelated to the APC application. Those changes may prompt adjustments to operating procedures and targets that have an economic impact on the plant. To separate the effect of implementing a new APC application from the effects of other changes, it may be desirable to limit the benefits to only include control-related changes.

The primary difference between this approach and the complete-project approach is the removal of any non-control effects. Some of these are obvious, such as institutional learning and personnel changes. It is less clear whether or not to include new instrumentation in the benefits estimates. In some cases, the instrumentation will be justified in a separate project that will claim the benefits. In other cases, the instrumentation will be part of the APC project because it is required for the proper operation of the APC application. In that case, any benefits resulting from the new instrumentation could be rolled into the APC benefits.

Advanced Process Control Benefits

Maintaining the performance of an APC application occasionally requires making significant changes to the structure. This might include adding new variables or updating the models to reflect changes in the process behavior. Most applications also require day-to-day monitoring and minor adjustments to deliver proper performance over a long period of time. In these cases it may be desirable to estimate the benefits associated with the APC application by itself, either to justify the effort to make a major update, to justify the cost of on-going maintenance, or to monitor the economic contribution of the controller over time.

This approach to defining APC benefits excludes everything except the actual changes made by the APC application. Excluded from the benefit estimate are any changes in the regulatory controls or instrumentation as well as changes in the operating procedures or economic drivers.

ESTIMATION OF APC BENEFITS

One of the challenges faced by APC implementers is the lack of standard tools for estimating the benefit of the application. Each implementer is forced to build custom calculations to estimate benefits. These calculations have no generally accepted approach. They are built around widely varying assumptions about what is being calculated, and even the meaning of "benefit" for an APC application.

Regardless of which benefit definition is selected, the benefit of an APC application is essentially a change in the profitability of the plant. The profitability of a process can be defined as the difference between the total value of the products produced and the cost of feed stocks and utilities consumed.

$$PP = \sum_i (S_i \times P_i) - \left(\sum_j (C_j \times F_j) + \sum_k (C_k \times U_k) \right) \tag{1}$$

where: PP = Process profitability
S_i = Selling price of product
P_i = Quantity of product produced
C_j = Cost of feedstock
F_j = Quantity of feedstock
C_k = Cost of utility
U_k = Quantity of utility

If we adopt the convention that positive values represent revenue while negative values represent cost, then equation (1) can be simplified by generalizing feeds, products and utilities as economically-significant process streams.

$$PP = \sum_i (C_i \times F_i) \tag{2}$$

where: PP = Process profitability
C_i = Value of economically-significant process stream
F_i = Quantity of economically-significant process stream

In general, the benefits estimate of interest includes only the effect of the APC application and not changes in the external environment. In this case, the value of each of the economically-significant process streams can be assumed to be constant. The change in profitability is then represented by:

$$\Delta PP = \sum_i (C_i \times \Delta F_i) \tag{3}$$

where: ΔPP = Change in process profitability
C_i = Value of process stream
ΔF_i = Change in quantity of process stream

Depending on the point in the project lifecycle that the benefits estimate is being made, the APC benefits (ΔPP) will result from a change in product/feed/utility quantity (ΔF_i) between the plant operation with the APC application running, without the application running, or with the application running in a different configuration.

Only one of these operating conditions can be directly observed at a time, so the benefits can never be calculated directly. At least one of the operating conditions will always need to be estimated, and thus requires some assumptions be made about the operation of the plant. The assumptions could be in the form of simple rule-of-thumb guidelines or detailed models of the process operation in the condition being estimated.

Several kinds of benefits estimation cases can be considered:

- The future benefits for the purpose of justifying an APC project
- The benefits currently being delivered by a running APC application, or that could be achieved if the operating limits were to be changed
- The benefits that would be delivered by a running APC application if its operation were to be improved by developing updated models or revising the application's design

Each of the "***Estimated***" entries in Table 1 requires the creation of a model. The models developed for these cases could be considered as part of a family of models that are developed in the lifecycle of an APC application.

Table 1: Benefit Estimation Cases

	Without APC	With APC	Changed APC
New Application Justification	Actual	***Estimated***	—
Running Application	***Estimated***	Actual	***Estimated***
Application Improvement	—	Actual	***Estimated***

Estimating Benefits for Justification of a New Application

During the justification phase of the project, a model can be developed of the process operation with the proposed APC application installed. The estimated benefits are the difference between the current operating conditions and the operating conditions predicted by the model. This kind of benefit estimate is the one in most common use because it mirrors the justification calculations that are done for most large expenditures.

One common approach is illustrated in Figure 1. The model used to predict the process behavior after the APC application is installed is based on the assumption of a reduction in process variability followed by a movement of the average operation closer to constraints (Latour, Sharpe, & Delaney, 1986) (Muske, 2003).

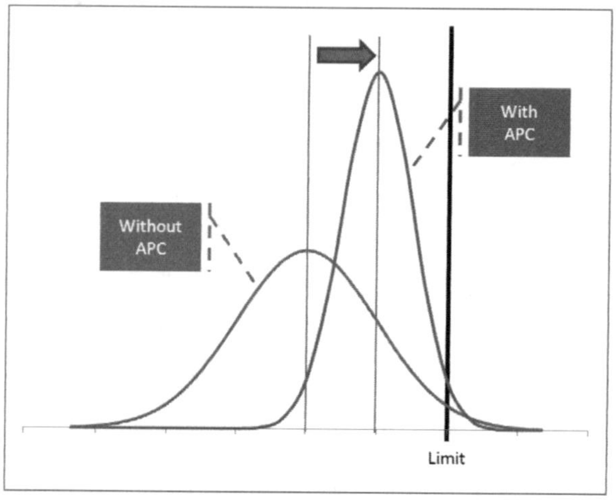

Figure 1: Justification by reduced variability

It is also necessary to estimate the change in the economically-significant process streams that corresponds to a change in the average of the process variable being moved towards a constraint.

In the simple column shown in Figure 2, the limit considered could be the overhead distillate impurity (x_D). If the economically significant streams in equation (2) were the distillate flow (D), the bottoms flow (B), and the reboiler duty (Q_R), then the benefit (ΔPP) would be calculated as shown in equation (4).

$$\Delta PP = C_D\left(\frac{\Delta D}{\Delta x_D}\right)\Delta x_D + C_B\left(\frac{\Delta B}{\Delta x_D}\right)\Delta x_D + C_Q\left(\frac{\Delta Q_R}{\Delta x_D}\right)\Delta x_D \qquad (4)$$

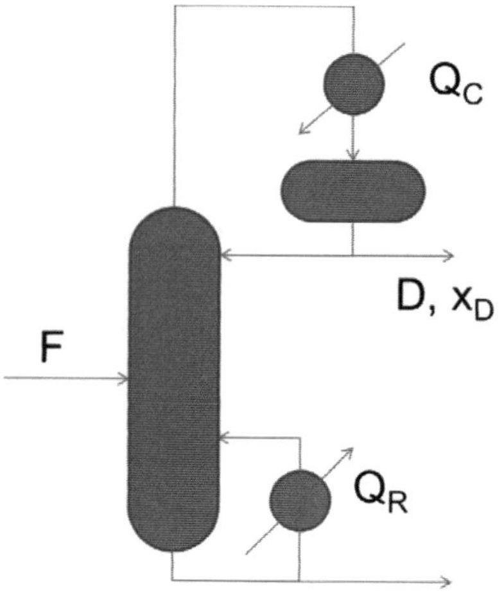

Figure 2: Distillation Column

The change in the distillate impurity (Δx_D) would be estimated by gathering historical data that represents the operation of the column. The variance reduction analysis would provide an estimate of the change in the distillate composition. Depending on the accuracy required in the benefit estimate, the change in distillate and bottoms flows and the reboiler duty could be modeled by a set of constants derived from historical process observations or they could be generated by a rigorous process model.

Other approaches for developing a model of the process operation with application are also possible. A simple model of the process could be used as a basis for an off-line optimizer calculation. This optimization would be applied to periodic historical data to estimate the "best" achievable operation at each point. Comparing this to the actual history data gives an estimate of the possible improvement. If the modeling approach is consistent with the APC application to be developed, the models can be used as a starting point for the application.

Estimating Benefits of a Running Application

One of the "take-home" messages from (Brisk, 2004) was the strong suggestion to implement proactive monitoring of APC application performance to sustain the benefits from the application. One of the critical performance measures for an APC application is the current benefit being delivered so it is important that this be estimated online in real time.

The benefit delivered by a running APC application can be estimated as the difference between the current operating conditions and those predicted by a model of the process operation with the pre-APC manual controls. This is the opposite kind of modeling problem to the one faced when estimating the benefits of a not-yet-installed APC application. The profitability of the process with the APC application running can be measured directly while the conditions without the application need to be estimated using a model of the process.

Fixed Benefit Weighted by On-Stream Factor

A common method to estimate the benefits of a running application is to assume that if it is turned on, then it is generating the benefits estimated in the justification of the project or in the post-project audit. Auditing the project is often required to estimate the actual benefits delivered by the APC application when it is commissioned. The audit is generally conducted as soon as possible after the application is running to minimize other changes to the system that might have an economic impact on the process.

This approach is only valid for a short period of time after the project is complete because other factors begin to change, and it becomes increasingly difficult to defend the assumption of a fixed benefit.

Regression of Historical Data

The model of the process operation with the pre-APC manual controls can be developed by regressing historical plant operating data (Bailey, Pedersen, & Hawkins, 1992). A regression model is developed to predict the economically-significant process streams (F_i) in equation (2) based on the process operating environment. The regression model is then used to predict the behavior of the process based on the current operating environment. The difference between this prediction and the current operation with the APC application running is then used in equation (3) to estimate the benefits. Figure 3 shows the real-time estimate of the steam consumed by a distillation column based on the column feed rate, pressure, and product quality targets. In this case, the reduction in steam flow is a direct indication of the controller benefit.

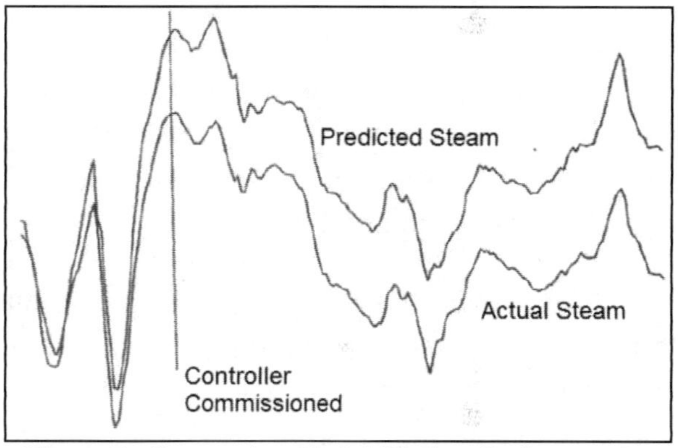

Figure 3 - Regression Model (Bailey, Pedersen, & Hawkins, 1992)

The advantage of this approach is that it reflects the real minute-to-minute changes in the process operation and directly estimates the pre-application values of the economically significant streams in equation (2). The model is generally easy to understand and easy to explain. This method works well for small applications with clearly defined economic objectives. As the APC application becomes more complex, it becomes difficult to design a regression model that is appropriate for all operating conditions. Significant changes in the underlying economics of the process can result in a change in basic operating strategy that invalidates the comparison with the historical operation.

Model the Operator's Behavior

As the application complexity increases, it becomes more difficult to directly model the process conditions that would be chosen by the operator in any given situation. However, it is possible to make some assumptions regarding the kind of strategy the operator would adopt and then adjust the current conditions based on those assumptions.

For example, a common assumption about operator behavior is the desire to leave a margin between the average operating conditions and the constraints, as shown in Figure 4. The size of the margin is based on the natural conservatism, limited measurement or inferential accuracy, and limited attention during manual control (Singh & Seto, 2002). The likely operating point for the plant can be estimated by selecting the size of the operating margin and then calculating the without-APC operating point for each constrained variable as an offset from the constraint. A model of the process itself is likely required to calculate some of the economically significant streams in equation (2), but this allows a real-time estimation of the process operating point without the APC application running.

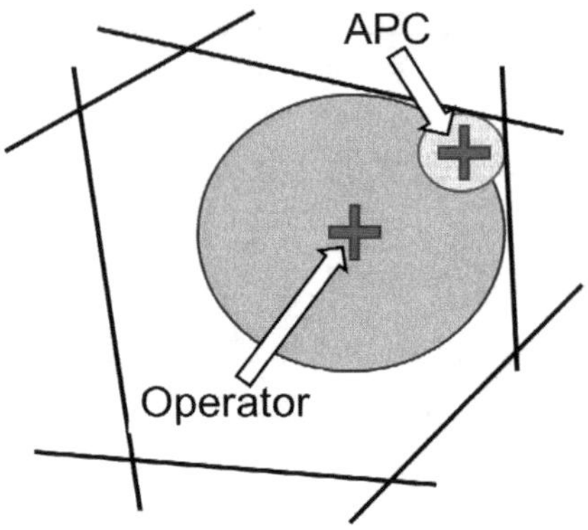

Figure 4 - Operating Margin

Another possible assumption about the operator's behavior is the tendency to make relatively infrequent adjustments. This assumption results in an operator strategy of making adjustments only when they are needed to keep the process within its constraints. This approach to modeling has the advantage that the application benefits drop to zero when the application is turned off, and only begin to increase as the application makes adjustments to correct for disturbances. It also captures the benefit of the continuous economic optimization that is a feature of many APC applications.

Estimating Potential Benefits of a Running Application

After an APC application is implemented, much of the attention focused on optimizing the performance of the application disappears. The operating limits tend to become more constrained than is necessary, typically as a result of the application's ability to handle large upsets in the process. The normal changes in the dynamic characteristics of the process slowly begin to introduce errors between the model used in the application and the actual process behavior. This causes the performance of the application and benefits delivered to deteriorate.

All of these changes happen very gradually and are often missed by the people responsible for maintaining the APC application. An online estimate of the potential benefit the application could achieve is a very useful performance indicator. The lost benefit associated with overly constrained limits can be estimated by running a "shadow" application that uses a relaxed set of operating limits. The benefit of improving the application's model is more difficult to estimate online because it requires an estimate of the error in the current model.

Conclusions and Summary

APC applications have moved into the every-day mainstream of process improvement projects and need to compete for resources on the basis of the actual benefits they deliver to the organization. It is necessary to have accurate and believable benefit estimates to invest resources in identifying and developing new applications, and in maintaining existing ones.

In order to present a clear picture of the on-going benefit being delivered by an APC application, it is necessary to make an online estimate of the benefits. This kind of benefit estimation provides an assessment of the application's performance and a means for ensuring that resources are allocated to maintaining the performance of the most important applications.

References

Bailey, J. K., Pedersen, C. C., & Hawkins, R. B. (1992). Estimate the Benefits of Advanced Controllers On-line. *Chemical Engineering Progress, June 1992*, 74-78.

Bauer, M., & Craig, I. K. (2008). Economic assessment of advanced process control – A survey and framework. *Journal of Process Control, Volume 18, Issue 1, January 2008*, 2-18.

Brisk, M. L. (2004). Process Control: Potential Benefits and Wasted Opportunities. *2004 5th Asian Control Conference, Volume 3 , Vol. 1*, 10-14.

Canney, W. M. (2003). The future of advanced process control promises more benefits and sustained value. *Oil & Gas Journal, April 21, 2003*.

Friedman, Y. Z. (2005). Has the advanced process control industry completely collapsed? *Hydrocarbon Processing, January 2005*, 15.

Friedman, Y. Z. (2006). Jim Ford's views on advanced process control. *Hydrocarbon Processing, November 2006*, 19-20.

Kern, A. G. (2005). Online monitoring of multivariable control utilization and benefits. *Hydrocarbon Processing, October 2005*, 43-47.

Kern, A. G. (2008). Outlook for multivariable predictive control. *Hydrocarbon Processing, October 2008*, 33.

Latour, P. R. (1996). Process control: clifftent shows it's more profitable than expected. *Hudrocarbon Processing Vol 75*, 75-80.

Latour, P. R., Sharpe, J. H., & Delaney, M. C. (1986). Estimating Benefits from Advanced Control. *ISA Transactions, Volume 25, Number 4*, 13.

Marlin, T. E., Barton, G. W., Brisk, M. L., & Perkins, J. D. (1991). Benefits from process control: results of a joint industry-university study,. *Journal of Process Control, Vol 1, June 1991*, 68-73.

Muske, K. R. (2003). Estimating the Economic Benefit from Improved Process Control. *Industrial & Engineering Chemistry Research, Vol. 42, Issue 20*, 4535-4544.

Singh, P., & Seto, K. (2002). Analyzing APC Performance. *Chemical Engineering Progress, August 2002*, 60-66.

Stout, T. M., & Kline, R. P. (1976). Control System Justification. *Instrumentation Technology (Intech), September 1976*, 51-58.

Wang, J. (2011). What is the outlook for advanced control engineering? *Hydrocarbon Processing, July 2011*, 79-84.

White, D. C. (2003). The Economics of Advanced Automation. *AIChE 2003 Spring Meeting, Paper 73c*.

(84c) Improved Closed-Loop Subspace Identification Technology for Adaptive Modeling and APC Sustained Value

Hong Zhao, Michael Harmse, Quinn Zheng and John Campbell, Aspen Technology, Inc., Houston, TX

Sustaining the value of Advanced Process Control (APC) applications has been a focus for many refinery and petrochemical companies over the past decade. Changes in plant operation such as equipment aging, lost bubble caps, fouling of packing in heat exchangers, process modifications and feedstock changes will deteriorate the performance of any APC applications and reduce the financial benefits of the controller. One of the necessary steps is to maintain the process model and keep it up to date. Frequent precision revamps of running controllers via a partial plant test and re-identification of a specific sub-model of the process is typically the most efficient and practical approach. New tools for model accuracy monitoring and model adaptation have been developed by exploiting the latest technology, such as automated closed-loop step testing (e.g. using the SmartStep™algorithm), online model quality monitoring and model re-identification. These new tools are designed to help APC users to diagnose and repair model-mismatch as often as is needed. The main benefit is to prevent the long delays occurring between model mismatch and model repair. Among the new tools that are utilized for this purpose, subspace identification plays a critical role.

Subspace identification emerged in the 1980s and eventually attracted a lot of academic attentions. AspenTech has first implemented it in DMCplus™(AMS 3.0) and this software was released in October 2000. In contrast to what is reported in most case studies found in academic papers, in industrial APC projects the users have to deal with many challenging problems that were not considered in the academic papers, such as the very large number of inputs and outputs. For example, the largest reported DMCplus controller has 283 manipulated variables (model inputs) and 603 controlled variables (model outputs), with thousands of native collinear subsystems, as well as extensive sparsity in the model. The main objective is to use cost-efficient ways to test the process unit in order to get more accurate models more quickly, since that will ensure better performance of the DMCplus controller. Control engineers are concerned with the test duration because the unit operates sub-optimally during this time period, and product give-away is often more expensive than the man-hours involved in the test. They are also concerned about the accuracy of the model's gains and gain-ratios (colinearity aspects). After more than 10 years of practical use in the process industry, the major features of subspace identification such as true multi-input multi-output (MIMO) identification and faster model convergence have been evident, resulting in substantial improvements in project execution efficiency. Recently, the properties of subspace identification algorithms in closed-loop environment have also been well studied and documented (e.g. Qin), and some improved algorithms have emerged to address these issues. With these developments, modern subspace identification technology has now matured to the point where it works in a synergistic way with innovative multivariable step testing technology like SmartStep. The combination of subspace identification and the multivariable step testing algorithms ensures that online adaptive modeling is becoming a reality. Control engineers doing APC maintenance work will be able to take full advantage of these technologies, ensuring the sustained value goal is met.

In this paper the recent progress of subspace identification technology is reviewed and several algorithms dealing with closed-loop data will be summarized in a unified Least-Squares framework developed by Qin (2006). Special issues in industrial MPC practice will be discussed and how that can be addressed will be demonstrated with actual project examples.

(84d) Successful Implementation of Advanced Process Control in Cryogenic Plants

Rafael Aguilar, Advanced Solutions, Honeywell, Houston, TX

Successful Implementation of Advanced Process Control in Cryogenic Plants.

Tom Lyndrup and Greg Rogers

Enterprise Products LLC

Houston, Texas

Rafael Aguilar and Duane Morningred

Honeywell Process Solutions

Houston, TX, and Camarillo, CA, USA

Paper to be presented at the 2012 AIChE Spring meeting, Houston, Texas

April 1-5, 2012

Abstract

Enterprise Products Partners L.P is anatural gasand crude oilpipelinecompany with headquarters inHouston, Texas. It has 18,368 miles of natural gas pipelines, 13,297 miles of NGL and petrochemical pipelines, 813 miles of Gulf of Mexico crude oil pipelines, 26 Gas Processing plants, 19 NGL and propylene fractionators along with Storage (Salt Dome), Drilling Platforms and NGL Terminals.

The Meeker Gas Plant 1 is a new process unit in Colorado's Piceance Basin, started up in late 2007 with a capacity to process 750 MMcf/d of natural gas and is capable of extracting up to 35,000 b/d of natural gas liquids (NGLs). Meeker, the largest natural gas plant in the continental United States, serves the Piceance Basin and other area deposits comprising the nation's fastest-growing natural gas field, increasing export capacity and providing gas to downstream markets through four major pipelines. Phase II of the complex (completed in the third quarter of 2008), doubled the processing capacity at Meeker to 1.5 Bcf/d of natural gas and 70,000 b/d of NGLs.

Enterprise Products wanted to maximize their Meeker, Colorado Cryogenic I plant's recoveries. This plant is subject to continuous changes in production plans, feed composition and pipeline constraints, which makes control difficult. The goal was to increase the recoveries by at least 1%.

The solution at Meeker Gas Plant utilized Honeywell's Multivariable Predictive Control applications that have steady state models. Step Testing was minimized due to the controller characteristics and the additional tools for online model ID used during the project implementation.

Since the initial commissioning, Enterprise Products has continued to enhance the applications finding new sets of variable limits and automatic update for optimization limits based on process conditions. In addition, the operator workflow has been improved, since the focus is now to understand the unit overall problem and search for a solution, instead of changing limits and setpoints to multiple controllers all the time. APC displays have been effective to show the current process constraints to increase production, as well as improve operational consistency across the multiple operation crews.

This paper summarizes the existing APC and Optimization solution and discusses the details of how the control strategy was recently implemented.

Improved Load Following of a Boiler with Advanced Process Control

Kevin R. Jensen and John D. Hedengren

Brigham Young University

Provo, UT 84602

Load following in power generation is a recent opportunity as time of day pricing and co-generation are adopted in refining, chemical, and power plants. Also, traditional sources of power generation are increasingly mixed with a growing fraction of emerging energy sources such as wind and solar. Wind and solar have the characteristic of being intermittently available in the case of cloud cover or as weather patterns move through an area. As of July 2011, 89% of base-load power generation in the U.S. came from coal, gas, oil, and nuclear plants (EIA, 2011). As non-traditional sources of energy integrate into the base-load, there is an opportunity to improve load-following to allow full utilization of the intermittent sources on a smart grid. The contribution of this study is to investigate the application of advanced process control for improved load following in traditional power generation, and in this case, boiler constraint control. Also, a new trajectory formulation is introduced for controlled variables in Nonlinear Model Predictive Control (NMPC). This new formulation allows rejection of process noise within a dead band while only introducing linear contributions to the objective function.

Base plant process control is typically deemed important from a safety and quality viewpoint. While these are two vitally important factors, many times, there are other aspects that greatly affect the success of a process unit. Controls are often developed from a working knowledge of a process or preconceived limitations. Although they have been very successful in the last 40 years since they were introduced, in many instances they "lack a systematic stability analysis and controller design" (Feng, 2006). It is based on operator knowledge, and adaptive algorithms (Hagan, Demuth, De Jesus, 2002). While operating a system in this manner may allow a process to proceed safely and meet quotas or quality assurance, there may be ways in which other elements of a process can be optimized (i.e. emissions, economics, and process unit life). Coal fired furnaces and boilers are an example of this. They are constrained by certain physical limits, such as rate of temperature change on tubes, allowing only restricted power cycling. Furthermore, energy from renewable resources has become increasingly popular. The National Renewable Energy Laboratory indicates that the US Department of Energy has a vision wherein wind energy contributions to total electricity production in the future is projected to increase. It is their aim that over the next 2 decades, wind will comprise 20% of US electricity production (Thresher Robinson, and Veers, 2008). However, coal-fired power plant cycling to accommodate renewable resources can actually increase wear and tear costs on coal-fired boilers. The longer the boilers sit idle, the greater the damage done to the boiler as it is ramped up after the idle time. This will decrease plant life and increase costs (Lew, Brinkman, Lefton, Piwko, 2011). It may seem as though these two factors cannot be optimized simultaneously. However, by using a robust controller, it may be possible to optimize multiple facets of a process and comply with multiple constraints.

In this study, we investigate the use of model-based control and PID controls in a coal fired furnace. By generating a differential algebraic model (DAE) of a coal-fired boiler, the constraints and parameters are explicitly modeled and controlled. Responses to process disturbances requiring power cycling are also optimized to increase profitability and process unit life.

The first step was to generate a model that could represent the cycling time and temperature changes of a coal-fired furnace. The model was created using first principles based on material and energy balances. The energy balance was built around the boiler, with appropriate heat-transfer terms for exchange between the bed, tubes, and high temperature water. A heat transfer term was incorporated into the model to represent the time delay of heating up and cooling down. The heat transfer was based on irradiative heating, as this is the dominant form of heating certain types of coal-fired boilers (Basu, Kefa, Jestin, 2000). The use of a lag variable allowed for the approximation of apparent dead time that is observed in a coal furnace. Dead time is often derived using step tests (Schnelle, Laungphairojana, Debelak, 2006). However, we used observations from operators and individuals knowledgeable about these systems. Load cycles were then simulated with control of the system through the APMonitor software. Control of the system was accomplished using trajectory tracking and constraints in Nonlinear Model Predictive Control (NMPC). For comparison the use of a PID controller on this system was explored. The results of the two control simulations were then compared.

NMPC was able to predict appropriate controller outputs in order to achieve the correct ramping rate and cycling. The model-based control was superior for several reasons. The physical constraints of the system may be set and the controller keeps the system explicitly within those bounds. For example, typical practice for changing furnace load is to increase heating output at a rate of about 1%/min. This rate constraint can be controlled explicitly in the multivariable controller along with trajectory load following. NMPC uses predictive values based on current measurements in order to achieve the set point within the dictated constraints. It was demonstrated over the entire range of operation, including transient and steady state conditions. It optimized load changes and achieved set points within the reliability constraints.

One of the major benefits of the PID is its simplicity. In simulations, it was able to achieve and maintain set point and reject disturbances. However, the controller had several shortcomings. In order to achieve the desired output temperature, the controller frequently saturated. The PID also had several challenges with start-up and ramping cycles. Typically, load changes are performed at a slower rate. The PID controller was unable to accommodate these constraints. In certain instances, especially those in which large disturbances were introduced, the PID controller violated the constraints on temperature gradient for the boiler temperature tube integrity. This situation can greatly affect safety, performance, and economic success, especially over the lifetime of the furnace. The PID controller was however very useful at steady state or for small disturbances. It was able to keep a set point, but not effective enough to be used in cases where large disturbances were encountered. It was not as robust as would be desirable in certain circumstances, especially those in which large disturbances or changing conditions are frequently observed.

Process data for this study was obtained from an operating coal-fired boiler facility. With current controls on the boiler, load changes can be performed at about 1%/min. A model-based controller challenges this restriction by driving to actual process constraints. Load changes can be performed faster if the constraints are better understood, or perhaps, in order to prolong boiler life, they should be done slower. The model-based approach enables an environment where constraints can be explicitly targeted in moving the boiler to the new power generation load. As a final statement, there are also some concerns that certain process units may be too complex to be accurately modeled. In these cases it may be beneficial to use a combination of empirical and first principles methods to obtain the best results. Future extensions to this work may also include forecasting of energy availability and load, time-of-day pricing, and anticipated peak power demands. This will enable feed forward information for improved load following optimization.

References

[1] EIA, U.S. Energy Information Administration (2011), Electric Power Monthly, URL: http://205.254.135.24/electricity/monthly/, Retrieved: 14 October 2011.

[2] Feng, G. (2006). A Survey on Analysis and Design of Model-Based Fuzzy Control Systems. *IEEE Transactions on Fuzzy Systems* 14(5) 676 – 697.

[3] Hagan, M., Demuth, H., and O. De Jesus (2002) An Introduction to the Use of Neural Networks in Control Systems, *International Journal of Robust and Nonlinear Control*, 12(11) 959-985.

[4] Thresher, R., Robinson, M., and P. Veers (2008). The Future of Wind Energy Technology in the United States. URL: http://www.nrel.gov/docs/fy09osti/43412.pdf. Retrieved: 14 October 14, 2011.

[5] Lew, D., Brinkman, G., Lefton, S., and D. Piwko (2011). How Does Wind Affect Coal? Cycling, Emissions, and Costs, URL: http://www.nrel.gov/docs/fy11osti/51579.pdf. Retrieved: 14 October 2011.

[6] Basu, P., Kefa, C., and L. Jestin (2000). *Boilers and Burners.* New York, NY: Springer

[7] Schnelle, K., Laugphairojana, A., and K. Debelak (2006). Emission reduction of NOx and CO by Optimization of the Automatic Control System in a Coal-Fired Stoker Boiler. *Environmental Progress* 25(2), 129-140.

(84f) APC2: Aligning People with Control

Raymond Coker, Automation, BP Texas City Refinery, Texas Ctiy, TX

Advanced Process Control (APC) can take on many flavors and levels of sophistication. On the digital control systems (DCS), advanced regulatory controls can create value using property predictions cascaded to DCS controllers or feed forward or decoupling signals can be added to improve stability and robustness of controls. In order to take on entire units of control, medium to large multivariable controllers have been developed to handle process interactions and to drive the unit to the most economical combination of constraints. Multiunit control and optimization in the form of larger linear program provides a tool to coordinate the flows and optimal targets across the refinery. More detailed effects can be modeled and included in the linear program solution as new knowledge is incorporated into the solution. Fuel gas, steam, hydrogen, amine, cooling water, tanks, shipments, electricity and process lineups create opportunities to synergize across multiple units.

Aligning People with Control (APC2) provides the mechanism to design controllers, commission controllers and sustain value long term. Critical resources include advanced control engineers, unit control engineers, process engineers, operations personnel and economics and planning personnel. Economics and planning provide the big picture view of the refinery and help identify the gold nuggets for control. Advanced control personnel help planning by providing more fidelity in the models and tradeoffs between quality and production and by commissioning model-based controllers to coordinate moves across each unit and to drive to the planning targets. Unit control personnel help with control system configurations, regulatory controls and advanced controls. Process and operations provide critical process and constraint information. The extended team includes management sponsors, instrument and analyzer personnel, environmental personnel, laboratory personnel, corporate technical leaders and sponsors, commercial managers, upstream suppliers and downstream customers.

Refining Energy Utilization and CO_2 Emissions Modeling

John J. Marano, JM Energy Consulting, Gibsonia, PA

William R. Morrow III, Jayant Sathaye, Eric Masanet, Tengfang Xu, Lawrence Berkley National Laboratory, CA

Abstract
Petroleum refining consumes a significant amount of energy to convert crude oil into transportation fuels and other products. Based on U.S. Department of Energy statistics, this industry accounted for roughly 9% of U.S. industrial energy usage and 14% of industrial CO_2 emissions in 2010. Each refinery is configured differently, processing varying quality crudes into a variety of specification products; thus, no two refineries are the same. Process-specific energy use and CO_2 emissions for petroleum refineries are not publically available, and may not even be adequately or accurately measured; making it difficult to understand current or to predict future refining energy requirements and emissions. A tool for estimating individual process energy requirements and emissions as a function of varying feedstocks, products and processing would be useful to policy makers and researchers focused on improving energy efficiency and reducing emissions. JM Energy Consulting has developed such a modeling tool to support a Lawrence Berkley National Laboratory (LBNL) project aimed at estimating costs and benefits of energy-efficiency improvements in the U.S. refining sector. In this paper, we present our methodology and some preliminary results from this modeling effort.

Introduction
Sources of CO_2 emissions from petroleum refining include: on-site combustion of fuels to provide process heating and to generate steam and electricity for use within the refinery; the production of hydrogen used in refinery operations; and coke burn-off from catalyst regeneration (primarily in fluid catalytic cracking) [1]. Secondary, but sometimes significant, refinery CO_2 sources include flaring of off-gases during start-up, shut-down and process up-sets; biotreatment plants, and other fugitive emissions [2]. Figure 1 shows a typical breakdown of the major CO_2 refinery sources. The majority of CO_2 emitted from refining derives from fuel combustion. Interestingly, the distribution of emissions, as well as, total emissions are similar regardless of whether the refinery employs fluid catalytic cracking (FCC) or hydrocracking for the conversion of heavy gas oil to light product (*i.e.* fuel and power requirements are about the same, carbon is rejected either in the FCC regenerator or the hydrogen production plant required for hydrocracking).

Approaches for reducing refinery emissions may be categorized as follows [1,3]:

- Energy Efficiency Improvements – reduction of specific-energy required to manufacture product by minimizing energy losses
- Process Yield Improvements – reduction of specific-energy required to manufacture product by improving process yields

- Fuels and Utilities Switching – substitution of low-carbon fuels and other utilities for existing high-carbon ones
- Crude Oil and Other Feedstock Switching – substitution of light-sweet crudes for heavy-sour crudes, or introduction of renewable bio-oils as feedstocks
- Product Switching – substitution of drop-in biofuels as blendstocks, reducing refinery crude oil throughput
- Product Slate Optimization – adjustment of product slate to reduce processing requirements and emissions
- Carbon Capture and Sequestration – CO_2 removal from flue gas or syngas used to produce refinery hydrogen, followed by geological, oceanic or terrestrial storage in perpetuity

Some of the options identified above are broader than normally encountered in refinery specific studies aimed at identifying short term operating strategies, and would require changes in existing fuel markets outside of refiners' control. In order to get a complete and accurate picture, life-cycle emissions outside of the refinery must also be considered when making choices among options, since it is total emissions that are important.

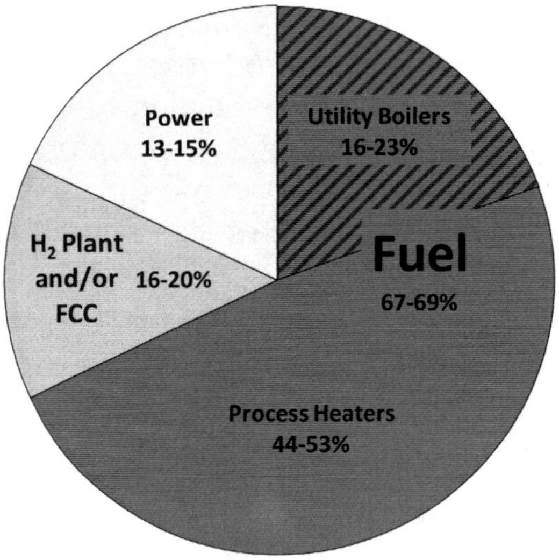

Figure 1. Typical Refinery CO_2 Emissions Distribution [4]

Adoption of efficient process technologies is an important approach to reducing CO_2 emissions, in particular those associated with combustion. In many cases, implementing energy efficiency measures is among the most cost-effective approaches that any refiner can make, improving productivity while reducing emissions. Therefore, careful analysis of the options and costs associated with efficiency measures is required to establish sound carbon policies addressing global climate change, and is the primary focus of LBNL's current petroleum refining sector analysis for the U.S. Environmental Protection Agency. The analysis is aimed at identifying efficiency-related CO_2 abatement measures and developing energy and CO_2 emissions abatement supply curves for the U.S. refining industry. A refinery model has been developed for this purpose that being general, also has the potential to be used to examine the costs and benefits of the other CO_2 mitigation options listed above. This model is described in more detail below.

Methodology - Background

Lawrence Berkley National Laboratory [5] and others [6-8] have been active in identifying possible measures for improving the efficiency of petroleum refining operations. However, it has been difficult to quantify the full impact of implementing these measures industry wide. One approach for doing this, attempts to quantify energy use and emissions at the process plant level within the refinery using average performance parameters for the individual process plants. This approach has been successful in establishing nominal baselines for energy use and emissions (*e.g.* [9]) for the refining industry as a whole. On the other hand, the approach has been less than definitive when it comes to quantifying potential benefits and costs from applying energy efficiency improvement measures [10]. This shortcoming results from the absence of predictive capabilities in the approach; as well as, issues associated with quantifying the inherent variability that exists between individual refineries.

Petroleum refineries in the U.S. are complex industrial facilities involving multiple processing plants configured both in parallel and series. The component plants are integrated through utility systems, which supply fuel, steam, cooling water, and electricity to the various processes. Individual operations within these plants are also extensively integrated. Due to the high degree of energy integration, both at the refinery and plant level, efficiency improvements to a single operation may have implications throughout the refinery. Efficiency improvements are in general not additive, with the possibility of the effectiveness of later modifications being dampened by those implemented earlier (*e.g.*, a 10% reduction in fuel consumption based on a revamped heat exchanger network design and a 5% reduction due to less furnace fouling does not necessarily result in a 15% total reduction). For this reason, it is essential to examine efficiency improvements in refining at the process level using a bottom-up, predictive approach to estimate energy usage on an operation-by-operation basis. A number of excellent case studies based on this approach can be found in the open literature [11-13]; however, these do not systematically address the state-of-play within the industry as a whole, and also require very sophisticated and detailed analyses (*e.g.*, heat exchanger network pinch analysis).

A predictive approach requires a certain degree of engineering to be incorporated into the analysis that may not be necessary when considering simpler manufacturing facilities. Technical descriptions of the various processes are required that relate process feed and operating conditions to process product streams and utility requirements on a step-by-step basis. There is precedence for this type of analysis within the refining industry for determining economically optimal refinery operation using Linear Programming (LP) models. A number of articles have appeared over the years that describe use of refinery LP models to examine possible CO_2 emissions reductions [1,4,14,15]. However, problems exist with standard approaches to refinery LP modeling that must be addressed when analyzing efficiency improvements.

Traditional LPs are formulated as an optimization of a refinery economic objective function and a system of constraints expressed as equalities and inequalities related to material and utility balances, along with operating and product quality specifications. For real refineries, models are very large with many variables being considered for optimization, which may be performed at a refinery on a day-by-day basis. Accuracy is achieved through the use of measured plant data or rigorous front-end process simulations (*i.e.* the LP model must be tuned-up). For the present application, interest is not in determining absolute economic optima, but rather *in sequentially organizing the cost hierarchy of a series of energy efficiency measures that might be applied across the industry*. Simplicity and flexibility is essential, and the primary concern is that the model mirrors correct refinery responses to changes of interest [16]. For this application, a detailed LP model may miss the big picture by focusing too much on the details (*e.g.*, multi-tiered product specifications, elaborate process models, multiple crude oils). This response to non-routine situations is sometimes referred to as model "constraint gridlock".

In addition, LP formulations are typically limited to the present or the near future, have difficulty handling capital investment decisions, and mass and energy balances may not be closed. The focus here is on short to midterm (twenty-year horizon) investments and operational changes that can improve efficiency, with attention being given to state-of-the-art and possible next generation technologies. For the latter in particular, costs are not well known and are dependent on the outcomes of on-going research, development and demonstration projects. When examining solutions to long range problems, accurately estimating production volumes may be more important than maintaining rigorous product specifications. There is a considerable degree of uncertainty with product quality requirements and other assumptions when looking as much as twenty years out into the future.

Special attention must be given to mass and energy balances. Early refinery LP models were neither mass nor energy balanced. Material balances were formulated on a volume basis (*e.g.*, barrels per day), and a total mass balance for the refinery model was not achieved without adjustments to stream densities. This situation has been mostly corrected in advanced LP models in use today; however, compositional balances for carbon and hydrogen are still not adequately represented in most LP models. Efforts to date have focused on methods to close these balances around the refinery model; however, these appear to be by means of forcing closure through arbitrary adjustments to specific stream compositions. Such methods are not adequate when one of the primary functions of the model is the prediction of CO_2 emissions. The best example of where this situation cannot be tolerated is when compositions of fuel gas streams are approximated by an externally specified and fixed fuel composition, leading to poor estimates of CO_2 emissions [1]. Likewise, early models and those still in use today perform only utility balances, that is the amount of steam, power, etc. used is equal to that generated or purchased. Once again, this is not adequate when considering issues directly related to energy efficiency. It is necessary to explicitly estimate energy use and losses throughout the refinery, instead of inferring them from utility balances. In order to do this, the energy content of refinery feed, product and internal streams must be known or predicted, and process-to-process heat exchange must be explicitly considered.

Methodology – Approach

To address the issues described above, standard methods for building refinery LP models have been adapted for the present application of generating refinery energy-usage abatement supply curves. A combination of empirical data correlations and semi-rigorous engineering models has been used to develop reduced-form refinery models to estimate refinery product yields and energy requirements. In many respects, these reduced-form models are similar to what might be used in a more traditional LP model applying "delta-based" sub-models. The refinery model developed has the following attributes:

- The model is formulated to maintain total mass and energy balance, as well as elemental balances for carbon, hydrogen, nitrogen and sulfur. This has been achieved by reformulating individual process yields data on a weight carbon basis (*e.g.*, ton of carbon in naphtha product per ton of carbon in feedstock). Hydrogen contents of process streams are based on correlated changes between the hydrogen-to-carbon ratio of feed and product streams, as are stream energy content (heating values). Nitrogen and sulfur removal efficiencies and distributions to liquid products are explicit input to the individual conversion processes represented in the model.
- Relatively simple sub-models are employed for conversion processes, relating yields and utility use to at most a few essential feed-quality and operating-severity parameters. Intermediate streams within the refinery are not aggregated (pooled) within the model, and streams of distinctly different character employ different yield correlations (*e.g.*, vacuum gas oil and coker gas oil FCC feeds). Utilities usage and emissions are disaggregated at the process level, and as much as possible, estimated at the operational level (*e.g.*, for the atmospheric crude unit, steam usage is estimated for individual column product draws, instead of being specified as a fixed proportion to crude feed rate).
- In general, individual product specifications are met by refining and not blending. What this means is the model does not optimize intermediate-stream pathways through the refinery, which are only varied when considering alternative crude oils and refining configurations (*e.g.*, the amount of light straight-run naphtha blended vs. that upgraded is input; whereas, the amount of gasoil sent either to hydro or cat cracking is calculated to achieve the desired gasoline-to-diesel ratio of the product slate specified). Sufficient processing severity is specified within the model in order to infer that product specifications are met. Individual refinery configurations are based on a statistical analysis of U.S. refinery process capacity, and an elementally-balanced slate of fourteen crude oil types is used to select a blend of crudes meeting the desired characteristics for any given refinery analysis.

Input to the refinery model includes production volumes for key products, in most cases these are transportation fuels: gasoline, jet, diesel, and marine fuel oil. Predicted outputs are crude and intermediate stream flows. Refinery energy usage and emissions are estimated on a process-by-process basis. The model was developed in MS Excel®; calculations are performed in the following sequence:

1) The refinery carbon balance is solved with Solver® using carbon-based yield correlations for individual processes;

2) Based on correlations, the hydrogen contents of all process streams are calculated, and the refinery hydrogen balance is closed by purchasing hydrogen or reforming natural gas on-site;
3) Electricity, fuel, steam and cooling water requirements for individual process operations are estimated, and utility balances for the refinery are closed by calculating the amounts of supplemental fuel (typically natural gas), electricity and make-up water required. Utilities may either be purchased or generated on-site;
4) Process energy usage and losses are calculated to achieve an overall refinery energy balance, and refinery CO_2 emissions are calculated;
5) The energy flows for the base refinery simulation are combined with a database of cost and performance efficiency-measures and used to sequentially rank the measures based on their return on investment.

The above procedure ensures that the refinery model is elementally and energy balanced without the need for ad hoc adjustments. While the models and methods used are less rigorous than those employed in sophisticated process simulators, they contain a more detailed engineering description than found in standard LP refinery models.

Sample Results

To demonstrate the robustness of the method described above, refinery CO_2 emissions were estimated and allocated to the various refinery intermediate streams that are blended into gasoline and distillate fuels. Only the results relating to distillate products are discussed here. The refinery was modeled as a "notional" refinery set up to approximate a typical U.S. Gulf Coast refining configuration.

Figure 2 compares the embodied refinery CO_2 emissions and combustion emissions for the following streams that may be blended into diesel fuel: straight-run kerosene, hydrocracker kerosene, straight-run light-gasoil, hydrocracker light gasoil, cat-cracker light-gasoil produced from straight-run vacuum gasoil, cat-cracker light-gasoil from heavy coker gasoil, and coker light-gasoil. The coker-derived streams are produced from delayed coking of vacuum resid. Other than the hydrocracker product streams, all of the distillate streams have been hydrotreated to meet product sulfur specifications. Also shown are the emissions for the final diesel product blend. The first thing to note is that the combustion emissions for all the diesel blendstocks are almost identical, varying by not more than 2% from the blended product value.

The embodied emissions are reported as a percentage of the combustion emissions. These are significantly less than the combustions emissions, ranging from a low of slightly less than 2% for hydrotreated straight-run kerosene to a high of slightly greater than 20% for the hydrotreated, cat gasoil produced from heavy coker gasoil. It is clear from Figure 2 that embodied CO_2 emissions are directly related to the extent and severity of processing that a stream experiences within the refinery. In particular, refinery conversion operations such as hydrocracking, cat cracking, and coking produce more emissions than hydrotreating operations alone. The embodied emissions for the diesel blend are about 9% of combustion emissions.

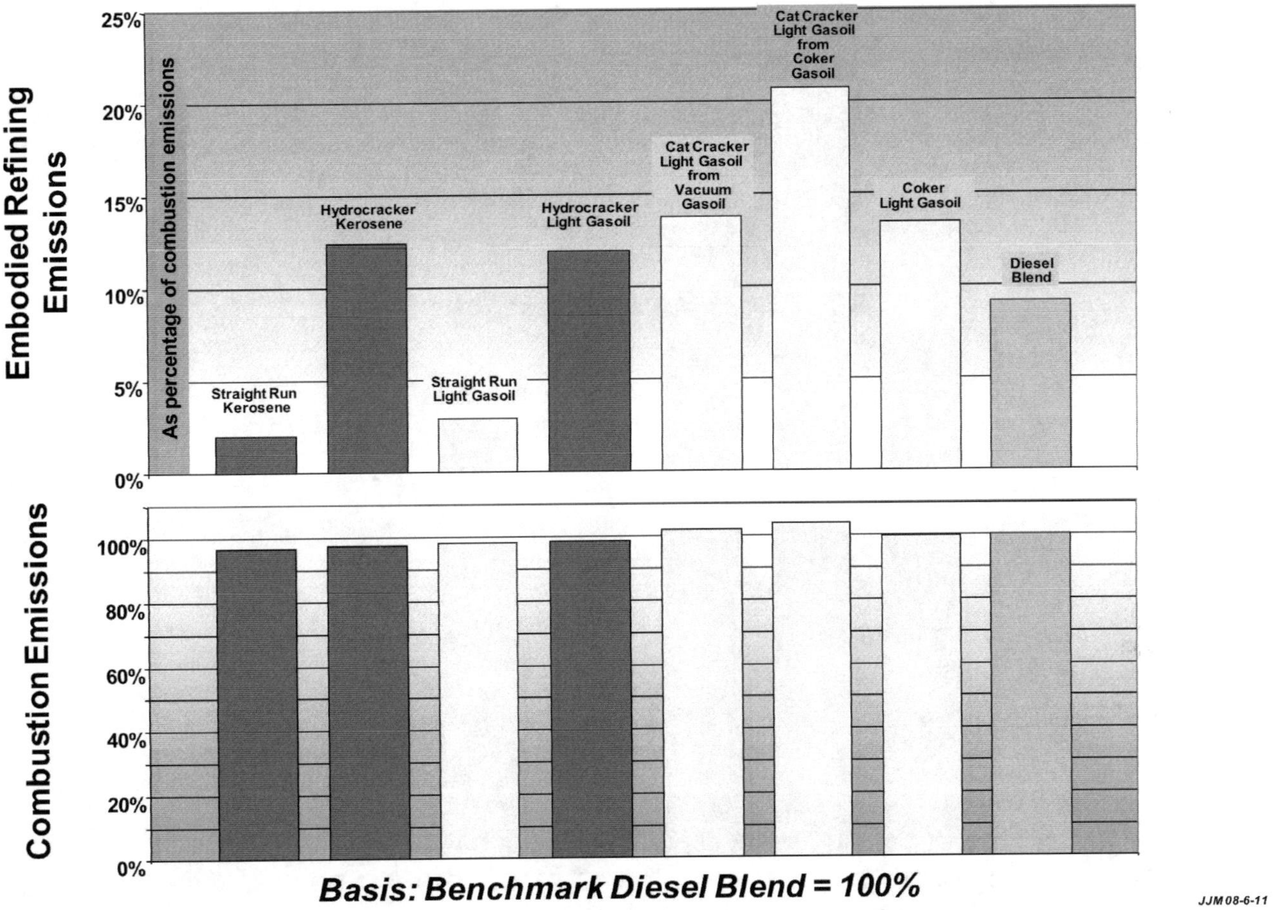

Figure 2. Relative CO_2 Emissions of Various Diesel Blendstocks [17]

Figure 3, compares CO_2 emissions for blended diesel produced from the refining of two different quality feedstocks, high-sulfur, light and high-sulfur, heavy crude oil. Embodied emissions from the heavy crude are about 20% higher than those from the light crude. This result is due more to the final distribution of distillate streams in the blend than to specific variations in the severity of refining seen by the individual blendstocks. This can be seen from the second row of pie charts in Figure 3. The heavy crude contains significantly less straight-run distillate, with low embodied emissions, than the light crude (47 vs. 56% of the diesel pool), and produces more FCC light gasoil (21 vs. 15%) due to the heavier crude's higher content of vacuum gasoil. Interestingly, the contribution of the coker to embodied emissions is relatively constant for both crude oils (45+14=59% vs. 39+21=60%).

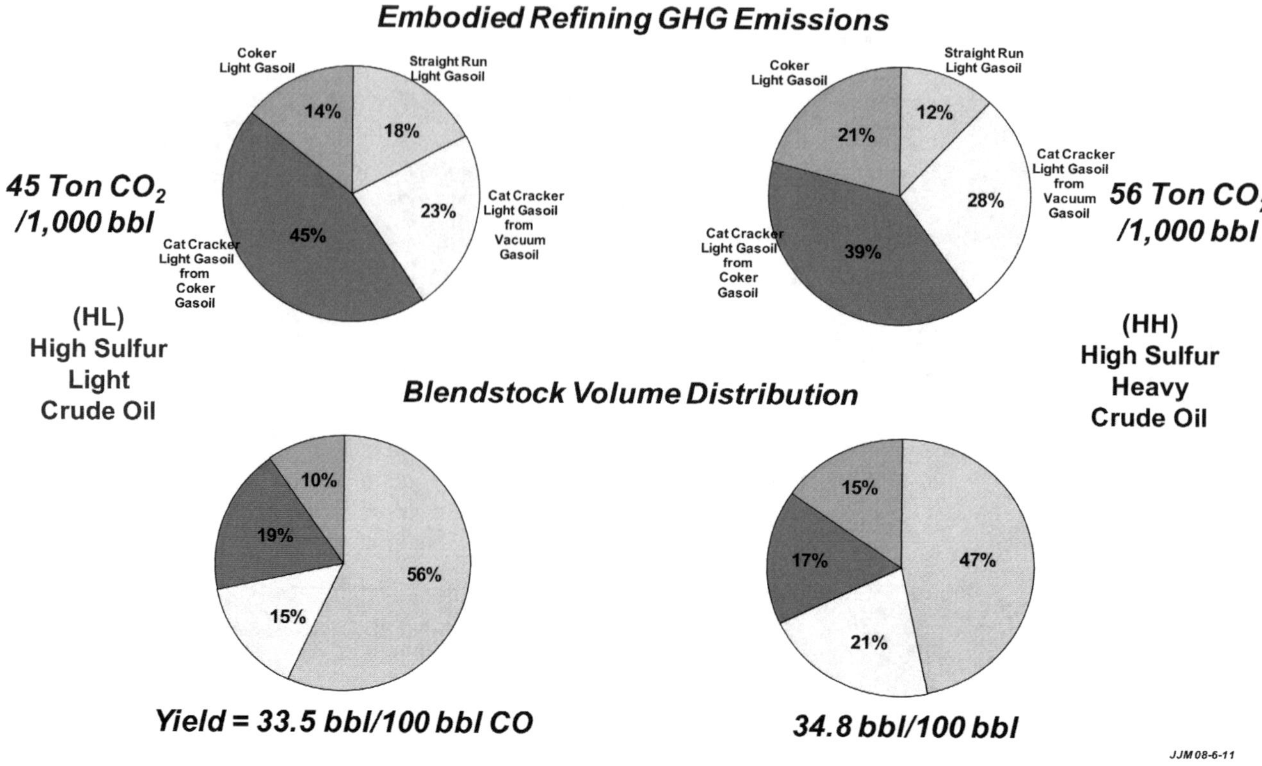

Figure 3. Comparison of CO_2 Emissions for Diesel Production from Light & Heavy Crudes [17]

Status and Next Steps

JM Energy and LBNL are continuing work to establish supply curves for refinery energy-usage and CO_2 emissions abatement. The current focus in on developing a database of energy abatement measures, efficiency and costs that can be applied to the refinery model results. In addition, a methodology for handling the variation between U.S. refineries is under development. The major challenge here, is establishing a baseline for the existing industry, since individual refineries do not publically report their energy use; though, there are now regulations in place requiring individual refineries to report their CO_2 emissions. A question still to be addressed is: How widely have any given efficiency measures been adopted by refiners? A clue to answering this question may be available from examining historical energy usage data for the industry, which clearly shows an improvement in efficiency over the last ten years. Data compiled on energy use will be used to update and tune the refinery model to match government-collected aggregate energy consumption in the industry. A final report on this analysis should be available from LBNL by late 2012.

To date, the current analysis has not considered the ramifications of current trends in petroleum refining related to technology, crude oil quality, fuel specifications, and product slate. These trends will have a significant impact on the future path of the U.S. refining industry. Future challenges that will likely affect the industry include: lower gasoline-to-distillate product ratios due to ethanol blending into gasoline, vehicle hybridization, further dieselization of the U.S., and increased air travel; internationally agreed to marine SOx reductions requiring low-sulfur bunker fuels; refinery crude slate changes due to increased production of domestic shale oils, and increased imports of Canadian synthetic crude oils and dilbit blends; and further implementation of renewable and/or low-carbon fuel standards, which may result in the long term to the introduction of truly drop-in biofuels. Future sensitivity analyses will be needed to examine impacts of these potential changes, since many of these could have negative ramifications for improving efficiency and lowering emissions, while some may be positive. The role of CO_2 capture and sequestration in petroleum refining will also need to be examined more completely if CO_2 emissions are to be drastically reduced over the next fifty years.

Acknowledgements
The authors would like to express their appreciation to the U.S. Environmental Protection Agency for providing financial support for this project.

Any conclusions, comments or opinions expressed in this paper are solely those of the authors and do not represent any official positions held by LBNL, EPA, or the U.S. Government. Information contained herein has been based on the best data available to the authors at the time of preparation.

Copies of the slides used for this *2012 AIChE Spring Annual Meeting* presentation may be obtained by contacting John Marano at *marano@zoominternet.net*

References

1. Stockle, M.; Bullen, T., "Integrate Refinery Carbon Dioxide Plans Plant-Wide," *Hydrocarbon Processing*, pp.106-111, Nov. 2008.
2. Spoor, R.M., "Low-Carbon Refinery: Dream or Reality," *Hydrocarbon Processing*, pp.113-117, Nov. 2008.
3. Gunaseelan, P.; Bueler, C.; Chan, W.R., "Greenhouse Gas Emissions: Characterization and Management," *Hydrocarbon Processing*, pp.57-70, Sept. 2009.
4. Clarke, S.C., "CO_2 Management – A Refiners Perspective," Foster Wheeler Energy, Ltd., available at http://www.fwc.com/publications/tech_papers/oil_gas/CO2.pdf
5. Worrell, E.; and Galitsky, C., *Energy Efficiency Improvement and Cost Saving Opportunities for Petroleum Refineries - An ENERGY STAR® Guide for Energy and Plant Managers*, Ernest Orlando Lawrence Berkeley National Laboratory, Report No. LBNL-56183, Feb. 2005.
6. White, D.C., "Trading Silicon for Carbon: How to Reduce Energy Usage through Automation," *Hydrocarbon Processing*, pp.61-70, June 2010.
7. Beaubien, R., "Efficiency Improvements Cut GHG Emissions, Help Profits," *Oil & Gas Journal*, pp.20-26, Sept. 28, 2009.
8. Tice, M., "Carbon and Energy Management: Innovative Approaches for a Responsible Future," *National Petrochemical & Refiners Association Annual Meeting*, San Diego, CA, Paper No. AM-08-37, March 9-11, 2008.
9. *Energy and Environmental Profile of the U.S. Petroleum Refining Industry*, prepared by Energetics Inc. for the U.S. DOE Office of Energy Efficiency and Renewable Energy, Industrial Technologies Program, Nov. 2007.

10. *Energy Bandwidth for Petroleum Refining Processes*, prepared by Energetics Inc. for the U.S. DOE Office of Energy Efficiency and Renewable Energy, Industrial Technologies Program, Oct. 2006.
11. Glasgow, I.M.; Polcar, S.; Davis, E.; Nguyen, T.; Price, J.; Stuecheli, C; Palmer, R.E., "How will Carbon Emissions Regulations Revise Energy Conservation Economics?," *Hydrocarbon Processing*, pp.35-41, June 2010.
12. Rossiter, A.P., "Improve Energy Efficiency via Heat Integration," *Chemical Engineering Progress*, pp.33-42, Dec. 2010.
13. Carbonetto, B.; Pecchi, P., "Going 'Green' with FCC Expander Technology," *Chemical Engineering Progress*, pp.79-83, Jan. 2011.
14. Marano, J.J.; Rogers, S.; Spath, P.L.; Mann, M.K., "Life-Cycle Assessment of Biomass-Derived Refinery Feedstocks for Reducing CO_2 Emissions," *Proceedings of the Third Biomass Conference of the Americas*, Montreal, Quebec, August 25-28, 1997, pp.325-337.
15. Hirshfeld, D.; Kolb, J., in *Comparison of North American and Imported Crude Oil Lifecycle GHG Emissions, Final Report*, prepared by Tiax LLC and MathPro Inc. for the Alberta Energy Research Institute, Tiax Report No. D5595, July 6, 2009.
16. Bredeson, L., "Quick Overview of Complexity Issues in Refinery Modeling," *Energy Information Administration Technical Workshop*, Sept. 30, 2009, available at http://www.eia.gov/oiaf/emdworkshop/pdf/lbredeson_lfmm.pdf
17. Marano, J.J., "GHG Profiles for Refining & Upgrading Fossil Distillate Fuels," *9th International Conference on Greenhouse Gas Control Technologies*, Washington, DC, November 16-20, 2008.

PROFITABILITY THROUGH LEAN SIX SIGMA
Save Energy - Save Environment

Author
Bhanumurthy, Maddala V,
Tel No.: 00966-43978540 /00966-506285745
FaxNo.:00966-43978118
BHANUMMV@ARAMCO.COM
MADDALA.BHANUMURTHY@ARAMCO.COM
Saudi Aramco/Yanbu Refinery,Yanbu
Co-Author
Mohammad A. Al-Malki
Tel No.: 00966-43978101

Biography:

Bhanumurthy, Maddala Joined Saudi Aramco as a Senior Process Engineer in 2007 and is currently working as Yanbu Refinery Energy Coordinator. He is a Chemical Engineering Graduate with MBA from Andhra University. He had more than 25 years refinery experience in reputed organizations like Hindustan Petroleum Corporation Limited, Qatar Petroleum. Before joining SA, he worked as a Deputy General Manager/FCC Head with Essar Oil Limited, India.

Mohammad A. Al-Malki is a Chemical Engineer graduated from King Fahad University of Petroleum and Mineral and joined Saudi Aramco in 2011 and currently working as Utilities Area Engineer.

Abstract Summary:

Energy performance is measured as Energy Intensity Index (EII) and it is ratio of Actual Energy/Standard Energy Consumption. Lower EII indicates lower energy consumption per unit of output and better performance. YR EII is 78.9 while World's best refinery EII is 73.5 One unit change in EII will result in a cost avoidance of 0.77 MM$. We can avoid up to $ 4.08 MM expenditure. Actual Energy Consumption is comprised of Fuel Gas (FG), Fuel Oil, and Electrical Consumption. Reduce energy consumption to improve EII. Standard energy consumption is calculated from unit throughputs. Reduction in throughputs increases EII.

LSS project is to identify and reduce defects in energy system causing high indices which will enable to meet best performing refineries EII. LSS drives speed, quality, and cost simultaneously. Lean focuses on the elimination of waste while Six Sigma focuses on reducing process variation and then on improving the process capability through the methodology DMAIC (Define-Measure-Analyze-Improve-Control).

Project started with team selection, Project scope finalization and identification of CTQ (critical to quality) factors. Brainstorming conducted for identified defects causing process deviations and reducing process capability. For each identified defect, Fishbone created, data collected, Failure Mode Effect Analysis (FMEA), detailed process maps and Pareto charts developed. Data analyzed using, hypothesis tests, and regression analysis to draw conclusions and developed solutions for short and long term .

Short-term measures like steam optimization, Boiler load management, Flue gas oxygen optimization were implemented. The LSS project implementation improved EII to 77.3 by end of 2011, while it was 78.1 by end of 2010 from 78.9 in year 2010 beginning with a cost avoidance of $1.4MM/year.

Yanbu Refinery, SA

Saudi Aramco's Yanbu Refinery Department (YRD) is a 235 MBD hydro skimming refinery located in western province of Saudi Arabia. YRD is committed to reduce energy costs as part of Saudi Aramco Energy conservation policy and reduce environmental emissions. Yanbu refinery (YR) adopted Lean Six Sigma (LSS) approach with Management and Organizational Consulting department support in its aim to reduce energy costs, improve energy key performance indicator (KPI) and reduced emissions.

Energy Intensity Index (EII)

Energy Intensity Index (EII) is ratio of Actual Energy/Standard Energy Consumption. Lower EII indicates lower energy consumption per unit of output. YRD uses Solomon methodology to calculate EII. Actual Energy Consumption is comprised of Fuel Gas (FG), Fuel Oil and Electrical Consumption. Energy consumption is to be reduced to improve EII. Standard energy consumption is calculated from formulas using unit throughputs. Any Reduction in throughputs will increase EII.

LSS Approach

LSS project is an ideal way to improve profitability by energy optimization, reduce EII and meet best performing refineries KPI. LSS drives speed, quality, and cost simultaneously. Lean focuses on the elimination of waste while Six Sigma focuses on reducing process variation and then on improving the process capability through the methodology DMAIC (Define-Measure-Analyze-Improve-Control) as shown in Fig1. LSS compliments to engineering solution where a problem is converted into a mathematical problem to obtain mathematical solution which will be applied to resolve the original problem.

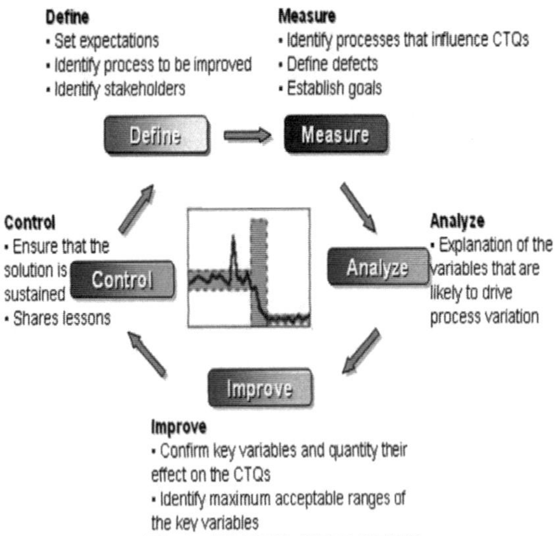

Fig1. DMAIC PROCESS

Features that set Six Sigma apart from other initiatives include a clear focus on achieving measurable and quantifiable financial returns, commitment to making decisions based on verifiable data, rather than assumptions and guesswork and an increased emphasis on strong management leadership and support. A special infrastructure of "Champion," "Master Black Belt and Green Belts", etc. will lead and implement the Six Sigma approach.

LSS Project

LSS project is to identify and reduce defects in energy system causing high indices which will enable to meet best performing refineries EII. Project started with selection of members from Process, Electrical, Mechanical Engineers, and Operations Foremen and supported by master black belt from Organization Consulting Department. Project scope clarified with problem and goal statements. Refinery EII is 78.9 while World's best refinery EII is 73.5 based on 2008 Solomon fuels study. One unit change in EII will result in a cost avoidance of 0.77 MM$. We can avoid up to $ 4.08 MM expenditure. To reduce EII by one unit, reduce total Fuel consumption by 97 BPD or Power by 65MW/Day.

Process boundaries identified using SIPOC (Supplier, Input, Process, Output, and Customer). Team identified CTQ (critical to quality) parameters as shown in figure and defects causing process deviations to meet customer requirements. For fuel gas system critical parameters impacting fuel gas composition include boilers load, boiler/heaters efficiencies, feed to heater inlet and outlet temperatures, stack temperatures, oxygen in stack flue gases etc, as shown in Fig2.

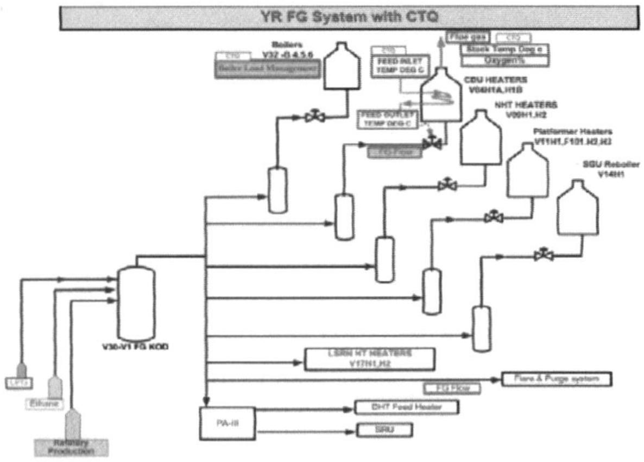

Fig2. Fuel Gas system CTQ

Pareto principle uses 80: 20 rule. This helps to identify major consumers/ areas (80%) to focus and minimize the defects in those areas for quick results. Pareto chart will be developed and used for all factors influencing Energy Consumption separately.

Pareto chart for factors influencing Standard Energy Consumption was prepared as shown in Fig 3. Refinery need to maintain high availability to maximize capacity utilization of the key units CDU, Diesel Hydrotreater and continuous catalytic regenerating reforming units. It also aids in maximizing refinery input by allowing blending operations.

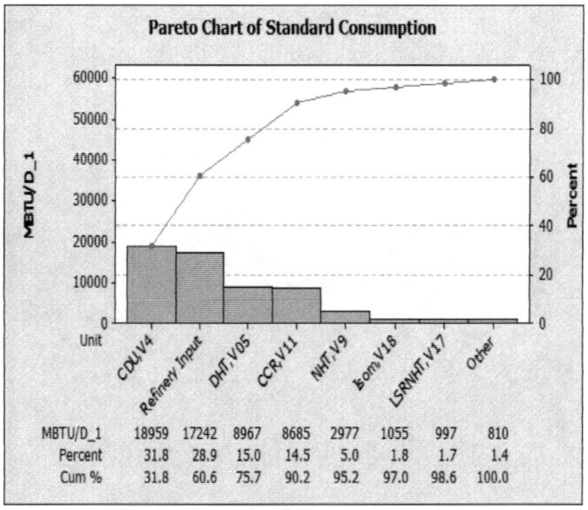

Fig3. Standard Energy Pareto

Data collected and created Pareto chart for factors influencing Actual Energy Consumption as shown in Fig4.

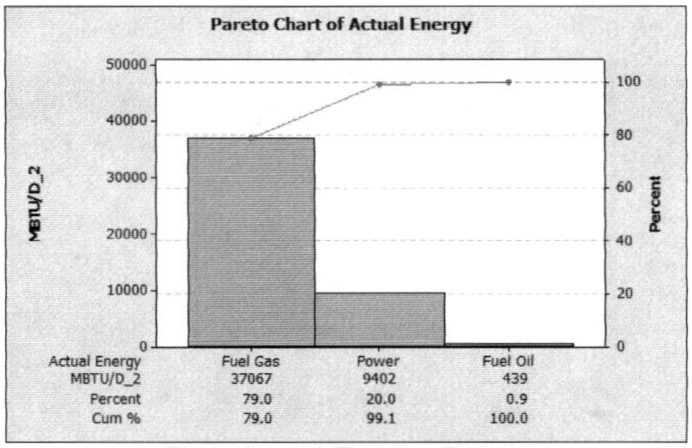

Fig4. Actual Energy Pareto

As can be seen from Pareto, refinery uses fuel oil (FO) only during emergencies as part of refinery environmental commitment and hence FO system is out of scope. Power consumption variation is mainly attributed to changes in ambient conditions as shown in fig 5 and will be excluded from further study as there is little control over ambient conditions.

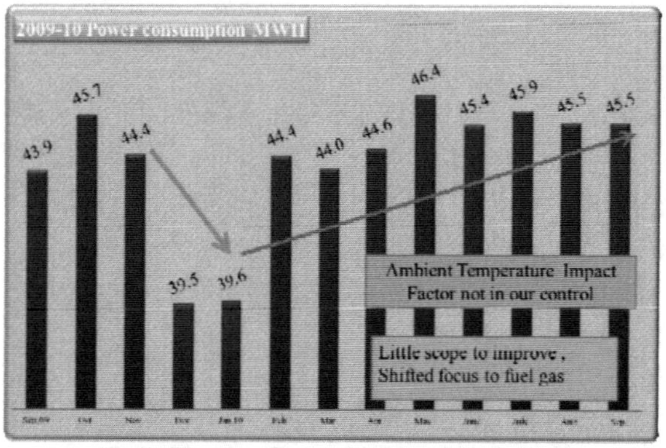

Fig5. Power Consumption Variation

Pareto chart for fuel gas (FG) Consumption prepared as shown in Fig6. CCR, Crude units and boilers consume 80% of fuel gas and will be focus areas for further study

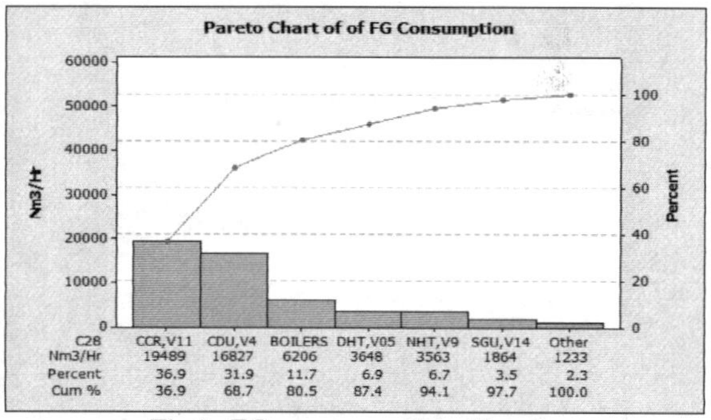

Fig6. FG consumption Pareto

Brainstorming was done for identifying defects for each actual energy factor and are represented in Fishbone chart as shown in Fig 7.

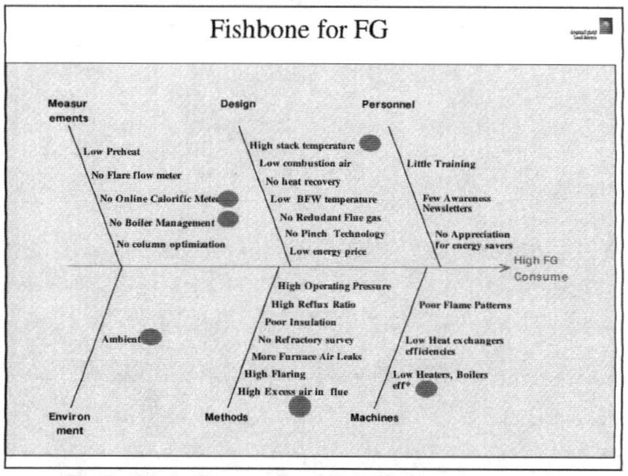

Fig7. Fish Bone for high FG consumption

Failure Mode Effect Analysis (FMEA) used for ranking the above factors based on severity, occurrence and determination. Detailed and specific process maps developed and analyzed for value and non-value added steps to identify opportunities.

Accordingly major fuel gas consuming heaters/boilers stack gases analyzed for oxygen. Prior to the project, heaters discharging flue gases to atmosphere with oxygen above 4% were 70% and were optimized to less than 40 % initially and finally to 10%. Accordingly the major fuel gas consumers, crude and platformer Heaters, oxygen in stack gases optimized. By optimizing these heaters a saving of $ 0.7MM/Year is realized.

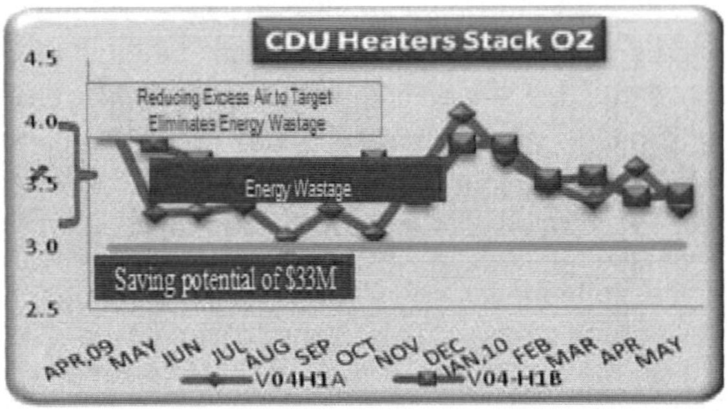

Fig8. Crude Heater flue gas oxygen

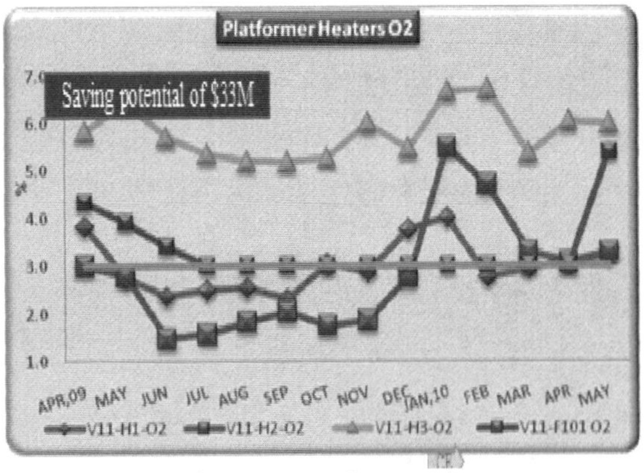

Fig9. CCR heaters flue gas oxygen

Flue gases generated after combustion are vented through stack and flue gas temperatures indicate amount of energy lost through stack gases. Flue gases exit temperatures can be reduced to 188-200 deg C, considering sox dew point impact, before venting through stack to minimize energy loss. Lean process recommends identifying and eliminating any wastage of energy in the refinery. Accordingly, Crude/CCR Heaters/Boilers, the three major FG consumers in the refinery stack flue gas temperatures data analyzed. Of the three major FG consumers, boilers and platformer Heaters, have waste heat recovery equipment on flue gases before being vented to atmosphere while Crude heaters consuming 31% of FG have no waste heat recovery equipment and discharge hot flue gases to atmosphere at above 480 deg C. Techno economical option will be evaluated and implemented to minimize energy loss through crude heaters.

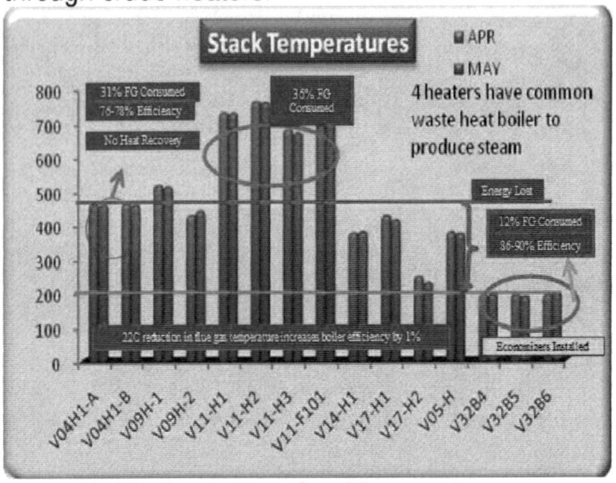

Fig10. Stack Flue Gas Temperatures

Refinery has three boilers with two boilers online always while the third one remains on hot standby condition to generate 150 psig(10.5 Kg/cm2) steam. Boiler loads analyzed as one of the major FG consumer utilizing hypothesis tests, and regression analysis to draw conclusions. It can be noted from the Fig 11 Boiler 5 operated at higher load than the other boilers while it had the lowest efficiency as compared to other two boilers.

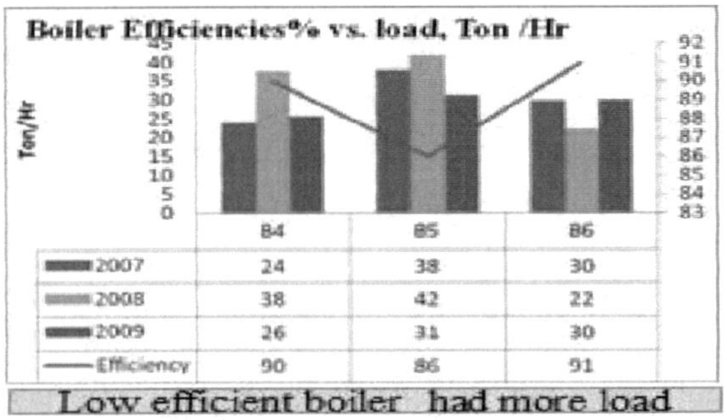

Fig11. Boilers Load & Efficiency

Six sigma tools were used to analyze data including boiler loads and fuel gas flow. In the first phase, fuel gas vs steam flow data analyzed using correlation and regression analysis. Strong positive Correlation is shown between fuel gas vs steam flow indicating high accuracy of the data as shown below.

Boiler FG versus steam flow Correlations:

Pearson correlation = 0.873

Then regression analysis performed between FG and steam flow for each boiler.

Regression Analysis:

Boiler 6 FG = 115 + 0.0605 steam flow

Boiler 5 FG = 696 + 0.0497 steam flow

From the regression analysis, boiler FG consumption fig12 was developed. It was obvious from the chart that Boiler#5 is consuming higher FG than Boiler#6 at normal operating loads.

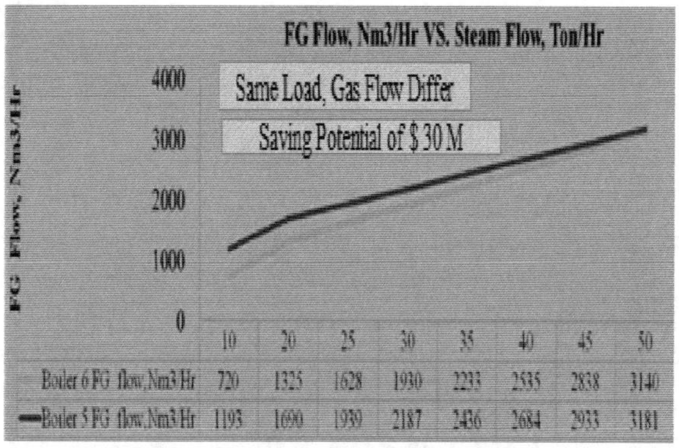

Fig12. Boiler FG consumption

It was recommended to operate high efficient Boiler#6&4 at higher loads than Boiler#5 to optimize boilers FG consumption. The regression analysis also allowed detecting meter errors and rectifying them early.

It was observed that boilers load is on increasing trend as shown in Fig13.

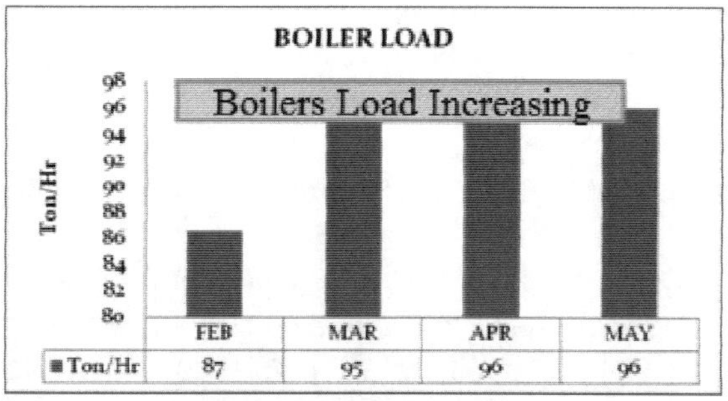

Fig13. Boilers load

Pareto chart developed for steam consumers in the refinery as shown below in Fig14.

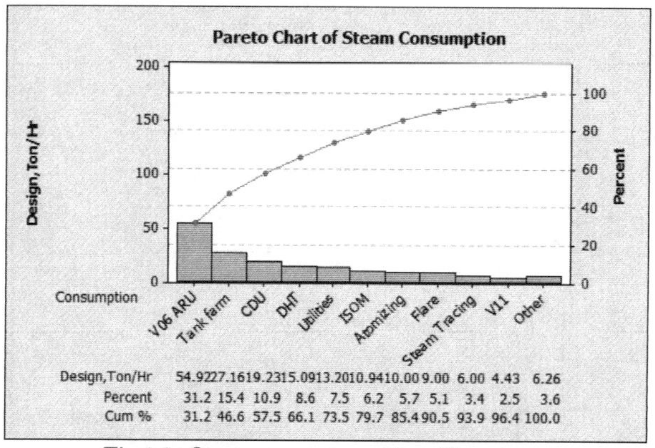

Fig14. Steam consumption Pareto

Amine regeneration unit (ARU) is the major consumer as can be seen From Fig 14. Accordingly ARU operations were optimized with the help of Area Engineers and Operations support as shown in Fig 15.

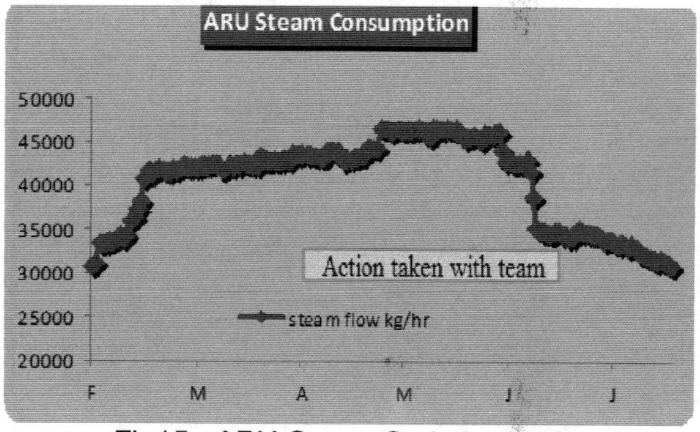

Fig15. ARU Steam Optimization

The Pareto chart can be also used to trouble shooting and restoring operations quickly during any partial steam failures/load shedding.

Various Process maps were developed during the project. Condensate system process map as shown in Fig16 also identified that refinery had excess 50 psig Low pressure steam (3.5 Kg/cm2) and air coolers are used as required to condense the excess steam to avoid venting. LSS project recommended shifting some steam users from medium pressure to low pressure steam. This will help to reduce steam generation from boilers using fuel gas firing and power consumption towards fin fans used to condense the steam.

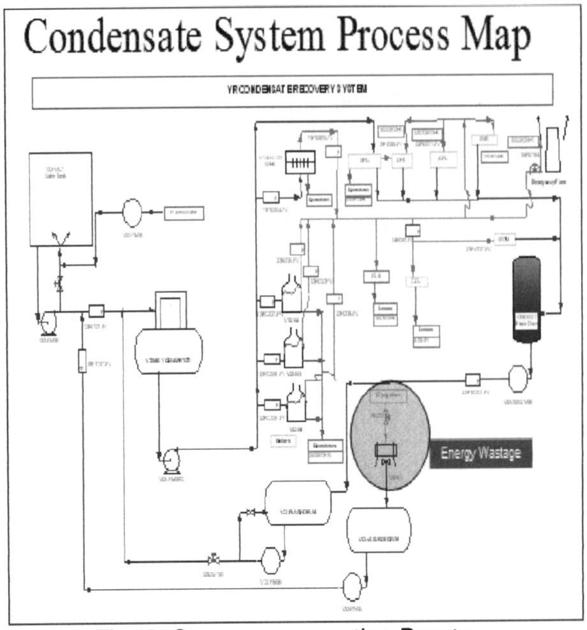

Fig16. Steam consumption Pareto

Save Energy –Save Environment;

Reduction of energy usage also reduces Harmful emissions of Sox,Nox,Co2 as shown in Fig 17.

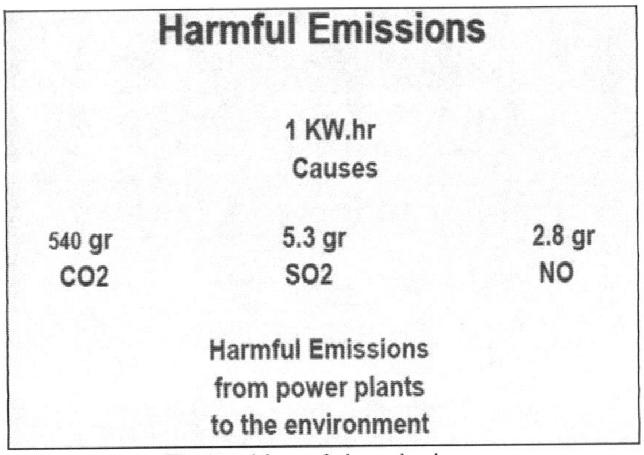

Fig17. Harmful emissions

Recommendations:

Project identified several short and long term recommendations to optimize energy consumption and reduce emissions. Short-term measures implemented include steam optimization, Boiler load management, Flue gas oxygen optimization, inaccurate flow meters identification and correction. Long-term measures include shifting users from Medium to low pressure steam and waste heat recovery from flues gases. Summary of key recommendations is shown in Table 1.

#	Root Cause	Data Output analysis	Recommendation
1	No Boiler management	Correlation proved	Increase load on high efficient boiler and reduce load on low efficient boiler
2	High steam consumption	Pareto chart helped in identifying consumer	Monitor and optimize steam consumption
3	Low condensate recovery	Lean	Enhance condensate recovery
4	High number of defective traps	Lean	Reduce defective traps percentage by forming special team
5	High oxygen in Heater Flue gases (At Upper limit of range 3-5%)	Relation	1. Reduce oxygen in flue gas to Minimum Target levels 2. Redundant analyzer like CO analyzer 3. Auto draft control
6	High Heater Flue gas temperature	Lean	Install heat recovery equipment for CDU heaters
7	YRD is long on LP steam and using fin-fans to condense LP steam.	Lean	Switch consumers from medium to low pressure steam
8	Fuel gas quality variation	Correlations affected	Install Online analyzer to measure and monitor fuel gas heating value

Table 1.LSS Project recommendations

Project Summary

Lean six sigma compliments to engineering solution with speed, quality, and cost simultaneously. Identify defects in major focua areas will yield quick results. Based on project recommendations, Refinery implemented several short-term measures like steam optimization, Boiler load management, Flue gases oxygen optimization. Long-term measures recommended include shifting users from Medium to low pressure steam and waste heat recovery from flues gases. Equipment efficiency, plant reliability play critical role and require continuous monitoring and enhancement. The LSS project implementation improved EII to 77.3 by end of 2011, while it was 78.1 by end of 2010 from 78.9 in year beginning with a cost avoidance of \$1.4MM/year. Energy conservation reduces costs and emissions.

Gratitude

Authors wishes to record their sincere regards for the YRD management and OCD of Saudi Aramco for extending their support for successful completion and implementation of project and presenting the paper. Authors also wishes to record sincere appreciation to the 2012 AIChE Spring meeting organizers.

AIChE Paper Number 247037

REDUCING CO_2 EMISSIONS FROM REFINERIES THROUGH AMINE CHEMISTRY UPGRADE

Michelle Valenzuela

Technical Service and Development Manager

Dow Oil & Gas

Ana Eisenhauer

Refinery Strategic Marketing Manager

Dow Oil & Gas

Linda Pirtle

Senior Technical Services and Development Specialist

Dow Oil & Gas

Prepared for Presentation at the 2012 Spring National Meeting

April 3, 2012

Houston, TX

REDUCING CO_2 EMISSIONS FROM REFINERIES THROUGH AMINE CHEMISTRY UPGRADE

Michelle Valenzuela

Technical Service and Development Manager

Dow Oil & Gas

Ana Eisenhauer

Refinery Strategic Marketing Manager

Dow Oil & Gas

Linda Pirtle

Senior Technical Services and Development Specialist

Dow Oil & Gas

Regional initiatives are underway to reduce and limit the emissions of greenhouse gases. With CO_2 emissions making up over 80% of the greenhouse gases, cap-and-trade programs will focus on those industries that consume the most energy. Refineries are among the top energy consumers and are seeking opportunities to reduce usage. With tightening margins, energy management programs will not only help refineries meet CO_2 emission regulations, but can also provide a competitive advantage.

With the trend towards heavier and higher-sulfur containing crudes, refineries are increasing processing capabilities, which can include capital-intensive projects and additional energy consumption. Energy conservation plans should include optimization of these processes. One area to consider includes the acid gas removal systems in refineries.

Through the selection and use of optimal solvents and implementation of energy efficiency techniques, which require minimal capital investment and expenditures, refineries can reduce energy usage, overall CO_2 emissions, and total cost in acid gas systems. This paper will discuss these approaches and share case studies detailing the implementation and results.

Introduction

Despite the lack of a global greenhouse gas (GHG) emission reduction agreement, regional schemes are being developed to limit CO_2 and other GHG emissions to the atmosphere. In the United States, the state of California is moving forward with regulations that will cap emissions from installations producing over 25,000 tons CO_2e/year. Equipment such as boilers and fired-heaters, commonly used in refineries for energy generation, are included in the law and will be obligated to reduce CO_2 emissions.

To reduce CO_2 emissions, refineries often rely on energy efficiency improvements. Three types of energy efficiency projects include:

1) Equipment Efficiency — The replacement of aging assets with high-efficiency technology

2) Operational Efficiency — The use of energy management and operating strategies to optimize energy-intensive assets

3) "Chemical," or Solvent Efficiency — The use of improved chemicals to reduce overall energy requirements of equipment

In reducing total energy consumption, it is also important to consider regulations and overall plant emissions. The trend towards higher sulfur-containing crude products affects energy consumption of refineries. To meet stringent regulations in the global market, refineries must increase processing capabilities to accommodate a heavier, sour crude slate. This includes upgrading hydrotreating capacity to address the additional sulfur load and adding acid gas removal systems to treat the resulting H_2S. The result is overall increased energy demand and CO_2 emissions. The question then becomes, "Which is more important: addressing energy consumption and CO_2 emissions, or environmental regulations and overall plant emissions?" The answer is "both."

Specific to acid gas removal systems, it is possible to reduce overall energy requirements while maintaining environmental compliance with the use of chemical and operational efficiency. Both require minimal capital investment, which is important for an industry already struggling with tightening margins.

Chemical efficiency includes upgrading the chemistry of the solvent to increase capacity and reduce required processing energy (i.e. reboiler duties). Often refineries design and operate acid gas removal systems with the sole objective of meeting permitted sulfur emission limits. Monoethanolamine (MEA) and diethanolamine (DEA) are amine solvents commonly used for removal of H_2S from sour gas streams. However, these solvents degrade to form corrosive products. As a result, refineries maintain low solvent concentrations to reduce degradation products and lessen the corrosion potential. Less amine in the system requires increased circulation rates and more aggressive stripping of the lean amine to meet stringent H_2S and SO_2 regulations. These practices result in excessive energy consumption and, subsequently, excessive CO_2 emissions.

Operational efficiency requires an understanding of the system performance. With the use of mass and energy balances, a plan is developed to optimize appropriate operating conditions to reduce energy consumption while maintaining environmental performance.

This paper will discuss these two approaches in more detail and will share field experiences and learnings from actual implementation of these practices.

Environmental Drivers for Energy Efficiency

Greenhouse gases, which include carbon dioxide, methane, and nitrous oxide, contribute to the global warming effect by trapping radiation on the earth's surface. Global and regional programs have been initiated to reduce the emission of GHG.

The United Nations Framework Convention on Climate Change (UNFCCC) is an international treaty established in 1992 with the intent of stabilizing GHG emissions as a global response to climate change. To further enforce efforts towards emission reduction, the UNFCCC adopted the Kyoto Protocol in 2007, legally binding developed countries to GHG reduction targets through 2012.

The United States signed the Kyoto Protocol but never ratified it. Subsequent legislation to enact a cap-and-trade program to control GHG in the U.S. also failed in 2009. In the absence of a market-based program, the United States Environmental Protection Agency (EPA) issued the Mandatory Reporting of Greenhouse Gases Rule (40 CFR Part 98, Greenhouse Gas Reporting program), requiring facilities that emit 25,000 metric tons or more per year of GHGs to submit annual reports to the EPA. The purpose of the rule is to collect accurate and timely GHG data to inform future policy decisions.

The EPA is currently working on rules to limit the emissions of GHG from large sources under the Clean Air Act but has yet to implement or enforce any regulations. Although a federal mandate has not been set for GHG emission control, several regional initiatives have been established. In 2007, a collaboration of seven U.S. states and four Canadian provinces formed the Western Climate initiative (WCI) to identify, evaluate, and implement policies to address climate change.

Through WCI, states and provinces can implement regional cap-and-trade programs with the intent of integrating the programs into national or federal programs as they are developed. Of the WCI partners, California is the only state that has enacted regulation to implement a GHG emissions cap-and-trade program.

On October 2011, the California Air Resource Board adopted the final cap-and-trade Resolution 11-32. The regulation covers 360 businesses representing 600 facilities and is divided into two phases: the first, beginning in 2013, includes all major industrial sources along with electricity utilities; the second, starting in 2015, brings in distributors of transportation fuels, natural gas and other fuels producing over 25,000 tons CO_2e/year. CO_2 emissions are calculated based on the consumption of fossil fuel by a stationary combustion facility (boiler, generator, fired heater, etc). California refineries will be required to comply with this resolution starting in January 2013. To reduce CO_2 emissions, refineries will need to focus on energy efficiency opportunities.

Opportunities for Energy Management

Although the trend towards energy efficiency seems relatively recent with the latest global and domestic implementation of GHG initiatives, refineries have long been accustomed to energy management and optimization programs as a means of staying competitive. As energy prices increased in the 1970s-1980s, refineries were forced to evaluate plant energy demand to find opportunities to reduce fuel, natural gas, steam, and electricity usage. Since the initiation of energy management in the 1970s, refineries have reduced energy usage by 30%. However, the increasing use of sour crudes, in combination with stricter control and accountability for overall emissions, has led the refining industry to revisit energy goals and strategies.

Refineries must also consider ultra-low-sulfur diesel (ULSD) requirements and environmental regulations when considering energy management plans and goals. Additional hydrotreating capability to remove more sulfur, especially from sour crudes, will result in more sulfur-containing waste or "acid" gases. Further processing of the acid gases is required before final disposition: emission, incineration, combustion, etc. Typically, refineries use amine-treating systems to process waste gases in refineries.

As with the other processing systems in the refinery, amine systems can require a significant amount of energy depending on asset age and condition, acid gas loads, solvent used, and operating parameters. This paper will include energy-reduction techniques for amine systems that use products, technologies, and services in the AMINE MANAGEMENT^{SM} Program from Dow.

Refinery Amine Systems

Refinery amine systems are required to effectively remove acid gases from a variety of gas and/or liquid streams to targeted levels. Most U.S. refineries use amine treating systems (Figure 1) consisting of the following equipment:

- Absorber: Used to "scrub" or remove acid gases from the inlet stream with an amine solvent

- Rich Amine Flash Tank: Used to separate co-absorbed components (i.e. hydrocarbons) from the rich amine
- Lean/Rich Cross Exchanger: Used to heat the rich amine solvent by exchange with the hot, lean amine from the regenerator
- Regenerator (including reboiler): Used to steam strip the amine to remove acid gases before recycling to absorber as lean amine
- Condenser: Used to cool the acid gas stripped from the rich amine; condensed water returns to top of regenerator
- Amine Cooler: Used to cool the lean amine before entering the absorber

Figure 1: Typical Refinery Amine System

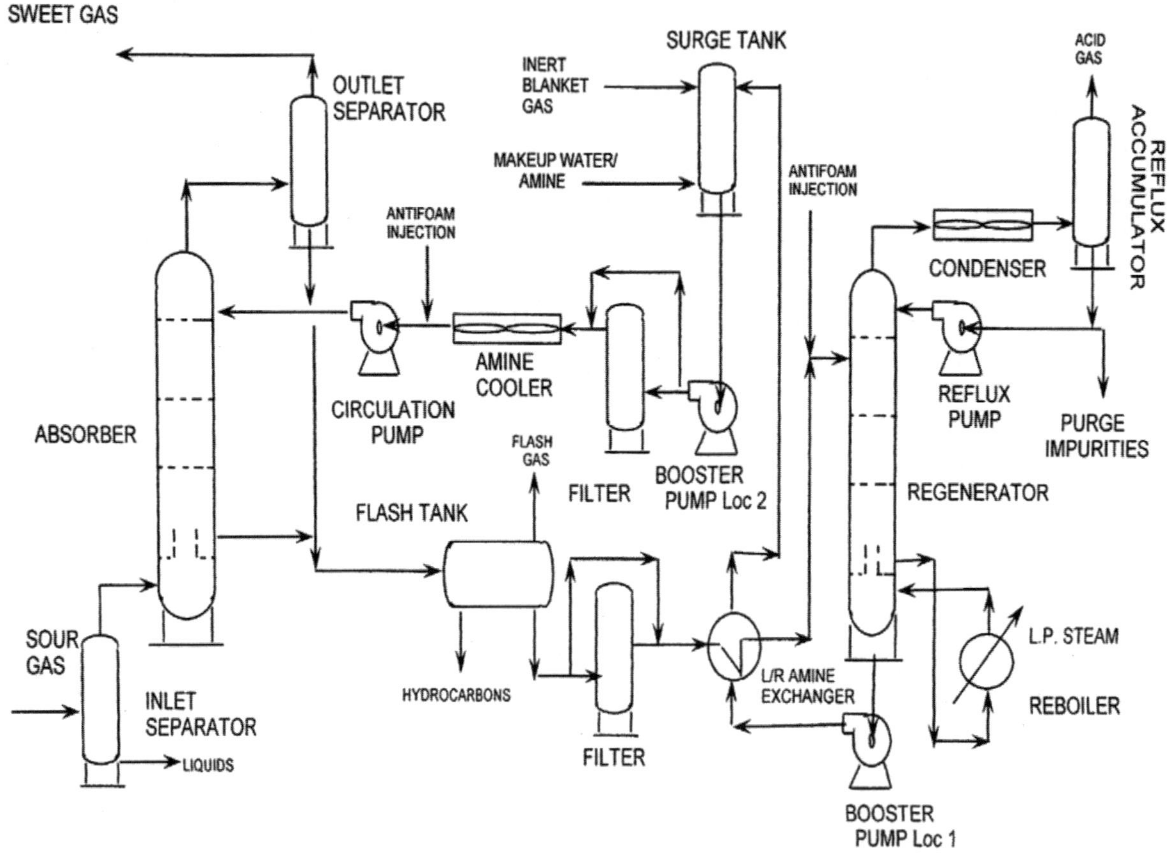

Solvent Selection

The development of amine solvents for acid gas treating has evolved from traditional generic amines, monoethanolamine (MEA), diethanolamine (DEA), and diglycolamine (DGA), to more selective gas treating amines that include diisopropanolamine (DIPA) and methyldiethanolamine (MDEA). The increased selectivity of the amine allows for preferential H_2S removal and CO_2 "slipping" into the treated gas stream, which is especially important for improved sulfur plant operations. MDEA, a tertiary amine, has become the most common selective solvent due to its resistance to degradation, low tendency for corrosion, higher acid gas loading capacity, and lower energy consumption.

The latest advances in amine technology have included the introduction of formulated specialty solvents developed to overcome the weaknesses inherent with generic amines. Proprietary additives are combined with the amine offering the greatest potential for system optimization. Initially, formulated solvents were based on MDEA to provide increased H_2S selectivity and energy savings. However, additional formulations and products have been developed to provide enhanced performance benefits in specific refinery applications. Performance solvents, for example, are now available to achieve the following objectives:

- Treat acid gas to meet <10ppm H_2S specifications in low pressure (i.e. tail gas units) applications while minimizing steam usage
- Reduce solvent losses in LPG service
- Reduce corrosion rates while increasing capacity
- Reduce energy consumption and, in turn, CO_2 emissions from utility generation

Table 1 below compares the overall performance of the gas treating solvents.

Table 1: Typical Amine Solvents Used in Refinery Acid Gas Treating Systems

Amine	Acid Gas Selectivity	Increasing Energy Savings
Formulated Solvent	Very high H_2S Removal	↑
MDEA	High H_2S Removal	
DIPA	Some H_2S Removal	
DGA	Total Acid Gas Removal	
DEA	Total Acid Gas Removal	
MEA	Total Acid Gas Removal	

The energy savings realized with formulated solvents stem from a reduction in sensible heat, a lower heat of reaction, and a reduction in necessary stripping steam. Tertiary amine-based solvents can operate at higher concentrations (up to 50wt% solution) without increased concern for corrosion. Therefore, circulation rates can be decreased, and the sensible heat requirement in the regenerator reboiler is reduced. The heat of reaction for the formulated solvents and tertiary amine is lower than primary and secondary amines. This reduces the amount of energy required to regenerate the solvent. Furthermore, formulated solvents can tolerate higher acid gas loading, thus reducing the stripping steam requirement and demand on the reboiler.

Identifying the optimal solvent is critical when reducing energy consumption. Many factors should be considered to ensure compliance with environmental requirements as well. Nevertheless, even with the best solvent selection, operation of the amine unit is equally important. To maximize the performance potential of the amine solvent, the operation of the amine unit should be optimized.

Amine System Performance and Management

A proactive approach to managing amine systems is required to provide both economic and performance benefits. The AMINE MANAGEMENT^{SM} Program from Dow is a comprehensive service program developed to provide sustainable results that reduce overall energy costs for amine treating systems while also meeting the rapidly changing demands of refinery unit operations.

As part of the program, an initial survey of the refinery amine system(s) is performed to gather key operating data (e.g. flows, temperatures, etc.), as well as equipment data (e.g. absorber and regenerator design data, reboiler duties, etc.). Amine samples are also collected and analyzed. The information collected is used to develop a fundamental understanding of current performance and identify opportunities for improvements. For instance, very low lean loadings (based on the amine analysis) and high overhead regenerator temperatures (based on the operating data collection) could indicate over-circulating or over-stripping of the solvent. This can result in unnecessary energy usage and increased amine losses, which can be remedied by reducing circulation or steam rates.

To complete the amine unit evaluation, technical tools are used to outline an improvement plan to meet refinery objectives. These tools include Dow proprietary simulation platforms, routine analytical analysis, pilot plant capabilities, heat stable amine salt management, and extensive knowledge and experience with more than 800 customers worldwide. An improvement plan is then presented to the refinery, implemented, validated, and sustained through routine visits and system reviews with Dow engineers.

The Use of Simulation Tools

Dow proprietary simulation platforms are used to develop operational solutions to meet energy reduction and/or other targets. State of the art simulation tools provide design and operating information to help achieve the following:

- Optimization of amine flow requirements and treatment results
- Gas treating solvent selection (including formulated solvents)
- Exploration of process limitations
- Process Flow Diagram
- Material balance with stream data tables
- Exchanger duties and other design information

The simulation output provides details on the current system performance, expected benefits from optimized performance, and any limitations of the amine system. An example of a temperature profile output is shown in Figure 2. Under ideal liquid-to-gas (L/G) ratios, the temperature profile of the absorber tower is shown with the "temperature bulge," or exotherm, at the bottom of the tower. However, in an effort to aggressively reduce steam usage, it is possible to reduce the L/G ratio such that the temperature bulge shifts to the middle or top of the absorber tower. This can compromise the treating capability and H_2S removal of the absorber system. When optimizing systems to reduce energy usage, it is important to understand limits of the process. An upset in the amine unit affecting a downstream unit or causing process downtime can quickly minimize the value of a poorly implemented energy savings plan.

Figure 2: Amine Absorber Temperature Profile with Normal L/G

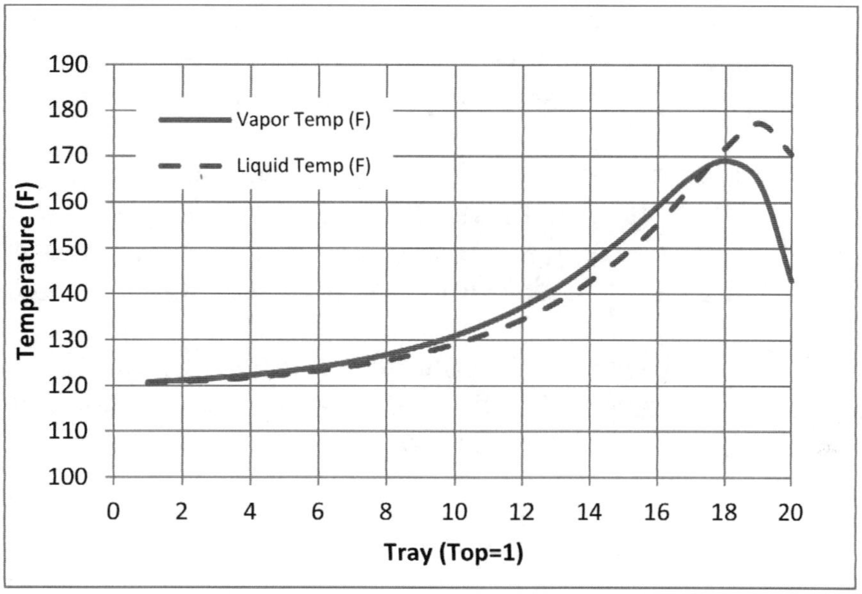

The Use of Pilot Plant Capabilities

In conjunction with the use of simulation tools, pilot-plant results are beneficial in modeling simulation output or plant performance, developing customer specific formulations, and validating operational improvements. As part of the AMINE MANAGEMENT^{SM} Program, pilot plant capabilities have helped:

- Demonstrate lower energy and CO_2 removal with solvent conversion
- Quantify organic sulfur removal performance for new gas treating applications
- Validate CO_2 slip and improve H_2S removal in low pressure applications (i.e. Tail Gas Units)

- Screen new additives and formulations for better acid gas removal with less energy usage

Solvent Hygiene

Often, energy management programs focus on equipment or process optimization. However, in amine systems, routine solvent analysis is important in identifying opportunities for energy savings. Monitoring amine solvent health is required to maintain optimal system reliability and operational performance. Ideally, routine sample analysis should include identification and quantification of all the components in the solvent. This would include amine concentration, acid gas content, individual degradation components, specific contaminants, and water content.

Contaminants in the system, such as formate, thiosulfate, thiocyanate, and other organic and inorganic acids, bind the amine to form Heat Stable Amine Salts (HSAS), which reduce functionality. High HSAS levels reduce system capacity, which can require increased steam usage and/or higher amine circulation rate to maintain system performance. Thus, it is important to monitor and maintain low levels of HSAS in the system.

Acid Gas loadings in the lean amine can serve as an indicator of stripping performance. In refinery systems, H_2S is present and measured as acid gas loading in the amine solvent. The absence of H_2S in the lean amine sample could indicate the amine is over-stripped. Operating data would then be used to confirm regenerator conditions and steam rates could be reduced.

As part of the AMINE MANAGEMENT^{SM} Program, a routine solvent analysis program is developed to monitor and trend solvent health as a means of determining system performance, anticipating and/or troubleshooting problems, and identifying opportunities for system improvement.

Case Studies

The following case studies will show how the implementation of the AMINE MANAGEMENT Program led to significant cost savings for each refinery. In all cases, solvent selection coupled with an optimization plan led to better treating performance, reduced steam consumption, and lower CO_2 emissions. The refineries below continue to use the AMINE MANAGEMENT Program from Dow to proactively identify opportunities for optimization and improved operational performance.

Case Study #1

A large-size, U.S. refinery completed a capacity expansion project to increase heavy crude processing capability. The expansion was successful but added strain on the utilities and downstream assets, increasing energy costs and limiting amine treating capacity. The shift in the crude slate increased the sulfur content and resulting acid gas volumes. Although the refinery was equipped with a complex amine treating system consisting of three regenerator trains and 18 gas and liquid contactors, the current solvent and system conditions were unable to maintain reliable acid gas removal from the process streams. A comprehensive, on-site evaluation of the refinery operations was conducted using the AMINE MANAGEMENT℠ Program, including detailed simulations of the existing amine systems and a comprehensive review of maintenance and operating procedures. The assessment showed that the refinery could achieve the following objectives:

- Address increased acid gas volumes without additional capital expenditure
- Decrease operating costs by reducing energy use
- Increase system reliability by reducing corrosion

The AMINE MANAGEMENT Program proposal recommended a solvent upgrade from the current solvent, DEA, to a high capacity, UCARSOL™ Specialty Solvent formulated for primary amine treating systems in refineries. Recommended operating concentration for DEA is less than 30% due to degradation that increases corrosion; however, the UCARSOL Solvent can target a much higher operating concentration, increasing the capacity available to remove acid gas. With the upgrade to the UCARSOL Solvent, the refinery increased solvent concentration from <30 wt% to an optimal target, allowing for reduced circulation rates and steam usage with the added benefit of reduced corrosion potential.

In addition to the solvent upgrade, process and solvent conditions were optimized to minimize environmental exceedances and improve system reliability. Amine loading targets and steam-to-lean amine (lb/gal) ratios were established to prevent over-stripping. The steam-to-lean ratio is a quick means of determining the amount of steam used to regenerate, or strip, the amine and should not be used as a replacement for reflux ratio. Steam-to-lean ratios were reduced by 20% indicating less steam usage. The lean and rich amine loadings were set at optimal targets allowing for reduced reboiler duty (or increased treating capacity).

When optimizing the system, targets and/or limits should account for occasional process swings (e.g., increase in H_2S content in inlet gas, increasing gas content, equipment fouling, etc). One of the regeneration systems at the refinery was able to aggressively reduce the steam-to-lean ratio by 25%, without affecting treating performance at existing conditions. However, to ensure safe operation and avoid H_2S and downstream SO_2 exceedances, the ratio was set at a conservative target.

The overall steam usage by the amine system regenerators is shown over a four-year period (2002-2006) in Figure 3. The solvent upgrade and optimization resulted in a significant energy reduction – 175,000 MMBTU/y – which translates to approximately $1.7 million/year in cost savings and a reduction of approximately 10,000 tons/year CO_2e emissions. Sulfur Recovery Plant energy usage was reduced by nine percent, and better utilization of the new solvent helped reduce amine losses by 17 percent.

Figure 3: Steam Usage for Refinery Amine Regenerators

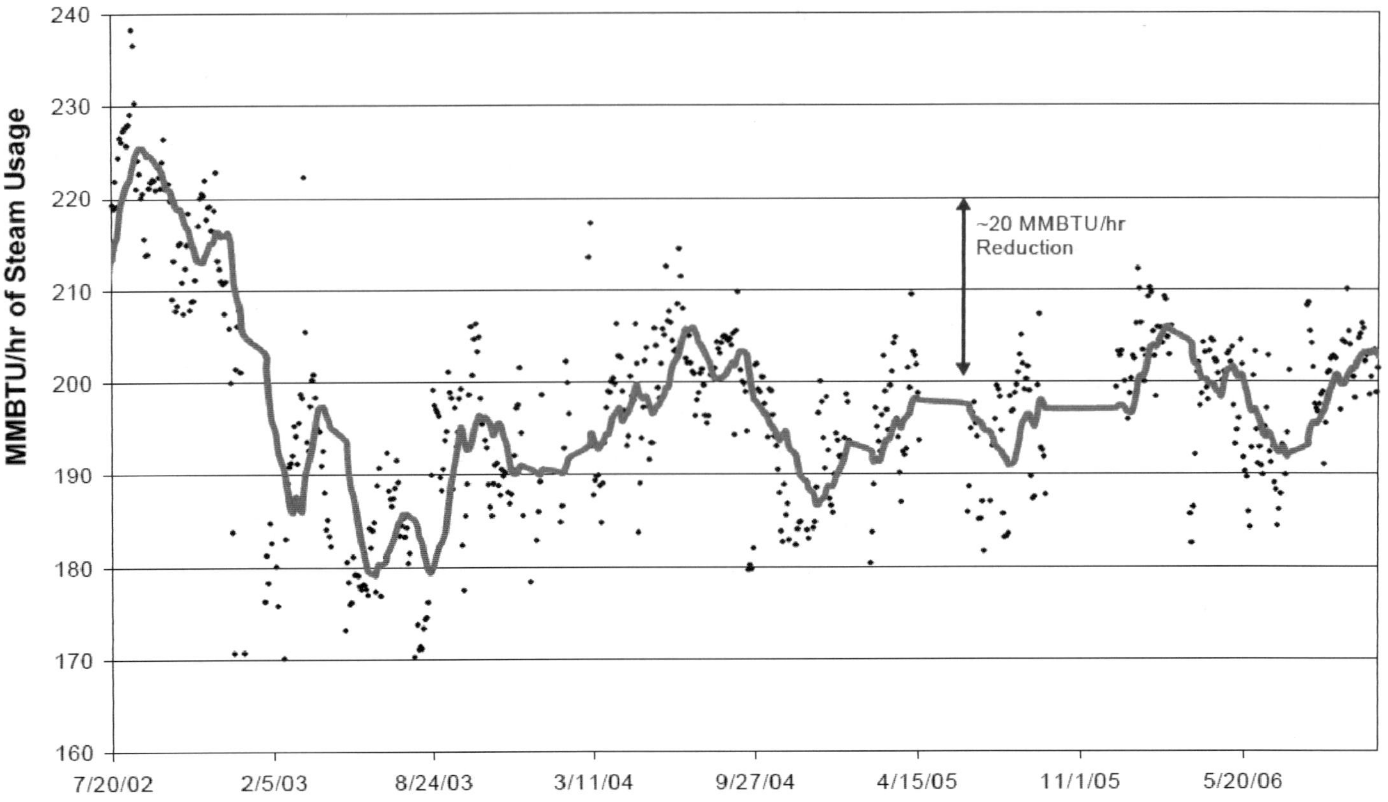

Case Study #2

This case study will show how a systematic approach using the AMINE MANAGEMENT[SM] Program from Dow was able to address the refinery needs. Initially, refinery objectives were focused on identifying a new solvent to enhance processing flexibility; however, through the years, opportunities to reduce solvent losses, enhance treating performance, and reduce energy usage have resulted in over 350,000 MMBTU/y savings and reduced CO_2 emissions.

A mid-size refinery in the U.S. used two generic solvents to treat process gas in the main and tail gas units. However, to increase system reliability and flexibility, the refinery moved to an integrated amine treating system requiring only one solvent for both primary and tail gas systems. The solvent would be required to meet treating needs for the contactors on the main system, including LPG liquid treaters, as well as provide CO_2 slipping and H_2S selectivity for the tail gas application. A comprehensive evaluation of the system was conducted using the AMINE MANAGEMENT Program to meet the following refinery objectives:

- Provide a single amine solvent to be used across the refinery
- Provide enhanced total sulfur removal
- Maintain adequate CO_2 slip, while removing meeting H_2S removal targets

A new specialty solvent was developed to meet the refinery needs. The refinery upgraded to a UCARSOL™ Solvent formulated for both primary and tail gas applications. Hesitant to increase solvent strength, the refinery operated at less than 30 wt% to minimize losses. However, with tools from the AMINE MANAGEMENT℠ Program, losses were tracked to reduce overall losses by 34% (as shown in Figure 4).

Figure 4: Refinery Amine Loss Reduction

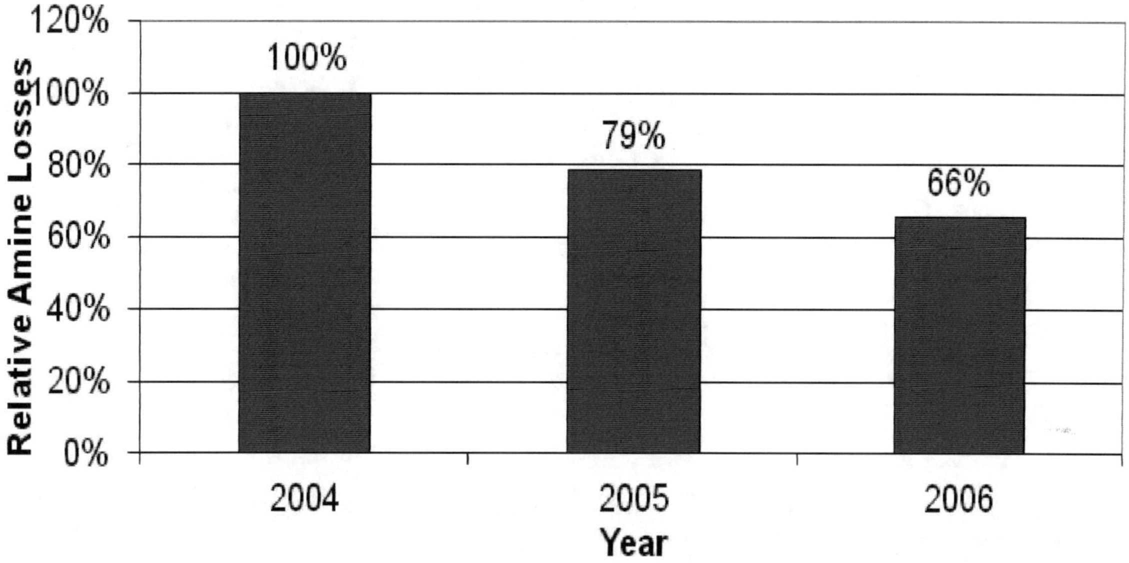

With losses optimized, another comprehensive evaluation by Dow engineers was performed and showed that increasing solvent concentration to greater than 40 wt% would result in a 26% decrease in energy costs, totaling a net savings of more than $1MM/year. Conversely, if energy usage was not a key objective, the concentration upgrade would increase overall treating capacity of the amine system by over 30%. Additional recommendations included reducing steam-to-lean ratios and adding improved process control logic to maintain optimal circulation rates.

The refinery increased the amine strength to 35wt% and implemented the optimization recommendations, which resulted in a reduced amine circulation rate. The lower rate led to a 30% decrease in regenerator duty (shown in Figure 5), and the energy reduction was approximately 40 MMBTU/h. These improvements translate to 350,000 MMBTU/y energy savings, 20,000 tons/y CO_2e emission reduction, and $1.1MM/year in total cost savings.

Figure 5: Refinery Trend in Total Regenerator Reboiler Duty

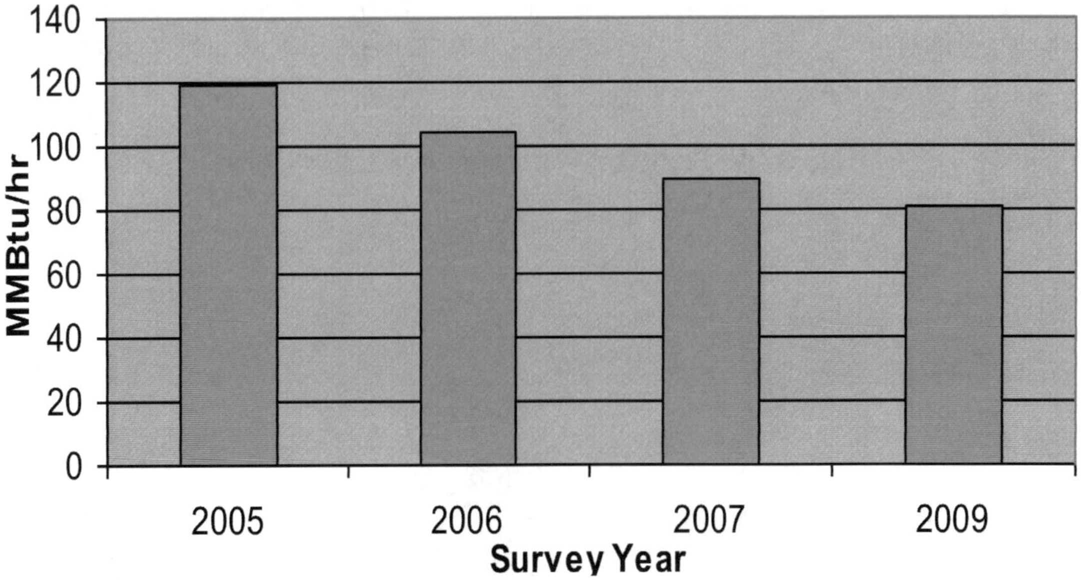

Case Study #3

With the flexibility to process large volumes of heavy crude, a large, U.S. refinery needed to achieve the following objectives to become more competitive:

- Increase gas treating capacity
- Improve system reliability
- Reduce total operating costs

The existing amine treating system used MEA. Although MEA is aggressive in total acid gas removal, the primary amine readily degrades, thus increasing the corrosion potential of the solvent. To minimize corrosion, the amine requires frequent and costly reclamation to remove degradation by-products. Operating concentrations are limited to 20wt%, and regeneration for MEA requires more stripping steam due to a higher heat of reaction as compared to formulated solvents. The use of MEA limited long-term capacity goals for the refinery.

An evaluation of the amine system was performed using the AMINE MANAGEMENT℠ Program. The evaluation determined that a solvent upgrade and optimization would help the refinery increase treating capacity by 48%, reduce amine losses by 50%, and reduce energy consumption by approximately 500,000 MMBTU/y. A two-step approach was developed and implemented to help the refinery achieve outlined objectives.

Phase 1: Solvent Upgrade The refinery upgraded from MEA to a UCARSOL™ Solvent designed to minimize energy usage and meet acid gas removal requirements in refinery applications. With the formulated solvent, the operating concentration was increased to an optimal range, and steam-to-lean ratios were reduced by 20%. Phase 1 of the optimization plan resulted in an immediate circulation rate reduction and an energy savings of over 400,000 MMBTU/y, translating to 22,000 tons CO_2e/y reduced emissions. Furthermore, the upgrade to the UCARSOL™ Solvent occurred with no shutdown or incidents.

A routine analytical plan was implemented to ensure optimal solvent and system performance. Solvent losses were also tracked, monitored, and improved. Overall, solvent health improved, solvent losses were reduced by 59%, and corrosion decreased.

Phase 2: Optimization The next step was to focus on optimization of the amine treating system. The individual gas contactors were evaluated using simulation techniques from the AMINE MANAGEMENT℠ Program. Proprietary equations were developed and implemented to optimize circulation rates, and rich loadings were increased to target an optimal range. This resulted in an additional 15% reduction in circulation rates and 300,000 MMBTU/y in energy savings, with a further reduction of 18,000 tons CO_2e/y.

The two-phase approach led to 700,000 MMBTU/y in energy savings, a reduction of 40,000 tons CO_2e/y emissions, a 59% decrease in solvent losses, and improved system performance with a total cost savings of over $3MM/year.

Summary

Energy conservation as a means of reducing CO_2 emissions and total operating costs is a target objective for refineries. As crude slates shift towards heavier, sour crudes, it is necessary to consider cost-efficient means of increasing processing flexibility while heeding environmental regulatory requirements. Specific to acid gas removal systems, this paper reviewed the implementation of AMINE MANAGEMENT℠ Program tools and techniques to reduce energy consumption and CO_2 emissions, resulting in a reduction in total operating costs.

In the case studies discussed, upgrading from generic solvents to formulated solvents improved overall system performance and increased gas treating capacity, meeting refinery objectives to address process expansion and flexibility needs. With the use of proprietary simulation tools, optimal solvents were selected based on the refineries' needs, and solvent upgrades were safely implemented with no environmental exceedances. Operating limitations were identified and targets for optimization were set specific to acid gas loadings and stripping steam ratios. Routine analytical plans were established to monitor solvent performance and proactively address system issues. Additional capacity gains and energy savings were realized with proprietary equations for individual absorbers and improved system process logic. The successful implementation of these optimization techniques, coupled with the solvent upgrade, led to 1,225,000 MMBTU/y in energy savings, a reduction of 70,000 tons CO2e/year emissions, and nearly $6 MM in operating cost savings combined for the three case studies discussed.

References

Ackman, R, Pirtle, L, "Amine Treating: A Review of the Technology", AICHE Spring National Meeting, April 2012.

"Annual Energy Outlook 2011", Available at http://www.eia.gov/oiaf/aeo/tablebrowser/, February 12, 2012.

Burns, D, Gregory, A, "Waste Minimization and Compliance Benefits To Be Gained From A Good Amine Management Program", NPRA, October 1995.

Echt, W, Wendt, C, "Reduce Sulfur Emissions From Claus Sulfur Recovery Unit Tail Gas Treaters", AICHE Spring National Meeting, March 1993.

"Energy Consumption", Available at www.need.org/needpdf/infobook_activities/IntInfo, February 12, 2012.

"Energy Efficieny Roadmap for Petroleum Refiners of California", California Energy Commission, Prepared by Energetics Incorporated, April 2004.

GPSA Engineering Data Book, Eleventh Edition (1998), Gas Processors Suppliers Association.

"Greenhouse Gas Emissions", Available at http://epa.gov/climatechange/emissions/index.html, February 12, 2012.

Gunn, Matt, "Amine System Upgrade to UCARSOL™ LE-801", Brimstone Symposium, September 2007.

"Kyoto Protocol", Available at http://unfccc.int/kyoto_protocol/items/2830.php, February 12, 2012.

"Manufacturing Energy and Carbon Footprint", Available at http://www1.eere.energy.gov/industry/rd/footprints.html, February 12, 2012.

"Petroleum Refining Industry Profile", Available at http://www1.eere.energy.gov/manufacturing/industries_technologies/petroleumrefining_profile.html, February 12, 2012.

"Refinery CO_2 Management Strategies", Available at http://www.hydrocarbonpublishing.com/ReportP/energy09.php, February 12, 2012.

"Western Climate Initiative", Available at http://www.westernclimateinitiative.org/index.php, February 12, 2012.

(86d) Carbon Capture Ready Heavy Oil Extraction and Upgrading

Eloy Flores III, Chemical Engineering Department, Division of Chemistry and Chemical Engineering, Southwest Research Institute, San Antonio, TX

Carbon dioxide emissions from heavy oil production are typically higher than that of producing conventional crudes. Ultimately, carbon dioxide emissions can make heavy oil production less competitive if regulated. Carbon dioxide emission regulations will affect the overall economics of heavy oil production and upgrading. The additional carbon dioxide gas emissions from heavy oil production, when compared to conventional crude oils, come from the additional energy that is typically needed for heavy oil extraction processes. A heavy oil upgrading process test facility, currently in operation at Southwest Research Institute (SwRI), has the capability to extract and upgrade heavy oil while producing a capture ready concentrated stream of carbon dioxide gas. The heavy oil upgrading process being demonstrated is a thermal cracking process, which is capable of directly providing the energy needed for extraction. Extraction will be done by steam assist gravity drain (SAGD). This process utilizes a fluidized bed process similar to that of a fluid catalytic cracking unit (FCC). Using oxygen combustion in the regeneration step, it is possible to generate a concentrated stream of capture ready carbon dioxide.

This presentation will examine and describe an alternative heavy oil upgrading process, currently being demonstrated, for carbon dioxide capture. It will also evaluate similar traditional heavy oil upgrading processes from literature that utilize oxygen combustion techniques.

(86e) Move to Atomic and Renewable Energy in An Oil Exporting Region?

Ahmed Khogeer, KAUST & Thuwal development Projects, Saudi Aramco, Thuwal, Saudi Arabia and **Hisham Bamufleh**, Chemical Engineering, King AbdulAziz University, Jeddah, Saudi Arabia

Move to Renewable energy in an oil exporting region?

With its abundant resources of oil and gas, and having 25 % of the world oil reserves and low cost of production, it was not the forecasted that Arab gulf countries will consider energy resources other than what they are exporting to the outer world. However, for many reasons, these countries are considering nuclear and renewable energies particularly solar for many residential and industrial uses. In this paper we will cover the reasons, economics, feasibility and some major announced projects in the region.

(86f) Process Integration for Dimethyl Ether Synthesis Via Catalytic Dehydration of Methanol

Mansour Emtir, Process Engineering, Harouge oil operations, Tripoli, Libya, Aymen Elbabour, Chemical Engineering Department, Academy of Graduate Studies, Tripoli, Libya and Elmahboub Edreder, Rsearch & Development, Libyan petroleum institute, Tripoli, Libya

Distillation is the most widely used separation technique in the process industry. Its main disadvantage is the high consumption of thermal energy: the heat requirements of the distillation columns often represent most part of the total energy cost of the plant. The energy saving in distillation is then an essential field of the chemical engineering research since the early 1970's; appropriate integration of the distillation columns with the overall process often results in a significant reduction of the energy consumption in many cases.

In this work the concept of the integration with the overall process has been applied by two integrated design approaches:

- Heat-integrated design; where the distillation system of the process will be heat-integrated with the background process (i.e. integration between the distillation columns and the heat exchangers).
- Reactive distillation design; where the reaction and distillation will be accomplished in the same unit operation.

The process of dimethyl ether (DME) synthesis by the catalytic dehydration of methanol has been considered as a case study.

Heat-integrated design of DME process showed a significant total annual cost saving of 41% compared to the conventional process, and the reactive DME distillation column design showed highest saving in total annual cost by 81%.

In addition to the cost saving of the two proposed designs, also the flue gas emissions have been reduced using process integration methodology, where 51% of the total emissions has been reduced by the heat-integrated design and 86% by the reactive distillation column design.

(93a) A Photo Safari Though a Petroleum Refinery

Matthew J. Reisdorf, Process, Fluor Corporation, Sugar Land, TX

Cut a crude assay, dehead a coke drum, and stalk wild naphtha in a 30-minute whirlwind tour of a petroleum refinery. This tour of refining products and processes is aimed at chemical engineers who want a better understanding of how and why the different parts of a refinery work together as an integrated system.

(93e) Planning for a Career in Refining - What Every Chemical Engineer Must Know

Shankar Vaidyanathanand Kevin Turini, Fluor Corporation, Sugar Land, TX

The paper will present an overview of how a chemical engineering graduate could put together a self managed development program to prepare for a career in the refining industry. It will cover the basic skills that are required and how to pursue them. It will suggest various resources that are available in the public domain and a suggested reading list. The paper will present an outline of a proposed action plan as well. The attempt is to enhance the knowledge and training of the individual chemical engineer by a structured learning program that can be adjusted based on individual needs.

(93b) From World Scale Refineries to State of the Art Nano Labs and Super Computers. From Coker to Clean Rooms, From KBPD to LPH. How to Translate Your Past Experience Into a Competitive Advantage?

Ahmed Khogeer, KAUST & Thuwal development Projects, Saudi Aramco, Thuwal, Saudi Arabia

In this paper we will cover some case studies on how a chemical engineer with a specialization in one area can find himself in a position where he is required to work on the design and construction in other chemical engineering discipline that he never practiced in the past. Some challenges, lesson learned and tips to translate what may be thought irrelevant experience into a competitive advantage.

Refinery-Wide Topics for Profitability, Reliability and Safety

Mike Siravo, Fluor Corporation, Sugar Land, TX

Introduction

The objective of this presentation is to raise your awareness of the multitude of problems lurking among the many activities performed in a complicated facility like an oil refinery. These problems take many forms and have varied causes. If not identified they can and do result in:

- Serious safety and environmental incidents
- Community relations breakdown, loss of reputation
- Unnecessary capital expenditure
- Damage to equipment and expense increases
- Loss of production and profitability

We will review several case studies of design and operational incidents to identify learnings. We will discuss what went wrong and why. It is beneficial to learn from other's mistakes, but hopefully the negative impact of these case studies on people, businesses and, possibly, careers will do more than provide specific learnings. Hopefully these negative impacts will instill in you the proper level of anxiety and accountability to provide the motivation to remain vigilant and alert as you perform your work.

Identifying and mitigating potential pitfalls in your work is not easy, especially when you may be encountering new assignments and the unfamiliarity that brings. Time constraints overlay everything we do and provide possibly the largest impediment to quality results. Although in these case studies you will encounter the bad news of things gone wrong, the good news is that the skills to root out problems before they occur can be developed. As we go through these case studies, look for the common themes of poor communication, lack of attention to detail and violation of procedures.

Case Study 1: Know Procedures and Follow Procedures

A refinery Crude Unit was nearing the end of its four week shutdown/turnaround. The lack of crude processing over the turnaround period had significantly reduced the rates of those refinery units that stayed on-line; units like the FCCU, Alky and Hydrocracker. These units and others were able to operate due to inventories of feedstock that were generated before the Crude Unit was shutdown. Gasoline blending was also partially reliant on inventories of certain blend components. These inventories were getting low as planned, but it was imperative that the Crude Unit start up on time.

An operating group can become extremely unpopular if they are unable to maintain schedule and achieve "oil-in" on the expected date. If delay were to occur, there would be some heated conversations and the plant manager would be making telephone calls to his boss, the President of Refining, at headquarters. Calls that he desperately wants to avoid, if possible. In summary the refinery staff was feeling a very strong sense of urgency for this startup.

One of the activities during the turnaround was to perform a light oil wash of some of the heavy oil equipment. Once the wash was completed, removal and disposal of this oil was necessary to proceed with commissioning of the unit for startup. Around that time, in the evening, it was discovered that the original plan for disposing of the contaminated wash oil was no longer an option. A scramble for an alternative ensued. Some ingenious soul discovered that a series of lines existed that could be used to connect the Crude Unit vessel holding the oil to a heavy oil tank in the Offsites area of the refinery. As often happens in an older refinery, sometime in the past, the need for flexibility had added these connections for a reason no one could now remember. This appeared to be very fortuitous.

Telephone conversations were initiated between the Crude Unit and the operations foreman in the tank farm area. The Offsites tank in question normally held a heavy oil used in blending asphalt. It was a fixed roof tank with a steam coil to keep the heavy oil fluid and a roof vent to the atmosphere. As luck would have it, the tank was essentially empty. The tank farm foreman was, correctly, not interested in receiving transfer of the wash oil. The foreman resisted the transfer because he knew that his operating procedures limited use of the tank to heavy oils with a high flash point. He did not know exactly what this oil was and the Crude Unit was not particularly forthcoming with information.

The standoff remained for the next few hours until the tank farm foreman relented. Much of his thinking was rooted in the culture of the refinery which held that the oil processing units was where the money was made. Any operations that were part of the refinery offsites and utilities were support functions that were never to be the cause of a problem in a place like the Crude Unit. This was a frequent source of friction between the two areas.

About three hours after the transfer was made, the manager of the refinery offsites area who lived about ten miles from the refinery was awakened by a loud, rumbling sound at about four AM. Not long after falling back to sleep, the manager's pager began buzzing. A return phone call to the refinery provided the news that one of the tanks in the manager's area had exploded. Miraculously, no one was injured. The roof had been launched some distance damaging another tank. The heat from the fire damaged the adjacent tanks. The ignition source was determined to be either pyrophoric iron in the tank or a thunderstorm that passed overhead around that time.

Putting everything else aside, the Offsites manager was going to spend the foreseeable future leading an incident investigation. The Plant Manager got to call his boss, the President of Refining, at headquarters, which is what he wanted to avoid in the first place.

Learnings:

- Have proper and clear procedures
- Train to the procedures
- Audit the understanding of the procedures
- Use weekly safety meetings to review and discuss procedures
- As often as necessary, using all reasonable means, management needs to reinforce to every person in the facility that they have the authority to shutdown any operation that poses an immediate threat to safety. Everyone needs to know that they will be lauded for such an action. Conversely, there should be no perception of reward for anyone that violates a procedure or takes risks to benefit production.

<u>Case Study 2: Poor Unit Monitoring</u>

During a shut down, a 30 MBPD diesel hydrotreater reactor was loaded with the latest generation of fresh catalyst. Startup proceeded smoothly and without issue. When catalyst is new the reactor temperature needed to meet the product sulfur specification is at its lowest point. As time goes on and the catalyst ages or deactivates, the reactor temperature is increased to overcome the lower activity of the catalyst to continue to meet spec. The predicted catalyst life was 2 years, meaning that at that time the reactor temperature will have been increased slowly over time until the maximum operating temperature is reached. Generally the reactor temperature is increased linearly for the first 75% of the 2 year run and the at an increasingly faster rate until end of run (EOR) temperature is reached.

About 5 months into the run, the operations supervisor noticed that to meet the product sulfur specification he needed to increase reactor temperature more often and in larger increments than was his experience with prior runs. Discussions were held with the process engineering department which agreed with the assessment. More intense monitoring of unit performance was initiated. Investigation showed that the catalyst deactivation rate (rate of required reactor temperature increase) was approximately three times higher than expected. At this deactivation rate the catalyst would reach EOR after only 8 months, 16 months earlier than planned.

Process engineering resources and support staff from the catalyst supplier began a concerted effort to understand the cause of the rapid deactivation. After several weeks three causes were identified:

- With this current run, the feed to the unit included a 15% increase in cracked stock. Cracked stock in this case included diesel boiling range material from the FCCU and Coker unit. Cracked stock has a higher hydrogen consumption than straight run diesel from the Crude unit.
- The higher activity catalyst increased hydrogen consumption compared to the catalyst used in prior runs due to an increase in aromatics saturation.
- The hydrogen content of the recycle gas in the diesel hydrotreater unit was designed to be 80 mole%. The centrifugal compressor, therefore, was designed for a recycle gas of 80% hydrogen. For some time, even in the prior run, the recycle gas purge rate was set too high. Make up hydrogen at 99 mole% hydrogen entered the unit on pressure control. So since the purge gas rate was high, the make up hydrogen rate was higher than it needed to be to maintain recycle gas purity at 80%. In fact the make up rate was so high that the recycle gas purity was running 90%. Generally, higher recycle gas hydrogen purity should be good for hydrotreater performance. However, as the hydrogen purity increased the molecular weight decreased significantly. Since the recycle gas compressor was a centrifugal design, this drop in molecular weight decreased discharge pressure and led to a reduction in recycle gas rate. The net effect was a reduced rate of hydrogen available for reaction.

Unfortunately, all three of the above combined to reduce the hydrogen partial pressure in the reactor by 35%. This greatly escalated coke formation on the catalyst causing rapid deactivation.

Indeed, as predicted, the catalyst lasted only 8 months although feed rate needed to be reduced to even achieve this. A fresh load of catalyst needed to be procured on short notice at a premium price. The unplanned shut down caused disruption to the refinery operating plan just prior to entering the profitable gasoline season. The total cost of the incident was several million dollars. The process engineering department, responsible for daily monitoring of the unit, had failed to do their job. And it was obvious to everyone.

Learnings:

- Specific and up to date operating guidelines should have been in place providing targets for recycle gas rate and hydrogen purity. In fact operating guidelines should be in place across the entire unit.
- The unit process engineer should have trended unit performance daily to identify abnormalities as early as possible. With proper monitoring the change in feed quality, low recycle gas rate and high hydrogen consumption all could have been

identified in the first month of operation. Corrective actions could have taken place to ensure the 2 year run length.

- Proper unit monitoring requires the process engineer to understand the reasoning behind all process conditions on a unit. In this case an understanding of the importance of hydrogen partial pressure and the variables that affect it were necessary. If you do not feel you have this level of understanding, discuss this with your supervisor and ask him or her to help you develop a plan to acquire the required competency.

Case Study 3: Installation Failure

An FCCU debutanizer was being revamped to increase capacity to allow for increased vapor and liquid traffic. This was necessary to improve separation between the bottoms product gasoline and the overhead product C3/C4 stream. The objective was to recover additional C4's from the bottoms product to reduce its blending vapor pressure and reduce the C5 content of the overhead product since C5's have a negative impact on the Alkylation Unit which is the ultimate destination for the C4's.

It was determined that the reboiler, condenser and reflux pumps all had the necessary spare capacity to allow for the increase in stripping required to improve the separation. An evaluation of the trays showed that a vapor-liquid bottleneck existed but only in the trays above the feed point or in the rectification section of the column.

A redesign and replacement of the trays was found to fall short of adding the required capacity as did the possibility of replacing the trays with random packing. The only possibility for gaining the needed capacity was replacing the trays with structured packing. Simulation and packing vendor data showed that structured packing would work. At the time of this revamp structured packing had been successfully used extensively in the refining industry but not in services handling C4's and lighter materials. In fact some industry literature discussed concerns with this application. Additional research including discussion with fractionation experts was successful alleviating enough of the perceived risk for the structured packing revamp to proceed.

In the final revamp design the eighteen trays above the feed point were to be replaced with two beds of structured packing. Above the top bed was a perforated tray that would take the reflux stream and drop it evenly across the top of the top packed bed through several hundred small (3/8" diameter) weep holes. Even distribution of liquid across the packing was essential, and this design would serve that purpose. Vapor leaving the top of the packing rose through this perforated tray through eight rectangular risers. Each riser was six inches wide and several feet in length. The risers extended one foot above the tray to ensure that liquid could only reach the packing below through the weep holes. Upon exiting the risers the vapor would leave the top of the column as normal on its way to the condenser.

Since there were two packed beds, there was a similar liquid distribution tray between the beds. This tray was to collect the liquid exiting the bottom of the upper packed bed and redistribute it evenly through the several hundred weep holes onto the lower packed bed. Vapor from the lower packed bed rises through eight vapor risers that were almost identical to those just described for the upper bed distributor. To eliminate the possibility that liquid falling from the upper bed could fall into the riser, each of the risers for the redistributor tray between the packed beds was equipped with a cover. The cover took the form of a gutter sitting five inches above the top of the riser and extending the full length of the riser. The gutter has side walls so that any liquid that lands on it can be routed to and run off its ends. This collected liquid then falls harmlessly onto the perforated tray deck. Since there are eight risers there are eight gutters. The five inch gap between the top edge of the riser and the bottom of the gutter provided the area for vapor to exit the riser.

Despite the concern of the new application of structured packing to such a revamp, the early phase of the column start up proceeded smoothly as the FCCU was brought on line. A problem developed, however, as the FCCU and column rate approached 60% of the design. The newly installed differential pressure instrument for the upper packed bed alarmed on high pressure drop. Overhead and bottoms product sampling confirmed the problem as the degree of separation was very poor.

For several days everything imaginable was done to determine what was causing the column flooding so it could be corrected. It was no small effort to start up the FCCU, and there was no desire to shut it back down so the column could be entered for inspection. On the other hand the economic impact of limiting the FCCU capacity to 60% of plan for much longer was also brutally painful to consider. A significant part of the thought process working against shutting down and entering the column was the fear that no problem would be found. And then what? It was unthinkable to go through the time, money, effort and embarrassment to enter the recolumn only to find later that it was not necessary to correct the problem.

Once all of the diagnostics and theories were exhausted, there was no other choice but to shutdown the FCCU, prep and enter the column for inspection. This took another day and a half, all the time the refinery was hemorrhaging money at the rate of about one half million dollars a day.

Once the column manways were opened and the air mover was in place. The FCCU unit process engineer and the process engineer from the project group responsible for the revamp design began inspecting what they could from the manways in the hope that something obvious could be found. As often is the case there was some good news and some bad news.

The good news was that the cause of the problem was painfully obvious. From one of the manways it was very easy to see the liquid redistributor tray that was located between the two packed beds. At first it seemed fine but something was not quite right. No part of it was loose or damaged, but it became clear that the space provided in the design between the vapor risers and the liquid collection gutters was somehow blocked. The gap between the two was slightly less than one inch instead of the design gap of five inches. How was all of the column vapor supposed to flow up through the risers when the cross sectional area for flow was restricted by more than 80%. How was this possible?

The explanation lay in a mistake made by the maintenance craft during the tray installation. The tray deck sections came with the risers already in place. During installation it was necessary for the crafts to mount the gutters above the risers using the supplied support strips and bolts. The crafts did not properly read the tray drawing. Thinking that the gutters were like a long hat sitting on top of the risers, they installed them upside down. This meant that instead of extending upward to collect liquid, the side walls of the gutters were extending downward, blocking off most of the area allotted for vapor flow.

As a result the pressure drop for vapor flow through the tray was high, much higher than design. With this high pressure drop the liquid level on the tray (designed to be only a few inches high) increased until it was above the top of the vapor risers. This further restricted vapor flow increasing pressure drop even more. Ultimately the liquid level rose into the top packed bed increasing bed pressure drop (thus the high pressure drop alarm) and destroying the degree of separation.

In a relatively short time the position of the gutters was reversed, the column was closed and the FCCU was restarted, reaching design rate in the typical amount of time.

The bad news was that the installation error could have been easily avoided by proper supervision of the crafts during installation. Certainly it is a reasonable expectation that the drawings be followed. But the responsibility for the final inspection prior to the column being closed lay with the process engineers who are generally tasked with this inspection since they are the most familiar with the details of the design intent.

Learnings:

- Since there were two process engineers involved, accountability for the final inspection became uncertain. The engineer involved in the revamp design had been reassigned to other project work before the column shut down and packing installation. The unit engineer was less experienced and was not directly informed of the responsibility to inspect all such changes to equipment in his area.
- In a perfect world one or both of the process engineers would have felt the need to perform the final inspection. One could argue that management had a responsibility to assign the task. Since none of this happened and luck didn't play out either, they all got to share the substantial amount of blame involved.

Modern Loop Tuning Tutorial

by

Tim Olsen
Emerson Process Management
Tim.Olsen@emerson.com

and

Jim Dunbar
Emerson Process Management
Jim.Dunbar@emerson.com

Prepared for Presentation at the

AIChE 2012 Spring Meeting
Houston, TX
April 1-5, 2012

Session 93, Young Professional Tutorial - Topics in Refining
Paper 93d

Abstract

The 2012 AIChE Spring Meeting presentation 93d is a tutorial on tuning PID process controllers. The course will explain the operation of the PID controller, typical variations of the PID controller, common process responses and how to identify the dynamics of the process. The instructor will demonstrate how to calculate PID tuning parameters from the process dynamics and the desired controller response.

Introduction

This paper is intended to encourage young professional engineers to understand the dynamics of their processes and use this tuning tutorial to improve process control performance. Modern loop tuning is a simple and inexpensive tool that any engineer can utilize for better operation.

Step 1 – Understand the Process and Operating Objectives

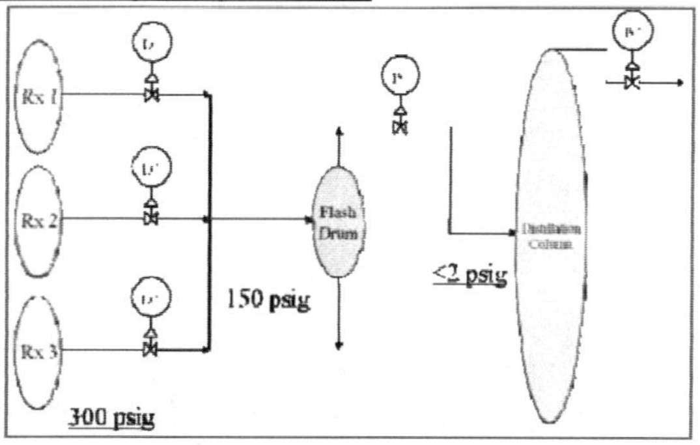

One of the first steps before actually starting to tune a loop is to understand the process and the objectives of the process. The process on the right, Figure 1, has several reactors on individual level control operating at 300 psig flowing into a flash drum at 150 psig and then vapors into a distillation column less than 2 psig. Each control loop was tuned individually, but the overall process was unstable and difficult to control the distillation column including product quality. Changes in the reactor to maintain level sent a disturbance to the flash drum, but the flash drum was tuned very aggressively (standard deviation of 0.1 psi from 150 psig setpoint) and the flash drum vapors significantly disrupted the performance to separate in the distillation column.

The objective of this process is not to have very accurate pressure control of the flash drum, but to maximize yield of desired products from the reactors and then effectively separate impurities from the product in the downstream distillation column.

Step 2 – Prioritize the Control Loops

Based on information gathered in step 1, you are now ready to prioritize the control loops to determine which should execute fast and which should execute slower. Process control loops interacting with each other will fight each other if given equal priority in tuning. This is easiest to understand in a cascade controller where steam flow is the inner loop and temperature is the outer loop (process temperature setpoint). The inner control loop is tuned 5-10 times faster than the outer loop to quickly compensate for changes in steam pressure. Since the setpoint of the inner loop is a function of the outer loop, you will always interact; if the inner and outer loop tuning is equal priority, this will cause instability or cycling because the inner loop cannot track the setpoint from the outer loop.

For the reactor-flash drum-column example in Figure 1 above, the primary objective is to maintain proper levels in the reactors to ensure maximum product yield. The downstream separation process is next in priority to ensure product quality at the lowest energy usage required for separation. The flash drum is just a vessel in-between the reactors and distillation column for a quick, basic separation. The 150 psig is not a critical value since the separation of flashing is simple separation based on pressure drop from the 300 psig reactors. The flash drum can be used as a variability buffer to absorb disturbances from the reactors maintaining level, so to minimize disturbances to the distillation column at less than 2 psig.

As good practice, one should utilize levels as variability buffers to absorb disturbances from upstream processes, thereby minimizing the impact to downstream processes. This can be a culture change for many facilities where console operators like seeing levels at 50% with a flat line process variable matching the setpoint. Surge drums, distillation column bottoms and overhead drums are good examples where the level can fluctuate from 40 and 60% (around the 50% setpoint) to maintain steady flow downstream with gradual changes as needed to rebalance the process flow. Note, some levels need to have minimal variability such as the example with the reactors in Step 1, so not all levels can be variability buffers, but many can.

Step 3 – Step Test each Loop and Identify Hardware issues

A step test is when you place a controller in manual and change the output a small percentage; the size of the output step should be large enough to observe the response above any process noise that may be present, yet small enough to minimize process upsets.

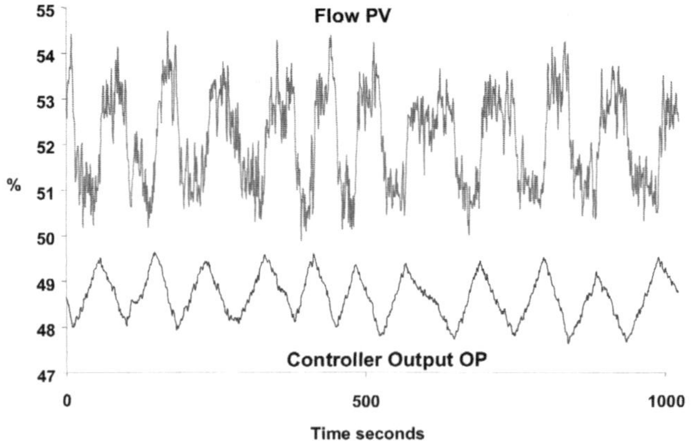

Figure 2: Limit cycle in a flow loop due to a sticky valve

Ensure you understand the process before executing step tests to prevent unwanted process upsets (for example, do not reduce hydrogen quench flow to cause a hydrocracker reactor temperature runaway; start with increasing hydrogen and then changing direction back to normal hydrogen use). Start with small steps such as 0.25%, 0.50% or 1% depending on sensitivity of the process. If you do not see a change in flow / temperature / pressure, continue making step changes until you do see a change. It is not uncommon to see control loops requiring an output change in excess of 2% to see changes from the final control element movement (Figure 2 above shows a limit cycle due to a sticky valve). Under-performing field devices add process constraints that cannot be tuned out or compensated with advanced process control.

Step 4 – Fix Hardware Issues

As you identify hardware issues like sticky control valves, hopefully you have the ability to make fixes by isolating the control valve to make repairs (bypass), or plan the repairs at the next upcoming turnaround. Smart positioners have made significant improvements over the past couple decades and help control valve performance for better response and positioning.

Step 5 – Re-Step Test Loops that had Hardware Fixes

Now that control valves move appropriately when the controller asks it to move, you can place a controller in manual and change the output a small percentage (same as step 3).

As shown on the right in Figure 3, the response is the most common type, a first-order plus dead time.. There are many types of responses that could be observed such as integrating, second order response, inverse response and others. This paper will address the first-order plus dead time response.

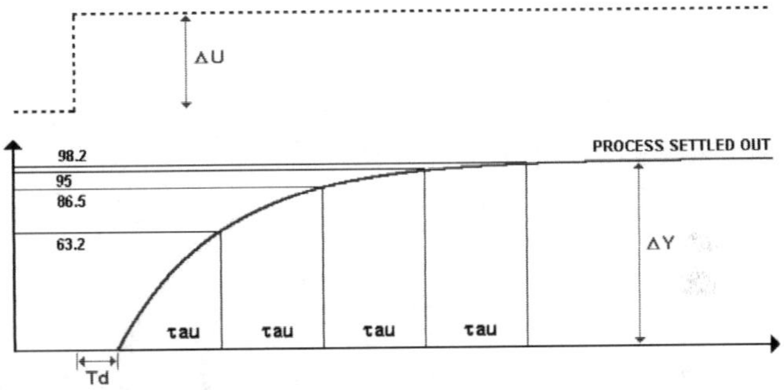

Figure 3: Typical first order step response

The primary process dynamic information attained from a step test is steady-state process gain, dead time, and the time constant. Typically these parameters are an average of several step tests around the normal operating range.

The steady-state gain can be calculated by taking the measured process value before and after the step test and then dividing by the output change (normally the gain is converted to units of %PV/%Output because most controllers calculate their output in dimensionless units). The time constant Tau is the time required to reach 63.2% of the final process change. The dead time is the amount of time after the step test is made until the process actually begins to move.

Step 6 – Calculate Tuning parameters

Once an engineer understands the process dynamics and control loop priorities, model-based tuning such as Lambda Tuning can be applied to set the speed of response for each controller based on operating objectives and priorities. "Lambda Tuning" is a tuning method where the control loop speed of response is a selectable tuning parameter; the closed loop time constant is referred to as "Lambda" (λ). Lambda Tuning originated with Dahlin [1] in 1968; it is based on the same IMC theory as MPC [2, 3]. It is model-based and uses pole-zero cancellation to achieve the desired closed loop performance. A recommended Lambda starting point to ensure robustness is three times the larger of Tau or

Deadtime. The time for the loop to reach setpoint is approximately four times the selected Lambda value.

Self-Regulating Process (First-order with deadtime model, Classical PID) [3, 4]

τ = time constant (from step test)

T_d = dead time (from step test)

K_p = steady state process gain (from step test)

λ = closed loop time constant (Lambda – user defined speed of response)

Controller Settings

$T_r = \tau$

$K_c = (1 / K_p) * T_r / (\lambda + T_d)$

Note, tuning parameters setup for normal operation may not work as effectively during turndown rates. If turndown rates continue for a long period of time, one should consider re-evaluating the required tuning parameters for the new operating conditions. Write down the old tuning parameters for when you return to normal operation.

Step 7 – Input New Tuning Parameters and Observe

There is an old saying that if you change the tuning parameters on a controller, you own it. If you make changes to the tuning, you need to observe the controller ability to minimize disturbances while maintaining the overall operating objectives effectively. A common way to test the speed of response is to install the new tuning parameters and then perform a small setpoint to confirm that the response is as expected.

Keep a log book for tuning parameters, changes including time and date, along with who made the changes. Some modern systems will electronically log changes and the user that made the tuning changes.

Step 8 – As needed, Follow-up on Tuning to Ensure Desired Performance

Control valves have moving parts and degrade over time, and the performance of these final control elements will impact the performance of tuning. A control loop that worked just fine in the past may not be working today due to hardware issues or a change in the operating parameters. If operating conditions change and will continue for awhile, a re-evaluation of the process and operating objectives along with new tuning parameters may be in order.

Conclusions

Young professional engineers have the ability to understand more about the dynamics of their processes and then use this information to improve process control performance. Modern loop tuning is a simple and inexpensive tool that any engineer can utilize. These steps listed above can be used to identify control valve deficiencies, understand overall process dynamics, and assist with tuning control loops. Additional examples and details will be shown in the presentation at the meeting.

REFERENCES

1. Dahlin E.B., *Designing and Tuning Digital Controllers,* Instr and Cont Syst, 41 (6), 77, 1968.
2. Morari M. and Zafiriou E., *Robust Process Control,* Prentice Hall, 1989.
3. Chien I-Lung and Fruehauf P.S., *Consider IMC Tuning to Improve Controller Performance*, Hydrocarbon Processing, Oct. 1990.
4. J. Martin, A.B. Corripio and C.L. Smith, *How to Select Controller Modes and Tuning Parameters from Simple Process Models*, ISA Transactions, Vol 15, April 1976, pp 314-319.

(106a) Synthesis of PVA Based Membranes Modified by Benzimidazole Group for Direct Methanol Fuel Cells

Pratima Gajbhiye, Chemical Engineering, Indian Institute of Technology, Kanpur, State of Uttar Pradesh, India, Jayant K. Singh, Chemical Engineering, Indian Institute of Technology, Kanpur, Ka, India and Anil Kumar, Department of Chemical Engineering, Indian Institute of Technology Kanpur, Kanpur, India

Abstract

In recent years, there has been a great effort to look for alternate membrane material and Polybenzimidazole (PBI) membranes have been synthesized where the conducting moiety of this polymer is benzimidazolium groups. The monomers used to synthesize these polymers are bis conjugated amines and diacids are expensive. In this work, we have synthesized grafted polymers of PVA as membrane materials where the grafts are either Poly Acrylic Acid, Poly maleic anhydride (PMA) or Poly Methacrylic acid (PMAA) and these materials were denoted by PVA-g-PAA, PVA-g-PMA, and PVA-g-PMAA. We then synthesized C1, C3 & C4 compounds (formula given at the bottom of Table S1) having bis conjugated amines whereas compound C2 having only one conjugated amine group. One of the steps in synthesizing C1 compound is nitration of phenyl ring, where there is a possibility of formation of two isomers out of which only one of them has strictly bis conjugated amines and can be polymerized with Isophthalic acid giving the formation of PBI. The other byproduct formed is an impurity and is minimized by carrying out the nitration step at low temperature but this cannot be completely removed from the product which is a mixture. Table S1 presents the summary of our experimental results on characterization of membrane, In the C1 to C4 compounds and in some cases (membranes 9, 10, 11), Sulfocuccinic acid were added to the casting polymer solution as additional crosslinking agent in order to achieve higher conductivity. We added the polymer, thus formed, to the casting solution, we first compared the results of the Fuel cell formed by adding C1 (C3 and C4) compound alone and heating at 180°C for some time. Table S1 has two sets of results, one on characterization (columns 1, 2, 3,4,5,6, and 7) of membranes while the other on Fuel cell (columns 8 to 11). For the latter columns 10 and 11 gave results on characterization on Fuel cell while columns 8 and 9 gave results when the Fuel cell was run on control mode for higher efficiency.

Experiments on membrane pore sizes are given in column 1 which lies between 8.88 to 9.52nm and they all are ultra filtration type. The values of the pore size are mostly governed by the nature of the graft chains, e.g. membranes 1,3,4,7 and 12 consist of PAA grafts and were found to have similar pore sizes. Similarly membranes 2,5,8,9 were of the similar pore size because the PVA have the same grafts of PMA. The next important thing is to determine the wall charges of these membranes and they were estimated by experimentally determining the rejection of Chromic (VI) acid solution through these membranes. Subsequently, a space charge model was proposed for the flow of the solution through their pores and a semi analytical solution was derived for the overall problem of separation of chromic acid solution. We observed that the high surface charge was because of the benzimidazolium groups present on the surface and they ranged from minimum of 465 to maximum of 525mV. The thermal degradation (TGA) experiments were carried out which showed that, through the modification of the polymer by benzimidazole groups, lead to stable membrane and the degradation temperature varies from 150°C for 10% weight loss and for approximately 80% weight reduction, a minimum of 300°C to 700°C.

The twelve membranes reported in Table S1 were first characterized completely in terms of pore size, wall charge, conductivity, IEC and thermal degradation. In fuel cell applications, higher temperatures give large current density and it is observed that some of the membranes (rows 1, 2, 5, 6, 7, 8 and 12) could easily be operated upto 150°C without the membrane degrading. However in our case, we insisted that the membrane of direct methanol fuel cell were operated at 90°C for which most of the fuel cell results for Nafion 117 are available. The catalysts used on the anode and cathode side were Pt-Ru and Pt loaded on carbon respectively, which were synthesized by following procedure given in literature. We find that in each case, the power density and the current density were higher and for some of our results (membranes 1,6,12) were atleast two to three times higher than that of Nafion 117 as seen in column 10 and 11.

It was recognized that the membrane material has hydrophilic PVA matrix and hydrophobic benzimidazolium groups at the surface formed by reaction of the C1 to C4 compounds with grafts (see Fig.S-1). When the reaction occurs (confirmed by IR and NMR), the hydrophobic benzimidazolium are formed which tends to separate as confirmed by the SEM results. In view of this we assumed that the pores of the membrane would have a thin hydrophobic layer with positive charge walls. This is confirmed by the contact angle measurements where the minimum contact angle is 51° and varied upto 89°. Even though most of the polymer matrix is PVA with its grafts, when the ions move through the pores (giving rise to current in Fuel cell), the pore walls get highly charged (as measured by column 2 of Table S1). Some of the membranes have Sulfo succinic acid SSA as part of the matrix (e.g. the membrane 9 to 11) which would tend to give sulfonate groups inside the PVA matrix as well as at the phase boundaries, this way giving rise to bipolar nature of the membrane (benzimidazolium group being strictly cationic in nature). In addition to this, some of the compounds such as C1 and C2 compounds have bonded sulfonate groups which would give the positive and negative charge separated by the molecular distance, once again producing a bipolar nature to the membrane. In remainder of the cases the charges would be induced on the phase boundaries as seen in Fig.S-1. The conductivity observed in our membranes is very high and could be attributed to the special bipolar structure of our polymers.

Table S1. Fuel cell and Membrane Characterization

Compound	Membrane	1	2	3	4	5	6	7	8	9	10	11

*** Formatting Issue - Best Available Paper/Graphic ***

			MWCO (pore size, nm)	Wall Charge	Conductivity (S/cm)	IEC	Thermal Degradation temp (°C) 10% - 80%		Water Uptake	Contact Angle	Average Voltage Gain (V)	Average time cycle (min)	Power Density (mW/cm)	Current Density (mA/cm)
C1	1	PVA-g-PAA-C1	9.1	488	0.04	2.7	199.732	268.70	7.5	72	0.523	55.75	50.47	74.88
	2	PVA-g-PMA-C1	9.35	465	1.23×10^{-2}	1.8	142.659	447.6	10.691	74.917	0.568	42.896	0.305	3.81
	3	PVA-g-PAA-modified	9.52	488	1.51×10^{-2}	2.12	115.123	706.53 (28.93%)	13.665	75.581	0.778	69.85	37.37	64.44
C2	4	PVA-g-PAA-C2	9.52	525	6.59×10^{-2}	2.38	119.558	691.93	13.4	51.2	0.704	39.83	23.0020	50.55
	5	PVA-g-PMA-C2	9.36	519	1.04×10^{-2}	2.17	181.59	562.348	14.9	50.24	0.77	161.93	34.72	62.11
	6	PVA-g-PMAA-C2	8.99	482	1.16×10^{-2}	2.18	201.055	472.568	13.8	51.8	0.6066	147.81	76.72	92.33
C3	7	PVA-g-PAA-C3	9.0	512	2.51×10^{-2}	2.23	204.908	453.03	8.624	70.368	0.6883	82.55	36.397	75
	8	PVA-g-PMA-C3	9.5	510	3.51×10^{-2}	2.27	166.819	505.293	13.513	62.944	0.726	37.557	36.11	64.381
	9	PVA-g-PAA-SSA-C3	9.36	497	1.71×10^{-2}	2.51	154.914	813.769 (24.32%)	10.5	54.8	0.8096	99.423	23.45	57.778
	10	PVA-g-PMA-SSA-C3	9.83	480	2.28×10^{-2}	2.46	115.12	428.713	8.5	76.2	0.663	224.22	28.561	58.358
	11	PVA-g-PMAA-SSA-C3	8.88	525	3.66×10^{-2}	1.85	139.772	681.715 (30.78%)	7.2	89.2	0.77	18.36	20.832	48.11
C4	12	PVA-g-PAA-C4	9.5	495	2.56×10^{-2}	1.62	189.87	451.86	8.927	59.132	0.704	74.19	41.1	71
	13	Nafion 117			0.1								21.01	48.33

Compound C1, [Bis-3-Amino-4-{3-(triethylammoniumsulfonato) phenylamino}-phenyl sulfone] hydrochloride

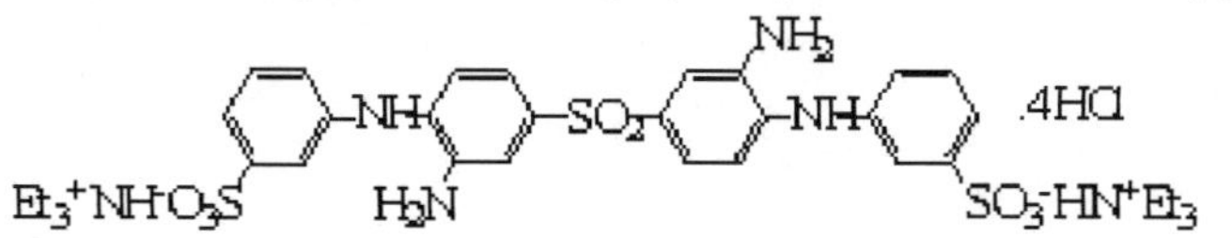

*** Formatting Issue - Best Available Paper/Graphic ***

Compound C2, 3-Amino -4-[3-(triethyl ammonium sulfonato) phenyl amino] phenylene hydrochloride

---

Compound C3, Bis -4 [1, 2 diamino phenyl]-phenyl sulfone) hydrochloride

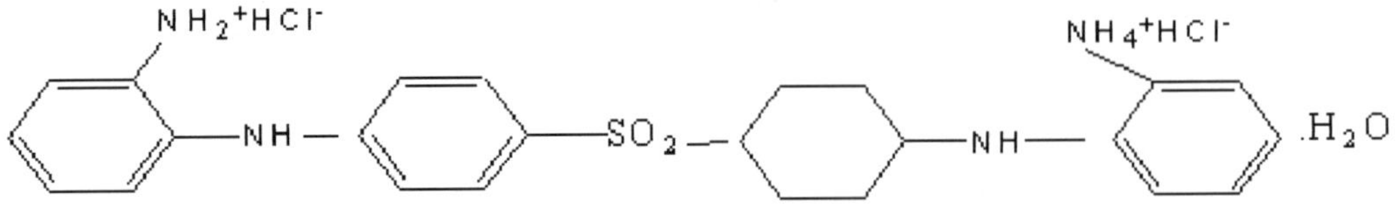

Compound C4, 1, 4 [1, 2 diamino phenyl] Phenylene hydrochloride

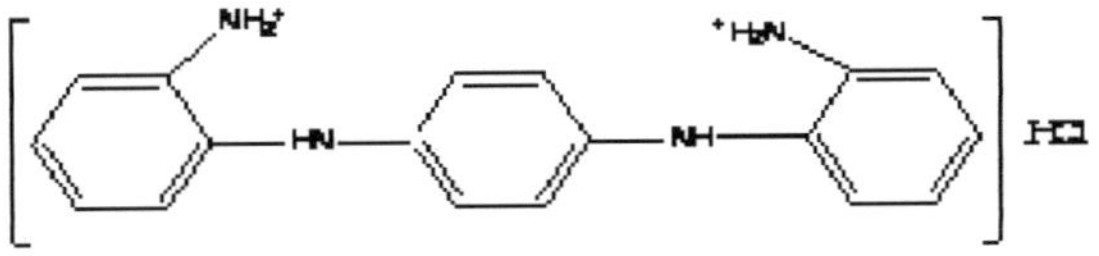

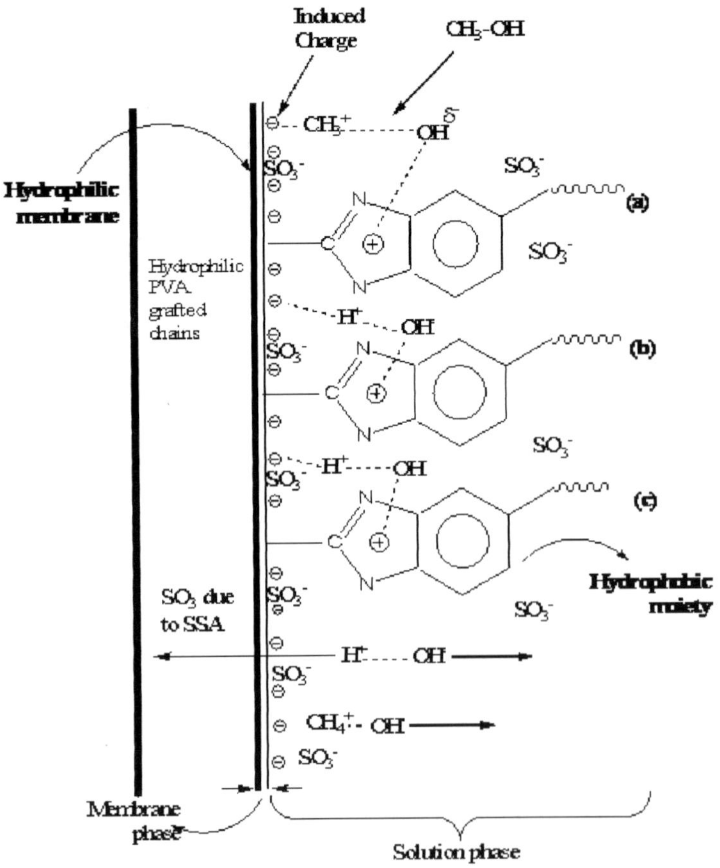

Figure S-1. The Bipolar nature of the membrane surface

*** Formatting Issue - Best Available Paper/Graphic ***

(106b) Operation and Controls of Operating Direct Methanol Fuel Cells with PVA Based Membranes Modified by Benzimidazolium Groups

Pratima Gajbhiye, Chemical Engineering, Indian Institute of Technology, Kanpur, State of Uttar Pradesh, India, Pranjal Joshi, Chemical Engg, IIT Kanpur, Kanpur, India, Jayant K. Singh, Chemical Engineering, Indian Institute of Technology, Kanpur, Ka, India and A. Kumar, Department of Chemical Engineering, Indian Institute of Technology Kanpur, Kanpur, India

Operation and Control of Direct Methanol Fuel cell with PVA based membranes modified with benzimidazolium groups

Pratima Gajbhiye, Pranjal Joshi, Jayant K.Singh, Anil Kumar

Department of Chemical Engineering, Indian Institute of Technology, Kanpur 208016, India

Abstract

In recent years, in an effort to find alternate membrane material (other than Nafion 117), Polybenzimidazole (PBI) membranes have been synthesized where the conducting moiety of this polymer is benzimidazolium groups. The monomers used to synthesize these polymers are bis conjugated amines and diacids are expensive. In this work, we have synthesized grafted polymers of PVA as membrane materials where the grafts are either Poly Acrylic Acid, Poly maleic anhydride (PMA) or Poly Methacrylic acid (PMAA) and these materials were denoted by PVA-g-PAA, PVA-g-PMA, and PVA-g-PMAA. We then synthesized C1, C3 & C4 compounds (formula given at the bottom of Table S1) having bis conjugated amines whereas compound C2 having only one conjugated amine group. One of the steps in synthesizing C1 compound is nitration of phenyl ring, where there is a possibility of formation of two isomers out of which only one of them has strictly bis conjugated amines and can be polymerized with Isophthalic acid giving the formation of PBI. The other byproduct formed is an impurity and is minimized by carrying out the nitration step at low temperature but this cannot be completely removed from the product which is a mixture. Table S1 presents the summary of our experimental results on characterization of membrane, In the C1 to C4 compounds and in some cases (membranes 9, 10, 11), Sulfocuccinic acid were added to the casting polymer solution as additional crosslinking agent in order to achieve higher conductivity. We added the polymer, thus formed, to the casting solution, we first compared the results of the Fuel cell formed by adding C1 (C3 and C4) compound alone and heating at 180°C for some time. Table S1 has two sets of results, one on characterization (columns 1, 2, 3,4,5,6, and 7) of membranes while the other on Fuel cell (columns 8 to 11). For the latter columns 10 and 11 gave results on characterization on Fuel cell while columns 8 and 9 gave results when the Fuel cell was run on control mode for higher efficiency.

Experiments on membrane pore sizes are given in column 1 which lies between 8.88 to 9.52nm and they all are ultra filtration type. The values of the pore size are mostly governed by the nature of the graft chains, e.g. membranes 1,3,4,7 and 12 consist of PAA grafts and were found to have similar pore sizes. Similarly membranes 2,5,8,9 were of the similar pore size because the PVA have the same grafts of PMA. The next important thing is to determine the wall charges of these membranes and they were estimated by experimentally determining the rejection of Chromic (VI) acid solution through these membranes. Subsequently, a space charge model was proposed for the flow of the solution through their pores and a semi analytical solution was derived for the overall problem of separation of chromic acid solution. We observed that the high surface charge was because of the benzimidazolium groups present on the surface and they ranged from minimum of 465 to maximum of 525mV. The thermal degradation (TGA) experiments were carried out which showed that, through the modification of the polymer by benzimidazole groups, lead to stable membrane and the degradation temperature varies from 150°C for 10% weight loss and for approximately 80% weight reduction, a minimum of 300°C to 700°C.

The twelve membranes reported in Table S1 were first characterized completely in terms of pore size, wall charge, conductivity, IEC and thermal degradation. In fuel cell applications, higher temperatures give large current density and it is observed that some of the membranes (rows 1, 2, 5, 6, 7, 8 and 12) could easily be operated upto 150°C without the membrane degrading. However in our case, we insisted that the membrane of direct methanol fuel cell were operated at 90°C for which most of the fuel cell results for Nafion 117 are available. The catalysts used on the anode and cathode side were Pt-Ru and Pt loaded on carbon respectively, which were synthesized by following procedure given in literature. We find that in each case, the power density and the current density were higher and for some of our results (membranes 1,6,12) were atleast two to three times higher than that of Nafion 117 as seen in column 10 and 11. In normal Fuel cell operations, there is decay in the cathode catalyst which is further increased by the methanol crossover. The voltage of the Fuel cell increases, the average volatage gained is reported in column 8 and it is found that the voltage gain depends on the nature of the membrane which is not the same for all of them. In column 9, we have measured the average time cycle and we found that some of the membranes (membranes 6, 7, 9, 10) gave extremely good results.

It was recognized that the membrane material has hydrophilic PVA matrix and hydrophobic benzimidazolium groups at the surface formed by reaction of the C1 to C4 compounds with grafts (see Fig.S-1). When the reaction occurs (confirmed by IR and NMR), the hydrophobic benzimidazolium are formed which tends to separate as confirmed by the SEM results. In view of this we assumed that the pores of the membrane would have a thin hydrophobic layer with positive charge walls. This is confirmed by the contact angle measurements where the minimum contact angle is 51° and varied upto 89°. Even though most of the polymer matrix is PVA with its grafts, when the ions move through the pores (giving rise to current in Fuel cell), the pore walls get highly charged (as measured by column 2 of Table S1). Some of the membranes have Sulfo succinic acid SSA as part of the matrix (e.g. the membrane 9 to 11) which would tend to give sulfonate groups inside the PVA matrix as well as at the phase boundaries, this way giving rise to bipolar nature of the membrane (benzimidazolium group being strictly cationic in nature). In addition to this, some of the compounds such as C1 and C2 compounds have bonded sulfonate groups which would give the positive and negative charge separated by the molecular distance, once again producing a bipolar nature to the membrane. In remainder of the cases the charges would be induced on the phase boundaries as seen in Fig.S-1. The conductivity observed in our membranes is very high and could be attributed to the special bipolar structure of our polymers.

Table S1. Fuel cell and Membrane Characterization

Compound	**Membrane**		**1**	**2**	**3**	**4**	**5**		**6**	**7**	**8**	**9**	**10**	**11**
			MWCO (pore size, nm)	**Wall Charge**	**Conductivity (S/cm)**	**IEC**	**Thermal Degradation temp (°C) 10% - 80%**		**Water Uptake**	**Contact Angle**	**Average Voltage Gain (V)**	**Average time cycle (min)**	**Power Density (mW/ cm)**	**Current Density (mA/ cm)**
C1	1	PVA-g-PAA-C1	9.1	488	0.04	2.7	199.732	268.70	7.5	72	0.523	55.75	50.47	74.88
	2	PVA-g-PMA-C1	9.35	465	1.23×10^{-2}	1.8	142.659	447.6	10.691	74.917	0.568	42.896	0.305	3.81
	3	PVA-g-PAA-modified	9.52	488	1.51×10^{-2}	2.12	115.123	706.53 (28.93%)	13.665	75.581	0.778	69.85	37.37	64.44
C2	4	PVA-g-PAA-C2	9.52	525	6.59×10^{-2}	2.38	119.558	691.93	13.4	51.2	0.704	39.83	23.0020	50.55
	5	PVA-g-PMA-C2	9.36	519	1.04×10^{-2}	2.17	181.59	562.348	14.9	50.24	0.77	161.93	34.72	62.11
	6	PVA-g-PMAA-C2	8.99	482	1.16×10^{-2}	2.18	201.055	472.568	13.8	51.8	0.6066	147.81	76.72	92.33
C3	7	PVA-g-PAA-C3	9.0	512	2.51×10^{-2}	2.23	204.908	453.03	8.624	70.368	0.6883	82.55	36.397	75
	8	PVA-g-PMA-C3	9.5	510	3.51×10^{-2}	2.27	166.819	505.293	13.513	62.944	0.726	37.557	36.11	64.381
	9	PVA-g-PAA-SSA-C3	9.36	497	1.71×10^{-2}	2.51	154.914	813.769 (24.32%)	10.5	54.8	0.8096	99.423	23.45	57.778
	10	PVA-g-PMA-SSA-C3	9.83	480	2.28×10^{-2}	2.46	115.12	428.713	8.5	76.2	0.663	224.22	28.561	58.358
	11	PVA-g-PMAA-	8.88	525	3.66×10^{-2}	1.85	139.772	681.715	7.2	89.2	0.77	18.36	20.832	48.11

*** Formatting Issue - Best Available Paper/Graphic ***

		SSA-C3						(30.78%)						
C4	12	PVA-g-PAA-C4	9.5	495	2.56×10^{-2}	1.62	189.87	451.86	8.927	59.132	0.704	74.19	41.1	71
	13	Nafion 117			0.1								21.01	48.33

Compound C1, [Bis-3-Amino-4-{3-(triethylammoniumsulfonato) phenylamino}-phenyl sulfone] hydrochloride

Compound C2, 3-Amino -4-[3-(triethyl ammonium sulfonato) phenyl amino] phenylene hydrochloride

Compound C3, Bis -4 [1, 2 diamino phenyl]-phenyl sulfone) hydrochloride

Compound C4, 1, 4 [1, 2 diamino phenyl] Phenylene hydrochloride

Figure S-1. The bipolar nature of PVA-g-PAA-BASPAPS membrane which reduces the methanol crossover due to the molecular dissociation. In (a), the imidazolium group is shown to separate out from the PVA moiety of the chain into a separate phase. In (b), the water is shown to complex with the SO and the imidazolium group. As a result of this, interaction in (d) and (e), the water and the methanol is shown to split into anion and cation and the anion permeates through the membrane giving bipolar effect

(106e) Numerical Investigation in Heat Conduction Effect on Heat Exchange for Mini Cross-Flow Heat Exchanger (MCHE)

Hongyue Duan[1], Wei Mu[1], **Xigang Yuan**[1] and Lingai Luo[2], (1)State Key Laboratory of Chemical Engineering, Tianjin University, Tianjin, China, (2)LOCIE Chambery, Savoie University, Savoie, France

In a traditional heat exchanger, the effect of heat transfer within the wall material driven by the temperature gradient in axial direction of the fluid flow is usually very small because the area of the cross section of the tube wall is much smaller than that of the tube for heat exchange. However, in a mini cross-flow heat exchanger(MCHE), where the heat conducting medium is typically a cubic-shaped block of metal and the hot and cold streams are put into heat exchange via two sets of channels in the block, the heat conduction in the wall material in the direction parallel to the fluid flow may have a significant effect on the performance of the heat exchanger.

In this paper, a 2-channel-layer MCHE was simulated and a periodic boundary condition was applied so that the results could be applicable to entire MCHE. The proposed model and the mesh grid size were proved by the good agreement of the simulated overall heat transfer coefficient with that found in the experimental reported in the literature[1]. Then the effect of heat conduction on heat transfer performance in a MCHE was studied numerically. The simulation results suggest that the performance of the MCHE can be improved by increasing the wall thickness between the adjoining channels in the same layer or by decreasing the wall thickness between the adjacent layers. The heat conduction effect can also be weakened by optimizing the material conductivity or by increasing the diameter of the channels. New correlations were developed based on the simulated results to predict the value of the overall heat transfer coefficient as a function of the operating parameter such as Re in the channel and the design parameters of the MCHE such as the diameter of the channel, wall thickness and thermal conductivity.

[1]Luo L A, Fan Y L, Zhang W W*et al.*Integration of constructal distributors to a mini crossflow heat exchanger and their assembly configuration optimization[J]. Chemical Engineering Science, 2007, 62 (13): 3605-3619.

(106g) The Effects of Out Gassing Through Cylindrical Walls on the Transport Coefficients

Luma. H. Yaseen[1], **Nada B. Nakkash**[2] and M. N. Latif[1], (1)Al-Nahrain University, Baghdad, Iraq, (2)Chemical Engineering, Al-Nahrain University, Baghdad, Iraq

Abstract

This work concerns with studying the influence of blowing on friction factor and mass transfer coefficient when the flow is normal to a tube bank with outgassing porous cylinders. The friction factor and the mass transfer coefficients are measured for a range of Reynold's number and blowing rate and expressed in terms of various dimensionless groups.

The test rig consists of a rectangular duct (20.3 × 11.3 cm), five rows of three porous tubes in staggered arrangements fixed across the duct with 1.5pitches (both longitudinal and transversal). The outside and the inside diameters of tubes were 25.4 and 21.4 mm, respectively.

The friction factor measurements was studied for Reynold's number range of

$3000 < Re < 17{,}000$ and blowing rate up to 4×10^{-3}. The results show that the blowing intothe boundary layer for the cylinders increases the overall friction factor from 10%to 50% depending on the blowing rate. The mass transfer coefficient was measured for Reynold's number range of $5000 < Re < 16{,}000$ and blowing rate of $10^{-5} < m/G_\infty < 10^{-3}$.

A new definition of the blowing rate was considered and expressed by the mass transfer driving force. Two successful formulas were obtained to account the outgassing effect on the mass transfer coefficient. The mass transfer coefficient for the outgassing boundary layer was obtained first by determining the non-blowing coefficient and multiplying it with the correction factor .

The first model was modified to account the density variation associated with the boundary layer change:

$$St_m Sc^{2/3} = \frac{0.51\, Re^{-0.5}}{(1 + \beta_m)^{4.7}} \left(\frac{\rho_s}{\rho_\infty}\right)^{0.4}$$

With the coefficient of determination R2 = 0.93.

The second model depends on the flat plate correction factor:

$$St_m Sc^{2/3} = 0.5\, Re^{-0.5} \left(\frac{Ln(1+ \beta_m)}{\beta_m}\right)^{9.94}$$

With the coefficient of determination R2 = 0.92.

These two models are found to be in agreement with the previous literature and a reduction of about 80% in the mass transfer coefficient has been obtained due to the outgassing.

This page intentionally left blank.

(106h) Down the Hubbert Curve

Ian Sutton, Sutton Technical Books, Houston, TX

Down the Hubbert Curve

In March 1956 M. King Hubbert, who was at that time Chief Geologist with the Shell Development Company published a seminal paper with the title "Nuclear Energy and the Fossil Fuels" (Hubbert 1956). In this lengthy and authoritative paper, which was presented before the American Petroleum Institute in San Antonio, Texas, Dr. Hubbert showed how key fossil fuel resources in the United States, particularly coal and oil, were of finite size and that their output would reach a maximum, and then decline. Dr. King's forecasts for the production of oil in the United States were accurate; he predicted that oil production would peak in the early 1970s, and that is what has happened. (The widely-used term "Peak Oil" is frequently accredited to Dr. Colin Campbell (2001)).

In the fifty years since the publication of Dr. King's paper great advances have been made in finding and extracting oil from what were considered to be marginal fields. But Dr. King's overall conclusion is still valid. Production of oil in the United States remains below the 1970's peak. Although Dr. Hubbert's analysis was confined to the United States, subsequent analysts applied his insights to other parts of the world. For example, the book *Twilight in the Desert: The Coming Saudi Oil Shock and the World Economy* (Simmons 2005) by the late Matthew Simmons achieved wide recognition by showing that the concept of peak oil most likely applies to reserves in Saudi Arabia, and other nations such as Iran and Kuwait.

The purpose of this paper is to examine how oil production and consumption have changed in the last five years, and to examine whether or not the concept of Peak Oil remains valid. Key to this discussion is an understanding of the concept of Energy Returned on Energy Invested (EROEI). It is concluded that the concept of Peak Oil, as first postulated by Dr. King, remains valid and current, and the production of oil peaked in the period 2005-2011 and is now set to move down the right hand side of the Hubbert Curve.

The title of this session is *Challenges in Instituting Energy Savings Project in the Present Market Conditions*. If the production of oil and has indeed already passed a peak then the challenge is not to do with "instituting" energy savings; those savings are going to occur whether we like it or not. The challenge will be to determine how the process industries can restructure themselves given that the energy resources available (both as direct energy and as petrochemical feedstock) will be in decline and that there will be intense demand for those same resources from other areas of the economy.

(106j) Statistical Decision Making for In-Service Inspection of Underground Waste Storage Tanks

Emily Mitchell, Emory University, Atlanta, GA and **Stephen P. Harris**, Savannah River National Laboratory, Aiken, SC

Pitting is a localized form of corrosion that occurs on passivated metal surfaces exposed to aggressive chemical environments. The pitting process involves the breakdown of the passive oxide film and subsequent anodic dissolution of the metal at a local site, while the remainder of the surface remains protected by the film.

Deep pits may result in lowering the maximum level of waste that made be stored in the tank and eventually lead to loss of confinement. Pit depths, established by ultrasonic testing, are subject to variables that cause uncertainty, including the repeatability of metric location, time between measurements, different transducers, and different inspectors.

The generalized extreme value (GEV) distribution is fit to the data gathered from ultrasonic testing and statistical projections are made for the full tank wall interior. A comparison is made with results from other well known statistical distributions. The estimated parameters are used in statistical tolerance limits for bounding pit depth in a waste tank.

The tolerance limits give an indication of extreme values of pit depth for the current condition of the tank walls. The upper tolerance limit for pit depth will then be compared to action levels for pitting. If the upper tolerance limit for pit depth exceeds the action level, more frequent examinations or increased tank surface area inspections are conducted.

(106k) Thermal Inbreathing Requirements of Low Pressure Storage Tanks At Elevated Temperatures

Donald E. Brooks, Pressure Relief Analysis, Lloyd's Register Energy Americas, Inc., Houston, TX

ISO 28300 (API 2000) "Petroleum, petrochemical and natural gas industries - Venting of atmospheric and low pressure storage tanks" provides guidance on determining the required thermal inbreathing flow rates to protect atmospheric and low pressure storage tanks against vacuum due to weather changes (e.g. a decrease in atmospheric temperature)." For heated, un-insulated tanks with vapor space temperatures above 58.9 °C (120 "F), ISO 28300 recommends an engineering review of thermal inbreathing requirements.

This paper describes one approach to evaluating thermal inbreathing requirements for un-insulated storage tanks with vapor space temperatures exceeding 58.8 °C (120 "F)." The tank is assumed to be exposed to the cooling effects of a sudden rain. A simplified unsteady state heat transfer analysis is performed to determine the tank wall and vapor space temperatures versus time. Initial gas mass and pressure are assumed to remain constant in the tank. Gas volume in the tank is calculated as a function of vapor space temperature using the ideal gas law. The rate of volume change is calculated for each time increment. The rate of volume change goes through a maximum value. This maximum value is selected as the thermal inbreathing requirement of the tank.

The calculation methodology described in this paper can be programmed in a spreadsheet for ease of use.

Dynamic Operational Risk Assessment with Bayesian Networks

Shubharthi Barua, Xiaodan Gao, M. Sam Mannan
Mary Kay O'Connor Process Safety Center
Texas A&M University, Jack E. Brown Engineering Bldg., 3122 TAMU, Room 244, College Station, TX 77843
shubharthi.barua@tamu.edu, xiaodan_gao08@mail.che.tamu.edu, mannan@tamu.edu

Extended Abstract Prepared for the Poster Presentation at the
8th Global Congress on Process Safety
Houston, TX
April 1 – 4, 2012

UNPUBLISHED

8TH GLOBAL CONGRESS ON PROCESS SAFETY

Dynamic Operational Risk Assessment with Bayesian Networks

Shubharthi Barua, Xiaodan Gao, M. Sam Mannan

Mary Kay O'Connor Process Safety Center
Texas A&M University, Jack E. Brown Engineering Bldg., 3122 TAMU Room 244, College Station, TX-77843
shubharthi.barua@tamu.edu, xiaodan_gao08@mail.che.tamu.edu, mannan@tamu.edu

Keywords: risk assessment, Bayesian network, dynamic Bayesian network, fault tree, event tree

Abstract

Oil/gas and petrochemical plants are complicated and dynamic in nature. Dynamic characteristics include ageing of equipment/components, season changes, stochastic processes, operator response times, inspection and testing time intervals, and timing of safety system operations, all of which are time dependent criteria that can influence dynamic processes. Conventional risk assessment methodologies can quantify dynamic changes in processes with limited capacity. Several methods are proposed for dynamic risk assessment such as C-DET, D-DET, DORA, etc., but all of them have deficiencies of some kind. Bayesian networks, a graphical inference technique, have some advantageous characteristics for risk analysis over other techniques. This research focuses on providing a framework for dynamic risk analysis based on Bayesian networks. A case study will be provided to illustrate this model.

1.0 Introduction

Oil/gas, chemical, and petrochemical process plants consist of numerous equipment and unit operations, thousands of control loops, and exhibit dynamic behavior. Different hazards and several types of risk are associated with these process facilities. For example, large process areas have the potential to stock a large amount of hazardous chemicals, which can result in the process area being highly congested. It is reported that process risk, i.e., risk in process facilities due to fires and explosions, greatly contributes to the total risk of offshore installations. Also, reactivity, toxic, and mechanical hazards have a significant effect on the total risk of these facilities. Therefore, it is very important to identify hazards, perform risk assessments, and take proper initiatives to minimize/remove hazards and risks; else a catastrophic accident may result.

From case histories, it has been observed that catastrophic accidents have a significant effect on people, environment, and society. Catastrophic accidents such as the Flixborough disaster, the Bhopal incident, and the Piper Alpha disaster caused fatalities and unbearable economic loss. In the past few years, several disastrous accidents in refineries, power plants, and offshore platforms have significantly affected people's perception, and consequently raised concern to emphasize process safety. A series of explosions in 2005 resulting from insufficient process safety and lack of risk reduction measures at the BP Texas City refinery accident resulted in 15 fatalities, 180 injuries, and $1.5 billion in losses [1]. The Kleen Energy power plant explosion in

Connecticut in 2010 is considered to be one of the worst industrial explosions in the USA [2]. In 2010, a fire and explosion resulting from a blowout at the Macondo well resulted in 11 deaths and 17 injuries. Also the continuous spill from the wellhead for 87 days had disastrous effects on the environment and wildlife surrounding the Gulf of Mexico. It has been reported that most of the accidents could be prevented or their severity could have been minimized by effective risk assessment and process safety management. But existing risk assessment methodologies cannot properly quantify various time-dependent effects. Therefore, the objective of this research is to develop a risk assessment methodology with the capability to capture variations due to different time-dependent effects.

2.0 Literature Review

N. Siu [3] summarized available methodologies for risk assessment. For example, the fault tree and event tree methods are two most common methodologies for risk assessment. But both techniques have limited ability to address different time-dependent effects such as equipment or component ageing, stochastic processes, operator response times, and inspection and testing time intervals. N. Siu [3] also described other methodology limitations while applying dynamic system analyses. Therefore, for dynamic systems, conventional techniques are not able to provide accurate results because of their inadequate ability to describe the variation of operational risks as time-dependent deviations, or the changes occurring in the process. Continuous event trees [4], discrete event trees [5], and Markov theory [6] methods have been proposed for dynamic reliability and safety analysis. But all of the proposed techniques have some deficiencies. For example, the Markov theory based model does not consider the effect of inspection on system-state transitions. X. Yang et al. [1] proposed the Dynamic Operational Risk Assessment (DORA) technique, which is able to take inspection and testing time intervals and repair time into account for evaluating operational failure.

Recently, Bayesian networks are receiving wide acceptance in reliability assessment, fault diagnosis, safety, risk, and maintenance analysis. Several authors have [7,8] compared some commonly used methods for accident analysis with Bayesian networks, and described several advantages of Bayesian networks over these methods. N. Khakzad et al. [2] illustrated parallelism between Bayesian networks and fault trees, and discussed the potential of Bayesian networks as a tool for process safety and accident analysis. H. Pasman and W. Rogers [9] described an approach to incorporate Bayesian networks into layer of protection analysis to make it more effective. These studies demonstrate the potential advantages of Bayesian networks application in risk assessment. This research provides a methodology for dynamic risk assessment based on Bayesian networks.

3.0 Methodology and Expected Result

3.01 Bayesian Networks

A Bayesian network, which is a probabilistic graphical model, describes causal influence relations among variables. A set of random variables and their conditional dependencies are represented by Bayesian networks via a directed acyclic graph (DAG). It has the ability to represent dependency among events clearly, accommodate multi mode and continuous random variables, and incorporate information i.e. generic, system specific and expert judgment to support optimum decision making.

A simple Bayesian network is shown in Figure 1. In a binary network, variables are represented by nodes and causal relationships among different nodes are shown by an arc. Conditional probability tables or defined probabilistic relations among nodes represent how one variable is linked with one or many variables. The nodes that influence another variable and have unconditional probability are called parent or root nodes. Nodes that are conditionally dependent on their direct parents are called intermediate nodes. The top node is defined as a leaf node, which is similar to the top event of a fault tree and initiating event for an event tree.

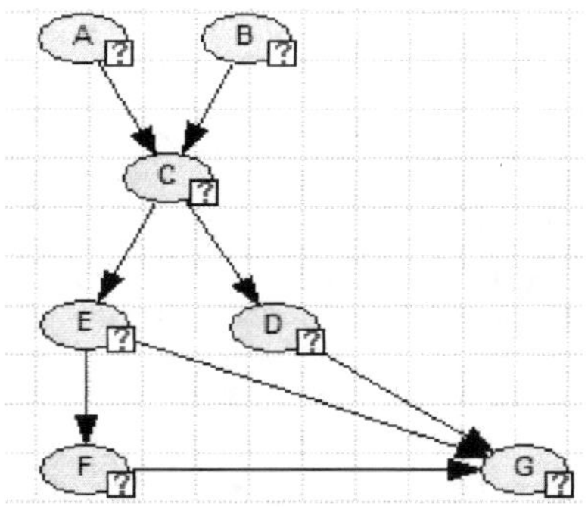

Figure 1: A simple Bayesian network

Let $N = (G, P)$ be a Bayesian network, where

$G = (V, E)$ is a directed acyclic graph; V (random variables) represents nodes; and E represents edges between pairs of nodes of DAG.

P represents probability distribution over V and $V = \{X_1, X_2, \dots, X_n\}$ is either discrete or continuous random variables [10]. These random variables are assigned to the nodes and the edges.

Bayesian networks can be represented by the joint probability distribution P(V);

$$P(V) = \sum_{X \in V} P\{X|pa(X)\} = P(X_1, X_2, \dots, X_N) = \sum_{i=1}^{n} P\{X_i|pa(X_i)\}$$

Here $pa(X_i)$ = parent nodes of X_i.

A main advantage of the application of Bayesian networks in risk analysis is the ability to update prior data using Bayes' theorem by incorporating new information as demonstrated by N. Khakzad et al. [2]. Also, Bayesian networks have an advantage of handling different types of uncertainty.

3.02 Dynamic Bayesian Networks

A general Bayesian network is a static model, i.e., the joint probability distribution is usually a representation of a fixed point or an interval of time [11]. A dynamic Bayesian network is a Bayesian network that has the ability to describe the evolution of joint probability distribution over time. Dean and Kanazawa [12] proposed modeling time in a discrete manner to represent the progression of time in a dynamic Bayesian network. In a dynamic Bayesian network, arcs links node/nodes from previous time slice to the node/nodes of the next time slice to represent temporal dependencies among them.

3.03 Methodology

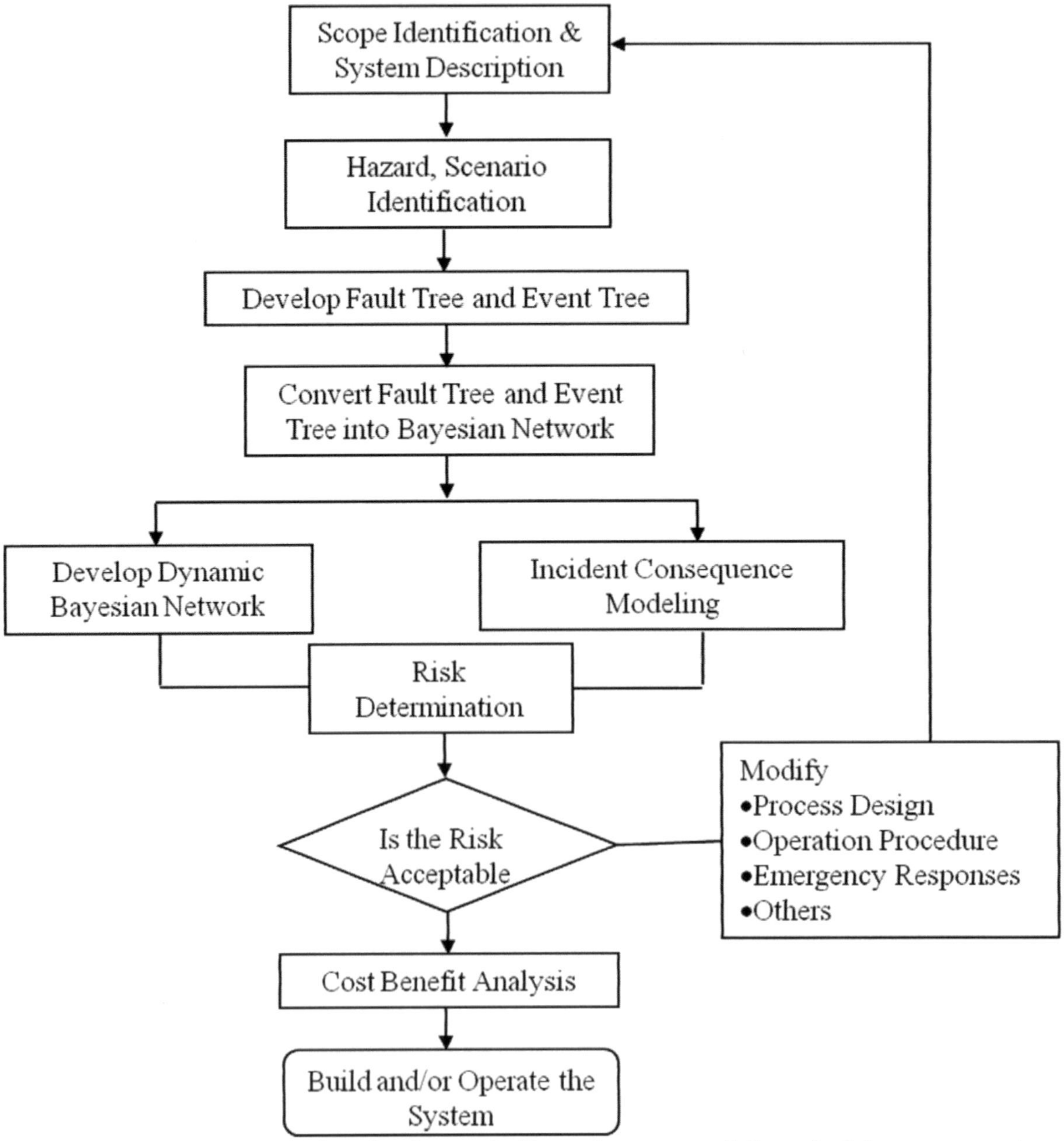

Figure 2: Framework for the dynamic Bayesian network based risk assessment

The objective of this research is to propose a dynamic risk assessment method based on dynamic Bayesian networks. According to the framework, a fault tree and event tree are developed for a system. The objective for following step is to convert the developed fault and event tree into a Bayesian network. Bobbio et al. [13] provided a methodology for conversion of fault trees into Bayesian networks. N. Khakzad et al. [2] showed equivalency between fault trees and Bayesian networks, and described the advantages of Bayesian networks for process safety analysis. Bearfield and March [14] proposed another methodology for generalizing event trees into Bayesian networks. The method has the ability to generalize simple event trees into Bayesian networks in a relatively easy manner, but the method becomes complicated for large event trees. Therefore, in the first phase, a methodology is proposed to convert fault trees and event trees into a static Bayesian network. In the second phase, a dynamic Bayesian network is developed from the static Bayesian network. The dynamic Bayesian network provides probability of a system at different time-slices for different states of the system. The probabilities are then combined with consequence modeling levels to calculate risk. GeNIe, which was developed at the Decision Systems Laboratory at the University of Pittsburgh, is used to create the model [15].

4.0 Conclusions

This article briefly describes a framework based on Bayesian network for dynamic risk assessment for the oil/gas, chemical, and petrochemical industries. This research illustrates the capability of Bayesian networks to represent all component interactions in complex systems, and to update all prior system probabilities in the presence of new information. The advantages of modeling flexibility and efficient computational mechanisms using dynamic Bayesian networks are also discussed.

4. References

[1] Yang, X. and Mannan, M.S., "The development and application of dynamic operational risk assessment in oil/gas and chemical process industry", Reliability Engineering and System Safety 95 (2010) 806-815

[2] Khakzad, N., Khan, F. and Amyotte,P., "Safety analysis in process facilities: Comparison of fault tree and Bayesian network approaches", Reliability Engineering and System Safety 96(2011) 925-932

[3] Siu,N. "Risk assessment for dynamic systems: an overview", Reliability Engineering and System Safety 43 (1994) 43-73

[4] Devooght, J. and Smidts,C. "Probabilistic reactor dynamics-I: the theory of continuous event trees", Nuclear Science and Engineering 1992; 111:229-40

[5] Amendola, A. "Accident sequence dynamic simulation versus event trees", Reliability Engineering and System Safety 1988; 22:3-25

[6] Howard, R. "Dynamic probability systems", Vols. I and II, New York, 1971

[7] Sklet, S. "Comparison of some selected methods for accident investigation", Journal of Hazardous Materials 2004; 111:29-37

[8] Nivolianitou, Z.S., Leopoulos, V.N. and Konstantinidou, M., "Comparison of techniques for accident scenario analysis in hazardous systems", Journal of Loss Prevention in the Process Industries 2004; 17"467-75

[9] Pasman, H. and Rogers, W.; "BBN, a tool to make LOPA more effective, QRA more transparent and flexible, and therefore to make safety more definable", 14th Annual Sysposium, Mary Kay O'Connor Process Safety Center, College Station, Texas, October 2011

[10] Donohue, S.K. and Dugan, J.B.; "Modeling the "Good enough to release" decision usin V&V preference structure and Bayesian belief networks", Reliability and Maintainability Symposium, 2003 Annual, 2003

[11] McNaught, K.R. and Zagorecki, A. "Using Dynamic Bayesian networks for prognostic modeling to inform maintenance decision making", IEEM 2009, Hond Kong, January 2010

[12] Dean, T. and Kanazawa, K., "A model for reasoning about persistence and causation", Computational Intelligence 1989; 5:142-150

[13] Bobbio A, Portinale L, Minichino M and Ciancamerla E., "Improving the analysis of dependable systems by mapping fault trees into Bayesian networks", Reliability Engineering and System Safety 71(3): 249-260

[14] Bearfield G and Marsh W, "Generalising event trees using Bayesian networks with a case study of train derailment", Available online at http://www.agenarisk.com/resources/white_papers/Generalising_Event_Trees.pdf , last modified: 23 February 2010, accessed: 7 January 2012

[15] Decision Systems Laboratory, University of Pittsburgh, GeNIe (Graphical Network Interface and SMILE (Structural Modeling, Inference, and Learning Enginer), Version 2.0 software, 2010. Available online at http://genie.sis.pitt.edu

(137b) Comparing the Performance of Compact Heat Exchange Reactor to CSTR for Fischer Tropsch Synthesis

Hani Gadalla[1], Zhijun Jia[1], Steven J. Vallee[1], Burtron H. Davis[2]and Dennis Sparks[2], (1)Chart Energy & Chemicals, La Crosse, WI, (2)Center for Applied Energy Research, University of Kentucky, Lexington, KY

A commercial Fischer Tropsch catalyst was studied in two Compact Heat Exchange Reactors (CHER) with different channel sizes, and in a continuous stir tank reactor (CSTR). The CHERs were constructed by Chart Energy and Chemicals with one reactor having a 5mm channels and the other having a 6mm channels, while maintaining the same catalyst volumeThe catalyst was crushed and sieved retaining particles ranging in size from 75-125µm. Testing was performed at the University of Kentucky Center for Applied Energy Research in Lexington, KY. The CO conversion, selectivity and hydrocarbon productivity were measured for each reactor over a wide range of operating conditions, as a function of temperature, pressure, H2:CO feed ratio, and flow rate. Both reactor types showed excellent temperature control, but the performances varied at different operating conditions. The larger channel CHER had a higher conversion than the 5mm channel, potentially due to slightly higher reactor bed temperature. The two CHERs approached and even exceeded the productivity of the CSTR for liquid hydrocarbons per gram of catalyst. At higher flow rates the methane selectivity increases at a lower rate for either of the CHERs than for the CSTR.

SIMPLE AND EFFECTIVE VAPOR FEED DISTRIBUTION IN A BOTTOM STRIPPING SECTION OF A CRUDE OIL COLUMN

Arlan L. de Souza[†], Silvia Waintraub[†], Vítor L. Pereira[*] and Karolline Ropelato[*]
[†]PETROBRAS – CENPES – Rio de Janeiro/RJ – Brazil
E-mail: arlan@petrobras.com.br, silviaw@petrobras.com.br
[*]ESSS – Engineering Simulation and Scientific Software – Rio de Janeiro/RJ – Brazil
E-mail: ropelato@esss.com.br, vitor.pereira@esss.com.br

Because of the relatively low vapor flow, the bottom stripping section of a crude column is usually designed with a smaller cross-sectional area than the flash zone's, either by using a smaller diameter or by installing an internal shell (shroud) inside the main column.

Injecting vapor in the region between the inner and the outer shell of a bottom stripping packing section of a crude oil distillation column has been found an easy, inexpensive and convenient way to achieve good vapor distribution at section entrance.

Different vapor distributors were evaluated using a 3D Computational Fluid Dynamic (CFD) analysis. A single-phase, fluid-porous domain approach was considered in all simulations. At the bottom of the simulated domain, a symmetry boundary condition was adopted to account for the shear effect on the reduced crude surface below.

Results show that a direct radial injection of vapor against the inner shell requires no additional column height and provides a better vapor distribution when compared to the perforated pipe sparger classical solution.

Keywords: annulus injection, annular space injection, CFD, vapor distribution, feed distributor, stripping section and crude oil distillation column.

INTRODUCTION

The initial breakup of crude oil into raw products is accomplished by fractional distillation in the Crude Distillation Unit (CDU). After preheated and partially vaporized, crude oil feeds the crude oil column in its flash zone, where vapor and liquid separate. The vapors (approximately 50-60%) move up the column, generating overhead vapor and side draw liquid products. The unvaporized portion of the crude, the residue, travels down the bottom stripping section of the column.

The basic function of a crude column bottom stripping section is to remove light-valuable components still contained on the residue. Steam is injected below the section (packing or trays) and rises through the descending residue, stripping out the lower boiling hydrocarbons and producing the reduced crude.

Bottom stripping section operates at low vapor and high liquid rate. While a high amount of reduced crude goes down through the stripping section, only stripped hydrocarbon and stripping steam compose the vapor phase. Because of this relatively low vapor flow, the bottom stripping section of a crude column has in general a smaller cross-sectional area when compared with the flash zone's.

Designers typically adopt for the bottom stripping section either a smaller diameter or install an internal shell (shroud) inside the main column. In either configuration, steam is injected below the section, most of the time through a perforated pipe distributor (sparger).

In this article, a new way of injecting vapor in a stripping section is presented. Instead of a sparger, vapor is injected in the annular space (annulus) between the inner and the outer shell of a bottom stripping section of a crude oil column. The quality of vapor distribution at the stripping section entrance is evaluated. The tendency of this vapor injection in generating turbulence on the residue's surface is also analyzed.

THE BOTTOM STRIPPING SECTION

In a typical design, the bottom stripping section includes a liquid distributor, four to six trays to facilitate good contact between liquid and vapor, and a vapor distributor below the trays. Packing is also used, most of time replacing the trays to increase the number of theoretical equilibrium stages, but the superior mechanical and fouling resistance of trays make them the first choice in a grassroots project.

The vapor distribution usually comprises a perforated pipe distributor, or sparger, installed beneath the stripping section [1]. The holes are normally directed downward to avoid to be plugged by the descending residue. Underneath the holes, an impingement baffle could be used to deflect the incoming steam sideways.

The design of the bottom stripping section is often neglected [2,3,4]. Stagnant areas, oversized downcomers and low weir loading are common problems that contribute to increase the fouling rate and plugging of the stripping trays. Excessive active area also is observed in bad designs. It reduces vapor velocity, causing weeping, and doesn't generate enough froth, making the contact of liquid and vapor very poor. This kind of misconceptions downgrade the already low efficiency observed in this high liquid low vapor section.

The carried light material on the residue coming from an inefficient stripping section increases the vacuum column load and its top (LVGO) pumparound duty. It can also overload the vacuum ejector system, increasing the vacuum column pressure and reducing vacuum gas oil recovery.

Increasing stripping steam rates, in order to compensate the low efficiency, doesn't achieve the expected or desired results. A higher stripping steam rate, besides increasing the condenser load, increases the column's pressure drop because of the vapor traffic increment and suppress the benefit of this operational change to some extent.

Poor performance in the bottom stripping section could be compensated by stretching the vacuum or any other downstream unit for recovery. This is true when the Vacuum Distillation Unit (VDU) has an extra available capacity and the vacuum column has fractionation capabilities. It is worth nothing, however, that the heat content of an atmospheric distillate stream is higher than the equivalent boiling range stream on the vacuum column. An energy optimization analysis should be done to set the proper amount of distillate recovered in each unit (atmospheric and vacuum) [2].

In a scenario of high competitiveness and tighter product specifications, many refiners have found in the redesign of the bottom stripping section a successful modification to improve refinery profitability [5,6,7,8,9].

A well designed stripping section helps to generate flash-zone vapor, which contributes to increase distillate yield and permits higher internal reflux, favoring diesel-AGO fractionation. An incremental volume of stripped vapor contains more diesel that the same volume generates with furnace temperature alone [5].

Since improved bottom stripping section allows a reduction in the crude heater outlet temperature at constant atmospheric distillation yield and lowers the quantity light material in the vacuum tower feed, overall crude unit performance is optimized [3].

Good distribution of the descending liquid and adequate vapor distribution at the bottom of the stripping section are the key to maintaining efficiency. Proper vapor distribution can materially improve the light material recovery in the residue for a given stripping steam rate, contributing to the refinery economics.

METHODOLOGY

Computational fluid dynamics (CFD) was employed to evaluate the quality of vapor distribution at the entrance of a crude column bottom stripping section. The stripping section is enclosed in a smaller diameter internal shell which contains a bed of packing. This work focused on the comparative analysis between two types of injection: (1) annulus injection, where vapor is fed radially in the gap between the inner and outer shell of a crude column, and (2) sparger injection, the classical solution, where the vapor enters the column through a perforated pipe distributor, as shown in Figure 1.

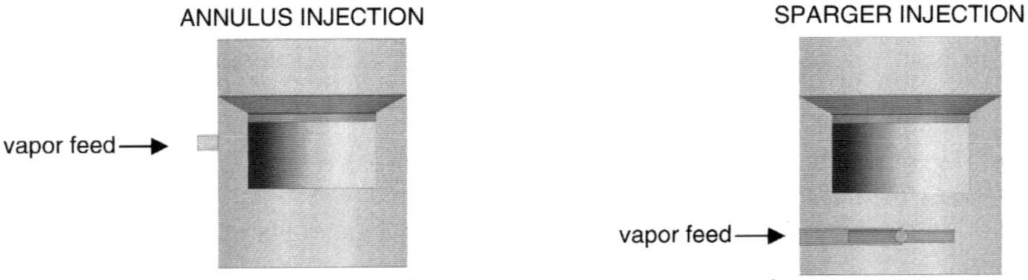

Figure 1 - Vapor injection configurations.

Single-phase, steady-state simulations were performed on ANSYS CFX 13.1.

The continuous phase (vapor) was considered incompressible and has been modeled using the k-ε turbulence model. It has been proved that this model is stable and numerically robust providing a well-established regime of predictive capability with accuracy [10]. A two-domain approach was employed: a fluid domain (vapor) and a porous domain (packing). The flow resistance on the porous zone is modeled according to Equation (1):

$$f = \frac{\Delta P}{h * U_y^2} \tag{1}$$

where ΔP represents the packing pressure drop; h the packing height and U_y the packing axial average velocity.

Applied boundary conditions (BCs) are shown on Table 1. The sparger distribution was carried-out through a 20 downward source-points set-up.

Table 1 - Boundary conditions.

Vapor inlet	Prescribed velocity
Bottom (residue surface)	Symmetry condition
Wall	No slip condition
Top (flash zone)	Prescribed pressure

Performance evaluation methodology for all simulated scenarios included flow analysis on two planes placed as shown on Figure 2. The vertical velocity was evaluated on a plane just below the packing bed (blue circle), and the shear strain rate on the residue's surface (red circle).

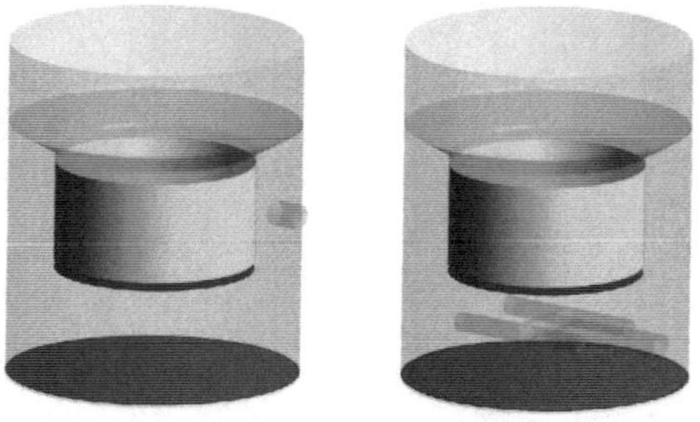

a) annulus injection b) sparger injection

Figure 2 - Evaluated planes.

Ascending flow homogeneity was evaluated through coefficient of variation (C_V) analysis for the axial velocity component [11,12,13]. A C_V approaching zero indicates an uniform flow situation. The coefficient of variation (C_V) is defined as:

$$C_V = \left[\frac{1}{a_t} \sum_{i=1}^{N} a_i \left(\frac{v_i - \bar{v}}{\bar{v}} \right)^2 \right]^{0.5} \tag{2}$$

And the average vapor vertical velocity can be defined by Equation (3):

$$\bar{v} = \frac{1}{a_t} \sum_{i-1}^{N} a_i v_i \tag{3}$$

where N is the total number of cells; a_t is the total column cross sectional area; a_i is the cross sectional area of a cell; v_i is the local vapor velocity.

Shear strain rate profiles were plotted in order to predict, qualitatively, the turbulence effects over the vapor-liquid interface existing at the column's bottom.

RESULTS AND DISCUSSION

PART 1 – SINGLE INLET

Figure 3 shows the vapor distribution at the packing bed entry when the vapor is injected radially in the annular space (a, b) and supplied via a sparger (c, d).

It can be observed from Figure 3b and d that the vertical velocity is more evenly distributed when vapor is injected in the annular space. For the annulus injection, vapor predominantly ascends with a velocity of 0.1 to 0.2 m/s, while via sparger the velocity profile has a more broad range, including negative (descending) velocities, meaning backmixed flow.

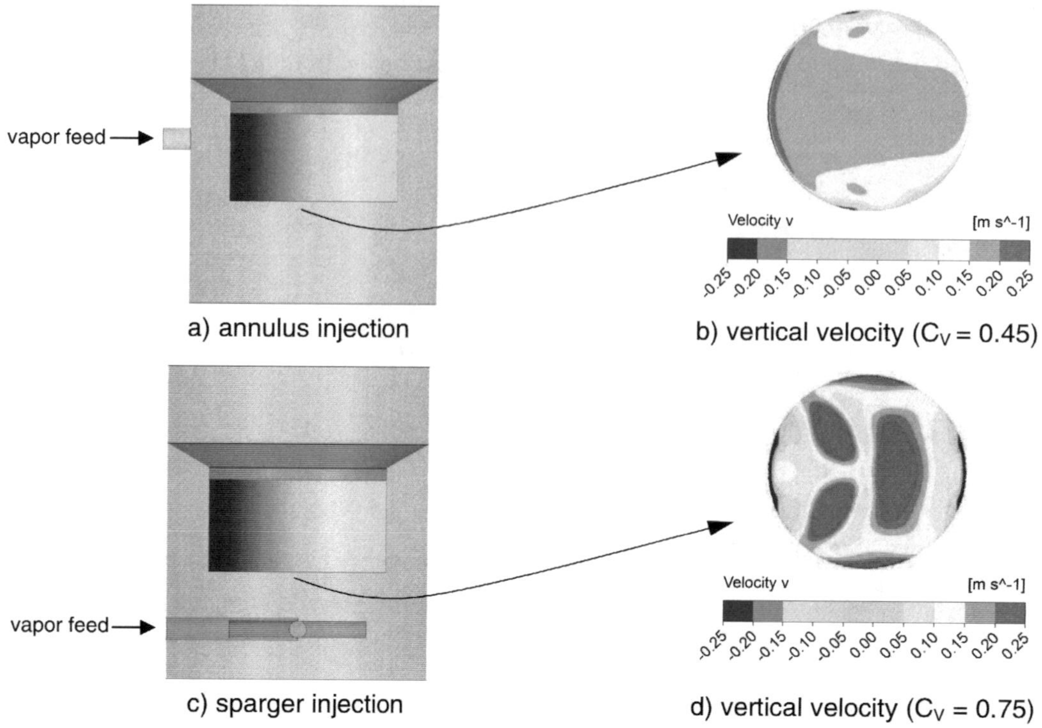

Figure 3 - Vapor vertical velocity below the stripping section bed.

When the vapor is fed by a sparger, high localized velocities appear between the sparger's pipe arms. Thus, the distance between sparger and the stripping section entry is an important parameter to be observed when using a sparger injection. In this case, a more uniform vapor distribution should be obtained when a larger gap is used.

The average of the coefficient of variation (C_V) is used to quantify the uniformity of the vapor velocity profile. It approaches zero as flow becomes more regular. The average values of C_V indicates that the flow via annulus injection (C_V = 0.45) is 60% more uniform than a sparger injection (C_V = 0.75).

The influence of the vapor injection on the liquid surface of the bottom crude column residue is also analyzed.

Kister [14] presents an interesting case of damage and unstable operation in a crude oil column caused by a sparger installed at a bottom stripping section. Steam left the sparger at 45° toward the liquid surfa ce, hitting the surface at high velocities. The level measurement started varying with aeration, becoming unreliable and misleading.

Figure 4 shows the shear strain rate at the residue's surface at the bottom of the column. Notice the different scales used in Figure 4b and d. The detail on Figure 4d shows the shear result in the same scale as in 4a.

The liquid under a sparger is significantly more affected by the vapor flow. A higher tendency of aeration is expected for the sparger because of vapor bubbling into the liquid. Not only the distance between sparger and the upper section should be a concern, as it was seen previously, but the distance between it and the lower section too.

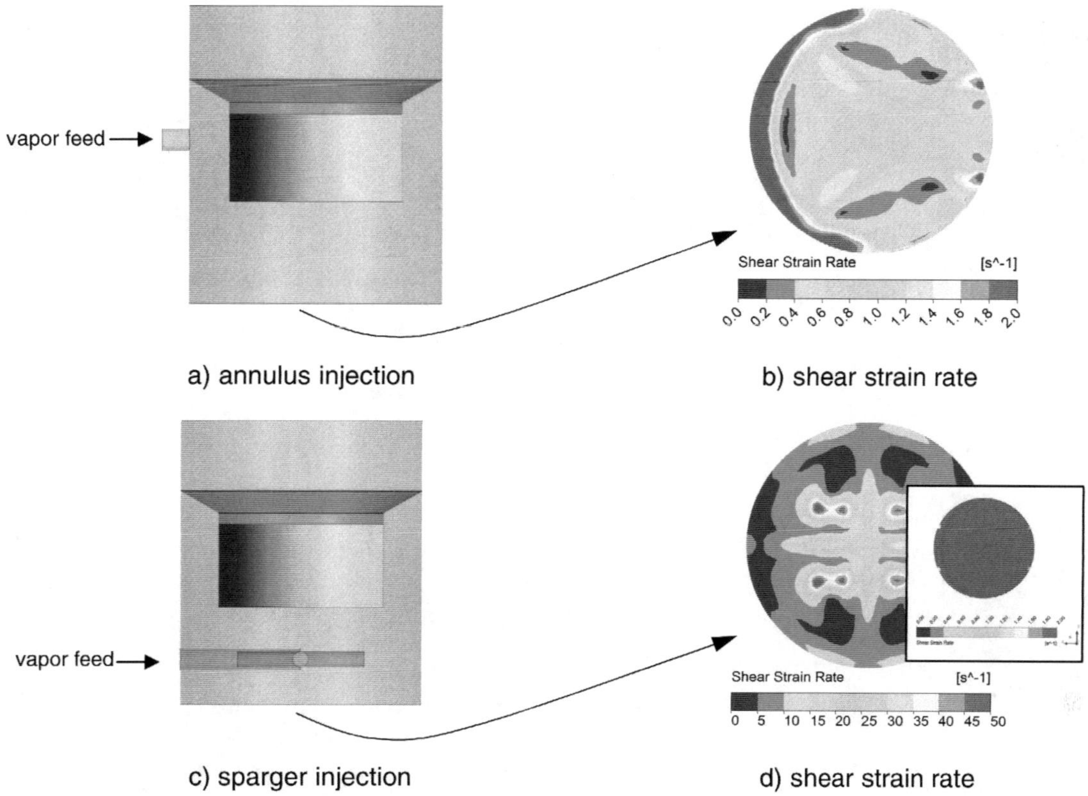

Figure 4 - Shear strain rate at residue surface.

PART 2 – TWO ALTERNATIVE INLETS (ONE OF THEM OUT OF SERVICE)

In the PART 2 of this work, the stripping section geometry was slightly modified. A pipe distributor was added to the PART 1 geometry, just below the packing bed, creating a two feed inlet stripping section (see Figure 5). This is the case for example when there are two sources of stripping fluid: steam and another inert gas [15], or even when steam is available at two different pressure levels. Nevertheless, the new one pipe distributor, just added to the geometry, is considered out of service (idle), so it was treated as wall (no flow across it) in our analysis.

Three elevations for the annular space nozzle were also considered: at the center of the shroud, shifted to an upper position and shifted to a lower position. In Figure 5a just the nozzle center position was showed. A tangential annular injection, in opposite to radial injection, was also evaluated. In order to allow an easy comparison, the results obtained in PART 1 are showed again (last column).

Figure 5b indicates that a radial annulus injection is much better in terms of vapor distribution than a tangential annular injection. There is no baffle to break the swirling motion of the vapor before it enters the stripping section above, so the tangential injection presents the worse vapor distribution. An upper nozzle position is good for the annulus radial injection. It seems that a higher elevation allows the vapor to primarily involve the shroud before taking the descending way, improving distribution. The presence of the idle distributor is not so critical to the annulus injection performance. It changes the average C_V from 0.45 (PART 1 annulus case) to 0.47 (upper offset), an increase of 4%. These C_V values are much better than the best sparger result (C_V = 0.75).

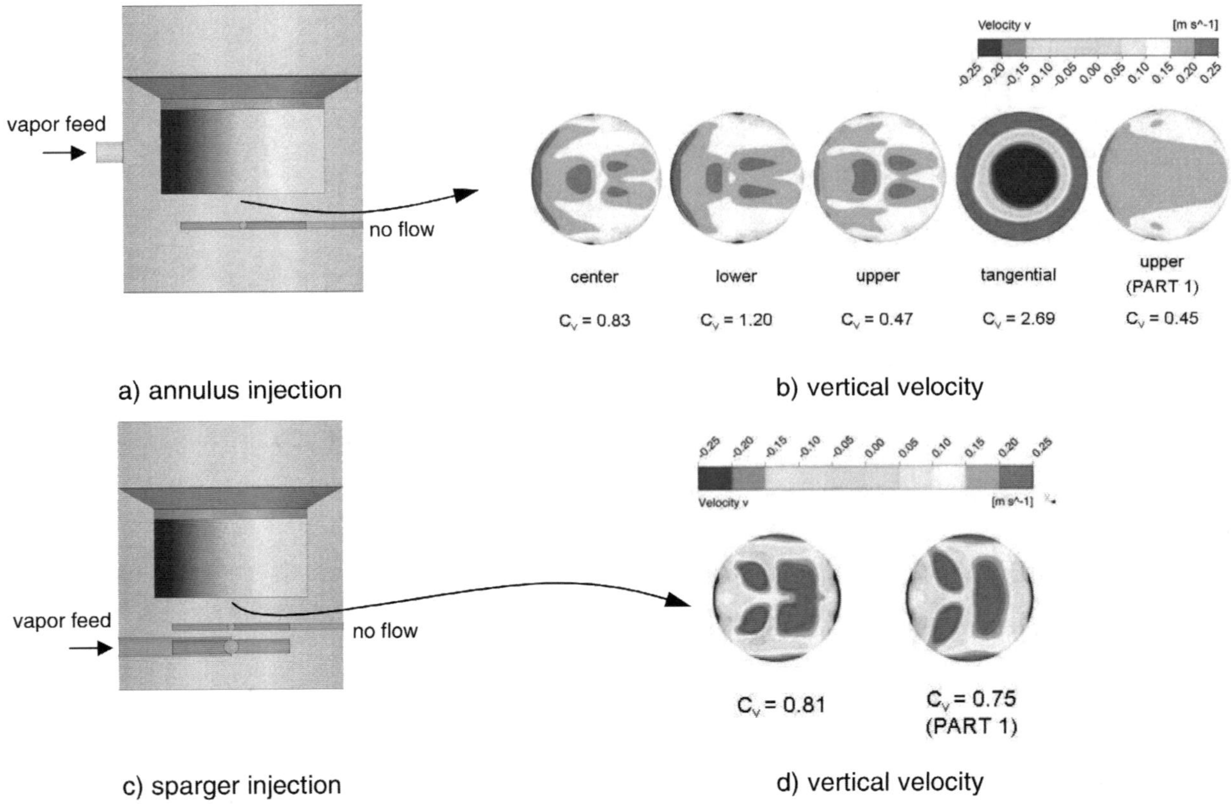

Figure 5 - Vapor vertical velocity below the stripping section bed.

The impact of the idle distributor on the vertical velocity sparger's results is a little bit higher. It is worth to notice in Figure 5c that the idle distributor is smaller than the original sparger, and the pipe arms of both distributors were vertically aligned. Despite of this, the average C_V changes from 0.75 (PART 1 sparger case) to 0.81 (PART 2 sparger case), an increase of 8%.

The shear strain results of PART 2 are quite similar to PART 1 (see Figures 6 and 7). Again, the liquid under a sparger is significantly more affected by the vapor flow.

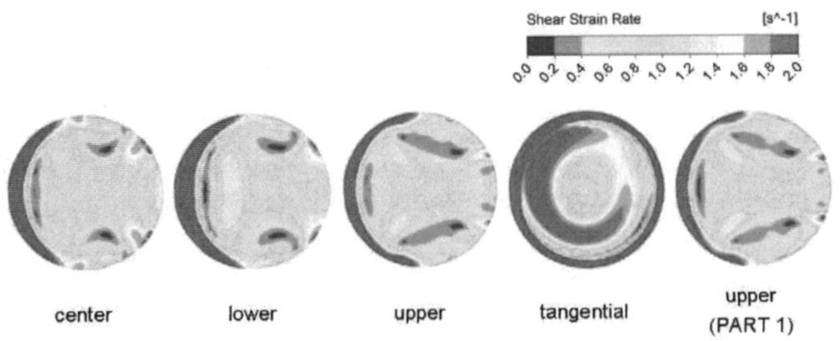

Figure 6 - Shear strain rate for annular injection at residue surface.

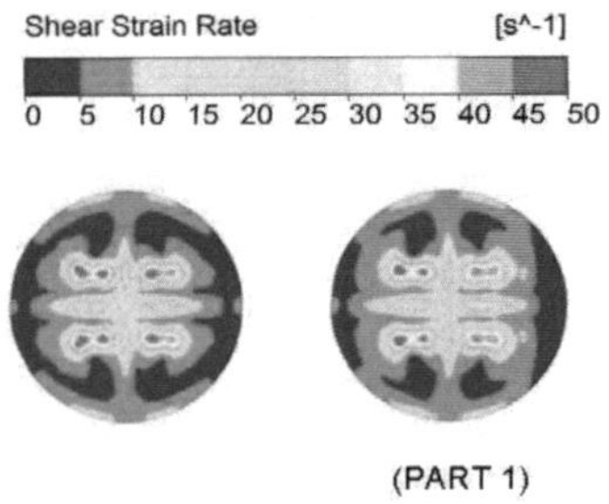

Figure 7 - Shear strain rate for sparger injection.

PRACTICAL RECOMMENDATIONS

The results outlined in the previous section point out the superior performance of a vapor injection in the annular space over the conventional vapor injection considering two different geometry configurations for the bottom stripping section.

A natural further modification in the geometry, in the direction of increasing complexity, would be an inclusion of beams, trusses and other mechanical structures necessary to support the stripping section. Complementary studies have indicated that the evenly annulus injection vapor distribution is minimal affected by the support beams. For better results, the vapor feed nozzle should be located in an elevation that allowed the vapor to primarily involve the shroud before taking the descending way.

The annulus injection requires a smaller nozzle than the required by a direct injection (just a nozzle below the stripping section) because the vapor has the whole annular space to reduce velocity before reaching the entry of the bottom stripping section. An impingement plate at the vapor inlet welded on the inner shell (shroud) should be considered to dissipate vapor kinetic energy and protect the shroud.

Water-induced pressure surges are common in bottom stripping sections. Undrained stripping steam rapidly vaporizes when entering into the tower, generating an explosion that rips trays or packing in great force. It is believed that the annulus injection configuration is more surge resistance than sparger or direct injection because the front of explosion firstly faces the heavier structures parts (shell, beam, trusses), rather than the trays or packing. Taking advantage of the annulus configuration, a considerable surge resistance improvement could be obtained installing on the blind annulus' cover a hinged panel or any other relief device capable of relieving small but sudden pressure surges.

CONCLUSION

Results show that a direct radial injection of vapor in the annular space formed between the inner and the outer shell of a bottom stripping section provides a better vapor distribution when compared to the perforated pipe sparger classical solution.

The vapor flow generated by the annulus injection does not directly impinge on the bottom liquid surface, so the possibility of erratic level control and liquid entrainment into the rising vapor is minimized.

The new way of feeding vapor in bottom stripping section is simple, effective and requires no additional column height, being a convenient way to improve the bottom stripping performance on revamps and grassroots designs.

REFERENCES

1. Kister, H.Z (1990) "Distillation operation," McGraw-Hill Company, pp. 81 and 87-89.
2. Golden, S. W. (2009), "Maximising diesel recovery from crude," *Petroleum Technology Quarterly,* Q1.
3. Hanson, D. W., T. Barletta and J. V. Bernickas (2005) "An atmospheric crude tower revamp," *Petroleum Technology Quarterly*, Q3.

4. Martin, G. R. and B. E. Cheatham (1999) "Keeping down the cost of revamp investment," *Petroleum Technology Quarterly*, summer, pp. 99-107.

5. Hanson, D. W., J. Langston, J. G. Keen, and C. Johnson (2003), "Low-capital crude unit revamp increases product yield," *Oil & Gas Journal,* v.101, pp. 65-69.

6. Singh, D (2011), "Meeting diesel specifications at sustained production," *Hydrocarbon Processing*, April.

7. Barletta, T., M. Smith and C. Macfarlane (2000) "Crude unit revamp to increase diesel yield," *Petroleum Technology Quarterly*, spring.

8. Lee, S.H and I. Buttridge (2009), "Optimizing crude unit design," *Petroleum Technology Quarterly,* Q2.

9. Spangler, R., D. Varraveto, R. Schoonover and T. Hanke (2006) "ConocoPhillips revamps crude unit to increase flexibility, profitability," *Oil & Gas Journal,* v.104, n. 31, pp. 49-56.

10. Wilcox, D. C. (2000) "Turbulence Modeling for CFD," *DCW Industries*, pp. 314.

11. Wehrli, M., S. Hirschberg and R. Schweizer (2003) "Influence of vapor feed design on the flow distribution," Trabs IChemE, v.81, Part A, January.

12. Billingham, J. F., D. P. Bonaquist and M. J. Lockett (1996) "Characterization of the Performance of Packed Distillation Column Liquid Distributors," *Praxair Technology Inc.* pp. 841-851.

13. Mohamed Ali, A., P. Jansens and Z. Olujic (2001) "The Use of Computational Fluid Dynamics to Model Gas Flow Distribution in Packed Columns," *6th World Congress of Chemical Engineering*, Melbourne, Australia 23-27 September.

14. Kister, H.Z (2006) "Distillation Troubleshooting", John Wiley & Sons, Case Study 8.4: Can Improved Level Monitoring Avoid High-Level Damage?, pp.149-152.

15. Plellis-Tsaltakis, C. and A.I. Lygeros (2011) "The Use of Fuel Gas as Stripping Medium in Atmospheric Distillation of Crude Oil," *Oil Gas Sci. Technol. – Rev. IFP Energies nouvelles,* v.66, n.6, November-December, pp. 1017–1024.

The Rayleigh Equation Revisited What to Do When Alpha (α) Isn't Constant

Robert G. Kunz

RGK Environmental Consulting, L.L.C.
Hillsborough, North Carolina

Paper 118 b

2012 AIChE Spring Meeting

April 1-5, 2012

Houston, Texas

Abstract

Integration of the Rayleigh Equation for batch distillation in closed analytical form has heretofore required that relative volatility (α) be assumed constant. A new technique presented in this paper produces a continuous analytical function for the Rayleigh Equation integral whether α is constant or not. The newly developed equation reduces algebraically to the traditional function when α is absolutely constant.

Both the traditional expression at constant α and the new equation are evaluated against numerical integration for a number of ideal and non-ideal binary vapor-liquid equilibrium (VLE) systems. The new method has proven especially useful for the non-ideal systems investigated, in which α varies widely with composition.

The method is then utilized successfully in a computer-simulated batch distillation of an ethanol-water solution at 1 atm pressure, a popular student laboratory exercise and one of the original systems studied by Lord Rayleigh. Results compare well with expected values.

A summary of VLE relationships for ideal and non-ideal systems plus temperature-dependent methods for analysis of constituent composition specific to the ethanol-water system are discussed in separate appendices. These latter include, among others, refractive index measurements assembled from various sources and estimates of liquid density for ethanol-water mixtures extended beyond the range of published data.

The paper consists of new material supported by background information tutorial in nature.

Keywords. Alcohol "Proof" Levels; Boiling Point Curve; Density of Ethanol-Water Solutions; Ethanol-Water Flash Points; Ethanol-Water Refractive Index; Gas Chromatography (GC); Numerical Integration; Rayleigh Equation; Simpson's Rule; Systems: Acetone-Water, Benzene-Toluene, Ethanol-Water, Ethylene Dichloride (EDC)-Toluene; Vapor-Liquid Equilibrium.

Disclaimer
("Some restrictions apply; batteries not included; your mileage may vary")

The following caveat applies: The information contained herein is offered in good faith but without guarantee, warranty, or representation of any kind (expressed or implied) as to its usefulness, correctness, completeness, or fitness for any particular purpose. The user assumes all risk for its implementation and should seek independent professional verification of its accuracy. The author assumes no responsibility and shall not be liable for any loss of profit nor any special, incidental, consequential, or other damages which may result from the use of any of the information contained in this presentation, be it oral or written. Any statements concerning design, construction, operation, what constitutes regulatory compliance, and/or how to achieve such compliance should not be construed as recommendations on the part of the author and/or his organization.

Introduction

Distillation is a separation technique utilizing the difference in composition of a liquid and the vapor in equilibrium with that liquid [**1,2**]. Simple, or differential, distillation employs a single equilibrium stage to effect that separation. A batch distillation operates on an unreplenished charge of material being boiled off from a still pot and then condensed as product in another vessel. It is an unsteady-state process, where the amount and composition of feed and product vary with time. The charge of feed material boils away and gradually becomes richer in the less volatile/higher boiling component(s). The condensed vapor product (distillate) increases in volume and is enhanced in the more volatile/lower boiling component(s). These are also known as heavy and light components, respectively.

This paper discusses the simple, differential batch distillation of a two-component (binary) system, in which none of the condensed vapor is returned, or refluxed, to the still pot, a configuration referred to by one author as a Rayleigh Distillation [**3**]. A typical laboratory setup is shown in Figure 1. It is a favorite student experiment in chemistry and engineering laboratories throughout the world.

Simple batch distillation is not practiced extensively in industry. It is used commercially for separating small quantities of high-value chemicals, when high purities are not required, as a preliminary step to be followed by further processing, for separations where the equipment requires frequent cleanout, or where the feed mixture is very easy to separate [**1,2**]. A rule of thumb for ease of separation is that the feed components differ in atmospheric boiling point by at least 25°C [**4**, pp.167-168], 30-40°C [**5**, p.61], or 75°C [**6**, p.89; **7**, p.89]. Whatever the precise numerical value, the intent is that the lighter component be much more easily vaporized in preference to the heavier component in order to distill a decent amount of high-purity product before too much of the feed material boils away.

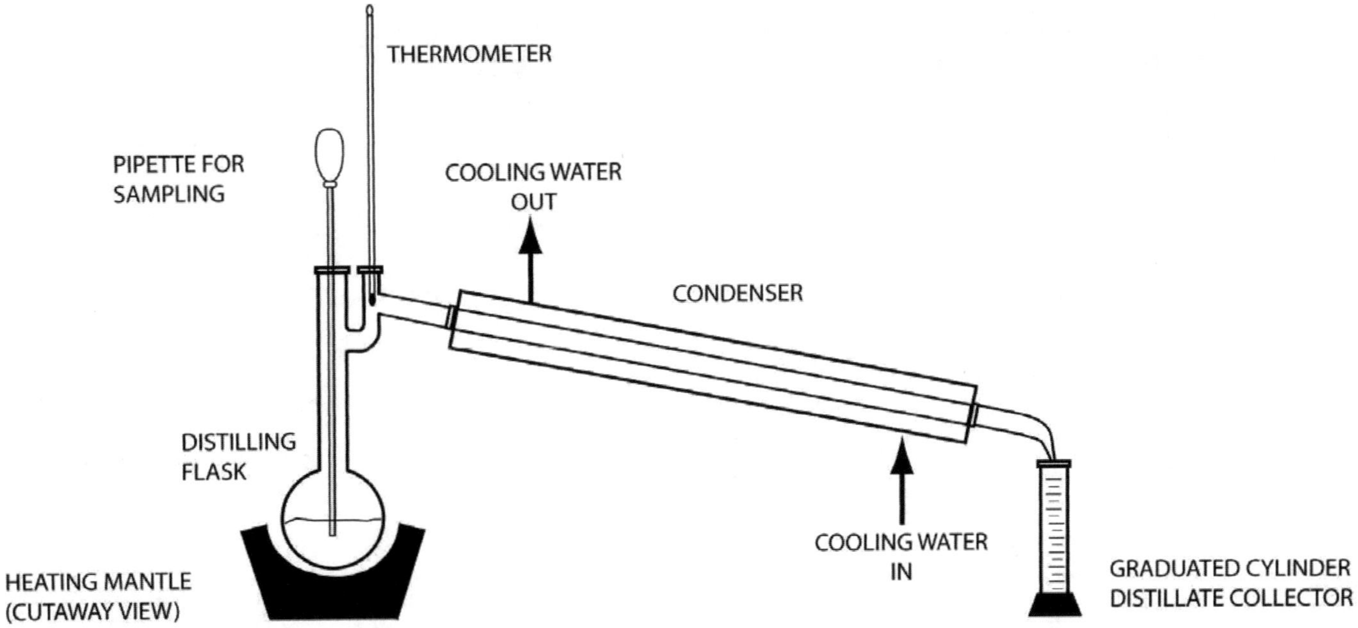

Figure 1. Typical Laboratory Setup for Simple, Differential Batch Distillation

The Rayleigh Equation

Simple batch distillation without reflux is described by the Rayleigh Equation, first advanced by Lord Rayleigh in 1902 [**8**]. In the derivation below [**2**; **9**, p.580-581; **10**; **11**], one begins with an overall mass balance:

$$F = W + D \tag{1}$$

where F = total moles of feed in the still pot/distilling flask at time (t) = 0
W = total moles left in still pot (bottoms)
and D = total moles boiled off and condensed into the distillate at the same point in time

and a component mass balance:

$$Fx_F = Wx_W + Dx_D \tag{2}$$

where x = the mol fraction of the more volatile component (MVC) in the feed (F), bottoms (W), and distillate (D).

Unsteady-state mass balance over a differential amount of time yields:

$$\underset{\text{initial amt in still}}{Wx} = \underset{\text{amt remaining in still}}{(W - dW)(x - dx)} + \underset{\text{amt being vaporized}}{ydW} \tag{3}$$

in which y denotes the mol fraction of the more volatile component in the vapor phase in equilibrium with its mol fraction in the liquid phase (x).

Expanding the equation above, one obtains:

$$Wx = Wx - xdW - Wdx + (dW)(dx) + ydW \tag{4}$$

Simplifying, rearranging, and neglecting the second order differential (dW) (dx) results in:

$$Wdx = ydW - xdW \tag{5}$$

From which:

$$\int_F^W dW/W = \int_{x_F}^{x_W} dx/(y - x) \tag{6}$$

and upon integration of the left-hand side between the stated limits, one obtains the Rayleigh Equation:

$$\ln(W/F) = \int_{x_F}^{x_W} dx/(y - x) \tag{7}$$

The integrand is not a simple function of x. At constant pressure, every value of x corresponds to a different temperature and hence vapor pressure. Except for ideal systems which follow Raoult's Law in the liquid and the Ideal Gas Law and Dalton's Law in the vapor, y is also not related to x in a simple manner.

The standard procedure is to evaluate the integral numerically or graphically,* or to come up with a suitable substitution in an attempt to produce an analytical solution in closed mathematical form. Numerical integration involves repeatedly fitting the numerical values of the integrand with a mathematical function over short ranges, integrating those functions mathematically, and then summing. This numerical fitting and summing process results in a series of discrete values, rather than a continuous function.

Relative Volatility

An analytical approach makes use of a parameter known as the *relative volatility* and denoted by the symbol *alpha* (α). It is defined as the ratio of the vapor-phase mol fraction of the more volatile component, denoted as *component 1*, to its liquid-phase mol fraction, this

* In previous times, in addition to counting boxes under the curve of the integrand or using a planimeter device [**12**], the curve was drawn carefully on high quality graph paper of uniform density and then was weighed and calibrated against a rectangular section of that same graph paper. This method is noted in Ref. [**13**, p.75].

ratio being divided by a similar ratio for the less volatile *component 2.* For a successful separation by distillation, α must be greater than 1.0. Mathematically:

$$\alpha = (y_1/x_1) / (y_2/x_2) = (y_1/x_1) / [(1 - y_1) / (1 - x_1)] \qquad (8)$$

The alternate expression above for α comes about since $y_2 = (1 - y_1)$ and $x_2 = (1 - x_1)$.

Rearranging Equation (8), one obtains for y_1:

$$y_1 = \alpha\, x_1 / [1 + x_1 (\alpha - 1)] \qquad (9)$$

and for $1 / (y_1 - x_1)$:

$$\frac{1}{(y_1 - x_1)} = \frac{1 + (\alpha - 1)\, x_1}{(\alpha - 1)\, x_1\, (1 - x_1)} = \frac{1}{(\alpha - 1)\, x_1\, (1 - x_1)} + \frac{1}{(1 - x_1)} \qquad (10)$$

The relative volatility is evaluated at every point along the vapor-liquid equilibrium x-y curve. In general, relative volatility is not absolutely constant across the entire range of x, even for the most ideal of systems. However, in those cases where α is considered to be reasonably constant, it is taken outside the integral sign to allow the Rayleigh Equation integral to be evaluated analytically by a closed-form mathematical function.

Now substituting the extreme right-hand side of Equation (10) into Equation (7) with the assumption of a constant α yields a new set of integrals for ln (W/F):

$$\ln (W/F) = \frac{1}{(\alpha - 1)} \int_{x_F}^{x_W} \frac{dx_1}{x_1 (1 - x_1)} + \int_{x_F}^{x_W} \frac{dx_1}{(1 - x_1)} \qquad (12)$$

in which each term can be integrated analytically [**14**, #40 and #29] to yield:

$$\ln (W/F) = \frac{1}{(\alpha - 1)} \left\{ \alpha \ln [(1 - x_F)/(1 - x_W)] + \ln (x_W/x_F) \right\} \qquad (13)$$

A mathematically equivalent form [**2**] is:

$$\ln (W/F) = \frac{1}{(\alpha - 1)} \left\{ \ln \left[\frac{(1 - x_F)\, x_W}{x_F\, (1 - x_W)} \right] + \ln \left[\frac{(1 - x_F)}{(1 - x_W)} \right] \right\} \qquad (14)$$

The percent of original charge remaining in the still pot is then given by:

$$100\,(W/F) = 100 \exp\,[\ln\,(W/F)] \tag{15}$$

Relative Volatility from Vapor-Liquid Equilibrium (VLE) Relationships

Equations for low-pressure binary vapor-liquid equilibrium (VLE) are summarized in Appendix A. Important variables are temperature (t), total pressure (P_T), partial pressure (p), vapor pressure (p^{vap}), liquid-phase (x) and vapor-phase (y) mol fractions of each component, plus so-called activity coefficients (γ's) when a given system cannot be considered to be ideal.

Pure-component vapor pressures are conveniently calculated using the Antoine Equation:

$$\log_{10} p^{vap}\,(\text{mmHg}) = A - B\,/\,[t\,(°C) + C] \tag{16}$$

Antoine constants are listed in Table 1 for chemicals of interest in this paper.

By manipulation of these equations, the relative volatility (α) in Equation (8) can be calculated at constant pressure for any temperature or the corresponding value of composition along the VLE curves from a ratio of vapor pressures and, if necessary, activity coefficients:

$$\alpha = (y_1/x_1)\,/\,(y_2/x_2) = (p_1^{vap}\,\gamma_1)\,/\,(p_2^{vap}\,\gamma_2) \tag{17}$$

in which the subscript *1* denotes the more volatile component (MVC).

A system can be considered to be ideal when $\gamma_1 = \gamma_2 = 1.0$ for all values of x. Relative volatilities for a number of systems – some ideal, some non-ideal – are listed in the second column of Table 2. Alpha is not perfectly constant for any of these systems across the range of interest. In general, α shows a wider variation in a system exhibiting non-ideal behavior than for an ideal system.

Other Quantities Are Determined by Material Balance

The Rayleigh Equation in whichever form – for example, Equation (7), Equation (13), or Equation (14) – describes the relationship between W and x_W for a given starting value of F. The other quantities of interest are determined by material balance from Equations (1) and (2) and Equation (18) below, derived from them:

$$x_D = (\,F x_F - W x_W\,)\,/\,(\,F - W\,) \tag{18}$$

The composition of the first drop of distillate, when W equals F and $x_W = x_F$ and the expression in Equation (18) becomes mathematically indeterminate, is provided by the x-y equilibrium curve, as is the instantaneous composition of the distillate at every moment during the distillation. The situation will become clearer by way of the following example.

Table 1. Antoine Equation Constants Used in the Analysis [(a)]

Chemical Name and Synonyms	Formula	MW [(b)]	BP (˚C) [(c)]	Applicable Range(˚C)	A	B	C
Acetone 2-Propanone Propanone Dimethyl Ketone	H_3CCOCH_3	58.08	56.2	—	7.02447	1161.0	224
Benzene Benzol	C_6H_6	78.12	80.1	8 to 103	6.90565	1211.033	220.790
1-Butanol n-Butyl Alcohol Butanol-1 n-Propyl Carbinol	C_4H_9OH	74.12	117.7	15 to 131	7.47680	1362.39	178.77
Ethanol Ethyl Alcohol Grain Alcohol Methyl Carbinol	C_2H_5OH	46.07	78.3	—	8.04494	1554.3	222.65
Ethyl Benzene Phenyl Ethane	$C_6H_5C_2H_5$	106.17	136.2	—	6.95719	1424.255	213.206
Ethylene Dichloride (EDC) Ethylene Chloride 1,2-Dichloroethane	$ClCH_2CH_2Cl$	98.96	83.5	—	7.18431	1358.46	232.2
Isobutyl Alcohol Isobutanol 2-methyl-1-propanol 2-methyl-propanol-1 Isopropyl Carbinol	$(CH_3)_2CHCH_2OH$	74.12	107.9	20 to 115	7.32705	1248.48	172.92
Methanol Methyl Alcohol Wood Alcohol Carbinol	CH_3OH	32.04	64.7	–20 to +140	7.87863	1473.11	230.0
Methyl Ethyl Ketone (MEK) 2-Butanone Butanone	$H_3CCOC_2H_5$	72.12	79.6	—	6.97421	1209.6	216

Table 1 (continued). Antoine Equation Constants Used in the Analysis [(a)]

Chemical Name and Synonyms	Formula	MW [(b)]	BP (˚C) [(c)]	Applicable Range(˚C)	A	B	C
1-Propanol Propanol Propanol-1 n-Propyl Alcohol Ethyl Carbinol	C_3H_7OH	60.11	97.2	—	7.99733	1569.70	209.5
2-Propanol Propanol-2 Isopropanol Isopropyl Alcohol Dimethyl Carbinol	$H_3CCH(OH)CH_3$	60.11	82.4	0 to 113	6.66040	813.055	132.93
n-Propyl Benzene Propyl Benzene	$C_6H_5C_3H_7$	120.20	159.2	—	6.95142	1491.297	207.140
Toluene Methyl Benzene Phenyl Methane	$C_6H_5CH_3$	92.14	110.6	6 to 137	6.95464	1344.800	219.482
Water Hydrogen Oxide	H_2O	18.02	100.00	0 to 60 60 to 150	8.10765 7.96681	1750.286 1668.21	235.0 228.0

Notes:

(a) Antoine Constants from "Lange's Handbook of Chemistry", 11th Edition [**15**, Table 10-10], except for toluene and the butanols, which are from the 12th Edition [**16**, Table 10-8].

(b) Molecular Weight (MW) from "Handbook of Chemistry and Physics," 62nd Edition [**17**, pp. C-65 to C-576], rounded for water [**17**, p.B-105] to 2 decimal places.

(c) Boiling point (BP) at 760 mmHg from "Lange's Handbook," 11th-13th Editions [**15**, p.4-59 and Table 7-4; **16**, p.4-59 and Table 7-4; **18**] or "Handbook of Chemistry and Physics," 62nd Edition [**17**, pp. C-65 to C-576, p.B105], chosen to be the most consistent with boiling point calculated using the Antoine Constants listed.

Illustrative Example. An illustrative example for batch/differential distillation is presented in McCabe and Smith [**11**]. In this example, an ideal solution of 50 mol percent benzene (MVC) and 50 mol percent toluene is subjected to batch distillation at 1 atm pressure. (Refer again to Figure 1.) Instructions are to take $1/\alpha$ as constant at 0.41 for this system over the entire range of the distillation and to plot the calculated results for temperature and benzene composition at various locations against the mol fraction of charge distilled (D/F). Instantaneous mol fraction benzene is desired in the still (x_W) and in the vapor leaving the still (y_1), along with the cumulative average benzene composition in the distillate (x_D).

Temperatures and compositions newly calculated here via electronic spreadsheet and plotted in Figures 2 and 3 reproduce the curves in the cited reference [**11**]. Calculations performed here assume the familiar benzene-toluene system to be ideal, that is, to follow Raoult's Law. Pure component vapor pressures were computed using the Antoine Equation with constants listed for benzene and toluene in Table 1. Any differences between the present computations and the tabulated values accompanying the plotted curves in Ref. [**11**] are Inconsequential and can be attributed to the graphical procedure employed there.

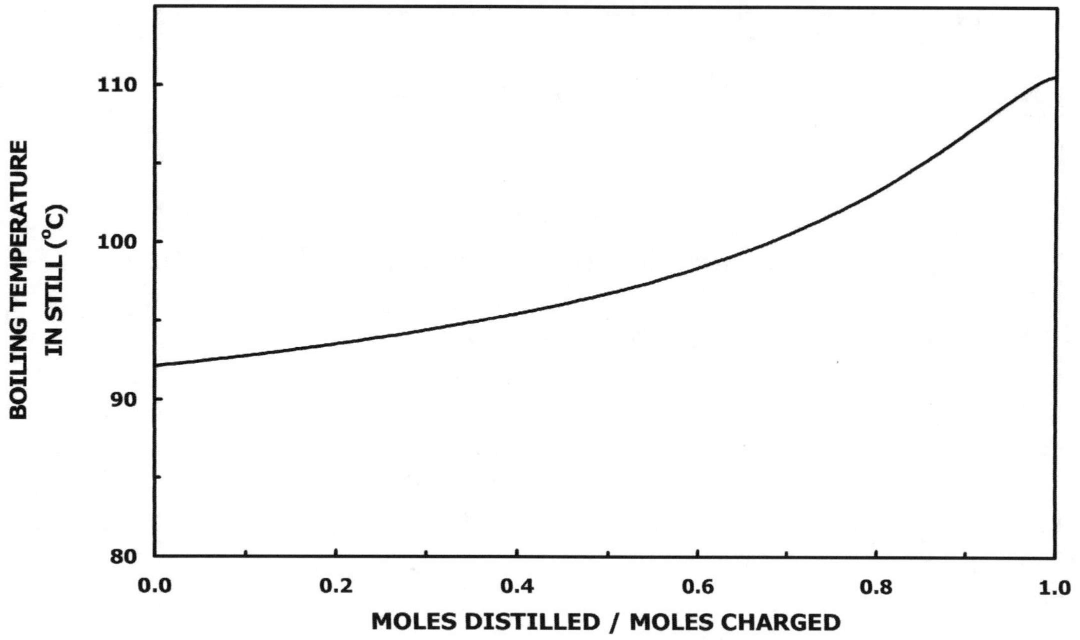

Figure 2. Calculated Temperature during Batch Distillation of 50-50 Mol Fraction Benzene-Toluene Solution at 1 Atm Pressure

Atmospheric boiling temperature for an ideal mixture depends on the vapor pressures and mol fractions of the pure components. The temperature in the still pot (Figure 2), initially just above 92 °C, approaches 110.6 °C, the boiling point of toluene present by itself in the still pot at the end of the distillation. Benzene concentration in the still pot (bottoms) approaches zero as the distillation proceeds (Figure 3). The last drop in the bottoms is pure toluene, although for safety reasons one would not want actually to distill the mixture to dryness. Benzene in the distillate, highest at its initial instantaneous value, decreases somewhat in the cumulative distillate as more toluene comes over. Instantaneous benzene concentration in the vapor condensing into the distillate declines more rapidly as the benzene is boiled off, leaving the still-pot composition richer in toluene.

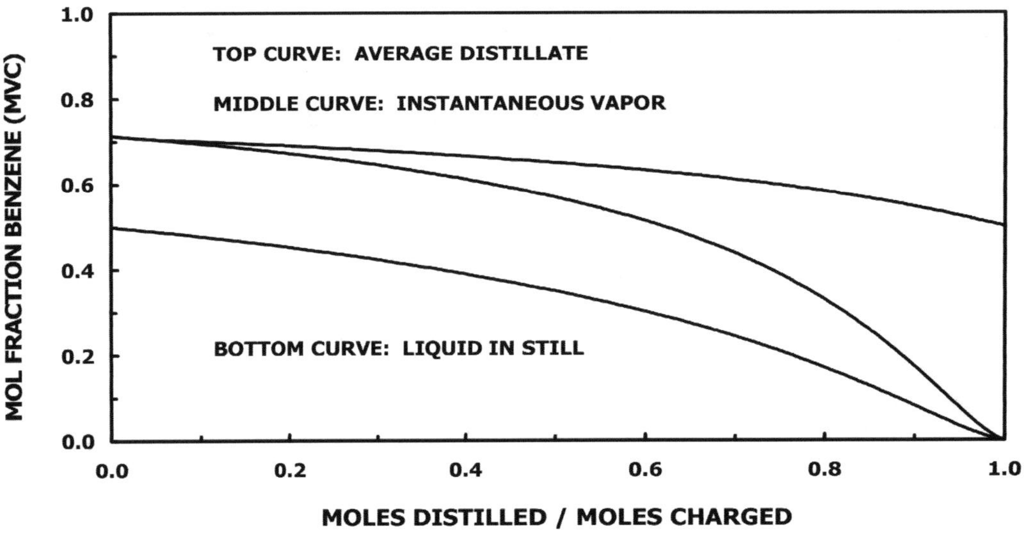

Figure 3. Calculated Compositions from Rayleigh Equation during Batch Distillation of 50-50 Mol Fraction Benzene-Toluene Solution at 1 Atm Pressure

Testing of Numerical Evaluation of the Integral

Using as the basis a starting value of 0.5 mol fraction in the distilling flask inspired by the illustrative example discussed above, results from numerical integration of Equation (12) by Simpson's Rule were compared with the exact values obtained from Equation (13). Since Equation (13) with an absolutely constant value of α is an exact analytical integration of Equation (12), it becomes a primary standard against which to test the accuracy of a numerical integration approximation.

Simpson's Rule for numerical integration [**12**]:

$$\text{Area under Curve} \approx (\Delta\ \text{abscissa} / 3)\ (\text{ordinate } 1 + 4 \times \text{ordinate } 2 + \text{ordinate } 3) \qquad (19)$$

approximates a small section of a curve with a parabola and computes the area under that parabola (an integral). Summing the areas from its repeated application to one adjacent section after another produces a series of values for cumulative area.† The process is terminated when the final desired abscissa is attained.

† In the present case, the area is negative since the integration is conducted from right (higher values of the abscissa) to left (lower values). That makes sense because logarithms are negative when the argument, as here, is a fraction less than 1.0.

It has been found here that repeated application of Simpson's Rule in steps of 0.02 mol fraction and an interval size of 0.01 mol fraction (the Δ abscissa above) reproduces the exact integral for ln (W/F), even when exponentiated, to a degree better than can be measured experimentally. This situation applies down to a mol fraction bordering on 0.02. Narrowing the interval size at the lower end of the mol fraction range, where the integrand begins to increase rapidly, maintains the accuracy of the numerical integration even further. Eventually, however, the integrand grows without bounds as mol fraction remaining in the still pot approaches zero.

Simpson's Rule numerical integration (the points) is compared with Equation (13) (the curves) in Figures 4 and 5 for typical values of constant α. The curves pass directly through the points, and one cannot detect any difference between points and curve at the scale of the figures. The same is true for even greater values of α, although those curves are not shown here. Note that the amount of bottoms remaining in the distilling flask drops steeply for the smallest value of α and begins to decline more slowly with each increase in α. The conclusion is that numerical integration with a properly chosen interval size is coincident with the results of Equation (13), the exact solution of Equation (12) with a constant α.

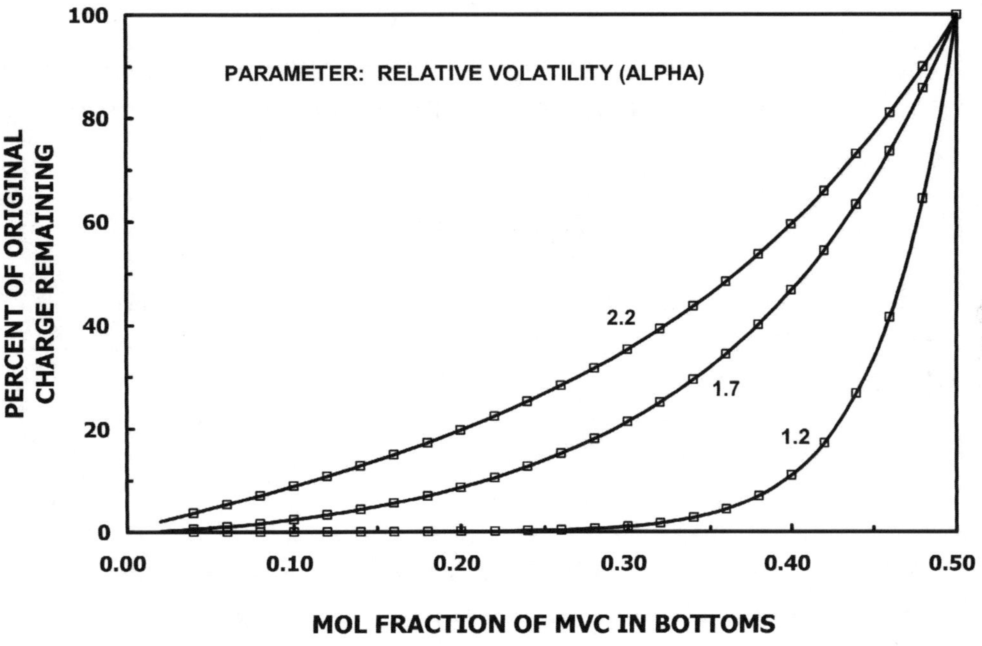

Figure 4. Comparison of Numerical Integration with Exact Solution of Rayleigh Equation at Constant α (Lines Analytical, Points Numerical) – Low Values of α

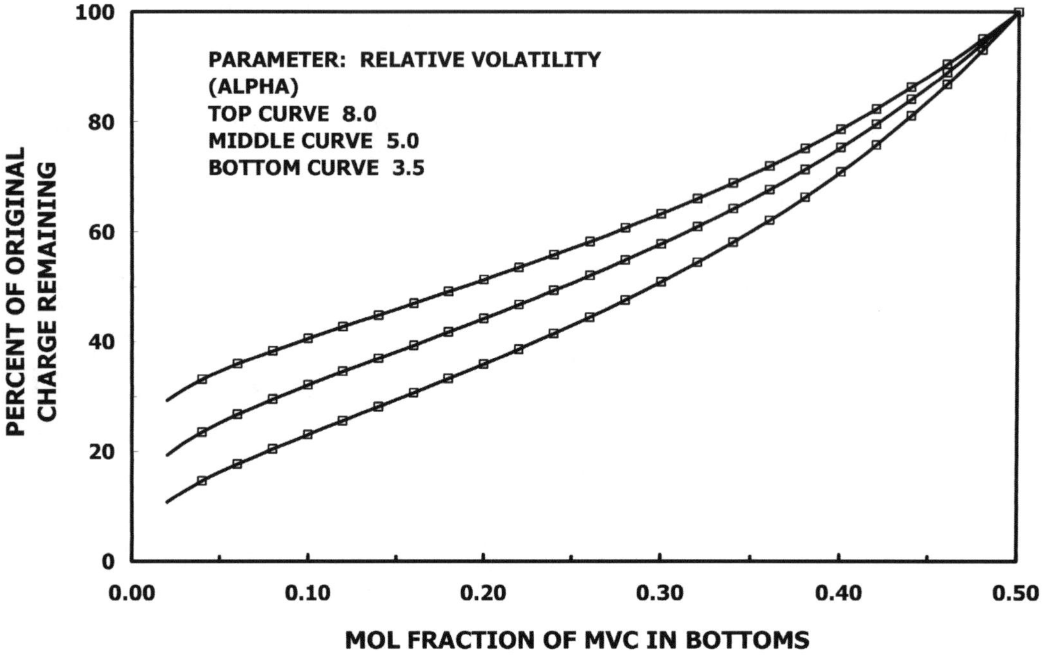

Figure 5. Comparison of Numerical Integration with Exact Solution of Rayleigh Equation at Constant α (Lines Analytical, Points Numerical) – Higher Values of α

A New Twist

Alpha need not necessarily be constant to come up with a solution to Equation (7) as a closed-form analytical expression. A less restrictive condition is that the factor $1/(\alpha - 1)$ can be fitted by a quadratic function of mol fraction (x):

$$1 / (\alpha - 1) = a\,x^2 + b\,x + c \qquad (20)$$

over the range of interest, in much the same way as the expression of gaseous heat capacities by an empirical power series in temperature. With $1/(\alpha - 1)$ so represented, the integrand of Equation (10) becomes a new series of terms in the second equation below, each having an analytical solution upon integration term by term [**14**, #32, #29, #40]:

$$\ln(W/F) = \int_{x_F}^{x_W} \frac{(a\,x_1^2 + b\,x_1 + c)\,dx_1}{X_1\,(1 - x_1)} + \int_{x_F}^{x_W} \frac{dx_1}{(1 - x_1)} \qquad (21)$$

$$\ln(W/F) = \int_{x_F}^{x_W} \frac{a\, x_1\, dx_1}{(1 - x_1)} + \int_{x_F}^{x_W} \frac{(1 + b)\, dx_1}{(1 - x_1)} + \int_{x_F}^{x_W} \frac{c\, dx_1}{x_1(1 - x_1)} \quad (22)$$

When the smoke clears and the dust settles, one is left with the following:

$$\ln(W/F) = a\,(x_F - x_W) + (1 + a + b + c)\ln[(1 - x_F) / (1 - x_W)] + c\ln[x_W / x_F] \quad (23)$$

This form too is an exact solution to Equation (7). The only source of error is the goodness of fit of the quadratic function in Equation (20). When α is constant for all values of composition, $a = b = 0$, $c = 1/(\alpha - 1)$ identically, and Equation (23) reduces to Equation (13).

Testing of the New Function against Numerical Integration

A number of real vapor-liquid equilibrium systems were chosen to test how well Equation (23) works in practice. These are broken up in Table 2 between Systems Taken to Be Ideal and Non-Ideal Systems, where coefficients are tabulated for Equation (20) over the range of α from $x_1 = 0.02$ to $x_1 = 0.5$. Chemical synonyms for the individual constituents of these systems can be found in Table 1.

Here, the previously validated Simpson's Rule integration is the standard, using the same interval size found suitable for the numerical integration of Equation (12) with constant α. The overall conclusion is that agreement of the newly derived Equation (23) with numerical integration is as good as in the previous graphs where α is constant (Figures 4 and 5). In addition, one has a continuous function in Equation (23) as opposed to numerical integration, which is valid only at discrete points. Results are discussed separately below for the Ideal and the Non-Ideal systems.

Ideal Systems. The system ethylene dichloride (EDC)-toluene, midway down the listing in Table 2, is representative of the ideal systems investigated. Percent of initial charge remaining in the still pot on a molar basis is plotted in Figure 6 against mol fraction EDC. The discrete points trace the numerical integration of Equation (7) or Equation (12) with a variable $1/(\alpha - 1)$ inside the integral sign. The continuous curve through the points is the integrated Equation (23) with values of *a, b,* and *c* for this system from Table 2.

The upper and lower curves represent the integrated Equation (13) with two different constant alphas bracketing the range of values in Table 2. The range has been expanded a bit to widen the difference between the curves. In this case, inserting the arithmetic average of the upper and lower alphas into the integrated Equation (13) for constant α would result in a curve virtually coincident with the middle curve calculated from Equation (23), passing through the points from numerical integration of Equation (21). Results from averaging of α are better for the Ideal Systems above EDC-toluene in Table 2 and not quite so good for those following. However, Equation (23) using proper values of *a, b,* and *c* was found to work regardless.

Table 2. Coefficients of $1/(\alpha - 1)$ Function

System	Range of α from x_1 = 0.02 to 0.5	$1/(\alpha-1) = ax_1^2 + bx_1 + c$ a	b	c
Systems Taken as Ideal:				
Benzene(1) – EDC(2)	1.108 – 1.110	0.0249847	–0.414753	9.26702
Ethanol(1) – 2-Propanol(2)	1.163 – 1.167	–0.00895551	–0.344138	6.15889
Methanol(1) – Ethanol(2)	1.66 – 1.71	0.077259	–0.251001	1.51560
Ethanol(1) – 1-Propanol(2)	2.03 – 2.11	0.0549304	–0.172904	0.977133
Acetone(1) – MEK(2)	2.09 – 2.17	0.0564870	–0.168884	0.924787
EDC(1) – Toluene(2)	2.20 – 2.26	0.0381021	–0.112977	0.838362
Benzene(1) – Toluene(2)	2.36 – 2.49	0.072534	–0.178451	0.741501
Methanol(1) – 1-Propanol(2)	3.16 – 3.65	0.125203	–0.241046	0.466673
Ethanol(1) – 1-Butanol(2) (a)	4.0 – 4.6	0.090100	–0.162840	0.338006
Benzene(1) – Ethyl Benzene(2)	4.3 – 5.3	0.138854	–0.210900	0.302701
Methanol(1) – Isobutyl Alcohol(2)	4.4 – 5.5	0.129731	–0.215647	0.297965
Benzene(1) – n-Propyl Benzene(2)	7.1 – 10.7	0.183439	–0.217725	0.167047
Non-ideal Systems: (b)				
Methanol(1) – Water(2)	7.6 – 3.6	0.228138	0.363189	0.144093
Ethanol(1) – Water(2)	9.8 – 2.0	3.5716	–0.07977	0.136641
Acetone(1) – Water(2)	28.6 – 5.0	0.91519	–0.04840	0.039525

Notes:

(a) May not be strictly ideal, but the ideal x-y curve is bracketed by the data of two different investigators [**19,20**].

(b) Coefficients $A_{1\text{-}2}$ and $A_{2\text{-}1}$ of van Laar Equations for Methanol-Water and Acetone-Water given in Perry's 3rd Edition [**9,** p.528] are 0.36, 0.22, and 0.89, 0.65, respectively ($\log_{10}$ based) or 0.83, 0.51 and 2.05, 1.50 (ln based). For the Ethanol-Water system, coefficients of 0.68381, 0.41724 ($\log_{10}$ based) or 1.5745, 0.9607 (ln based) were determined [**21**] by fitting the azeotrope of 89.43 mol % ethanol and 10.57 mol % water at 78.15°C [**9**, pp.631,633] using vapor pressures calculated from the ethanol and water Antoine Constants of Table 1.

Non-Ideal Systems. Of the three non-ideal systems evaluated, acetone-water displays the widest variation in α. Accordingly, this system has been picked to illustrate how Equation (23) can deal with a widely varying α. Amount of solution remaining in the distilling flask (W) is shown in Figure 7 as a function of acetone concentration (x_W). As in Figure 6, the points represent the results of numerical integration.

The middle curve drawn through those points is calculated from Equation (23) using the *a, b,* and *c* constants from Table 2 for this system. The curve provides an excellent fit to the numerically integrated points even though the $1/(\alpha - 1)$ fit itself is not quite so good as any of those obtained for the ideal systems. The fit of $1/(\alpha - 1)$ tends to improve for shorter ranges of x_W. If a better fit of $1/(\alpha - 1)$ is needed, the range of x_W can be subdivided into smaller sections to piece together the integration all the way from the starting point to the desired final composition. Alternatively, a cubic term can be added in the fit of $1/(\alpha - 1)$ against composition. In that case, Equation (22) would then contain an additional term, whose integrated form can be found in Ref. [**14**, #36]. That exercise is left to the reader.

The upper and lower curves in the figure depict the analytically integrated function of Equation (13) with α constant at each of its extreme values. Clearly neither one of these curves falls anywhere near the middle curve and the points from numerical integration, and it is not clear *a priori* how to average the α values to achieve such a fit. In addition, with values of α so far apart, it turns out that even the best "average" α utilized in Equation (13) does not reproduce the correct track of the middle curve and its associated points. Once again, use of Equation (23) with proper values of *a, b,* and *c* is the method of choice to obtain an analytical expression relating total moles remaining in the distilling flask with molar composition.

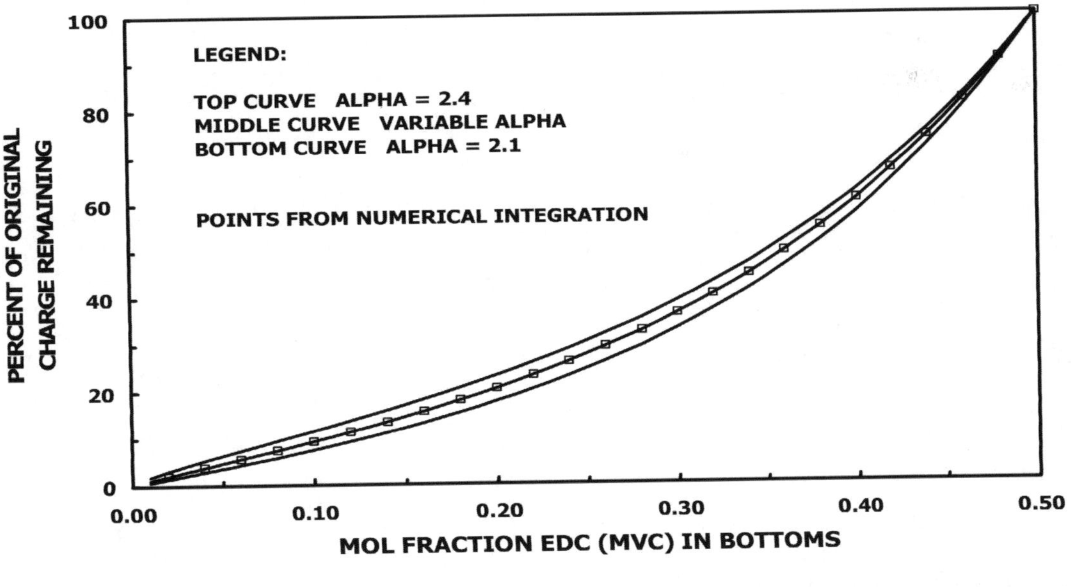

Figure 6. Newly Derived Function vs. Numerical Integration – EDC-Toluene at 1 Atm

Testing of the New Function in Simulated Laboratory Experiment

To demonstrate the ease of use of the newly developed technique, batch distillation of a pure ethanol-water solution was simulated at a standard atmospheric pressure of 760 mmHg (1.013 bar). This is a favorite system employed in batch distillation as a student laboratory exercise and one of the systems studied by Lord Rayleigh [**8**].

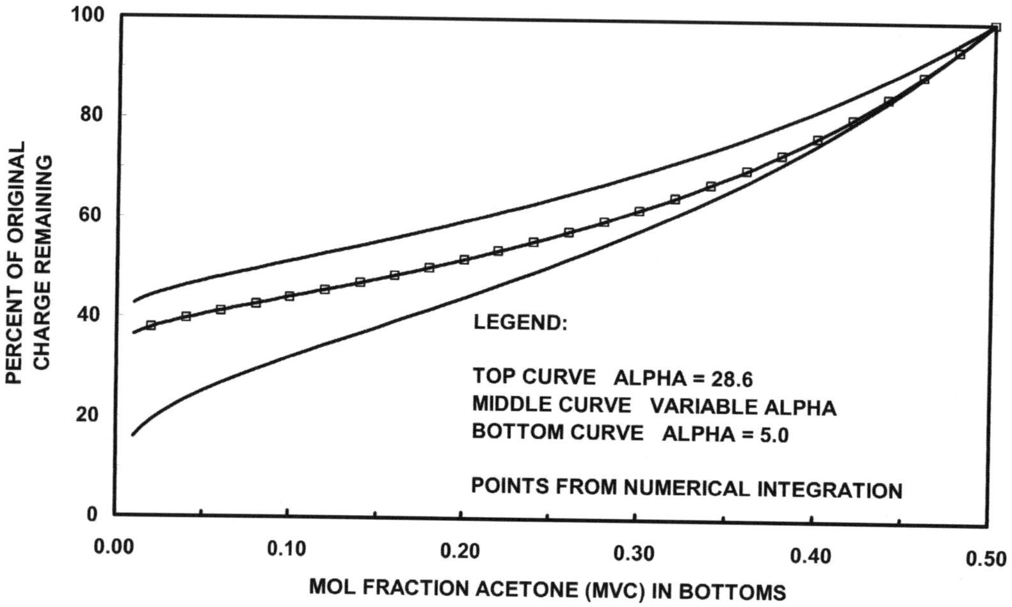

Figure 7. Newly Derived Function vs. Numerical Integration – Acetone-Water at 1 Atm

The Laboratory Experiment. An initial 200-mL charge of a pure ethanol-water solution, 20% ethanol by volume (at 60 °F, 15.56 °C) in an apparatus similar to Figure 1 at standard atmospheric pressure (760 mmHg, 1.013 bar) is typical [**6,7,22**]. Volume of the distilling flask is some 500 millimeters (mL). Boiling stones are added to prevent bumping [**4**, pp.146,170].

Samples of distillate and perhaps still bottoms are taken as every 5 mL of distillate is collected, and temperature is read at such times. The experiment is over when 50 mL of distillate has been collected. Sometimes the total collected distillate is subjected to further purification in a second step using a separate refluxed fractionating column. This experiment has also been conducted using denatured alcohol and water, fermented sugar solutions, and wine being distilled into brandy.

The student is cautioned to wait to read the temperature until droplets condense on the thermometer bulb [**4**, p.172]. The experiment then proceeds drop by drop. The recommended distilling rate varies, but numbers in the range of 2 [**6,7**] to 10 [**4**, p.171] drops per minute are stated. At an average 6 drops per minute (1 drop every 10 seconds), and 20 drops to the millimeter, collection of 50 mL would take just under 3 hours, exclusive of analytical time. This would allow completion of the first step of the experiment within a typical standard laboratory period.

Quantitative analysis is done by refractive index, density, or possibly gas chromatography. A qualitative analysis for ethanol in the distillate is often conducted by attempting to ignite several drops of solution on a watch glass, since ethanol and its aqueous solutions greater than about 50 % by volume are combustible at room temperature [**23**]. These methods are reviewed in some detail in Appendix B. Whatever the method of analysis, it is good practice to prepare one's own calibration curve(s) using the materials and analytical equipment employed rather than relying exclusively on literature data.

The Simulation. To demonstrate the ease of use of the technique, a computer simulation of the batch distillation laboratory experiment described above was conducted. The simulated experiment is valid only for a binary mixture of pure ethanol and water, with an x-y VLE curve as in Figure A3. It does not apply quantitatively to denatured alcohol containing methanol and other ingredients, fermented sugar solutions, or wine being distilled into brandy.

The simulation begins when equilibrium becomes established as the temperature in the laboratory experiment stabilizes and the first drop of liquid falls into the receiving vessel. The simulation also assumes that the distillation takes place slowly enough so as to maintain true vapor-liquid equilibrium with a drop of liquid condensed on the thermometer at all times.

Results are summarized in Table 3. Distilling flask temperature, amount of bottoms remaining and distillate formed, plus composition of bottoms and distillate are recorded at every 5 mL of liquid distillate collected at 20 °C. There are two ways to conduct such an experiment: save each 5-mL increment separately for individual analysis, or let the distillate accumulate as a composite sample. In the latter case, the distillate composition would follow the trend of the relationship depicted in Figure 3 for benzene-toluene in the illustrative example above, but with different chemicals. Results from the simulation are listed both ways in Table 3.

Gram moles of bottoms (W) and mol fraction of ethanol in the bottoms (x_W) are related by Equation (23), using a function for $1/(\alpha - 1)$ from Equation (20) fitted from x_W of 0 and 0.2 ($a = 1.73317$, $b = 0.511067$, $c = 0.103302$). As one might expect, these different constants provide a better fit for$1/(\alpha - 1)$ over the limited range of x_W in this "experiment" than the coefficients listed in Table 2 for a larger span of x_W.

Moles of distillate (D) and mol fraction of ethanol in the distillate (x_D) are computed from material balances [Equations (1) and (2), rearranged]. Moles are converted to grams using a molecular weight of 46.07 for ethanol and 18.02 for water from Table 1:

$$\text{Grams ethanol} = 46.07 \times \text{moles of ethanol} \tag{24}$$

$$\text{Grams water} = 18.02 \times \text{moles of water} \tag{25}$$

Mol fraction (m.f.) ethanol is converted to wt % ethanol by

$$\text{wt \%} = 100 \times (\text{m.f.})\,(46.07) \,/\, [(\text{m.f.})\,(46.07) + (1 - \text{m.f.})\,(18.02)] \tag{26}$$

to make use of weight-percent dependent density relationships for ethanol-water mixtures, either at an assumed laboratory temperature of 20 °C or as estimated for the solution remaining in the distilling flask (Appendix B).

Mass is converted to volume by means of density at known temperature and composition. Tables in Ref. [**24**, Tables 6.22 and 6.36], interpolated if necessary, relate wt % and % by volume at the standard temperature of 60 °F (15.56 °C).

Table 3. Summary of Simulated Batch Distillation Experiment

Temp (°C)	**Bottoms (W)** **g moles** **mass (g)** **mL @ 20 °C** ***mL @ temp***	**Distillate (D)** **g moles** **mass (g)** **mL @ 20 °C**	**Ethanol in Bottoms (x_W)** **mol fract** wt % *% vol*	**Ethanol in Distillate (x_D)**		
				Cumulative **mol fract** wt % *% vol*	**Individual** **mol fract** wt % *% vol*	**Instantaneous** **mol fract** wt % *% vol*
89.1	9.7367 194.6994 **200.0 *209.5***	0.0000 0.0000 **0**	0.0706 16.269 ***20.0***	0.3704 60.07 ***67.8***	0.3704 60.07 ***67.8***	0.3704 60.07 ***67.8***
89.6	9.5787 190.2358 **195.2 *204.2***	0.1581 4.4636 **5**	0.0658 15.26 ***18.8***	0.3645 59.46 ***67.2***	0.3645 59.46 ***67.2***	0.3577 58.74 ***66.5***
90.0	9.4180 185.7561 **190.3 *199.0***	0.3188 8.9433 **10**	0.0609 14.22 ***17.5***	0.3578 58.76 ***66.5***	0.3513 58.07 ***65.8***	0.3439 57.27 ***65.0***
90.5	9.2542 181.2583 **185.4 *193.7***	0.4825 13.4410 **15**	0.0560 13.17 ***16.3***	0.3508 58.04 ***65.8***	0.3370 56.51 ***64.3***	0.3289 55.63 ***63.5***
91.0	9.0871 176.7406 **180.5 *188.5***	0.6496 17.9588 **20**	0.0511 12.11 ***15.0***	0.3432 57.20 ***65.0***	0.3214 54.78 ***62.6***	0.3128 53.79 ***61.6***
91.6	8.9161 172.2004 **175.6 *183.3***	0.8206 22.4989 **25**	0.0463 11.04 ***13.7***	0.3352 56.32 ***64.1***	0.3045 52.83 ***60.7***	0.2952 51.72 ***59.6***
92.2	8.7409 167.6352 **170.7 *178.0***	0.9959 27.0641 **30**	0.0415 9.96 ***12.3***	0.3265 55.36 ***63.2***	0.2863 50.63 ***58.5***	0.2763 49.40 ***57.2***
92.8	8.5607 163.0420 165.8 *172.8*	1.1760 31.6573 **35**	0.0367 8.88 ***11.0***	0.3174 54.31 ***62.2***	0.2665 48.17 ***56.0***	0.2559 46.79 ***54.5***
93.5	8.3750 158.4176 **160.8 *167.5***	1.3618 36.2818 **40**	0.0321 7.82 ***9.7***	0.3075 53.10 ***61.0***	0.2454 45.40 ***53.1***	0.2341 43.87 ***51.5***
94.2	8.1830 153.7584 **155.8 *162.3***	1.5538 40.9410 **45**	0.0276 6.77 ***8.3***	0.2971 51.94 ***59.8***	0.2228 42.29 ***49.8***	0.2109 40.61 ***48.0***
94.9	7.9839 149.0609 **150.8 *157.0***	1.7529 45.6385 **50**	0.0233 5.76 ***7.2***	0.2859 50.59 ***58.4***	0.1909 38.84 ***46.1***	0.1867 36.99 ***44.0***

The boiling temperature curve relates still-pot composition and the temperature at which it boils. The boiling temperature curve is predicted from the VLE relationship described in Appendix A. The complete curve, including the entries in Table 3, Columns 1 and 4, is plotted in Figure 8 against weight fraction (wt %/100) ethanol as in Ref. [**24**, p.251]. Experimental data from three different sources judged to be reliable [**24**, p.251; **25**,**26**] are shown for comparison. Although the data are better correlated by the broken-line curve in the figure, agreement with the predicted (solid) curve is within several tenths of a degree Celsius for the vast majority of the 78 data points plotted. Difference equals or exceeds 1.0 °C for only 3 of those points, with a maximum temperature difference of 1.3 °C. Maximum deviations occur in the range of 0.05 to 0.3 weight fraction ethanol. These differences in the boiling-point curve highlight the minor imperfections in the activity-coefficient model fitted solely on the azeotropic point (Appendix A).

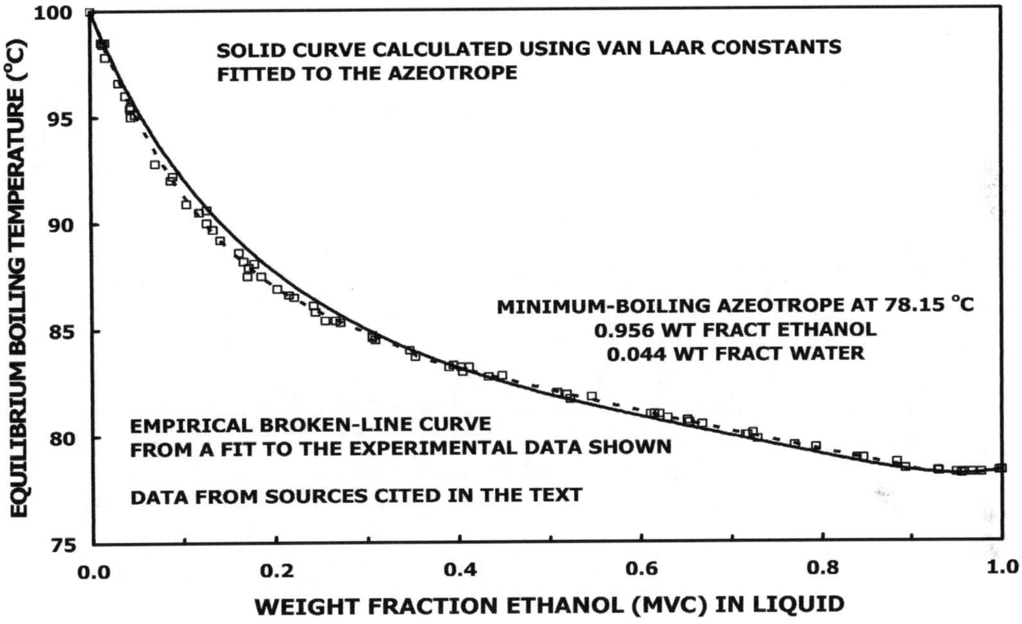

Figure 8. Boiling Temperature Curve for Ethanol-Water at 1 Atm Pressure

Plotting the temperature values in Table 3 against cumulative distillate collected results in the expected s-shaped curve of Fig. 5.8 of Refs. [**6** & **7**]. The simulated experiment has been extended to collect an extra 50 mL of distillate and provide additional temperature data points in Figure 9. Those temperatures are not listed in Table 3.

The density function developed to prepare Figure B1 in Appendix B can be used to estimate the hot liquid remaining in the still pot during distillation. Estimated bottoms volume is listed in Table 3, Column 2. Liquid volumes shown in the table at each boiling temperature are 4-5 % greater than the volume of the same liquid at 20 °C. Change in volume remaining in the distilling flask is therefore not accounted for exactly by subtracting the volume of distillate removed, measured at 20 °C. Even when bottoms and distillate volumes are both expressed at 20 °C, those volumes are not additive because of the difference in density with ethanol concentration.

Density of the residual liquid at each temperature and composition for the simulated experiment summarized in Table 3 is plotted against temperature as the points in Figure 10. As before, additional points were obtained by extending the simulation to collect another 50 mL of distillate (again not shown in the table). The point at the extreme upper right is the liquid density of pure water boiling at 100 °C and 1 atm. "Experimental" points extrapolate smoothly to this datum along the curve drawn in the figure.

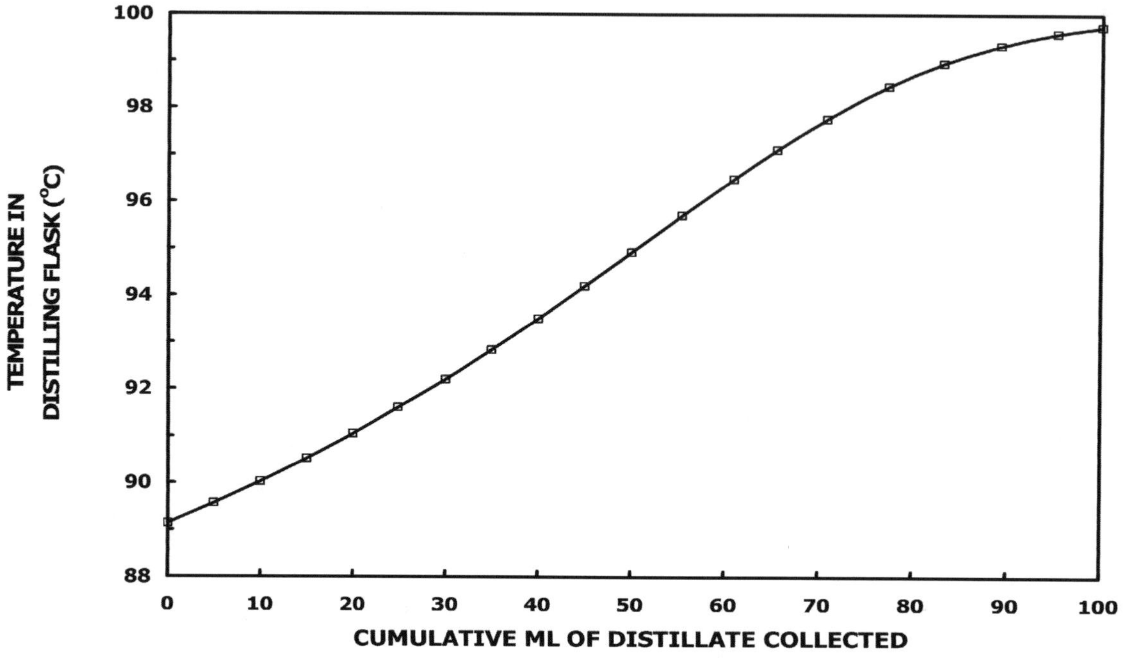

Figure 9. Still-Pot Temperature vs. Cumulative Volume of Distillate for Simulated Ethanol-Water Batch Distillation at 1 Atm Pressure

The empirical curve-fit function, forced to pass through the 100 °C point:

$$\rho_{\text{BOTTOMS}} = 0.958380 + 1.35444 \times 10^{-3}\,(t - 100\ °C) - 8.1550 \times 10^{-5}\,[(t - 100\ °C)]^{2} - 3.23156 \times 10^{-7}\,[(t - 100\ °C)]^{4} \qquad (27)$$

reproduces the points of the simulation with a correlation coefficient (r^2) of 100 % and a standard deviation (s^2) of 0.00002. This function allows one to estimate the bottoms density (and therefore the volume of solution left in the distilling flask), from temperature alone when batch-distilling without reflux a 20 % by volume (60 °F, 15.56 °C) solution of pure ethanol and water. Division of mass by density gives volume.

Composition of the distillate is enumerated in the last three columns of Table 3. Three versions of distillate collection are shown. In Column 5, data are recorded at every 5 mL as the distillate is allowed to accumulate in the receiving vessel. In Column 6, the receiving vessel is changed as soon as each 5 mL of distillate is collected, and the individual distillate samples are kept separate for "analysis." Column 7 represents the instantaneous composition at each temperature and distillate volume noted. For the first drop of distillate (0 mL accumulated), all three columns are exactly the same. Composition is also the same, but at a different value, in Columns 5 and 6 for the first 5 mL of accumulated sample. For other entries, compositions show a regular progression from top to bottom and from left to right in Columns 5, 6, and 7.

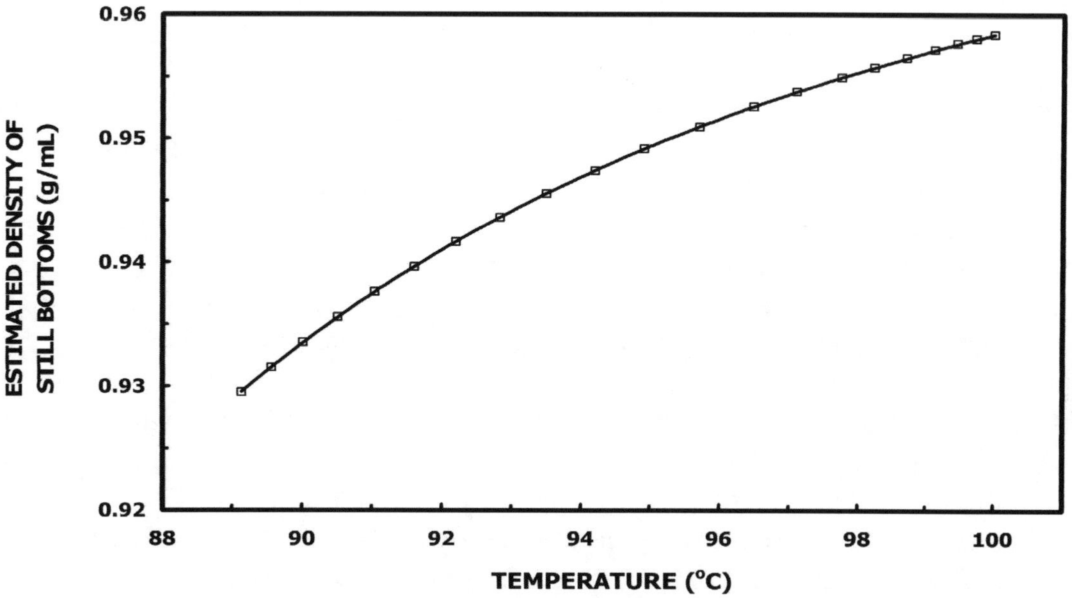

Figure 10. Estimated Density of Solution Remaining in the Still Pot at Boiling Temperature for Simulated Ethanol-Water Batch Distillation at 1 Atm Pressure

In the laboratory, distillate solutions cooled to a standard temperature of perhaps 20 °C would be analyzed quantitatively by any of the methods discussed in Appendix B. For example, densities at laboratory temperatures are well known. Any samples extracted from the distilling flask should also be cooled to laboratory temperature before analysis by density.

On a qualitative basis, unheated distillate samples of about 50 % by volume or more would light off at laboratory temperature with a small flame and leave no residue on the watch glass (Appendix B). This qualitative ignition test should produce positive results at every 5 mL increment for the cumulative samples and for most of the individual samples.

In some versions of the experiment, the composite ethanol-rich distillate is saved and further purified using a separate refluxed fractionating column. In a properly designed column, composition will approach pure water in the bottoms and the azeotrope in the overhead. A sample of that distillate would ignite on the watch glass as well, burning more vigorously with a brighter flame and again leaving no visible residue.

Wrap-up. The newly developed equation has successfully facilitated the simulation of an unrefluxed batch distillation experiment; however, modeling of any further purification of distillate samples in a fractionating column is beyond the scope of the present investigation.

Summary and Conclusions

- Batch, or differential, distillation without reflux is described by the Rayleigh Equation:
  - + Relates composition and amount of material remaining in distilling flask
  - + Other quantities determined by material balance
  - + Illustrative example from the literature presented.
- Numerical integration is required for the Rayleigh Equation in its basic form.
- Substitution of relative volatility (α) allows analytical integration for constant α.
- New equation derived here allows analytical integration whether α is constant or not.
- Analytical integration utilizing α has been evaluated here against numerical integration:
  - + For constant or nearly constant α
  - + For a number of real vapor-liquid systems, where α varies with composition.
- For ideal systems, with α not varying widely, use of an "average" α is adequate.
- For non-ideal systems, only the new equation follows the track of numerical integration.
- The new function was used in simulation of batch distillation of pure ethanol and water.
- Ethanol-water results compare favorably with expectations, for example:
  - + Boiling point vs. composition curve
  - + Still pot temperature vs. cumulative volume of distillate collected
  - + Density of solution remaining in still pot.
- Methods to analyze liquid composition are discussed for the ethanol-water system:
  - + Gas chromatography (GC)
  - + Refractive index data compiled from various sources
  - + Densities of ethanol-water solutions extended beyond range of published data
  - + Qualitative ignition test / flash points of ethanol-water solutions.

About The Author

Robert G. Kunz was an environmental engineering manager at Air Products and Chemicals, Inc., Allentown, PA before retiring after 26+ years of service. He then joined Cormetech, Inc., Durham, NC as Technical Project Manager in support of sales and marketing efforts in the petroleum refining and petrochemical industries. He held engineering positions previously at Esso Research and Engineering Company, Florham Park, NJ and The M.W. Kellogg Company, New York, NY and is currently an independent environmental consultant. "Dr. Bob" has earned a BChE degree in Chemical Engineering from Manhattan College, a PhD in Chemical Engineering from Rensselaer Polytechnic Institute, an MS in Environmental Engineering from Newark College of Engineering, and an MBA from Temple University. He has contributed numerous publications to the technical literature, including topics in phase equilibria, and is a recipient of the Water Pollution Control Federation's Harrison Prescott Eddy Medal for noteworthy research in wastewater treatment. He is author or co-author of two books and holder of one U.S. patent. He is a member of the American Institute of Chemical Engineers (AIChE), the American Chemical Society (ACS), and the Air & Waste Management Association (A&WMA) and is a licensed professional engineer in several states.

Acknowledgement

The author is grateful to Roberta Kunz Fox, AIA, of fox^2 design for help with Figure 1.

References

1. Kroschwitz, J.I. and M. Howe-Grant, editors, "Kirk-Othmer Encyclopedia of Chemical Technology," 4 ed., Vol. 8, pp.311-322, Wiley, New York (1993).

2. Coulson, J.M. and J.F. Richardson with J.R. Backhurst and J.H. Harker, "Chemical Engineering," 4 ed., Vol. 2, pp.436-440, Butterworth-Heinemann, Oxford (1999).

3. Treybal, R.E., "A Simple Method for Batch Distillation," *Chem. Eng.*, **77(21)**, 95-98 (Oct. 5, 1970).

4. Zubrick, J.W., "The Organic Chem Lab Survival Manual," 7 ed., 350 pp., Wiley, Hoboken, NJ (2008).

5. Mayo, D.W., R.M. Pike, and D.C. Forbes, "Microscale Organic Laboratory with Multistep and Multiscale Syntheses," 5 ed., p.61, Wiley, Hoboken, NJ (2011).

6. Williamson, K.L., R. Minard, and K.M. Peters, "Macroscale and Microscale Organic Experiments," 5 ed., pp.88-104, Houghton Mifflin Company, Boston, MA (2007).

7. Williamson, K.L., R. Minard, and K.M. Peters, "Macroscale and Microscale Organic Experiments," Selected Material Special Edition for University of Massachusetts, Amherst, 5 ed., pp.88-104, Houghton Mifflin Company, Boston, MA (2007).

8. Lord Rayleigh, *Phil. Mag.*, "On the Distillation of Binary Mixtures," Series 6, **4(23)**, 521-537 (Nov. 1902).

9. Perry, J.H., editor, "Chemical Engineers' Handbook," 3 ed., 1942 pp., McGraw-Hill, New York (1950).

10. Badger, W.L. and J.T. Banchero, "Introduction to Chemical Engineering," pp. 245-258, McGraw-Hill, New York (1955).

11. McCabe, W.L. and J.C. Smith, "Unit Operations of Chemical Engineering," 3 ed., pp. 484-487, 535-540, 606, McGraw-Hill, New York (1976).

12. Thomas, G.B., Jr., "Calculus and Analytic Geometry," 3 ed., pp.385-387, Addison-Wesley Publishing Company, Inc., Reading, MA (1960).

13. Shoemaker, D.P., C.W. Garland, and J.W. Nibler, "Experiments in Physical Chemistry," 5 ed., 890 pp., McGraw-Hill, New York (1989).

14. Hodgman, C.D., editor-in chief, "Mathematical Tables from Handbook of Chemistry and Physics," p.249, Chemical Rubber Publishing Co., Cleveland, OH (1958).

15. Dean, J.A., editor, "Lange's Handbook of Chemistry," 11 ed., McGraw-Hill, New York (1973).

16. Dean, J.A., editor, "Lange's Handbook of Chemistry," 12 ed., McGraw-Hill, New York (1978).

17. Weast, R. C. and M. J. Astle, "CRC Handbook of Chemistry and Physics," CRC Press, Inc., Boca Raton, FL (1981).

18. Dean, J.A., editor, "Lange's Handbook of Chemistry," 13 ed., p.4-59 (water) and Table 7-15 (organics), McGraw-Hill, New York (1985).

19. Brunjes, A.S. and M.J.P. Bogart, "Vapor-Liquid Equilibria for Commercially Important Systems of Organic Solvents," *Ind. Eng. Chem.*, **35**(**2**), 255-260 (Feb. 1943).

20. Hellwig, L.R. and M. Van Winkle, "Vapor-Liquid Equilibrium for Ethyl Alcohol Binary Systems," *Ind. Eng. Chem.*, **45**(**3**), 624-629 (Mar. 1953).

21. Kunz, R.G., "Environmental Calculations A Multimedia Approach," pp.41-52, Wiley, Hoboken, NJ (2009).

22. Fleser, L.F. and K.L. Williamson, "Organic Experiments," 7 ed, pp.64-68, D.C. Heath and Company, Lexington, MA (1992).

23. Anon., "Ethanol," www.worldlingo.com/ma/enwiki/en/Ethanol (accessed on 3/8/2011).

24. Flick, E.W., editor, "Industrial Solvents Handbook," 5 ed., 994 pp., Noyes Data Corporation, Westwood, NJ (1998).

25. Jones, C.A. E.M. Schoenborn, and A.P. Colburn, "Equilibrium Still for Miscible Liquids," *Ind. Eng. Chem.*, **35**(**6**), 666-672 (June 1943).

26. Otsuki, H. and F.C. Williams, "Effect of Pressure on Vapor-Liquid Equilibria for the System Ethyl Alcohol-Water," *CEP Symposium Series No. 6*, **49**, 55-67 (1953).

27. Ballard, L.H. and M. Van Winkle, "Vapor-Liquid Equilibria at 760 Mm. Pressure," *Ind. Eng. Chem.*, **44**(**10**), 2450-2453 (Oct. 1952).

28. Othmer, D.F., M.M. Chudgar, and S.L. Levy, "Binary and Ternary Systems of Acetone, Methyl Ethyl Ketone, and Water," *Ind. Eng. Chem.*, **44**(**8**), 1872-1881 (Aug. 1952).

29. Kesselman, W.D., G.E. Hollenbach, A.L. Meyers, and A..E. Humphrey, "Vapor-Liquid Equilibrium Data for Benzene-Alkylbenzene Systems," *J. Chem. Eng. Data*, **13**(**1**), 34-36 (Jan.1968).

30. Albert, N. and P.J. Elving, "Vapor-Liquid Equilibria in Binary Systems," *Ind. Eng. Chem.*, **41**(**12**), 2864-2867 (Dec. 1949).

31. Gmehling, J., U. Omken., and W. Arlt "Vapor-Liquid Equilibrium Data Collection Chemistry Data Series," Vol. 1, Part 7 (Aromatic Hydrocarbons), p.382, DECHEMA, Frankfurt (1980), citing Rollet, A.P., P. Toledano, G. Elkaim, and M. Senez, *Alger Sci. Phys.,* **2**, 403 (1956).

32. Gmehling, J. and U. Omken., "Vapor-Liquid Equilibrium Data Collection Chemistry Data Series," Vol. 1, Part 7a (Aromatic Hydrocarbons Supplement 1), p.241, DECHEMA, Frankfurt (2000), citing Rivenq, F., *Bull. Soc. Chim. Fr.*, **0**, 2427 (1974).

33. Carlson, H.C. and A.P. Colburn, "Vapor-Liquid Equilibrium of Nonideal Solutions – Utilization of Theoretical Methods to Extend Data," *Ind. Eng. Chem.*, **34**(**5**), 581-589 (May 1942).

34. Eduljee, H.E., V.N. Kumarkrishnarao, and M.N. Rao, "Correlation of Vapor-Liquid Equilibrium Data for Acetone-Water System," *Ind. Eng. Chem. Chem. Eng. Data Series*, **3**(**1**), 44-50 (1958).

35. York, R., Jr., and R.C. Holmes, "Vapor-Liquid Equilibria Data of the System Acetone-Acetic Acid-Water," *Ind. Eng. Chem.*, **34**(**3**), 345-350 (Mar. 1942).

36. Othmer, D.F. and R.F. Benenati, "Compositions of Vapors from Boiling Binary Solutions," *Ind. Eng. Chem.*, **37**(**3**), 299-303 (Mar. 1945).

37. Ju Chin Chu, "Distillation Equilibrium Data, p.24, Reinhold, New York (1950), citing unpublished paper by D.F. Othmer, D. Friedland, and E.G. Schiebel.

38. Gmehling, J. and U. Omken., "Vapor-Liquid Equilibrium Data Collection Chemistry Data Series," Vol. 1, Part 1c (Aqueous Systems Supplement 3), p.338, DECHEMA, Frankfurt (2003), citing Sugiyama, K. and K. Saka, *Kagaku Kogaku*, **33**, 470 (1969).

39. Gmehling, J. and U. Omken., "Vapor-Liquid Equilibrium Data Collection Chemistry Data Series," Vol. 1, Part 1c (Aqueous Systems Supplement 3), p.345, DECHEMA, Frankfurt (2003), citing Wohland, R. and T. Roscher, *Fiz Report*, 4273 (1973).

40. Beebe, A.H., Jr., K.E. Coulter, R.A. Lindsay, and E.M. Baker, "Equilibria in Ethanol-Water System at pressures Less Than Atmospheric," *Ind. Eng. Chem.*, **34**(**12**), 1501-1504 (Dec. 1942).

41. Othmer, D.F., W.P. Moeller, S.W. Englund, and R.G. Christopher, "Composition of Vapors from Boiling Binary Solutions," *Ind. Eng. Chem.*, **43**(**3**), 707-711 (Mar. 1951).

42. Dean, J.A., editor, "Analytical Chemistry Handbook," p.9.1, McGraw-Hill, New York (1995).

43. Andrews, L.W., "The Refractive Indices of Alcohol-Water Mixtures," *J. Am. Chem. Soc.*, **30**(**3**), 353-360 (1908).

44. Doroschewsky and Dvorsehantschik, *J. Russ. Phys. Chem. Soc.*, **40**, 908-931 (1908); *Chem. Zentr.*, **2**, 1569-1571 (1908), as cited in Simmonds, C., "Alcohol, Its Production, Properties, Chemistry, and Industrial Applications," Macmillan, London (1919).

45. Speight, J.G., editor, "Lange's Handbook of Chemistry," 16 ed., p.1.95 and p.2.294, McGraw-Hill, New York (2005).

46. Department of Commerce, Bureau of Standards, "Standard Density and Volumetric Tables – Circular No. 19," 6 ed., 72 pp., U.S. Government Printing Office, Washington, DC (Oct. 31, 1924).

47. Washburn, E.W., editor-in-chief, "International Critical Tables of Numerical Data, Physics, Chemistry and Technology," 1 ed., Vol. III, 444 pp., McGraw-Hill, New York (1928).

48. "The New Encyclopædia Britannica," Vol. 13, p.525, Encyclopædia, Inc., Chicago, IL (2002).

49. Schweppe, J.L. and J.R. Lorah, "Ternary System Ethyl Alcohol – *n*-Heptane – Water at 30 °C," *Ind. Eng. Chem.*, **46**(**11**), 2391-2392 (Nov. 1954).

50. Gray, D.E., editor, "American Institute of Physics Handbook, pp.2-142,2-152,2-153,2-157, McGraw-Hill, New York (1957).

51. "Ethyl Alcohol Handbook, 6 ed., Lyondell Chemical Company, Houston. TX, www.itecref.com/pdf/**Ethyl_Alcohol**_Handbook_Equistar.pdf (accessed on 3/8/2011).

52. Anon., "Alcoholic Proof," http://www.bookrags.com/wiki/proof-alcohol (accessed on 3/3/2011).

53. U.S. Census Bureau, Foreign Trade Statistics, "2011 Schedule B – Chapter 22," http://www.census.gov/foreign-trade/schedules/b/2011/c22.html (accessed 3/8/2011).

54. Anon., "At What 'Proof' Will Spirits Burn? – Alcohol Fire Chemistry," http://ask.metafilter.com/95484/At-what-proof-will-spirits-burn (accessed on 3/8 2011).

55. Spencer, A.B. and G.R. Colonna, editors, "Fire Protection Guide to Hazardous Materials," 13 ed., p.325-57 (ethanol-water), p.497-8 (pure ethanol), NFPA, Quincy, MA (2002).

APPENDIX A

Summary of Vapor-Liquid Equilibrium (VLE) Relationships

Ideal Systems

Equations. Low-pressure binary vapor-liquid equilibrium for an ideal system is described by the following relationships involving partial pressure (p), vapor pressure (p^{vap}), and liquid-phase mol fraction of each component. The total pressure is denoted as P_T.

$$p_1 = p_1^{vap} x_1 \tag{A1}$$

$$p_2 = p_2^{vap} x_2 \tag{A2}$$

and

$$P_T = p_1 + p_2 \tag{A3}$$

with

$$x_1 + x_2 = 1 \tag{A4}$$

Equations (A1) and (A2) are known as *Raoult's Law*.

If the Ideal Gas Law and Dalton's Law are assumed to apply to the vapor phase, vapor-phase mol fractions (y_1 and y_2) can be calculated from:

$$y_1 = p_1 / P_T = p_1^{vap} x_1 / P_T \tag{A5}$$

$$y_2 = p_2 / P_T = p_2^{vap} x_2 / P_T \tag{A6}$$

with

$$y_1 + y_2 = 1 \tag{A7}$$

Vapor pressures for any system, ideal or non-ideal, are conveniently calculated using the Antoine Equation:

$$\log_{10} p^{vap}\,(\text{mmHg}) = A - B / [t\,(^\circ\text{C}) + C] \tag{16}$$

Antoine constants are listed in Table 1 of the main text for chemicals of interest in this paper.

By combining and rearranging the equations above, one obtains:

$$x_1 = (P_T - p_2^{vap}) / (p_1^{vap} - p_2^{vap}) \tag{A8}$$

Chiefly by means of this equation, the entire x-y equilibrium curve and temperature-composition curves at constant pressure for ideal systems can be mapped out in a straightforward manner. One first chooses a temperature, computes vapor pressures from Equation (16), then calculates x_1 from Equation (A8) and y_1 from Equation (A5). Only if a specific value of x_1 is desired, does the calculation become trial and error, in which a new value of t is repeatedly chosen until the desired value of x_1 is obtained to within whatever tolerance is satisfactory.

Relative Volatility. The relative volatility (α) in Equation (8) can be calculated for any temperature or the corresponding value of x_1 along the VLE curves from a ratio of vapor pressures:

$$\alpha = (y_1/x_1) / (y_2/x_2) = p_1^{vap} / p_2^{vap} \tag{A9}$$

Relative volatilities for a number of systems – some ideal, some non-ideal – are listed in the second column of Table 2. Alpha is not perfectly constant for any of these systems across the range of interest. In general, α shows a wider variation in a system exhibiting non-ideal behavior than for an ideal system following Raoult's Law.

The systems considered as ideal [**9**, p.526;**25**;**27**-**29**] are listed in the top section of the table in order of increasing α. The higher the α, the more the bulge between the x-y equilibrium curve and the y=x, 45° line. EDC-toluene, midway down the listing, is representative of the ideal systems investigated, and its x-y curve is drawn in Figure A1 along with experimental data from several investigators [**25**,**30**-**32**]. Similar plots for methanol-ethanol and benzene-toluene have been published previously [**21**]. The other ideal systems noted here have also been verified with real data. However, those additional data-enhanced x-y diagrams are not shown since the curves are easily calculated and look largely the same.

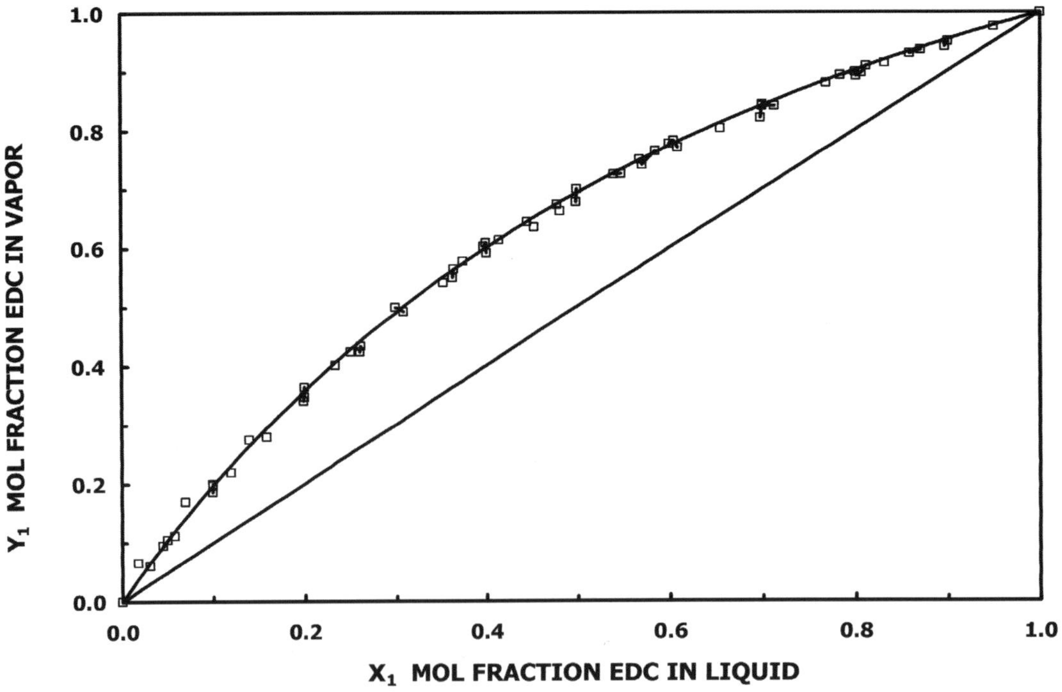

Figure A1. Vapor-Liquid Equilibrium Diagram for EDC-Toluene at 1 Atm Pressure

Because of the close similarity in α between EDC-toluene and the adjacent benzene-toluene, the resulting material balance relationships for EDC-toluene would closely resemble the curves of the illustrative example above for benzene-toluene (Figure 3). Since the pure component atmospheric boiling point is not quite the same for EDC and for benzene, the temperature graph analogous to Figure 2 would differ.

Non-Ideal Systems

Equations. For these systems, a so-called activity coefficient (γ) is introduced into Raoult's Law to yield:

$$p_1 = p_1^{vap}\,\gamma_1\,x_1 \qquad \text{(A10)}$$

$$p_2 = p_2^{vap}\,\gamma_2\,x_2 \qquad \text{(A11)}$$

and

$$P_T = p_1 + p_2 \qquad \text{(A3)}$$

with

$$x_1 + x_2 = 1 \qquad \text{(A4)}$$

Under the Ideal-Gas and Dalton's Law assumptions, vapor phase mol fractions are given by:

$$y_1 = p_1 / P_T = p_1^{vap}\,\gamma_1\,x_1 / P_T \qquad \text{(A12)}$$

$$y_2 = p_2 / P_T = p_2^{vap}\,\gamma_2 x_2 / P_T \qquad \text{(A13)}$$

again with

$$y_1 + y_2 = 1 \qquad \text{(A7)}$$

As in ideal systems, the Antoine Equation [Equation (16)] can be used to calculate vapor pressures. Selected Antoine constants are listed in Table 1.

Activity Coefficients. Activity coefficients (γ's) account for deviations from ideality. They are back calculated as point values from experimental data and then fitted to a special type of continuous function for use in VLE calculations. There are several empirical activity-coefficient formulations in widespread use; for any given application, some fit the deviations better than others across the range of data. The van Laar equations [**1**; **9**, p.527, **33**], for example:

$$\log_{10}\gamma_1 = \frac{A_{1\text{-}2}}{[1 + (A_{1\text{-}2}\,x_1)/(A_{2\text{-}1}\,x_2)]^2} \qquad \text{(A14)}$$

and

$$\log_{10}\gamma_2 = \frac{A_{2\text{-}1}}{[1 + (A_{2\text{-}1}\,x_2)/(A_{1\text{-}2}\,x_1)]^2} \qquad \text{(A15)}$$

fit the data well for the non-ideal systems considered here. These particular equations are also written in terms of natural logarithms (base e), using constants 2.302585... × $A_{1\text{-}2}$ and $A_{2\text{-}1}$.

Relative Volatility. Examples of non-ideal systems are contained in the last three entries of Table 2. These systems were selected to demonstrate the effect on α as the behavior becomes more and more non-ideal. The relative volatility (α) in Equation (8) for these and other non-ideal systems can be calculated by adding activity coefficients to the vapor-pressure ratio in Equation (A9) as shown below:

$$\alpha = (y_1/x_1) / (y_2/x_2) = (p_1^{vap}\gamma_1) / (p_2^{vap}\gamma_2) \tag{17}$$

Additional comments on the three example systems follow.

Methanol-Water and Acetone-Water. Activity coefficients for methanol-water and acetone-water at 1 atm pressure derived from experimental data are given in Ref. [**9**, p.528] and reproduced here in a footnote to Table 2 of the main text. The methanol-water x-y curve with accompanying experimental data from a number of investigators is shown elsewhere [**21**]. It resembles the curves for the ideal systems. It uses the van Laar constants listed in the Table 2 footnote, Antoine constants from Table 1, and a trial- and-error‡ procedure to compute each point on the continuous curve at the chosen total pressure.

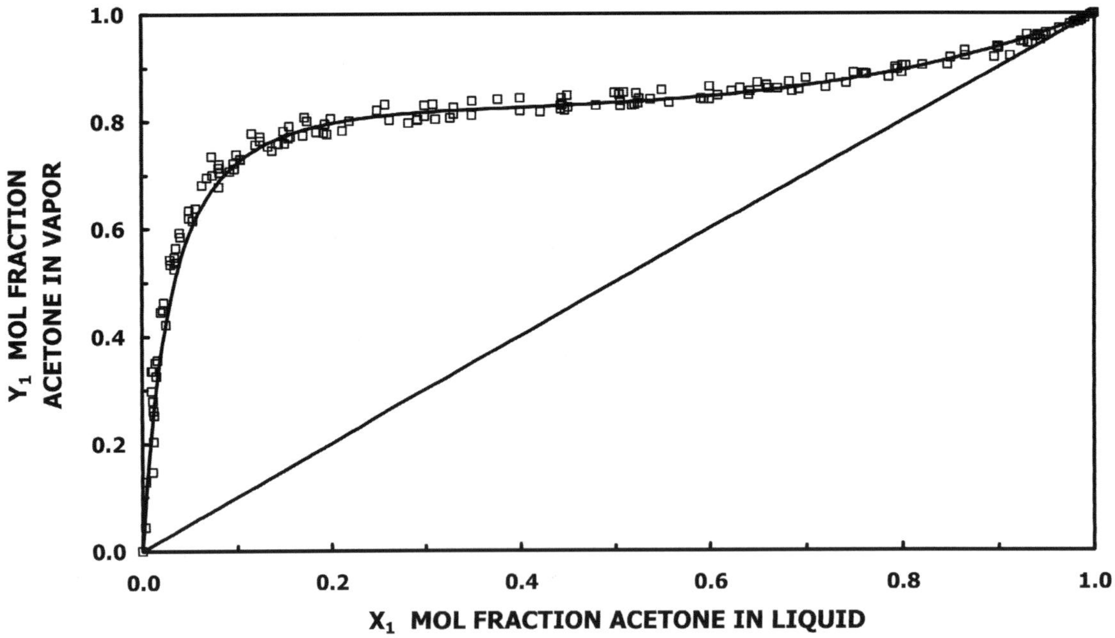

Figure A2. Vapor-Liquid Equilibrium Diagram for Acetone-Water at 1 Atm Pressure

‡ Constant-pressure calculations for non-ideal systems are always trial and error since the activity coefficients are functions of liquid composition. An efficient method to map out the x-y equilibrium curve is to select a value of x_1 and total pressure and then vary the temperature until the calculated total pressure from Equation (18) agrees as closely as desired with the total pressure selected.

The acetone-water system (Figure A2) is distorted somewhat but does not exhibit azeotropic behavior. The curve in Figure A2, prepared in a similar manner to the x-y curve for methanol-water, correlates the data of several investigators [**19**,**28**,**34**-**39**] reasonably well.

Ethanol-Water. The classic example frequently cited to illustrate an azeotrope is the ethanol-water system. This system is well investigated in the literature and is a favorite student exercise in college/university laboratories throughout the world. Its predicted x-y diagram at atmospheric pressure, shown without data points (Figure A3), was drawn by using the same trial-and-error procedure described above for methanol-water and acetone-water.

It exhibits an azeotrope composed of 95.6 wt % ethanol and 4.4 wt % water at 1 atm pressure (760 mmHg, 1.013 bar) and 78.15 °C [**9**, pp.631,633]. This corresponds to 89.43 mol % ethanol and 10.57 mol % water [**9**, p.633]. The azeotropic point is different at other total pressure levels [**9**, p.631]; water content decreases with decreasing pressure, and at least in theory becomes zero when distillation pressure is reduced low enough [**9**, p.631;**40**].

In the special case of an azeotrope, one can obtain the van Laar constants for use in Equations (A14) and (A15) from the composition of the azeotropic point alone [**21**]. Since the fitting technique to obtain them involves only the azeotropic point, agreement with other experimental values along the curve is not guaranteed. Although the fit is not perfect, the myriad of accompanying data points from multiple investigators (shown elsewhere [**21**]) are, however, so numerous as to obscure the calculated x-y equilibrium curve almost completely.

Furthermore, the ethanol-water system at constant pressure has been investigated from below atmospheric pressure [**40**] up to at least 300 lb_f/in^2 absolute (psia) [**26**,**41**]. It is well described in Ref. [**26**] by van Laar constants only slightly different from those obtained here and differing among themselves but little for huge changes in pressure. Therefore, small day-to-day variations in barometric pressure encountered in the laboratory would alter only P_T in simulation calculations, and not require modification of the 760-mmHg van Laar constants.

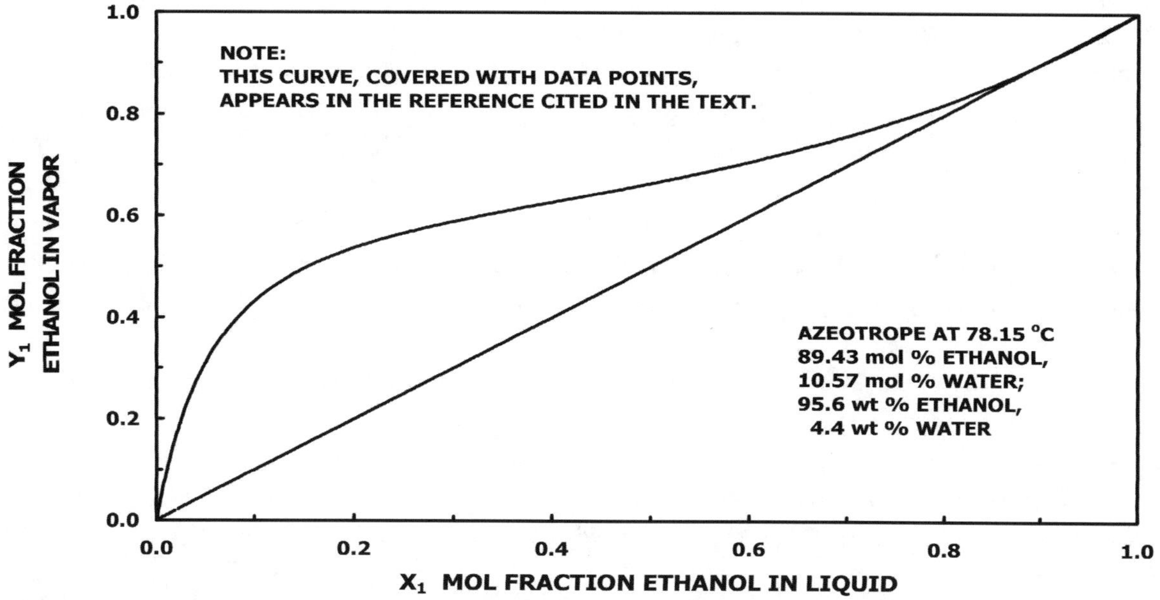

Figure A3. Predicted Vapor-Liquid Equilibrium Diagram for Ethanol-Water at 1 Atm Pressure

APPENDIX B

ANALYTICAL METHODS FOR ETHANOL-WATER SOLUTIONS

This appendix reviews several ways to analyze for chemical composition in the ethanol-water system. These include at least three quantitative methods and one qualitative technique. Quantitative analysis is done by refractive index, density, or possibly gas chromatography (GC). A qualitative analysis for ethanol in the distillate is often conducted by attempting to ignite several drops of solution on a watch glass since ethanol and its aqueous solutions greater than about 50 % by volume are combustible at room temperature [**33**]. These are discussed in turn below.

Analysis Using Refractive Index

Refractive index data for mixtures of ethanol and water at various temperatures (n_D^t) are summarized in Table B1, assembled from a number of sources. Refractive index can theoretically be determined to 1 part in 10,000 [**4**, p.319; **6**, p.299] but is more likely to agree with literature values to 1 part in 1,000 because of the presence of impurities [**6**, p.279].

Refractive index is commonly measured at 20 °C (n_D^{20}) [**17**, pp.D-200,D207]. Minor excursions from 20 °C are corrected for by adding a mean value of 0.00045 refractive index units for every °C above 20 °C [**6**, p.299;**42**].

$$n_D^{20} = n_D^t + 0.00045\,(t - 20\text{ °C}) \qquad \text{(B1)}$$

As seen in Table B1, differences in refractive index for more appreciable differences in temperature are not exactly accounted for by the general factor.

Refractive index for ethanol-water solutions passes through a maximum point when plotted against composition. This occurs at 79.3 wt % ethanol and 25 °C, as reported by Andrews [**43**], and in the vicinity of 79-80 wt % ethanol at 60 °F (15.56 °C), as determined by curve-fitting the data in Ref. [**24**, p.250] combined with differential calculus. A less rigorous inspection of the other entries in Table B1 detects maxima occurring somewhere in a wider range between 70-80 wt % ethanol. In the affected range, therefore, more than one value of ethanol composition corresponds to the same refractive index when used as the independent variable in a calibration curve.

Determination of refractive index requires only two or three drops of sample [**6**, p.300]. Once the procedure is set up and the proper technique is developed [**4**, pp.247-251], it is possible to analyze a new sample every minute [**13**, p.237]. Refractive index should be measured as soon as possible after sampling lest the samples decompose on standing [**6**, p.300]. Analysis by refractive index does, however, require specialized equipment [**6**, p.300].

Table B1. Refractive Index (n_D) Values of Ethanol-Water Solutions at Various Temperatures (°C)

Ethanol (wt %)	n_D^{15}	$n_D^{15.56}$	n_D^{20}	n_D^{25}	n_D^{30}	n_D^{40}	n_D^{50}	n_D^{55}
0	1.33345	1.33336	1.33300	1.33250	1.3318	1.3306	1.3290	1.3281
10	1.34020	1.3402	1.3395	1.3389	1.3384	1.3368	1.3349	1.3339
20	1.34778	1.3479	1.3469	1.3462	1.3450	1.3429	1.3406	1.3393
30	1.35470	1.3542	1.3535	1.3520	1.3510	1.3481	1.3452	1.3435
40	1.35948	1.3590	1.3583	1.3565	1.3550	1.3518	1.3484	1.3468
50	1.36290	1.3626	1.3616	1.3598	1.3578	1.3543	1.3506	1.3488
60	1.36505	1.3650	1.3638	1.3620	1.3597	1.3560	1.3522	1.3501
70	1.36645	1.3662	1.3652	1.363038	1.3608	1.3570	1.3528	1.3505
80	1.36690	1.3665	1.3658	1.36331	1.3611	1.3569	1.3525	1.3502
90	1.36626	1.3659	1.3650	1.36239	1.3603	1.3561	1.3515	1.3491
100	1.36332	1.36316	1.3614	1.35941	1.3573	1.3531	1.3487	1.3465

Notes:

(a) Values at 15°C, 30°C, 40°C, 50°C, and 55°C from Ref. [**44**].

(b) Values at 15.56°C (60°F) for pure water and ethanol from Ref. [**24**, p.250]. Other entries at 15.56°C from regression of Ref. [**24**, p.250] values at other compositions.

(c) Values at 20°C from Ref. [**17**, p.D-200 (water) and p.D207].

(d) Values at 25°C either reported by Andrews [**43**], assembled from entries on p.1.95, p.2.294, and Tables 10-71 and 10-72 of Lange's Handbook [**45**], or curve-fitted to interpolate between those values.

Analysis by Density

Published densities of ethanol-water solutions [**9**,**15**-**18**, **45**-**47**] facilitate ethanol analysis by careful measurement of density at constant temperature. Unlike the refractive index of ethanol-water solutions, for the most part, solution density at a given temperature corresponds to a single value for composition. The simplest technique employs a pycnometer, or specific gravity bottle, a piece of calibrated glassware of precisely known volume [**13**, pp.191-193]. It is weighed first empty and then full of solution, and density is calculated by dividing solution weight by solution volume. Pycnometer volume is typically about 10 mL [**48**]. Related studies used a 10-mL [**25**,**40**] or 5-mL [**49**] pycnometer to analyze by density.

Analysis by density has the advantage of using simple laboratory equipment in widespread use. It takes a minimum of time and provides a unique value for ethanol composition at a constant known temperature. However, in some cases, required sample size may be greater than the amount of distillate collected.

Density-Composition Plot. Density data for ethanol-water mixtures across the entire gamut of compositions and sample temperatures are difficult to find. This is not so much of a problem for analysis by density, but rather for converting from mass to volume at any given temperature and vice versa. Density data (g/mL) for liquid water (0-102 °C) and ethanol-water mixtures are tabulated in NBS Circular 19 [**46**] and handbooks reproducing said tables. For pure ethanol, density or its reciprocal, specific volume, is listed in various tables up to 80 °C [**47**, p.41], and tabular entries and correlating equations are provided for density from 0-78 °C [**50**]. Density data in lb_m/gal is plotted either against °C [**24**, p.286] or °F [**51**, p.62] from 5 to 60 °C, parametric in vol % ethanol. Density data for liquid solutions outside this temperature range could not be found in the published literature.

However, the parametric plots [**24**,p.286; **51**,p.62] appear to be well behaved, consistently spaced, and fairly linear at the upper temperatures, suggesting that missing values might be interpolated or extrapolated with some degree of success. To that end, the density plot has been reconstructed here in units of g/mL between 0 and 100 °C, with wt % ethanol as parameter (Figure B1).

The newly developed curves of Figure B1 are based on the data from NBS Circular 19 [**46**], the equations of Ref. [**50**], and the aforementioned graph [**24**, p.286]. The fit is exact for all values for pure water, for pure ethanol up to 78 °C, and for tabulated mixture data from 10-40 °C [**46**], and in reasonable agreement with values read with difficulty from the curves of Ref. [**24**, p.286] from 5 to 60 °C. Two different methods produce the same extrapolated pure ethanol curve from 78 to 100 °C.

The top curve (0 % ethanol, 100 % water) contains the actual maximum point in the density for pure water at about 4 °C. Curve fitting predicts a maximum in density for 20 wt % ethanol at about 2 °C, which may or may not be the case. As drawn in Figure B1, the other curves exhibit a continuous decrease in density with increasing temperature.

Estimated densities for pure ethanol beyond 80 °C and for mixtures below 5 °C and above 60 °C are not based on data and should be used with caution, since extrapolated densities are most likely not exact. Furthermore, although values are shown in Figure B1 for pure ethanol and for some mixture compositions above their atmospheric boiling points, these solutions do not exist as stable liquids at 1 atm pressure. In that case, the predicted densities correspond to whatever slightly elevated pressure is necessary to maintain the solution in the liquid state. This slight difference in pressure should have a negligible effect on density since enormous pressure increases are necessary to effect a significant change in liquid density [**47**, pp.40-42]. Finally, while estimated values of density may be useful to calculate volume from mass or vice versa, density values outside the range of data should <u>not</u> be used for analytical purposes to determine solution composition.

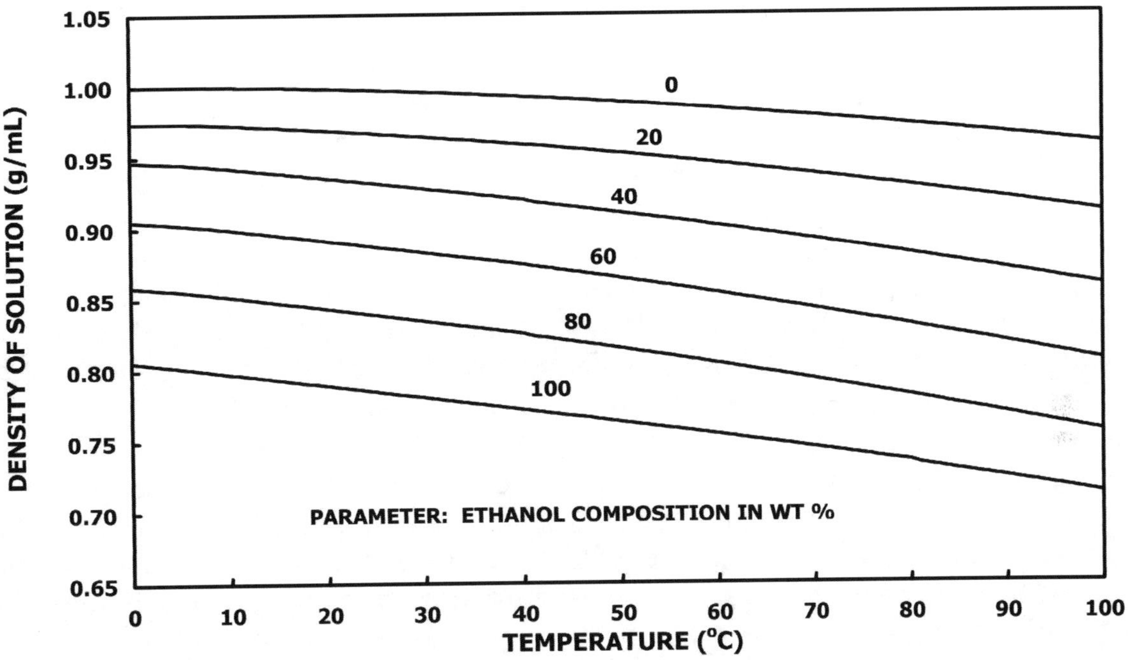

Figure B1. Estimated Liquid Density of Ethanol-Water Solutions

Analysis by Gas Chromatography (GC)

Gas chromatography is a well-documented separation and analysis technique used in a chemical laboratory. When a liquid sample is injected into the gas chromatograph instrument, it is picked up by a carrier gas, usually helium but possibly nitrogen, and passed through a heated column packed with adsorbent material, where different chemicals are adsorbed and desorbed at different rates. The chemicals being analyzed for arrive at a detector downstream as a series of peaks traced on a strip chart recorder. The time for the arrival of a peak is indicative of each particular chemical in the mixture. The width of each peak, or more properly the area under the peak, represents the amount of that chemical present. More information to supplement this highly simplified explanation is contained in the cited references [**4**, pp.255-274; **6**, pp.215-223; **7**, pp.165-173; **13**, pp.149-151].

Advantages: A gas chromatograph can detect extremely low concentrations in an unknown sample. It requires a sample size on the order of 1-25 micro liters (µL), smaller or much smaller than the several drops (20 drops = 1 mL) needed for a refractive index determination.

Disadvantages: The GC instrument is exceptionally expensive to purchase and may have to be shared with other investigators, if it is available at all. It needs an hour or so warm-up time, and requires experience it setting its temperature controls for the particular system being analyzed and some technique/practice in handling the micro-liter syringe to pierce the septum at the instrument's injection port. Residence time for each sample to pass through the column and show up as peaks at the detector may be several minutes or longer.

Analytical-Qualitative

A qualitative analysis for ethanol in the distillate is often conducted by attempting to ignite several drops of solution on a watch glass. Like hydrogen, ethanol burns with a bluish flame not easily visible in ordinary daylight [**51**, p.101]. Ethanol and its aqueous solutions greater than about 50 % by volume are combustible at room temperature [**23**]. These will ignite and burn to CO_2 and water without leaving any residue.

Alcohol Proof. In former times, this phenomenon formed the basis of a test of the alcoholic spirits given to members of the British Royal Navy as part of their rations [**52**]. In that test, a sample lighting off with gunpowder constituted "proof" that the alcohol had not been watered down. In the U.S. today, *proof* is defined as exactly twice the alcoholic content by volume, measured at 60 °F (15.56 °C) [**24**, Table 6.36; **51**, p. 108; **53**].

Anecdotal evidence suggests that a certain brand of vodka at 100 proof (50 % ethanol by volume) lights off, whereas this same vodka at 80 proof (40 % ethanol by volume) does not [**54**]. That this should be so is explained by the flash point of ethanol-water solutions (Table B2, prepared from a single source [**55**]); flash point is plotted vs. temperature in Ref. [**24**, p.254]. The flash point is the lowest temperature at which just enough liquid is evaporated at the surface to create a combustible concentration in air of gas that will catch fire if an ignition source is present.[§] The table shows that an ethanol-water solution is combustible at "room temperature" at about the stated 50 % by volume. Heating the solution first, which generates additional combustible ethanol vapor (and thereby lowers the flash point), allows solutions of lesser ethanol content to ignite at the same room temperature [**54**].

Table B2. Flash Points of Ethanol-Water Solutions [**55**]

Ethanol Volume %	Temperature		Ethanol Volume %	Temperature	
	°F	°C		°F	°C
5	144	62	60	72	22
10	120	49	70	70	21
20	97	36	80	68	20
30	85	29	95	63	17
40	79	26	96	62	17
50	75	24	100	55	13

[§] The autoignition temperature, for which no external ignition source is necessary, is much higher (for example, 363 °C for 100 % ethanol [**55**]). Degrees F for pure ethanol in table and footnote calculated from entries in °C.

(118c) Separation of the Cresol Isomers by a New Technique Combining Distillation and Crystallization

Lie-Ding Shiau, Department of Chemical and Materials Engineering, Chang Gung University, Taoyuan, Taiwan and Chih-Hao Huang, Chemical and Materials Engineering, Chang Gung University, Taoyuan, Taiwan

A new separation technology, called stripping crystallization (SC) or distillative freezing (DF), has been recently developed in our laboratory to separate the mixtures with close boiling temperatures, including the mixed xylenes [1-3] and the benzene/cyclohexane mixture [4]. In principle, SC is operated at a triple-point condition, in which the liquid mixture is simultaneously vaporized and crystallized due to the three-phase equilibrium. Thus, SC combines distillation and crystallization to produce pure crystals. Due to the close boiling points of p-cresol (201.98°C) and m-cresol (202.27°C), it is very difficult to separate them by conventional distillation. The experiments show that the SC operation can be applied to purify p-cresol from the cresol isomer while it is rather difficult to purify m-cresol. In the purification of p-cresol by SC, the recovery rate increases with increasing operating pressure while the final purity of the product decreases with increasing operating pressure.

References:

[1] L.D. Shiau, C.C. Wen, B.S. Lin,*Ind. Eng. Chem. Res.,***2005**; 44, 2258-2265.

[2] L.D. Shiau, C.C. Wen, B.S. Lin,*AIChE J.,***2006**; 52, 1962-1967.

[3] L.D. Shiau, C.C. Wen, B.S. Lin,*AIChE J.,***2008**; 54, 337-342.

[4] L.D. Shiau, C.C. Yu,*Sep. Purif. Technol.,***2009**; 66, 422-426.

An Engineering Approach to the Synthesis of Integrated Distillation Schemes for Systems Involving Difficult Separations

D. Jantes Jaramillo, G.T. Polley

Dept. of Chemical Engineering, University of Guanajuato, Mexico

The separations of iso-pentane and n-pentane and of iso-butane and n-butane are examples of what may be termed a difficult separation. They each not only require a large number of distillation stages but operation at elevated pressure if cooling water is to be used in the overhead condenser.

A "standard" problem that included both of these separations was proposed by Heaven (1969). This problem has been studied by Rathore et al (1974), Andrecovich & Westerberg (1982), Morari & Faith (1980), Meszaros & Fonyo (1986), Rajah & Polley (1995), Jobson (1997), Sobocam & Glavic (2002) and others. The details of the separation are given in Table 1.

Table 1. Heaven's Problem

Component	Mole Fraction	Molal Flow (kmol/hr)
A: Propane	0.05	45.36
B: i-butane	0.15	136.08
C: n-butane	0.25	226.80
D: i-pentane	0.20	181.46
E: n-pentane	0.35	317.52
Total	1.00	907.20

Given the attention directed at this problem, the obvious questions are "why look at it again?" The answer is that all of the work published to date appears to have involved the examination of all of the possible separation schemes. Consideration of the engineering needs ahead of synthesis reduces this number from 14 to 2. Given this situation there is no need for sophisticated software to solve this problem. Standard flow-sheeting packages can be used.

In solving this problem the reflux ratios are initially set at 1.33 times the minimum value in order to reduce the column height for the difficult separations. In final refinement of the design the more common criterion of 1.1 times minimum value can be investigated.

Engineering Heuristics

The number of theoretical stages required for a given separation increases with increasing pressure. Therefore, for difficult separations the operating pressure should (in the first instance) be minimised.

If high utility costs are to be avoided heat should be removed from the overhead condenser using cooling water (rather than chilled water or refrigerant). So, the minimum operating pressure for a separation should be set n the basis of a minimum overhead temperature (in this study we have chosen 50 C, obviously it is a constraint that is set by local ambient conditions).

Step 1. Identify the most difficult separation.

In this problem the hardest separations are those involving splitting n-pentane (component E) from iso-pentane (component D). The number of theoretical plates required for this separation (for the possible ways in which it can be undertaken) is shown in Figure 1. The result suggests that operating pressure for a column having a bottom product of component E should operate at the lowest possible pressure.

The relationship between operating pressure and overheads temperature for the various columns providing component E as bottom product is shown in Figure 2. This plot indicates that the D/E column provides both the lowest operating pressure (2 bar) and the lowest vapour traffic.

It is concluded that separation schemes incorporating this particular separation are favoured and the initial design component is a D/E column operating at 2 bar. The column requires 91 theoretical stages. The re-boiler load is 20.1 MW and the re-boiler temperature is 72.9 °C.

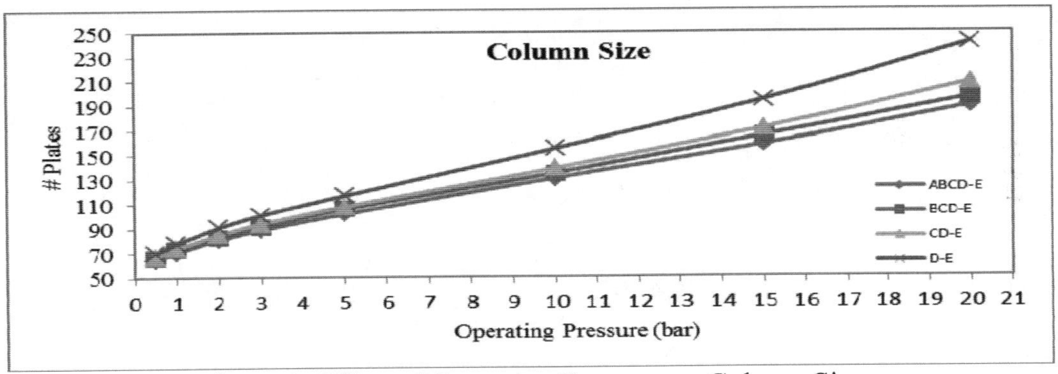

Figure 1. Effect of Operating Pressure on Column Size

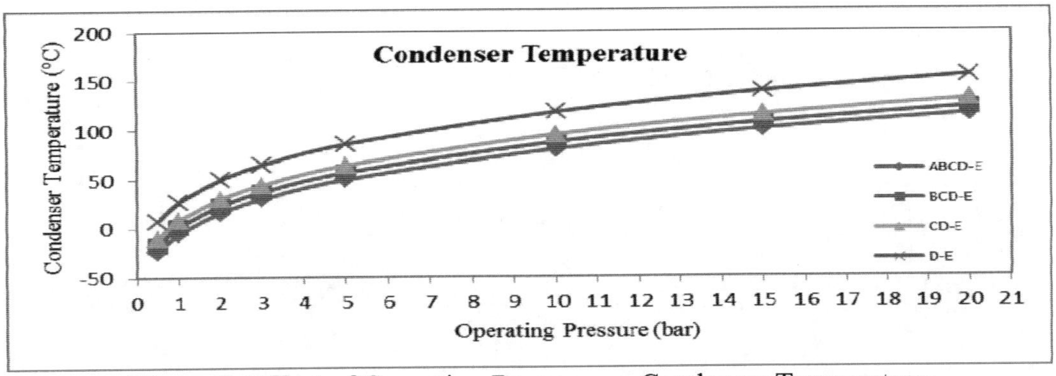

Figure 2. Effect of Operating Pressure on Condenser Temperature

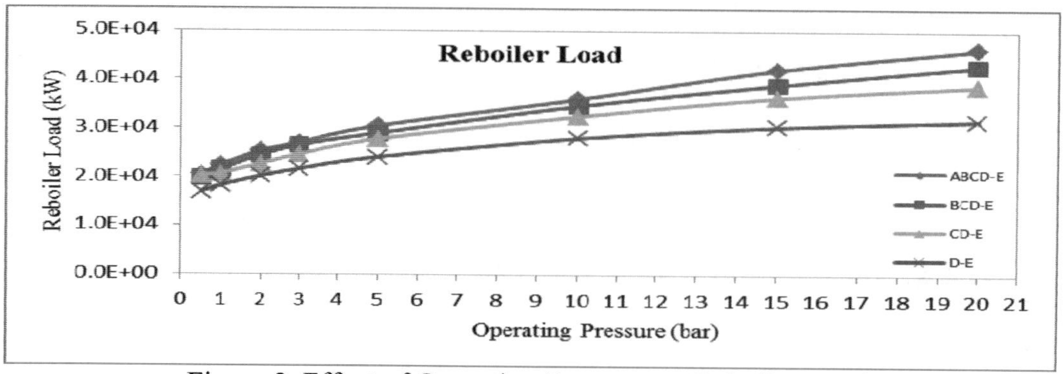

Figure 3. Effect of Operating Pressure on Reboiler Load

Step 2. Identify any other difficult separation.

The splitting of iso-butane and n-butane (components B and C) is also observed to be difficult. The effect of pressure on column size is shown in Figure 4; on condenser temperature is shown in Figure 5 and on reboiler load in Figure 6.

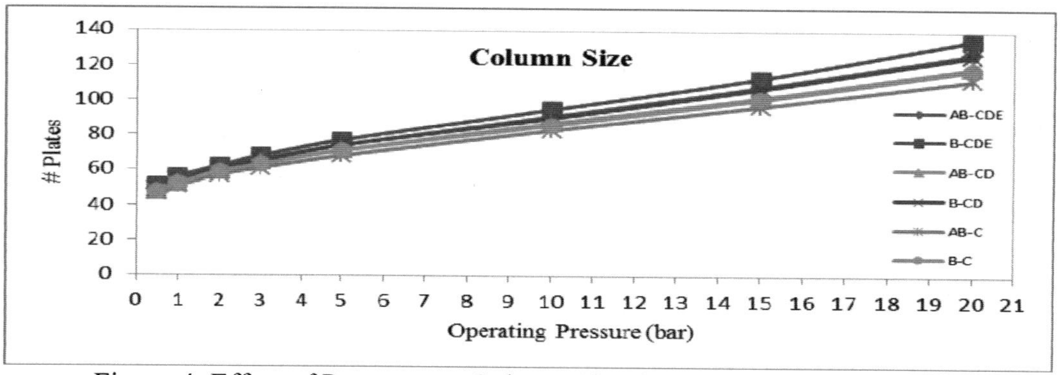

Figure 4. Effect of Pressure on Column Size for splitting i-C4 and n-C4

Examination of these figures indicates that considerations of capital cost favour the B/C column. The minimum operating pressure for this column is 6.5 bar. The reboiler temperature will be 65.8 °C with a load of 9.04 MW.

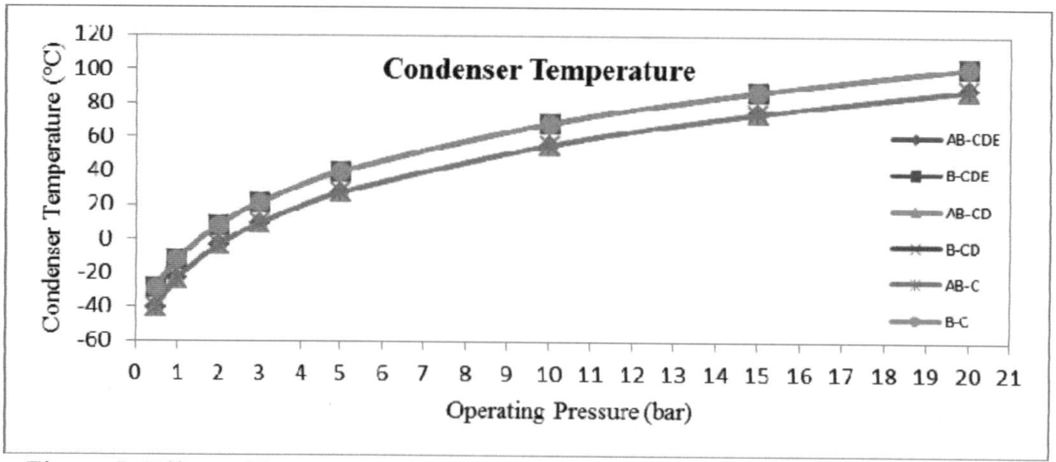

Figure 5. Effect of Pressure on Condenser Temperature for splitting i-C4 and n-C4

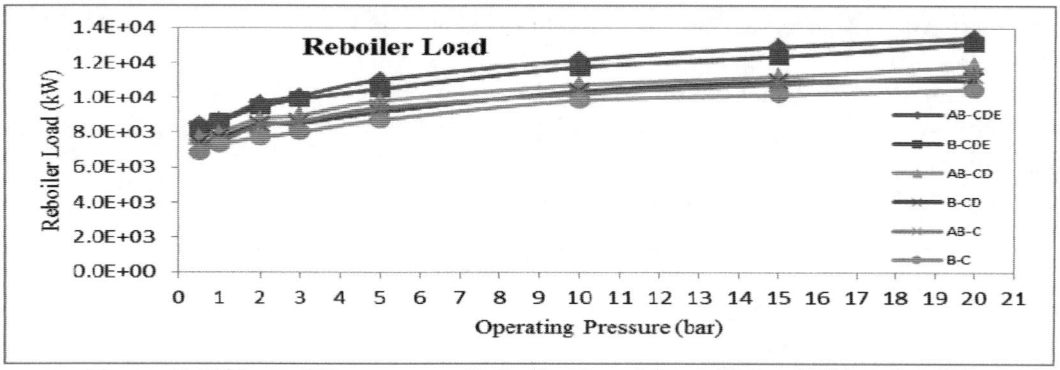

Figure 6. Effect of Pressure on Reboiler Load for splitting i-C4 and n-C4

Step 3. Identify Possible Schemes

The decision to have a scheme that incorporate columns handling D/E and B/C separations leads to a significant reduction in the number of schemes that can be employed. All of the schemes that start with the separations of ABCD/E and AB/CDE are eliminated, whilst for the schemes starting with the separation ABC/DE only one remains (shown in Figure 7). And, finally for the schemes starting with the A/BCDE separation (Figure 8) only one is valid. Rather than having 14 schemes to evaluate there are now only 2. Attention is now directed towards identifying how these schemes can be thermally integrated.

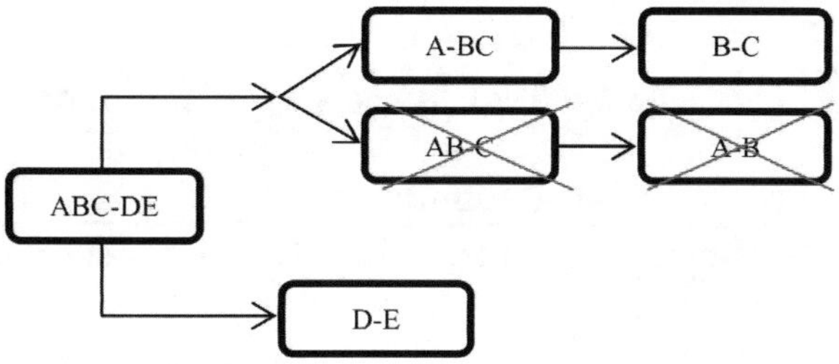

Figure 7. Schemes starting with the ABC/DE separation

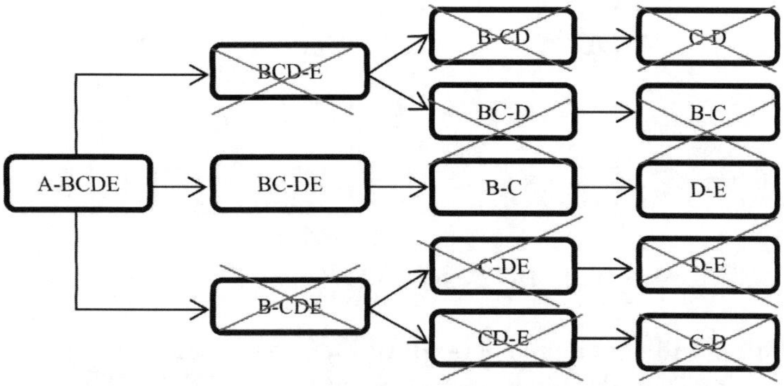

Figure 8. Schemes starting with the separation A/BCDE

Step 4. Identify Opportunities for Thermal Integration

It is assumed that the maximum column operating pressure is 20 bar. The condenser temperatures and loads at this operating pressure for the potential heat sources are listed in Table 2.

Table 2. Potential Heat Sources

Scheme	Separation	Condenser Temp.	Condenser Load
1	ABC/DE	102.4	7.78
	A/BC	59.6	2.48
2	A/BCDE	59.6	4.09
	BC/DE	109.7	7.80

The identified heat sinks are listed in Table 3.

Table 3. Identified Heat Sinks

B/C	65.8 C	9.04 MW
D/E	72.9 C	20.1 MW

The temperature at which component A condenses is too low for the stream to provide heat to satisfy either of the identified heat sinks. Heat from either of the nC/Dn columns can be integrated.

The two design options are listed in Table 4 and shown in Figures 9 and 10.

Table 4. Summary Options

Scheme	Column	Pressure	Heat Load	No. Plates
1	ABC/DE	20	8.16	46
	A/BC	16.0	2.60	33
	B/C	6.5	9.04	76
	D/E	2	20.1	91
	Q SAVING		-7.78	
	Summations		**32.12**	**246**
2	A/BCDE	15.6	4.25	34
	BC/DE	20	8.12	46
	B/C	6.5	9.04	76
	D/E	2	20.1	91
	Q SAVING		-7.8	
	Summations		**33.71**	**247**

Comparison with Result from the Literature

Jobson undertook a full cost analysis of all of the possible non-integrated schemes, the two schemes identified in this work were ranked first and third. The scheme ranked second commenced with the ABC/DE separation followed by AB/C and A/B. In the analysis undertaken in this work the decision was taken to favour the B/C separation over the A/BC separation.

Sobocam & Glavic (2002) solved this problem and found an "optimal" solution that had the same structure as scheme 1 but contained one further integrated re-boiler. This re-boiler used overheads from the column separating components B and C to provide part of the boil-up on the column separating D and E.

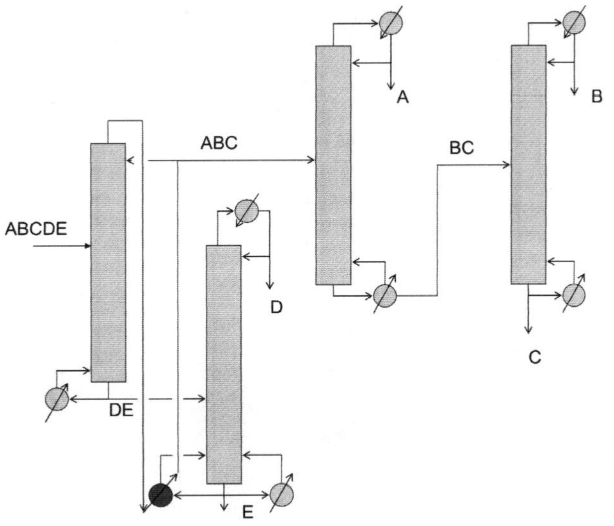

Figure 9. Scheme 1

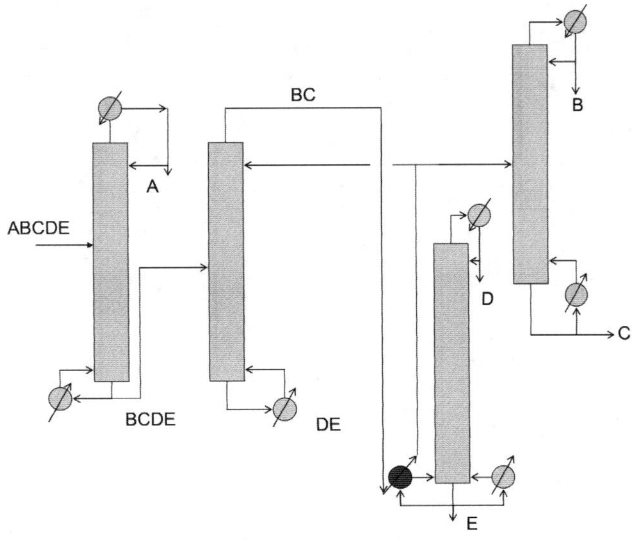

Figure 10. Scheme 2

In order to achieve this second integration the temperature of the overheads from the B/C column would need to have a temperature in excess of 88 C (based on a minimum temperature difference of 15 C). Therefore, the pressure of the B/C column needs to be in excess of 15 bar and the number of theoretical plates in excess of 100 (the current size being 76).

Since, it is probable that a separation requiring 76 theoretical plates would require two columns operating in series (because of height restrictions) the additional cost associated with this level of increase in number of plates could be justified through energy savings. So, a design involving a B/C column operating at a higher pressure and thermally linked to the D/E column should be examined.

Step 5. Examination of Assumptions and Refinement of Design

The effect of column pressure on condenser load for the separation of iso-butane from n-butane is shown in Figure 11.

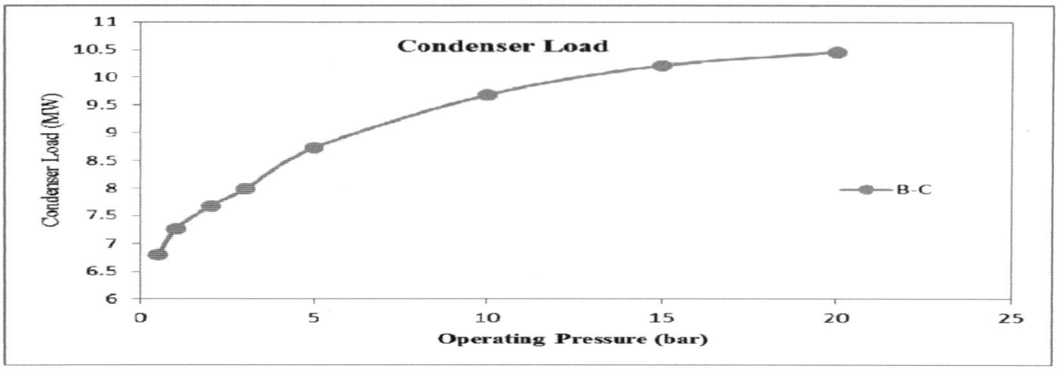

Figure 11. Effect of Pressure on Condenser Load for B/C separation

It is observed that there is little change once the column pressure exceeds 15 bar. Operation of this column at 15 and using the condenser load to provide part of the boil-up on the D/E column would save 10.21 MW.

This separation appears in both candidate structures. So, the final options are:

Table 5. Summary Options

Scheme	Column	Pressure	Heat Load	No. Plates
1	ABC/DE	20	8.16	46
	A/BC	16.0	2.60	33
	B/C	15	10.23	101
	D/E	2	20.1	91
	Q SAVING		-17.99	
	Summations		**23.1**	**271**
2	A/BCDE	15.6	4.25	34
	BC/DE	20	8.12	46
	B/C	15	10.23	101
	D/E	2	20.1	91
	Q SAVING		-18.03	
	Summations		**24.67**	**272**

Further practical considerations need to be considered (such as control strategy and flexibility) before the design is complete. Such study may well find that direct integration of B/C condenser with D/E re-boiler is unacceptable because both separations are very sensitive to changes in feed composition and the combined system is difficult to control.

However, two viable candidates do appear to have been identified without the need to use any specialist software or numerical techniques.

Conclusions

In this paper an examination of how existing engineering tools can be used to identify thermally integrated distillation schemes for problems involving difficult separations.

The presence of difficult separations greatly reduces the number of separation schemes that can be considered to be viable.

References

Andrecovich M.J. & Westerberg A.W. 'A simple synthesis method based on utility bounding for heat integrated distillation sequences', AIChEJ, 1985,31(3), 363-375

Jobson M. "Short-cut Evaluation of Distillation Sequences", Comp. & Chem. Engng., 1997, 21. Suppl. S553-S557

Heaven D.L. M.S. Thesis, University of California, Berkeley, 1969

Meszaros I. & Fonyo Z. "A new bounding strategy for synthsizing distillation schemes with energy integration", Comp. & Chem. Engng. 1986,10(6),545-550

Morari M. & Faith D.C. "The synthesis of distillation trains with heat integration", AIChEJ, 1980, 26,916-928

Rathore R.N.S., Van Kormer K.A. & Powers G.J. "Synthesis strategies for multi-component separation systems with energy integration", AIChEJ, 1974,20,491-501

Rajah W. & Polley G.T. "Synthesis of practical distillation schemes", Trans.I.Chem.E. 1995, 73A,953-964

Sobocan G. & Glavic P. "A simple method for systematic synthesis of thermally integrated distillation sequences", Chem. Eng. J, 2002,89,155-172

An Engineering Approach to the Synthesis of Integrated Distillation Schemes where there is a Significant Number of Options

D. Jantes Jaramillo and G.T. Polley

Department of Chemical Engineering, University of Guanajuato, Mexico

In a previous paper (Jantes Jaramillo et al, 2012a) we examined a how a multi-component mixture that had difficult separation tasks could be separated into its component parts. It was argued with mixtures having one or more difficult individual separation the optimisation was dominated by capital cost considerations. Consequently, the way in which such separations should be conducted needed to be identified first. This then led to a significant reduction in the number of viable schemes and identification of the better options was simplified.

In this paper attention is directed at systems that cannot be simplified in this manner and where a large number of options are available.

Malone et al (1985) have pointed out that the vapour traffic in a separation scheme is a guide to the cost of the scheme. There is a direct relationship between the vapour traffic and energy demands and it also influences column diameters and heat exchanger sizes. The better schemes should be those that minimise the vapour traffic.

Porter & Momoh (1991) subsequently derived an equation for the identification of the total vapour traffic involved in a given separation scheme. They called this the ROTE equation (emphasising that it was a "rule of thumb").

Jobson (1997) has provided a critique on this equation warning that it can lead to incorrect conclusions.

Alternative Method for Calculating Vapour Traffic

Porter & Momoh based the development of their equation upon the following equation for minimum reflux ratio:

$$R_{\min} = \frac{1}{\left(\alpha_{LK,HK} - 1\left[\frac{x_{D,LK}}{x_{F,LK}} - \alpha_{LK,HK}\frac{x_{D,HK}}{x_{F,HK}}\right]\right)} \qquad (1)$$

where subcripts F refer to feed and D to distillate. They state that this assumption was the most difficult to justify.

Scheibel & Montross (1946) have presented an empirical equation for the determination of the minimum reflux ratio in a column separating a multi-component mixture. They tested their method against plate-by-plate calculations covering wide separations.

Twenty three of the cases studied involved liquid feeds that contained components heavier than the heavy key. The relative volatilities of the keys ranged from 1.1 to 6.0. The relative volatilities of the heavy components varied from 0.9 to 0.2. They claimed that comparison between the reflux ratio determined from their equation and that determined using plate-by-plate calculation was very good.

A further ten case studies involved vapour feed in which 72% was a component that was either heavier or lighter than the keys. Agreement between prediction and plate-by-plate calculation was reported to be excellent.

Given these comparisons, the empirical equation of Scheibel & Montross appears to be a good starting point for the development of an alternative to the ROTE equation of Porter & Momoh.

The equation of Scheibel & Montross is:

$$R_{\min} = \frac{1}{\sum x_{LK-i}} \left[x_{LK} R^{*}_{\min} + \sum_{i=0}^{\substack{Heavier \\ Component}} x_{HK+i} \sum_{j=1}^{\substack{Heavier \\ Component}} \frac{x_{HK+i}}{\alpha_{LK/HK+i} - 1} + \sum_{k=1}^{\substack{Lighter \\ Component}} \frac{x_{LK-i}}{\alpha_{LK-i/HK}} \left(1 + \alpha_{LK/LK-1}\right) \right] \quad (2)$$

and involves the adjustment of a pseudo-minimum reflux ratio that is dependent on the relative volatility of the two key components and the mole fraction liquid mole fraction at which the feed line intersects with the equilibrium curve, and is given by:

$$R^{*}_{\min} = \frac{1}{\left(\alpha_{LK/HK} - 1\right) x_i} \quad (3)$$

The feed line is dependent on the ratio of liquid to vapour molar flows adjusted for non-key components and is given by:

$$y = x_F + m\left(x_F - x\right) \quad (4)$$

The ratio, m, is given by:

$$m = \frac{q - \sum_{i=1}^{\substack{Heavier \\ Component}} x_{HK+i}}{(1-q) - \sum_{j=1}^{\substack{Lighter \\ Component}} x_{LK-j}} \quad (5)$$

With q being the heat to vaporize one mole of a feed divided by the molar latent heat of the feed.

The difficulty in applying this approach is the determination of the value of x_i. Scheibel & Montross presented a quadratic equation for this term and took the solution lying between zero and unity.

Observation of values reported for this parameter indicates that it can be calculated from the simple expression:

$$x_i = \left(1 + \frac{0.18}{abs(1-m)}\right) x_f \quad (6)$$

Where

$$x_f = \frac{x_{LK}}{x_{LK} + x_{HK}} \quad (7)$$

So, the approach adopted here is to solve equations (7), (5), (6), (3) and (2) and then calculate the vapour traffic within the column using this minimum reflux ratio.

Example Problem

One major advantage that the above approach has over previous methods is that it allows feed conditioning to be examined through changes in the value of q (see Equation 4). Such facility is of particular benefit when integrated systems are examined for it allows feed conditioning can be conducted through thermal integration.

Different values of q can be specified for the individual separations. However, in this example it is assumed that q is 0.5 for all separations.

The system studied is the mix of alcohols outlined in Table 1.

Table 1. Example Problem

Component		Mole Fraction
A	Ethanol	0.25
B	Iso-propanol	0.15
C	n-Propanol	0.35
D	Iso-Butanol	0.10
E	n-Butanol	0.15

The individual schemes available for this separation are listed in Table 2.

The volumetric flows for the 14 schemes determined using the ROTE equation, the equations described above and rigorous simulation of the individual schemes are listed and ranked in Table 3.

The new methodology generates predictions that are generally within 10% of those derived using rigorous simulation. It predicts four out of the top five schemes having the lower vapour traffic.

The approach developed in this work indicates that scheme 14 is the better option. So, this is the scheme for which an integration scheme will be developed.

Table 2. Available Schemes

Scheme	First column	Second Column	Third Column	Final Column
1	ABCD/E	ABC/D	AB/C	A/B
2			A/BC	B/C
3		AB/CD	A/B	C/D
4		A/BCD	B/CD	C/D
5			BC/D	B/C
6	ABC/DE	A/BC	B/C	D/E
7		AB/C	A/B	D/E
8	AB/CDE	A/B	CD/E	C/D
9			C/DE	D/E
10	A/BCDE	BCD/E	BC/D	B/C
11			B/CD	C/D
12		BC/DE	B/C	B/E
13		B/CDE	CD/E	C/D
14			C/DE	D/E

Table 3. Vapour Traffic and Ranking of Schemes

Scheme	ROTE	RANK	NEW EQUATION	RANK	RIGOROUS	RANK
1	10.06	6	7.69	14	7.84	13
2	11.71	13	6.95	12	7.86	14
3	8.97	4	7.08	13	7.22	11
4	11.27	10	6.23	7	7.2	10
5	11.59	12	6.51	9	7.42	12
6	9.5	5	5.84	4	6.88	7
7	7.84	3	6.58	10	6.87	6
8	7.83	2	6.75	11	6.79	4
9	6.96	1	5.84	2	6.35	1
10	11.72	14	6.41	8	7.18	9
11	11.4	11	6.13	6	6.97	8
12	10.35	8	5.87	3	6.53	3
13	11.11	9	6.07	5	6.82	5
14	10.24	7	5.16	1	6.38	2

Thermal Integration

Following the procedure described in the earlier paper, we now develop the heat load table. This is reported in Table 4. It is derived from rigorous simulation of the individual separations conducted at differing pressures.

Heat from the condenser on the D/E column can be used to provide part of the heat load of either the C/DE re-boiler or the B/CDE re-boiler. Through suitable pressure manipulation, the overheads of both of the separations can be used to provide boil-up the A/BCDE separation. The larger temperature difference would be a match with the B/CDE separation. Table 5 shows a suitable integration system with the D/E condenser operating at 20 bar, partially driving a B/CDE column operating at around 8.5 bar which then partially drives a A/BCDE column operating a 2 bar.

Table 4. Heat Load Table for Chosen Scheme

Separation	Re-boiler Temperature	Load	Plates	Separation	Condenser Temperature	Load	Plates
				D/E 20	214	5.13	34
B/CDE 20	213.5	16.4	37				
C/DE 15	210.1	23.0	95				
				C/DE 20	207.2	31.3	135
A/BCDE 20	205.6	25.2	83				
D/E 10	202.9	5.8	40				
				D/E 15	201.3	5.8	37
B/CDE 15	199.3	16.2	36				
				C/DE 15	192.8	25.6	95
C/DE 10	192.6	18.7	71				
A/BCDE 15	191.4	29.2	87				
				B/CDE 20	187.7	19.4	37
				D/E 10	184.4	6.9	40
B/CDE 10	181.4	16.3	34				
				A/BCDE 20	180	29.1	83
				C/DE 10	174.3	21.5	71
				B/CDE 15	174	19.5	36
D/E 5	173.8	7.3	45				
A/BCDE 10	173.3	33.7	91				
				A/BCDE 15	167.1	33.4	87
C/DE 5	166.1	15.4	54				
				D/E 5	158.3	8.55	45
				B/CDE 10	156.3	19.7	34
D/E 3	155.8	8.1	47				
B/CDE 5	154.2	16.4	31				
				A/BCDE 10	150.4	38.2	91
C/DE 3	149.2	14.4	48				
				C/DE 5	146.8	18.4	54
A/BCDE 5	146.6	37.8	90				
D/E 2	143.5	8.4	48				
				D/E 3	141	9.46	47
C/DE 2	137.4	13.7	46				
B/CDE 3	137.4	16.1	30				
				B/CDE 5	130.9	20.5	31
A/BCDE 3	130.1	37.4	85				
				C/DE 3	129.2	47.6	48
				D/E 2	128	9.8	48
D/E 1	125.9	8,6	46				
B/CDE 2	125.7	15.9	29				
				A/BCDE 5	108.5	42.6	85
C/DE 1	120.3	13.3	43				
A/BCDE 2	118.7	35.5	80				

Table 5. Integration of D/E Condenser

	Temperatures					
	Condenser			Re-boiler		
Operating Pressure	D/E	C/DE	B/CDE	C/DE	B/CDE	A/BCDE
1	108	97	83.5	120	109	102
2	128	116	102	137	126	**119**
3	141	129	114	149	137	130
5	158	147	131	166	154	147
8.5	--	--	**149.5**	--	**174.2**	--
10	184	174	156	192	181	173
15	201	193	174	210	199	191
20	**214**	207	188	223	214	206

This scheme is the supplement with integration of overheads from a C/DE separation at 5 bar with the A/BCDE column operating at 2 bar (see Table 6). An auxiliary condenser must be installed in column B/CDE operating with 2.9 MW. This excess of heat can be used for partially conditioning the feed stream entering to this column. The energy consumption in reboilers and heaters (for feed conditioning) are shown in Table 7. Energy savings are shown in this table as well. The total number of stages required for this sequence is 201.

Table 6. Integration of the C/DE Column

	Temperatures					
	Condenser			Re-boiler		
Operating Pressure	D/E	C/DE	B/CDE	C/DE	B/CDE	A/BCDE
1	108	97	83.5	120	109	102
2	128	116	102	137	126	**119**
3	141	129	114	149	137	130
5	158	147	131	166	154	147
8.5	--	--	**149.5**	--	**174.2**	--
10	184	174	156	192	181	173
15	201	193	174	210	199	191
20	**214**	207	188	223	214	206

Table 7. Energy Consumption in reboilers and heaters and integrated condenser loads

Separation	Reboiler Loads	Feed Heating Loads	Integrated Loads
A/BCDE	35.5	5.5	35.5
B/CDE	16.43	5.9	16.43+2.9
C/DE	15.4	2.2	0
D/E	4.21	2.2	0
Total	**71.54**	**15.8**	**54.83**

The total energy consumption after integration is 32.51 MW.

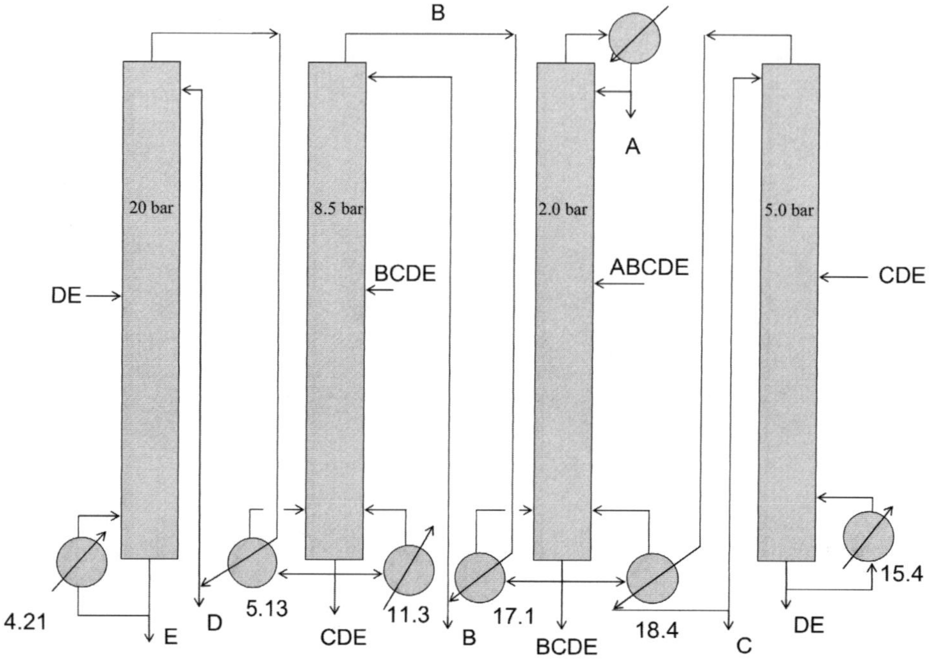

Figure 1. Integrated Scheme

We finally examine if rather than use column simulation to derive the heat load table one determined from the minimum reflux ratio can be used. The resulting table is given below:

Table 8. Heat Load Table for Chosen Scheme based on empirical equation

Separation	Re-boiler Temperature	Load	Separation	Condenser Temperature	Load
			D/E 20	214	8.93
B/CDE 8.5	174.2	8.41			
C/DE 5	166.1	8.52			
			B/CDE 8.5	149.5	8.92
			C/DE 5	146.8	12.07
A/BCDE 2	118.7	34.44			

The resulting loads are:

Table 9. Energy Consumption in reboilers and heaters and integrated condenser loads based on empirical equation

Separation	Reboiler Loads	Feed Heating Loads	Integrated Loads
A/BCDE	34.44	5.5	34.44
B/CDE	8.41	5.9	8.41+0.51
C/DE	8.52	2.2	0
D/E	7.78	2.2	0
Total	**59.15**	**15.8**	**43.36**

Total energy consumption after integration 31.59MW as compared with 32.51 MW identified using column simulations. However, there is a significant difference in some of the individual loads. The technique is being further exmined.

Conclusions

A simple approach based on a simplified empirical equation for the minimum reflux ratio is proposed for the calculation of the vapour flow rate. Once feed conditions (i.e. mole fractions and relative volatilities) and percentage of recovery for the light component are fixed, this procedure can evaluate all the sequences in order to find the one with the lowest total vapour flow rate. Once the sequence with the lowest vapour flow-rate is identified, the heat load table for such scheme is used to identify all heat integration opportunities.

In the example problem the Heat Load Table was derived using the HYSIS program for column simulation. In order to determine whether the number of simulations undertaken could be reduced by using a table derived from the identified vapour traffic. The initial results are encouraging.

References

Jantes Jaramillo D., Polley G.T., An Engineering Approach to the Synthesis of Integrated Distillation Schemes for Systems Involving Difficult Separations, AIChE Spring Meeting, Houston, 2012

Jobson, M., Short-cut evaluation of distillation sequences. Comput. Chem. Eng. (1997), 21, pp. 553-557

Malone M.F., Glinos K., Marquez F.E. and Douglas J.M., Simple Analytical Criteria for the Sequencing of Distillation Columns, AIChEJ, 1985, 31, 683-689

K.E. Porter and S.O. Momoh, Finding the optimum sequence of distillation columns - an equation to replace the "rules of thumb" (heuristics). The Chem. Eng. J., 46 (1991), pp. 97-108.

Rong, B.-G., Kraslawski, A., Nystrom, L., Design and synthesis of multicomponent thermally coupled distillation flowsheets. Comput. Chem. Eng. (2001), 25, pp. 807-820.

Scheibel, E. G., and Montross, C. F., Empirical equation for theoretical minimum reflux, Ind. Eng. Chem., Vol. 38, No. 3, (1946), pp. 268-272.

(118f) Synthesis and Optimisation of Demethaniser Flowsheets for NGL Recovery

Muneeb Nawaz, Costain Energy and Process, Manchester, United Kingdom, Megan Jobson, Centre for Process Integration, University of Manchester, Manchester and Adrian Finn, Energy & Process, Costain, Manchester, United Kingdom

Synthesis and Optimisation of Demethaniser Flowsheets for NGL Recovery

Muneeb Nawaz[1], Megan Jobson[2], Adrian Finn[1]

1.*Costain - Energy & Process, Costain House, Styal Road, Manchester, M22 5WN*

2.*Centre for Process Integration, School of Chemical Engineering and Analytical Science, University of Manchester, Manchester, United Kingdom*

A demethaniser process is characterised by interactions between the complex distillation column and other flowsheet units, including the turbo-expander, flash units, multistream exchangers and refrigeration system. This research presents a systematic approach for demethaniser flowsheet synthesis to generate cost-effective designs.

A demethaniser column has many degrees of freedom, including the operating pressure, multiple feeds, the number and duty of side reboilers and the flow rate of the external reflux stream. The additional feed and side reboiler streams enhance the efficiency of the process, but complicate process modelling. The number of design variables is also augmented by additional degrees of freedom such as the location and the order of feeds, the number of stages and the reflux ratio in the column. The complexity of the demethaniser column precludes the use of the Fenske–Underwood–Gilliland shortcut design method. A semi-rigorous boundary value method is proposed for the design of complex demethaniser columns for application within an optimisation framework for process synthesis and evaluation.

A simplified flowsheet simulation model based on shortcut models is developed that is able to account for various configurations and inter-connections in the demethaniser process. A methodology accounting for heat integration in multistream exchangers is proposed. A generalised superstructure has been proposed for demethaniser flowsheet synthesis that includes various structural combinations in addition to the operational parameters. A stochastic optimisation technique, simulated annealing, is applied to optimise the superstructure and generate energy-efficient and cost-effective flowsheets. The application of the developed synthesis methodology is illustrated by a case study of relevance to natural gas processing. The results allow insights to be obtained into the important trade-offs and interactions and indicate that the synthesis methodology can be employed as a tool for quantitative evaluation of preliminary designs as well as to facilitate evaluation, selection and optimisation of licensed demethaniser flowsheets.

Simple Opportunities for Improving Energy Efficiency

Alan P. Rossiter
Rossiter & Associates, Bellaire, Texas
alan@rossiters.org; www.rossiters.org/associates

Abstract: Energy efficiency has become an important theme in the process industries, and many companies have implemented sophisticated strategies to improve both the design and the operation of their plants. However, efficiency improvements are often possible simply by doing basic things right. This article presents a number of examples that show how operators and process designers frequently overlook simple changes that can lead to dramatic gains, and provides guidance on how to avoid some of the pitfalls.

Introduction

Great stride shave been made in energy efficiency in recent years. Many factors have contributed to this progress, including the following:

- Step-change technological improvements, such as new catalysts that increase yields and reduce operating pressure. These changes are fundamental, and they can lower the inherent energy demand of a process.
- Equipment advances, including novel distillation systems and high efficiency pumps, compressors, turbines and motors. Improvements to components like these increase energy efficiency by eliminating avoidable losses without changing the underlying process.
- Improvements to process design procedures. These include the global improvements resulting from advances in simulation and physical properties, and also more specific gains in energy efficiency from design methods such as hierarchical decision procedures (1), heat integration using pinch analysis (2) and numerical optimization (3). These methodologies do not rely on new types of equipment or fundamentally different chemistry. Rather, they provide insights that allow engineers and designers to make better use of existing technologies.
- Process control and real time optimization. Advances in these areas allow us to get more out of our facilities and to operate at tighter tolerances. This often results in significant energy savings.
- A better focus on energy use in plant operations. Energy management is increasingly being highlighted in operator training. Many companies are also deploying operations support systems, such as "energy dashboards" (4) that provide real time information on energy performance and guidance when corrective action is needed.
- Focused maintenance activities, such as steam trap and steam leak programs, heat exchanger cleaning and furnace tuning. Keeping equipment in good condition is an important part of achieving optimum performance.
- Changes in corporate policies. For example, many companies have instituted corporate goals for energy efficiency, and these are often coupled to specific

directives – e.g., incorporating energy efficiency into the procurement requirements for new equipment, or setting funds aside for energy efficiency investments.

However, in the midst of these sophisticated technologies and programs many basic opportunities are often missed. This article illustrates this fact by presenting a series of examples that have been observed by the author while conducting energy management projects at refineries and chemical plants in various parts of the world.

Example 1: One of the most frequent inefficiencies encountered is the unnecessary cooling of process streams that should be kept hot. Figure 1 shows an example of this (5), based on a study at a petrochemical facility.

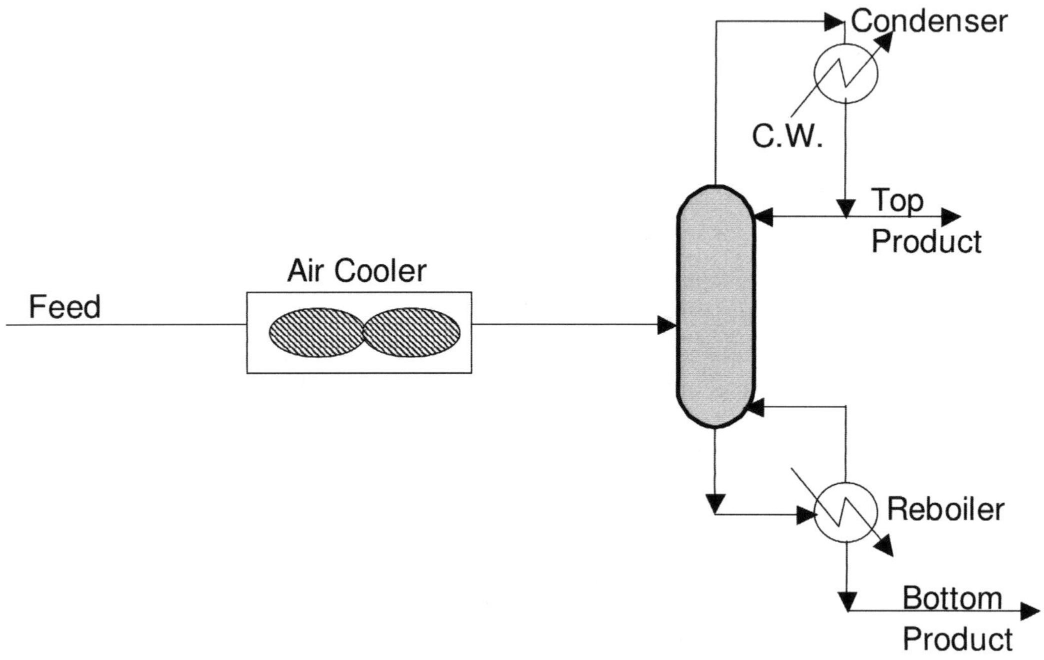

Figure 1: Inappropriate Cooling of Distillation Column Feed

In this example, the design of the distillation column includes an airfin cooler on the feed. This was installed to prevent overloading of the overhead condenser under certain abnormal operating conditions. However, the air cooler was operated continuously during normal operations, too. This removed heat from the feed, and consequently the heat load in the reboiler increased.

As a result of the study the procedures were changed and the air cooler fan was shut off during normal operations. This reduced the reboiler duty by more than 30%, a saving worth more than $1,000,000/year, at no cost to the facility.

Even with the fan off there was still a significant loss of heat due to convection in the air cooler. This loss can be eliminated by installing a bypass around the air cooler (Figure 2), a small project that saves an additional $200,000/year with a very good payback.

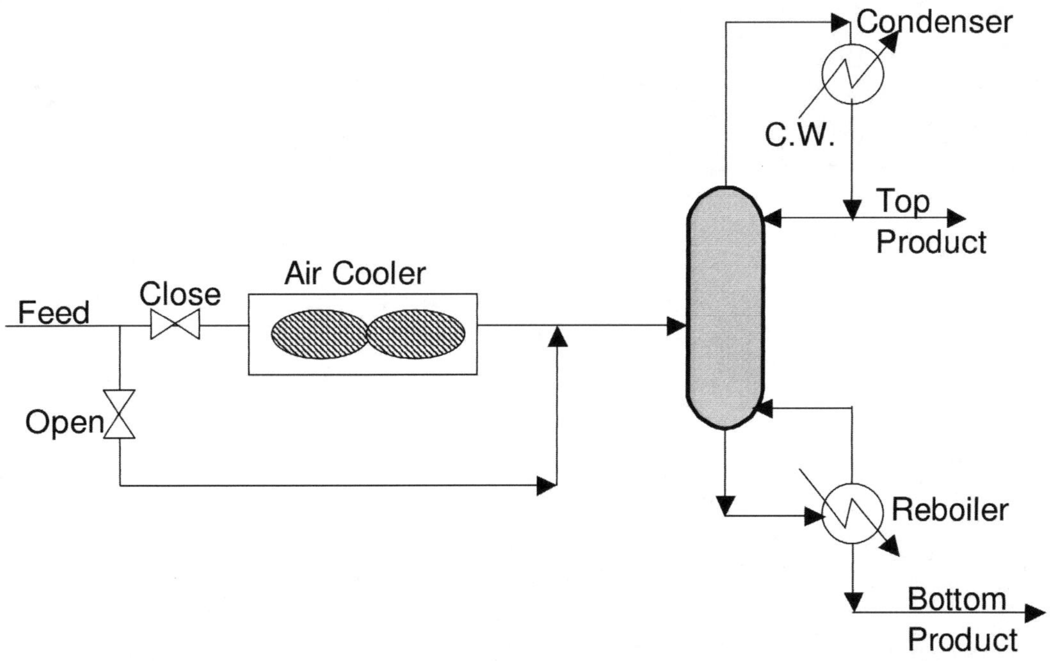

Figure 2: Inappropriate Cooling Corrected by Bypassing Cooler on Distillation Column Feed

This example illustrates the need to understand the purpose of all the equipment in a process. Operators often place equipment in service inappropriately, and once inappropriate operating norms have been established they can sometimes stay in place for years. Good operating procedures, together with training, including frequent refresher courses or continuing education, can go a long way to minimizing this type of misuse of equipment.

Example 2: The second example also illustrates an inefficiency caused by poor operation. However unlike the previous example, where placing equipment in service inappropriately caused a loss of energy efficiency, the converse is true in the present example.

Figure 3 shows a pumparound on one of the distillation columns at an oil refinery (5). It has two heat exchangers, the first to transfer heat to distillation column feed, and the second to reject excess heat to cooling water. However, the operators opened a bypass around the feed-preheat exchanger because the pumparound duty became too high during the winter, and they did not subsequently close the bypass in the summer. As a result, heat was unnecessarily rejected to cooling water for an extended period time.

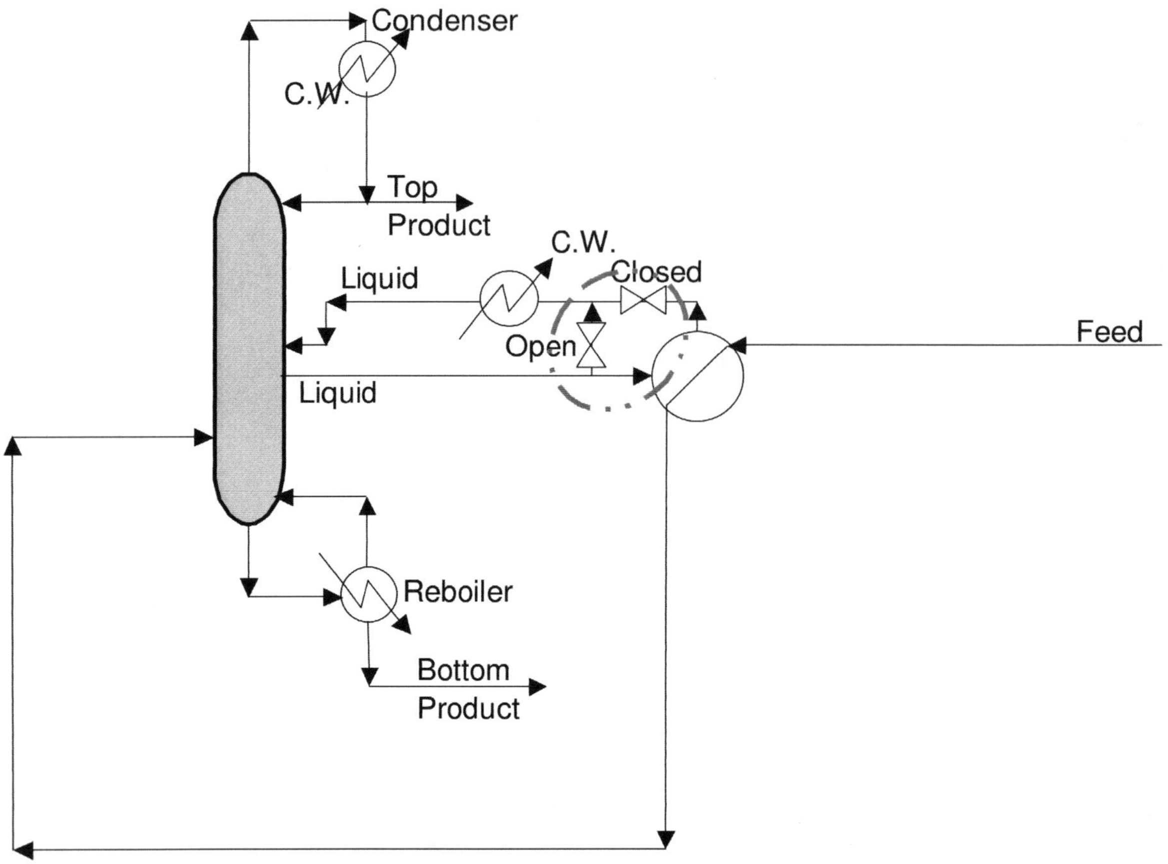

Figure 3: Unnecessary Bypassing of Heat Recovery Equipment

This inefficiency was readily corrected by closing the bypass, saving $600,000/year (Figure 4). In addition, a change was made to the operating procedures to ensure that the bypass valve position is routinely monitored. Further energy savings are possible with improved control systems.

In this example it is the unnecessary bypassing of heat recovery equipment that resulted in a lost opportunity. Permanent changes in operating procedures, together with operator training, help prevent the plant reverting to the incorrect operating practices. Automation, improved process control, and monitoring via DCS and data historian systems can also facilitate operating with lower tolerances. This can lead to significant energy savings.

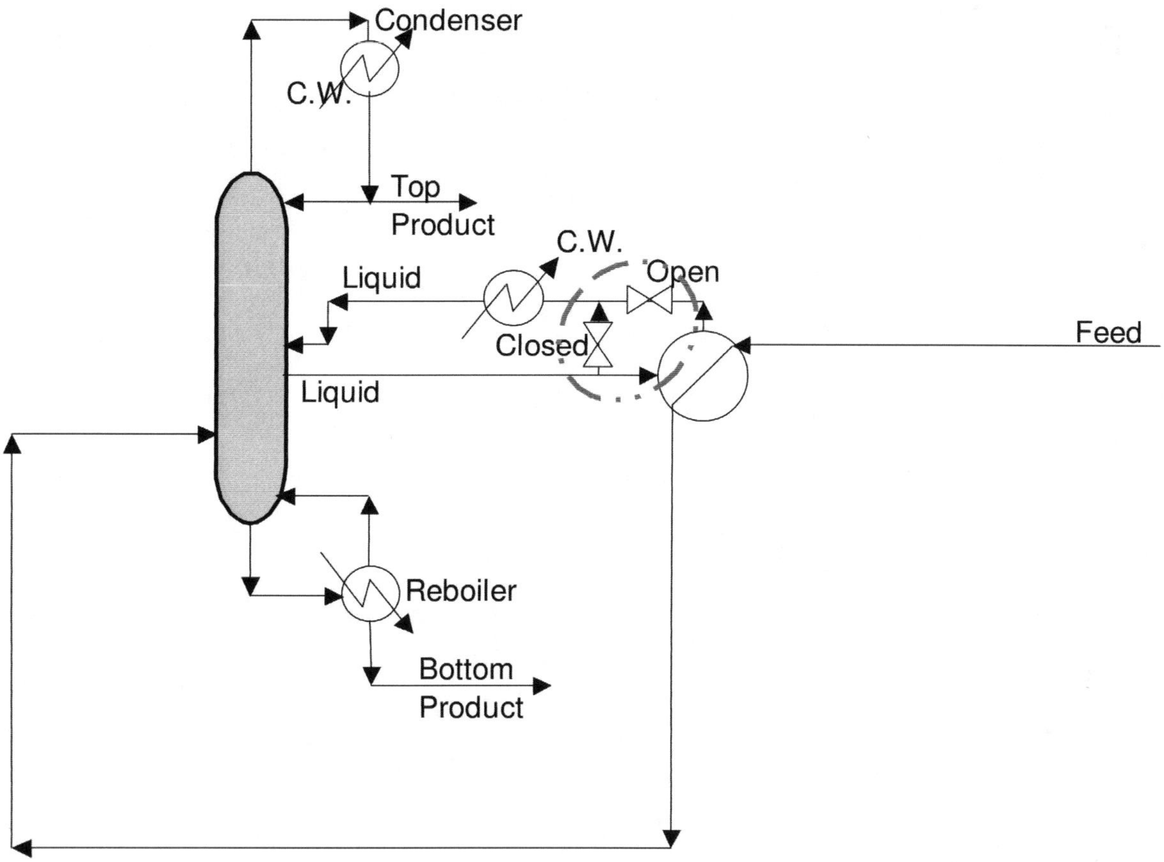

Figure 4: Closing the Bypass Restores Heat Recovery

Example 3: The third example relates to a pump in a plastics facility (6). A major high pressure feed pump is operated with spillback control (Figure 5). In this arrangement the pump itself operates at constant flow and constant delivery pressure, and any change in the process demand is satisfied by adjusting the opening of the spillback valve. This facilitates stable operation of the process, and it also ensures that the pump is never shut in, a condition that could damage to the equipment. A cooler is incorporated in the spillback line to avoid overheating.

Although the spillback control system offers very stable process conditions, this is accomplished at the cost of significant energy inefficiency. In normal operation the pump runs at high throughput and relatively low head, with a large recycle flow (see Figure 6).

There are numerous options for reducing pump power requirements. These include variable speed drives, high efficiency pumps and motors, trimming or replacing impellers, and adding a smaller pump for use during periods of low throughput. For various reasons none of these options was viable in the present case. However throttle control (Figure 7) does offer a simple alternative.

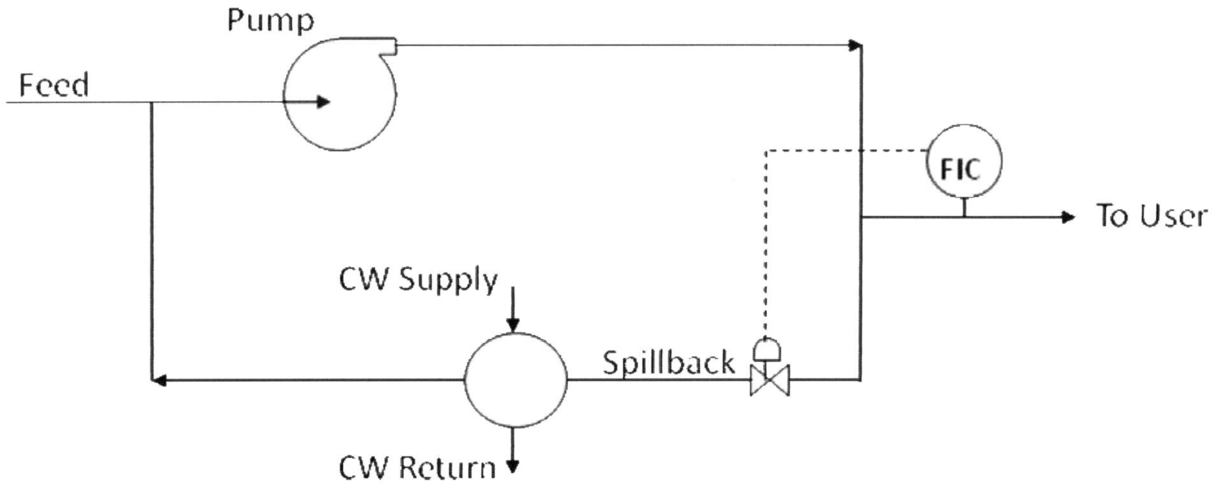

Figure 5: High Pressure Feed Pump with Spillback Control

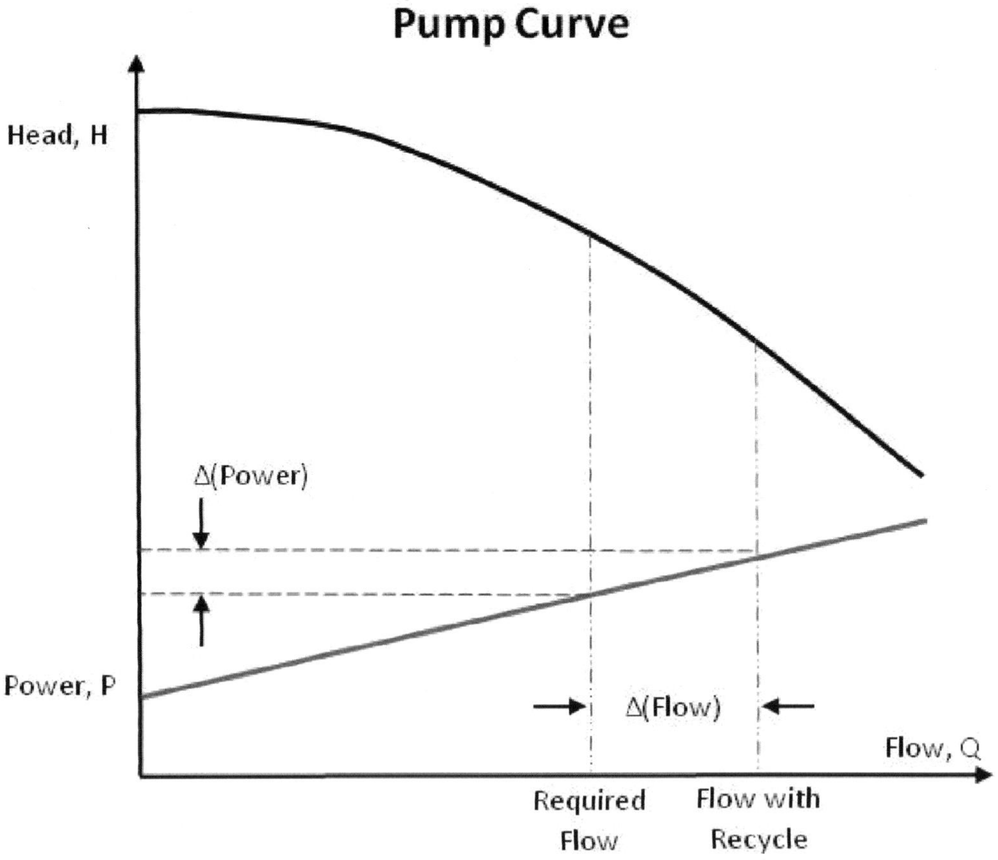

Figure 6: Pump Curve Comparing Spillback and Throttle Control

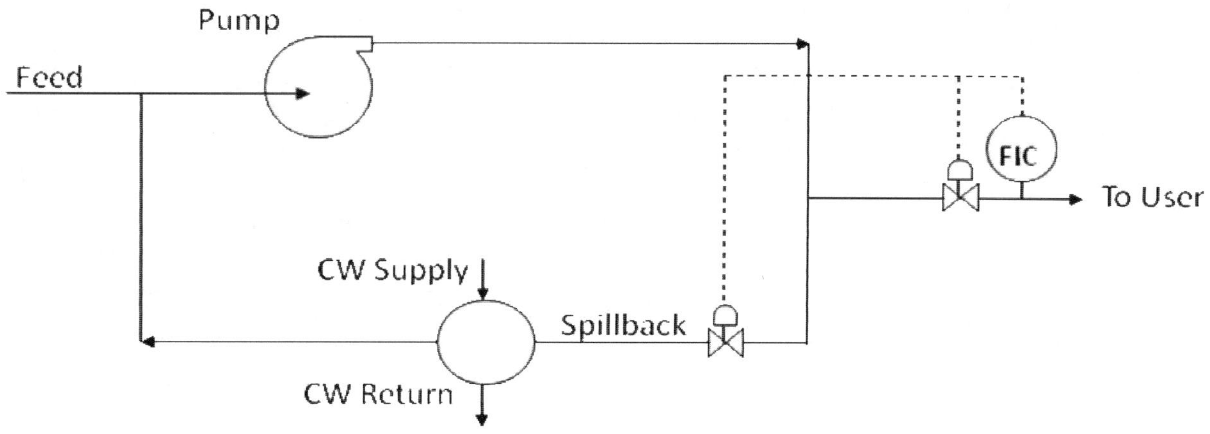

Figure 7: High Pressure Feed Pump with Throttle Control

With the throttle arrangement the flow through the pump in normal operation is equal to the amount of feed required by the process user ("Required Flow" on Figure 6). This is significantly less [by Δ(Flow)] than the normal flow through the pump with recycle using spillback control. Throttling also introduces a backpressure on the pump and this, together with the reduced flow rate, moves the operating point upwards and to the left on the pump curve (Figure 6). From the pump power curve (also on Figure 6) it is clear that the pump power is reduced by Δ(Power). It is still important to ensure that the pump is never shut in, so the spillback valve is retained and opens if the process feed requirement falls below a predetermined minimum allowable flow.

The only new facilities in this example are the throttle valve and associated equipment, together with reprogramming of the control system. The pump power is reduced by roughly 10%, worth $150,000/year.

Engineers and operators tend to accept the design of their control systems as long as the equipment functions reliably and safely. Reliability and safety are, of course, top priorities. However, many control systems are designed without regard to the energy losses they create. It is well worth reassessing these systems and challenging the losses.

Example 4. Heat exchanger cleaning programs are important in maintaining the performance of heat recovery systems, most notably crude unit preheat trains in refineries. Refinery management is becoming increasingly sophisticated in this area, with both improved cleaning techniques and better tools for assessing appropriate cleaning intervals for the heat exchangers in the circuit. However, the best cleaning methods and the most elegant optimization of cleaning intervals are of little use when communication fails.

In the course of a project in a refinery the author noted that a heat exchanger was out of service – not unusual, as heat exchangers often require maintenance. However, the records showed that this particular heat exchanger had been idle for more than three months. Further investigation revealed that the heat exchanger had been taken out of service for cleaning. The work had been completed within a couple of weeks, and the maintenance supervisor then notified the shift supervisor that the heat exchanger was ready for use. However, due to other activities the heat exchanger could not be brought back into service immediately. The shift supervisor failed to pass the information on to the next shift, and there was no follow up action until the author drew attention to the idle equipment.

When the unit manager was informed of the situation it required only a few hours to bring the heat exchanger back in service. The energy loss during the two and a half months that the heat exchanger had been left idle after the cleaning was worth over $100,000 (5).

In this example, better systems were needed for tracking the status of maintenance jobs on the unit. A simple electronic reminder system, for example, could have alerted the operators to the need to bring the heat exchanger back on-line.

Example 5. Steam is commonly used as a heat transfer medium in oil refineries and chemical plants. Most steam systems use thermal deaerators (Figure 8) to drive off oxygen and other dissolved gases from boiler feed water. In principle only a small amount of steam is needed to do this. However, as the incoming water is often far below the saturation temperature in the deaerator a substantial amount of additional steam is consumed in preheating the water. Because of this it is not uncommon to consume 10% or even 15% of the total steam make in deaerators. Not surprisingly, many facilities install deaerator feed water preheat projects to reduce this steam load.

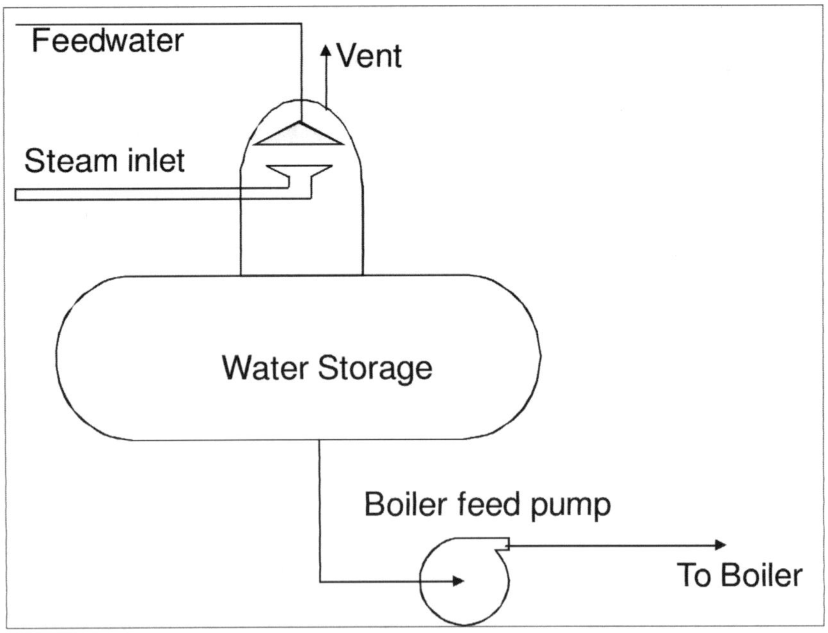

Figure 8: Thermal Deaerator – Main Components

An energy management project at a petrochemicals complex included a new heat exchanger to preheat deaerator feed water, using a product rundown stream as the heat source (5). The deaerator operates at 10 psig (240°F saturation temperature). The rundown stream was available at 350°F, and it went directly to an air cooler which cooled it to around 90°F. The project was intended to reduce the heat load in the air cooler, and at the same time reduce the steam demand in the deaerator.

In order for the deaerator to operate properly it was necessary to limit the deaerator feed water temperature to 230ºF. The design therefore included temperature control, as shown in Figure 9.

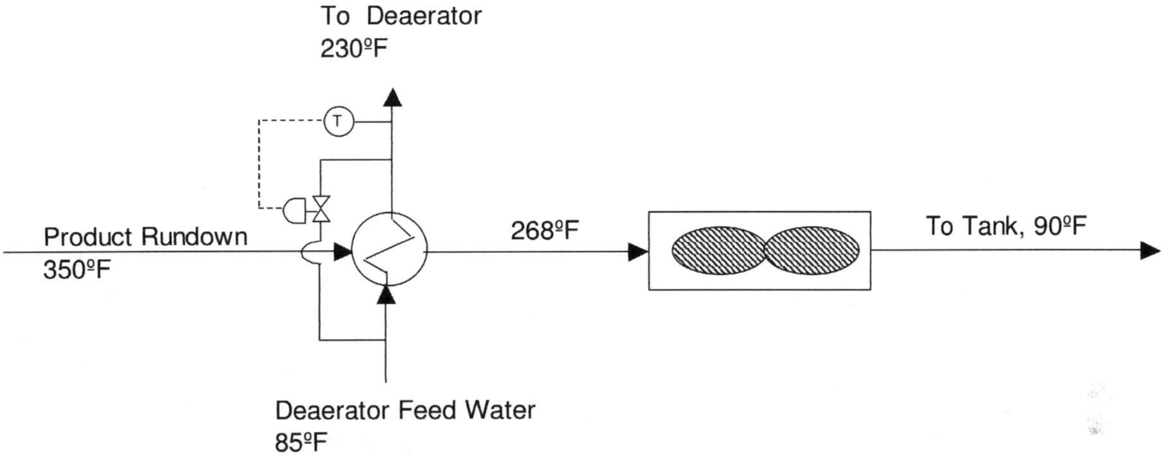

Figure 9: Deaerator Feed Water Preheat Project, as Originally Implemented

Shortly after the project came on line operators started reporting problems. The water preheat temperature frequently exceeded the 230ºF limit, and the pressure drop in the deaerator feed line was excessive due to vapor locking in the new heat exchanger. After numerous attempts to fix the problem by modifying plant operations the heat exchanger was taken out of service, using existing manual bypasses and isolation valves.

The underlying problem was in the control scheme. Bypassing deaerator feed water around a heat exchanger is effective in reducing heat pick-up. However as the heat pick-up goes down, so does the amount of water passing through the heat exchanger. Consequently the temperature of the water leaving the heat exchanger rises as the bypass opens, and it can easily reach its boiling point. This accounts for the observed difficulty in controlling temperature and the associated vapor locking.

The simplest interim solution to this problem is to open the existing manual bypass on the product rundown stream, as shown in Figure 10. This reduces the flow of the rundown stream through the heat exchanger while maintaining the water flow, thus ensuring that the water does not overheat. However, the manual bypass valve has to be about half open to ensure that the water does not boil in the heat exchanger under any of the anticipated operating conditions. This means that in some situations the temperature of the water going to the deaerator is much less than the target value of 230ºF. However, this simple strategy does allow the plant to obtain a significant percentage of the potential benefits with no additional investment.

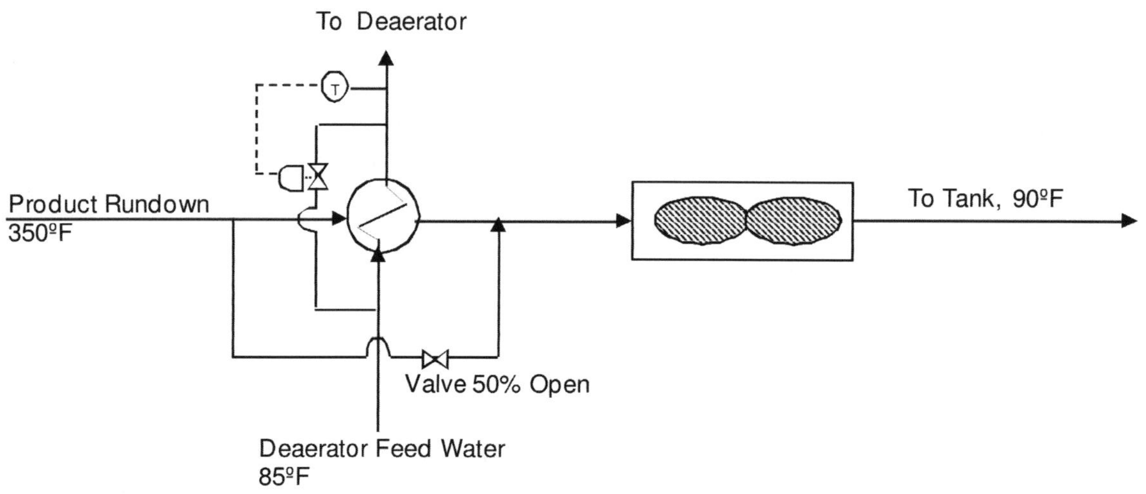

<u>Figure 10:</u> Deaerator Feed Water Preheat Project, with Manual Bypass on Product Rundown

A more complete solution requires the temperature control valve to be relocated on the product rundown bypass (Figure 11). With this arrangement the target deaerator feed water temperature is achievable in all of the anticipated operating conditions.

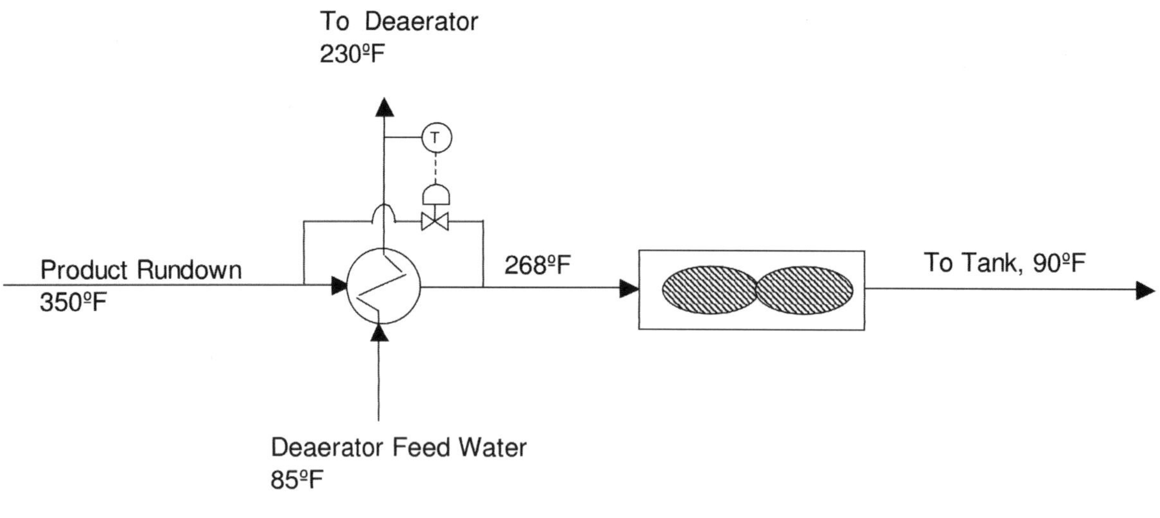

<u>Figure 11:</u> Deaerator Feed Water Preheat Project, with Automatic Temperature Control on Product Rundown Bypass

Heat integration schemes are generally developed from an evaluation of available heat sources and heat sinks at steady state conditions, often using pinch analysis. This is generally a very effective way of identifying opportunities. However, the final design must account for the inevitable variations in process conditions and how they will affect the operation of the equipment. In this particular example the variation in flows, temperatures and heat loads were recognized, but a basic physical property of water (it boils at 240°F under 10 psig pressure) was not. The key message is that we must never lose sight of basic principles, whatever we do.

Conclusions

No matter how sophisticated our energy management strategies become it is imperative that we continue to pay attention to basic principles. We must understand the purpose and limitations of the equipment in our processes, and also the physical properties of the materials we handle. Key principles of human interactions – especially communication – are also essential to all of our activities. This article has illustrated how these basics are often overlooked, how these oversights can undermine our best efforts at energy efficiency. It also provided insights on how to correct these problems.

References

1. Douglas, J. M., 'A Hierarchical Decision Procedure for Process Synthesis', AIChEJ, Vol. 31, No. 3, pp. 353-362, March 1985.
2. B. Linnhoff, D. W. Townsend, D. Boland, G. F. Hewitt, B. E. A. Thomas, A. R. Guy, R. H. Marsland, *A User Guide on Process Integration for the Efficient Use of Energy, Revised First Edition* I.Chem.E., Rugby, England, 1994.
3. El-Halwagi, M. M., 'Introduction to Numerical Optimization Approaches to Pollution Prevention' in *Waste Minimization through Process Design*, A. P. Rossiter (ed.), McGraw-Hill, New York, pp. 199-208, 1995.
4. Jagen, P., 'DuPont's Energy Dashboard', *Texas Industrial Energy Management Forum*, Brady's Landing, Houston, TX, April 6, 2006.
5. Rossiter, A. P., 'Back to the Basics', Hydrocarbon Engineering, Vol. 12, No. 9, pp. 69-73, September 2007.
6. Rossiter, A. P., *Energy Management in the Twenty-First Century Workshop Notes*, Rossiter & Associates, Bellaire, TX, 2011.

Capital Cost Targeting of Total Site Heat Recovery

Andreja Nemet, Petar Sabev Varbanov*, Jiří Jaromír Klemeš

Centre for Process Integration and Intensification – CPI[2], Research Institute of Chemical and Process Engineering, Faculty of Information Technology, University of Pannonia, Egyetem u. 10, H-8200 Veszprém, Hungary
varbanov@cpi.uni-pannon.hu

A single uniform ΔT_{min} specification had been mostly considered for the evaluation of the trade-off between the heat recovery and the capital cost for individual processes. However, exploiting heat recovery on Total Site level offers additional potential for energy saving (Klemeš et al., 1997). The current work deals with the evaluation of the capital cost for the generation and use of site utilities (e.g. steam, hot water, cooling water), which enables the evaluation of the trade-off between heat recovery and capital cost targets for Total Sites. The procedure involves the construction of Total Site Profiles and Site Utility Composite Curves and the further identification of the various utility generation and use regions at the profile-utility interfaces. This is followed by the identification of the relevant Enthalpy Intervals. The lower bound of the required heat transfer area can be then estimated which enables its further use in optimisation procedures.

1. Introduction

Methods for targeting capital and total cost of Heat Exchanger Networks (HENs) were initially developed by Townsend and Linnhoff (1984) and further elaborated (Ahmad et al., 1990). A trade–off between the rate of heat recovery and the involved capital cost for an individual process, accepting a single ΔT_{min} specification has been described (Serna-González et al., 2007). However, it still receives considerable attention (Serna-González and Ponce-Ortega, 2011). In a recent work (Varbanov et al., 2012) Total Site heat recovery targeting using multiple ΔT_{min} specifications for the site processes and process-utility interfaces has been explored. It is also possible to define and use the ΔT_{min} contributions of individual process streams in a process (Kravanja et al., 1997). The aim of this work has been to determine the lower bound on the heat transfer area for meeting the targeted heat recovery on the Total Site. This can be used in further work for finding the optimal configuration of ΔT_{min} specifications for heat recovery inside the processes and between them through the utility system.

2. Methodology

Total Site is a set of processes linked with a central utility system (Fig 1). The first step for Total Site targeting is to maximise the heat recovery within the processes. Total Site Profiles (TSPs) are then constructed to evaluate the heat recovery potential between the processes through the utility system. The procedure is described next and illustrated in Fig. 2.

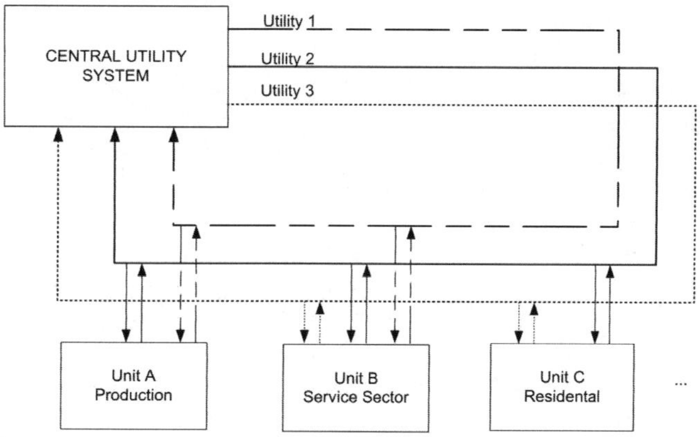

Fig 1: A Total Site (after Perry et al, 2008)

Step 1. Process heat recovery (Fig 2). The process-level utility targets and Grand Composite Curves (GCCs) are obtained using the Problem Table Algorithm (PTA). The heat exchanger area depends, besides others, on the heat transfer coefficients. However, the GCC segments generally represent several streams. To be able to determine the area of heat exchangers without resorting to detailed HEN design, a representative heat transfer coefficient should be selected. The aim of the current method is to set the lower bound of the heat exchange area required. According to this, a representative stream with the highest heat transfer coefficient should be selected. By this selection the lower bound is ensured – the heat exchanger area of any network cannot be smaller.

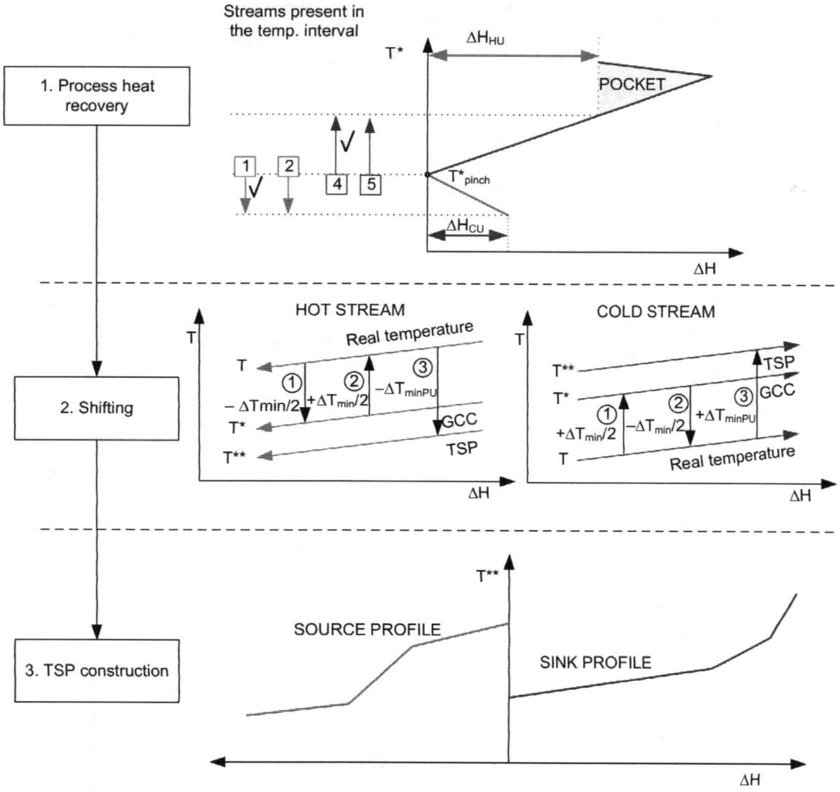

Fig 2: Constructing TSPs

Step 2. Shifting (Fig 2). In this step the segments identified at the previous step are shifted using the procedure by Varbanov et al. (2011), using individual ΔT_{min} specifications for heat exchange between process streams as well as between process streams and utility – for each of the site processes. Two shifts are performed for each GCC segment: (i) Back to the process stream real temperatures and after (ii) Forward by ΔT_{minPU}, which is the minimum temperature difference required for a feasible heat exchange between process streams and the utility.

Step 3. TSP construction (Fig 2). The GCC heat source segments are combined on the left hand side of the Y–axis. The heat sink segments are combined on the right hand side of the Y-axis. The construction of each profile (heat sources and sinks) is performed using the technique for building process CC, but in this case using the segments extracted from the GCC for each process.
As a result, the constructed plot consists of two parts. On the left-hand side is the Heat Source Profile and on the right-hand part is the Sink Profile. In this way the problem is partitioned into utility generation (Site Source Composite Curve) and utility use (Site Sink Composite Curve). Heat recovery can be performed through intermediate utilities/media (Fig 3).The Site Utility Composite Curves are constructed to evaluate the maximum site heat recovery. The Enthalpy Intervals (EIs) are identified, as illustrated in Fig.3, by selecting the enthalpies corresponding to the changes in the slopes of the Site Profiles and the Site Composite Curves.

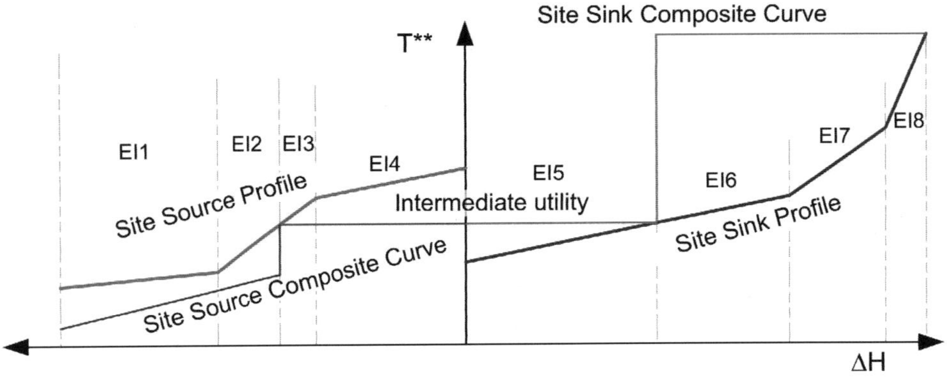

Fig 3: TSP with EI

The heat exchange areas are determined in each EI separately using the general equation for heat transfer area evaluation (Fig 4). In the TSP plot, the utilities are represented at their real temperatures while the Site Profiles are at temperatures shifted by whole ΔT_{minPU} with respect to the initial process streams.

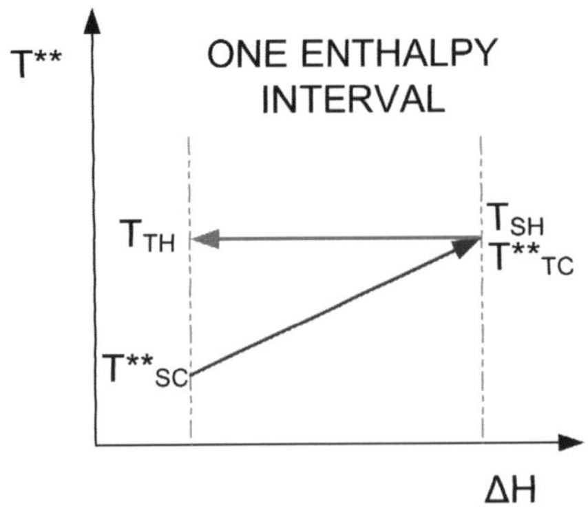

$$A = \frac{Q}{U \cdot \Delta T_{LM}}$$

$$\frac{1}{U} = \frac{1}{h_U} + \frac{1}{h_{PR}}$$

$$\Delta T_{LM} = \frac{(T_{SH} - T_{TC}) - (T_{TH} - T_{SC})}{\ln \frac{(T_{SH} - T_{TC})}{(T_{TH} - T_{SC})}}$$

Fig 4: Determining the heat transfer area in one EI

When the profiles for the process heat source and the hot water generation in Fig.4 touch each other they still have sufficient temperature difference equal to ΔT_{min}. When determining the heat exchanger area the temperatures of the process stream should be shifted back to their real temperatures. To determine the overall area for heat transfer between utility and process stream the areas from each EIs are summed up.

3. Case study

The input data for the case study are listed in Table 1. They are considered two processes A and B.

Table 1: Input data for case study

Process	Stream	Supply temperature [°C]	Target temperature [°C]	CP [MW/°C]	ΔH [MW]	Type of medium	H [W/(m² °C]
Process A	A1, cold	50	110	0.05	3.0	Liquid	800
	A2, hot	100	30	0.06	4.2	Liquid	800
	A3, cold	100	140	0.02	0.8	Gas	350
Process B	B1, hot	190	120	0.06	4.2	Gas	350
	B2, cold	100	240	0.04	5.6	Gas	350
	B3, hot	80	60	0.02	0.4	Liquid	800

They are three utilities available: cooling water with the input 30 °C and output 20 °C, h (water) = 800 W/(m^2 °C). The intermediate utility is steam at 120 °C, h (intermediate utility) = 10,000 W/(m^2 °C) and the utility with the highest temperature is available at 250 °C, h (steam) =11,000 W/(m^2 °C).

3.1 Results

The GCCs of the processes and the TSPs are shown in Fig 5. The next step is to determine the area of heat exchangers between utility and process streams by constructing the TSPs (Fig 5).

Table 2: Calculation of the heat exchanger area

Enthalpy Interval	Hot stream	Cold stream	Q [MW]	U [W / (m² °C)]	ΔT_{LM} [°C]	A [m²]
EI1	A2	CW	1.8	400	19.3	233.2
EI2	A2+B3*	CW	0.6	400	41.3	36.3
EI3	A2	CW	2.0	400	59.9	8.4
EI4	B1	IU	1.0	338.2	39.9	74.1
EI5	IU	A1	0.5	740.7	24.7	27.4
EI6	IU	A1	0.5	740.7	16.2	41.8
EI7	ST	A1	0.2	745.8	141.4	1.9
EI8	ST	A3	0.06	339.2	124.4	1.4
EI9	ST	B3	24	339.2	30.8	229.5
					TOTAL	**653.8**

**Both streams A2 and B3 media are liquid and as a consequence the overall heat transfer coefficients are assumed equal, and in the area calculation it is not important which one is the representative stream.*

From Table 2 is visible that by heat recovery 1 MW occurring in EI4, EI5 and EI6 (Fig 5) can be recovered through the intermediate utility. Required heat transfer area is143.3 m^2. This is determined by summation of areas required in those EIs. If the intermediate steam level (at 120 °C) is removed then the same part of the profiles would require larger heat transfer area amounting to 171.6 m^2.

By applying the heat recovery via the intermediate steam level both the utility targets and the capital cost targets are reduced.

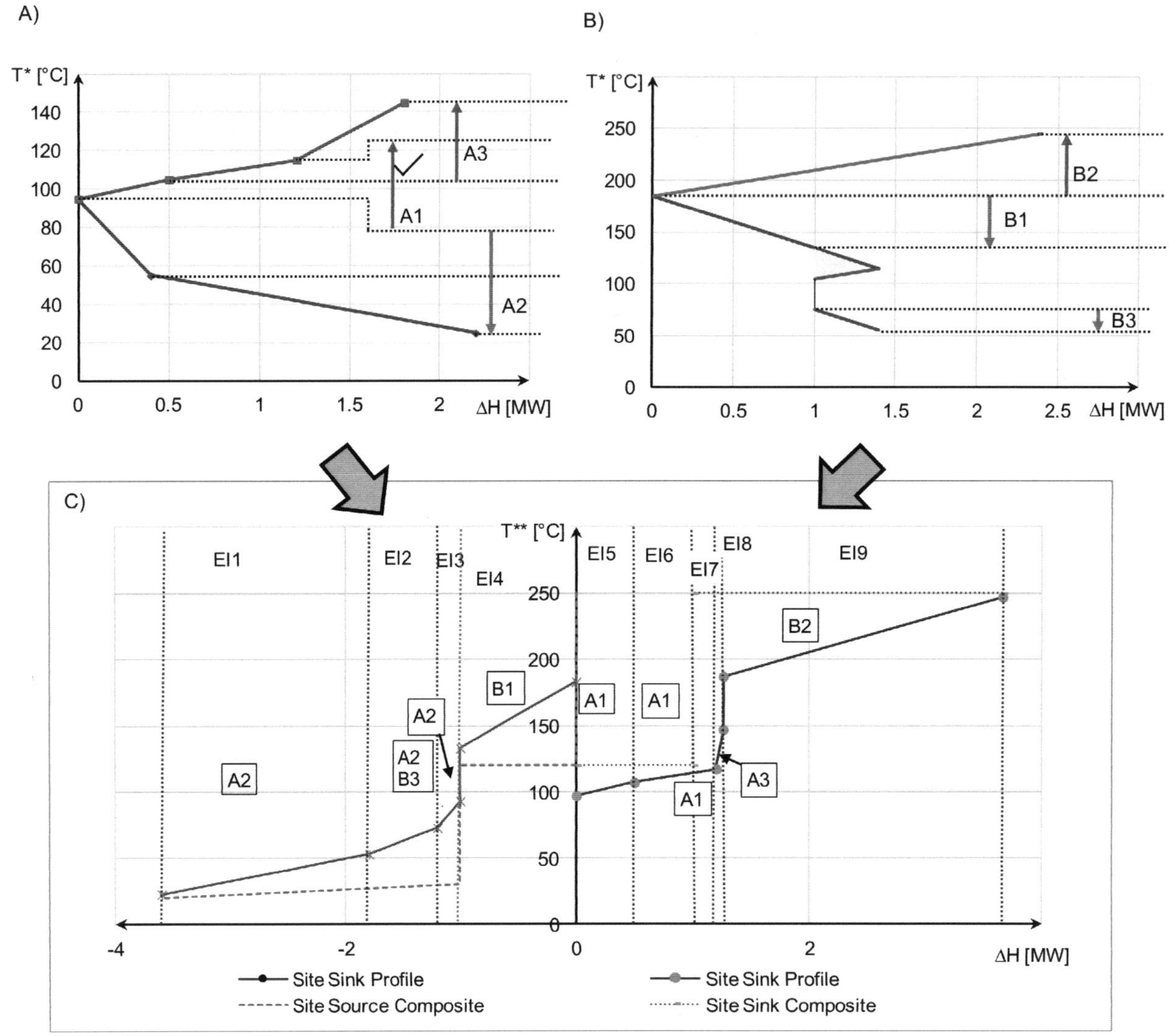

Fig 5: GCC of A) Process A and B) Process B and the c) TSPs

4. Conclusions

A procedure for evaluating the lower bound of the heat transfer area for a heat exchange between utility and process streams on a Total Sites has been developed and demonstrated. This methodology enables a preliminary analysis of the trade-off between the amount of recovered heat and the needed investment cost. In the case study presented 1 MW of heat can be recovered through the central utility system for which 143.3 m^2 heat exchanger area is required. In comparison if just a high pressure steam and cooling water are used the heat exchanger area is increased by 28.3 m^2. It indicates that the additional investment can be economically viable.

The developed model and the obtained results lay out the ground for a procedure evaluating the capital cost targets for all heat transfer units on a Total Site in a future work – also including the heat recovery at the process level. Based on this, the fundamental capital energy trade-off can be evaluated and an optimisation of the minimum allowed temperature difference specifications for whole Total Sites can be performed.

Acknowledgement

The financial support is gratefully acknowledged from the project TÁMOP-4.2.2/B-10/1-2010-0025 with the National Development Agency of Hungary.

Nomenclature

ΔT_{min}	minimal temperature difference between two process streams, °C
ΔT_{minPU}	minimal temperature difference between process stream and utility, °C
EI	enthalpy interval, MW
PTA	Problem Table Algorithm
CP	heat capacity flowrate, MW/°C
A	area of heat exchanger, m^2
ΔT_{LM}	Logarithmic mean temperature, °C
Q	heat, MW
Q_k	heat exchanged in enthalpy interval k, MW
h	heat transfer coefficient, W/(m^2 °C)
h_U	heat transfer coefficient of the utility stream, W/(m^2 °C)
h_{jPR}	heat transfer coefficient of process stream, W/(m^2 °C)
ΔH	enthalpy, MW
T	temperature, °C
T*	shifted temperature, °C
T**	twice shifted temperature, °C
U	overall heat transfer coefficient, W/(m^2 °C)

References

Ahmad, S., Linnhoff, B., Smith, R., 1990. Cost optimum heat exchanger networks: 2. Targets and design for detailed capital cost models. Computers & Chemical Engineering, 14(7), 751–767.

Varbanov PS, Fodor Z, Klemeš JJ, 2012, Total Site Targeting with Process Specific ΔT_{min}, Energy. Submitted.

Klemeš, J., Dhole, V.R., Raissi, K., Perry, S.J., Puigjaner, L. 1997. Targeting and Design Methodology for Reduction of Fuel, Power and CO_2 on Total Sites. Applied Thermal Engineering, 17 (8-10), pp. 993-1003.

Klemeš J., Friedler F., Bulatov I., Varbanov, P. 2010. Sustainability in the process industry – Integration and optimization. New York: McGraw-Hill., USA, 362 p.

Kravanja Z., Glavič P., 1997. Cost Targeting for HEN through Simultaneous Optimization Approach: A Unified Pinch Technology and Mathematical Programming Design of Large HEN. Computers Chemical Engineering, 21 (8), 833–853.

Perry S, Klemeš J, Bulatov I (2008) Integrating waste and renewable energy to reduce the carbon footprint of locally integrated energy sectors. Energy 33:1489-1497.

Serna-González M and Ponce-Ortega JM, 2011, Total cost target for heat exchanger networks considering simultaneously pumping power and area effects. Applied Thermal Engineering, 31, 1964-1975.

Serna-González M.,Jiménez-Gutiérrez A., Ponce-Ortega J. M., 2007, Targets for Heat Exchanger Network Synthesis with Different Heat Transfer Coefficients and Non-uniform Exchanger Specification. Chemical Engineering Research and Design 85, 1447-1457.

Townsend D.W., Linnhoff B., 1984. Surface area targets for heat exchanger networks. IChemE 11th Annual Research Meeting, Bath, UK, Lecture 7a.

APPROPRIATE PLACEMENT OF COMPRESSORS AND EXPANDERS

Truls Gundersen* and Danahe Marmolejo-Correa
Department of Energy and Process Engineering
Norwegian University of Science and Technology
Kolbjoern Hejes vei 1.B, NO-7491, Trondheim, Norway
*Corresponding author: truls.gundersen@ntnu.no

Extended Abstract

The term "Appropriate Placement" was introduced in Pinch Analysis during the 1980s to indicate how special unit operations should be correctly integrated with the background process to ensure maximum energy savings, or in fact energy savings at all. In complex processes and without using a "systems" (holistic) approach, it is indeed possible to heat integrate in such a way that no energy savings result. Pinch Analysis has provided guidelines for the design of heat exchanger networks as well as the Appropriate Placement of distillation columns, evaporators, heat engines and heat pumps. These guidelines are based on insight about the process Pinch as a bottleneck for heat recovery, and a fundamental decomposition effect by temperature level into a heat deficit region above Pinch and a heat surplus region below Pinch. Once an Appropriate Placement scheme is identified, Pinch Analysis also provides graphical (i.e. the Grand Composite Curve) and numerical/algorithmic (i.e. the based on the Heat Cascade) tools that can establish how much heat that can be correctly integrated.

Glavič and co-workers, while considering the Appropriate Placement of chemical reactors, also made statements about operation of compressors, however, without completing the discussion and without developing a methodology based on insight from thermodynamics. This paper will conclude on this issue and provide very accurate guidelines for the correct integration of compressors and expanders, such as optimal inlet temperatures.

The discussion in the paper assumes that streams subject to heat integration also may undergo pressure changes. In subambient processes, even streams with the same supply and target pressure may be utilized as utility streams by a sequence of compression and expansion stages to provide refrigeration for the process. If a process stream needs to be compressed, the investment in a pump (liquid phase) or a compressor (vapor phase) has to be made, and that opens for a holistic view on compression and heat integration. Likewise, if a process stream needs a pressure reduction, then the choice between a throttling process (valve) and a work producing process (expander) has to be made. In the latter case, one should again take a holistic view on expansion and heat integration.

Established design guidelines for rotating machinery suggest to operate compressors at low temperature (possibly with inter-stage cooling) to reduce work consumption and to operate expanders at high temperature (possibly with inter-stage re-heating) to increase work production. Based on insight developed in this paper that in turn relates to principles from Pinch Analysis, design guidelines for compression and expansion will be presented that are in conflict with the established ones. Compressors provide additional heating to the heat recovery system and should be placed above the Heat Recovery Pinch. Likewise, expanders provide additional cooling and should be placed below Pinch.

The extended problem definition (an extension of the classical heat recovery problem) for systems of process streams that may undergo changes in both temperature and pressure is the following:

Given a set of process streams with a supply state (temperature, pressure, and the resulting phase) and a target state, as well as utilities for power, heating and cooling; design a system of heat exchangers, expanders, pumps and compressors in such a way that an objective function is minimized.

The objective function could be based on economics (such as minimum Total Annual Cost) or thermodynamics (such as minimum irreversibilities or exergy losses). The operating cost is related to external heating and cooling (both at different temperature levels) and net shaftwork required by the process (total consumed minus total produced). An evaluation of or comparison between design alternatives will have to deal with both mechanical and thermal utility systems, and the obvious solution is to use a parameter that measures the quality of different energy forms. Thus there is a motivation for using exergy, and in this paper a newly developed Exergetic Transfer Effectiveness (ETE) is used to measure exergy efficiency of the design alternatives.

As indicated above, compression and expansion are used deliberately in below ambient processes to provide refrigeration as cold utility. For these types of processes, and in developing the ETE, the true exergy sources and sinks of a process were identified through a proper classification of exergy types and the decomposition of these exergy forms into exergy components.

In addition, the paper will discuss the importance of the Heat Recovery Pinch concept when process streams not only change temperature but also pressure. With changes in both these state variables, the path from supply to target state is not fixed and will be determined as part of the search for a system that satisfies the objective of the design exercise.

The paper will use a case study to conclude its discussion by identifying the optimal inlet temperature for compressors and expanders that are part of a heat recovery system. It should be no surprise that these optimal inlet conditions are the hot or cold Pinch temperatures depending on whether the actual process stream that is subject to pressure change is a hot process stream (to be cooled and/or condensed) or a cold process stream (to be heated and/or evaporated).

Evolutionary design of an LNG process using the ExPAnD method and a new graphical representation of exergy

Danahe Marmolejo Correa[1], Truls Gundersen[2]
Department of Energy and Process Engineering, Norwegian University of Science and Technology,
Kolbjoern Hejes v 1B; NO-7491, Trondheim, Norway.
[1]corresponding author: danahe.m.correa@ntnu.no; danahe.marmolejo@gmail.com.
[2]truls.gundersen@ntnu.no

Abstract

This paper presents a novel design methodology for low temperature processes that suits particularly well for natural gas liquefaction processes. The methodology combines similar graphical tools as the ones provided by Pinch Analysis (PA) Exergy Analysis (EA). The use of EA in the design procedure is justified by the fact that processes operating below ambient temperature require shaftwork for the compressor stations in the refrigeration cycle; mechanical work is pure exergy. The paper uses a new diagram for exergy and energy targeting which is based on a new energy quality parameter called *exergetic temperature*.

Introduction

The proposed design procedure uses the heuristic rules of the Extended Pinch Analysis and Design (ExPAnD) method for designing low temperature processes presented by Aspelund *et al.* (2007). The same heuristics were applied by Wechsung *et al.* (2011) in an alternative approach using Mathematical Programming and a novel superstructure where heating, cooling, compression and expansion operations to optimize an LNG process. The ExPAnD method combines the classical Pinch Analysis with Exergy Analysis and is applicable for the development of processes where both temperature and pressure are important design variables such as in processes operating below ambient temperature (T_0). The design of such processes represents several challenges. First, the design must be customized to match the climate conditions (i.e. ambient temperature and relative humidity) of the location where the plant will be built. Second, the plant size and location may constrain the number of units and equipment size, possibly reducing the process efficiency (i.e. offshore natural gas liquefaction processes have lower energy efficiency, fewer unit operations and smaller equipment size than onshore plants). Third, the number of degrees of freedom in the design can be quite large because of the manipulation of temperature and pressure. Forth, processes below ambient temperature have large energy consumption in the refrigeration cycles (low temperature level results in high energy consumption). For low temperature heat exchanger networks, the temperature difference between process streams should be small and the refrigeration loads minimized.

Exergetic Temperatures

One option to achieve low energy consumption is to reduce avoidable irreversibilities. Exergy Analysis is a powerful tool to identify and calculate such irreversibilites; however, because exergy tends to have rather complex relations (non-linear) with respect to the process variables (such as, temperature, pressure and composition), the calculations can be quite tedious. In addition, some graphical representations of exergy as a function of process variables or other energy quality parameters require multiple calculations for their construction (e.g.. the Exergy Composite Curves, ECCs, and Exergy Grand Composite Curve, EGCC, introduced by Linnhoff and Dhole (1992)) due to the non-linear relations. For these reasons, exergy analysis is commonly used as a post-design tool and not integrated in the design procedure. Exergy calculations, however, may be simplified by assuming ideal gas conditions and constant heat capacity. These assumptions do not lead to very large deviations for processes where gases such as methane and nitrogen are used (e.g. in natural gas processing). Moreover, the exergy content of a process stream can be decomposed into temperature based exergy ($\dot{E}^T$), pressure based exergy ($\dot{E}^p$) and chemical exergy. The sum of the first two components is the so-called thermo-mechanical exergy ($\dot{E}^{TM}$). If both temperature and pressure based exergies can be calculated using the above described assumptions, then the decomposition of the thermo-mechanical exergy (Eq. 1) can be done in such a manner that temperature based exergy only depends on temperature and pressure based exergy is a function of pressure only as shown by Eqs. 2 and 3.

$$\dot{E}^{TM}(T,p)=\dot{m}c_p\left[T-T_0-T_0\ln\frac{T}{T_0}+T_0\ln\left(\frac{p}{p_0}\right)^{\frac{\kappa-1}{\kappa}}\right] \tag{1}$$

$$\dot{E}^T=\dot{m}c_p\left[T-T_0-T_0\ln\frac{T}{T_0}\right]=\dot{m}c_pT^{E^T} \tag{2}$$

$$\dot{E}^p=\dot{m}c_p\left[T_0\ln\left(\frac{p}{p_0}\right)^{\frac{\kappa-1}{\kappa}}\right]=\dot{m}c_pT^{E^p} \tag{3}$$

The factors in brackets of Eqs. 2 and 3 are the so-called *exergetic temperatures*. For $\dot{E}^T$, the exergetic temperature (T^{E^T}) is only a function of temperature, and for $\dot{E}^p$, the corresponding exergetic temperature (T^{E^p}) is only a function of pressure. Moreover, pressure based exergy can be transformed into temperature based exergy and vice versa. For example, a process stream expanded below ambient temperature will produce work and reduce the stream temperature. This means that temperature based exergy actually increases, and this can be seen as a partial transformation of pressure based exergy to temperature based exergy. The rest of the pressure based exergy is transformed into work, except for the thermodynamic losses (exergy destruction due to irreversiblilities) in the expansion process.

A novel diagram for exergy targeting

The new representation of exergy targeting and exergy exchange utilized in this paper incorporates the simplified relations of exergy given by Eqs. 2 and 3. The simplifications allow having a linear relation between exergy and a new energy quality parameter referred to as *exergetic temperature*. The exergetic temperature for temperature based exergy is only a function of temperature and the exergetic temperature for pressure based exergy is only a function of pressure. The change in temperature or pressure based exergy is then equal to the product of a constant heat capacity flow rate and a difference in the corresponding *exergetic temperatures*. These changes in the energy quality parameter and changes in exergy for all the process streams can be plotted in a composite manner similar to the changes in enthalpy for the Composite Curves used in PA (Fig. *1a*). Fig. 1*b* illustrates the novel diagram for exergy targeting.

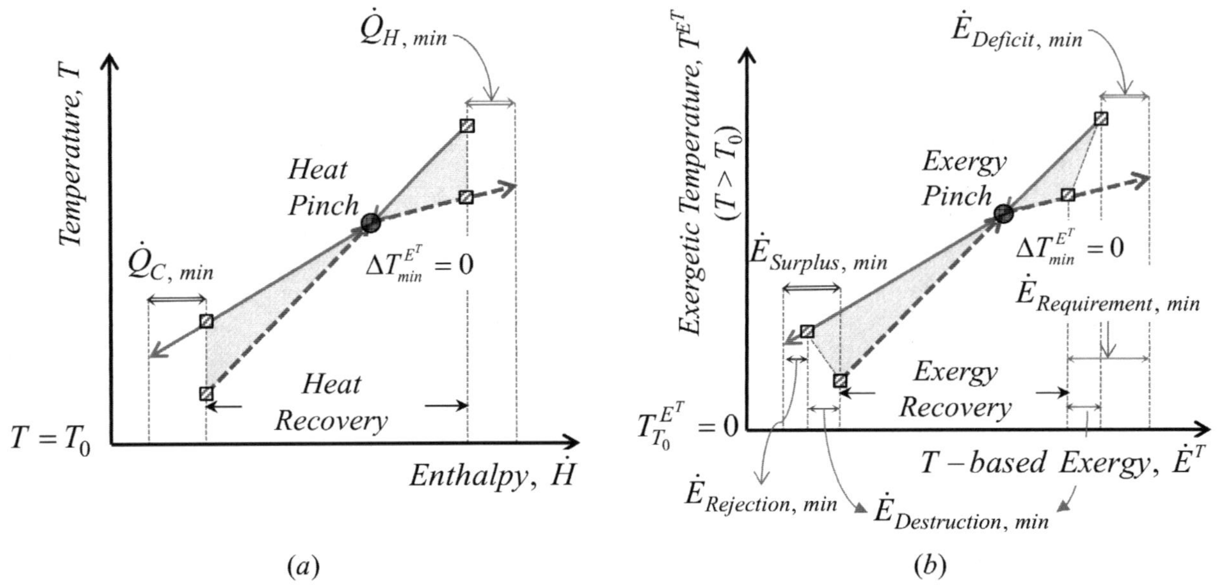

Fig. 1. (a) the Composite Curves and (b) the $T^{E^T} - \dot{E}^T$ diagram above ambient temperature

Some characteristics of the new diagram are:

- The minimum value of T^{E^T} (zero) is at ambient temperature. Thus, T^{E^T} is positive for the entire temperature range, which is in contrast to the Carnot factor in the ECCs. The stream with the condition that has the highest exergy value is always placed at the top of the curves while the lowest (closest to T_0) is at the bottom. *For streams below T_0, the coldest temperature corresponds to the highest exergy content and the largest exergetic temperature.*
- When the minimum approach temperature (ΔT_{min}) between the CCs is fixed, then a corresponding minimum exergetic approach temperature ($\Delta T^{E^T}_{min}$) is obtained. As a consequence:
  - The exergy deficit and surplus are calculated ahead of any design actions from the exergy cascade by only using the stream data, similar to how minimum external heating and cooling demands are obtained from the traditional heat cascade.
  - The *heat pinch* and *exergy pinch* are placed in corresponding enthalpy and exergy intervals, thus if the pinch rules are followed, exergy is neither imported below pinch nor exported above pinch, and exergy is not transferred across pinch.
- The exergy destruction is graphically shown by transforming the boundary temperatures of the overlapping region obtained in *the traditional CCs* into exergetic temperatures (squares in Figs. 1*a* and 1*b*).
- The maximum exergy recovery for a given ΔT_{min} is the maximum amount of exergy that the exergy sink curve can accept in the overlapping region, and the minimum process-to-process exergy destruction is the difference between the process-to-process exergy transferred from the exergy source curve and the maximum exergy recovered. The calculation of exergy targets is given by (4) and (5).

Above *exergy pinch*: $$\dot{E}_{Requirement,\,min} = \dot{E}_{Deficit,\,min} + \dot{E}_{Destruction,\,min} \quad (4)$$

Below *exergy pinch*: $$\dot{E}_{Rejection,\,min} = \dot{E}_{Surplus,\,min} - \dot{E}_{Destruction,\,min} \quad (5)$$

Manipulation of stream pressures

In the special case of isentropic pressure change (Eq. 6), the power required (generated) by the compressor (expander) as well as the changes in pressure and temperature based exergies are given by Eqs. 7, 8 and 9. By substituting Eq. 8 into Eq. 9 and realizing that $T_{\Delta s=0} - T_1 = \Delta T_{\Delta s=0}$, the relationship between the changes in exergetic temperatures reduces to the very simple form given in Eq. 10. This relation aids to the inclusion of pressure change effects in the graphical representation of exergy exchange. (Eq. 11) is the equivalent relation for exergy components and shaftwork for isentropic changes of pressure.

$$\frac{T_{\Delta s=0}}{T_1} = \left(\frac{p_2}{p_1}\right)^{\frac{\kappa-1}{\kappa}} \tag{6}$$

$$-\dot{W}_{\Delta s=0} = \dot{m}c_p\left(T_{\Delta s=0} - T_1\right) = \dot{m}c_p T_1\left(\left(\frac{p_2}{p_1}\right)^{\frac{\kappa-1}{\kappa}} - 1\right) \tag{7}$$

$$\Delta\dot{E}^{P}_{\Delta s=0} = \dot{m}c_p\left[T_0 \ln\left(\frac{T_{\Delta s=0}}{T_1}\right)\right]\Bigg|_{p_1\to p_2} = \dot{m}c_p\,\Delta T^{E^P}_{\Delta s=0}\Big|_{p_1\to p_2} \tag{8}$$

$$\Delta\dot{E}^{T}_{\Delta s=0} = \dot{m}c_p\left[T_0\left(\frac{T_{\Delta s=0} - T_1}{T_0} - \ln\frac{T_{\Delta s=0}}{T_1}\right)\right]\Bigg|_{p_1\to p_2} = \dot{m}c_p\,\Delta T^{E^T}_{\Delta s=0}\Big|_{p_1\to p_2} \tag{9}$$

$$\Delta T_{\Delta s=0} = \Delta T^{E^T}_{\Delta s=0} + \Delta T^{E^P}_{\Delta s=0} \tag{10}$$

$$-\dot{W}_{\Delta s=0} = \Delta\dot{E}^{T}_{\Delta s=0} + \Delta\dot{E}^{P}_{\Delta s=0} \tag{11}$$

Case study

The reverse Brayton process for the liquefaction of natural gas is used as a case study. The reverse Brayton process liquefies natural gas utilizing only nitrogen as working fluid in the refrigeration cycle. The exergy efficiency of the reverse Brayton referred in the literature (Venkatarathnam 2008) is close to 34% (defined as change in thermo-mechanical exergy of natural gas divided by net compressor work) with 6 compression stages in the refrigeration cycle (Fig. 2). In the example shown in this paper, only the final outcomes of the targeting and design stages are presented. The procedure for obtaining these results are explained in detail by Marmolejo-Correa and Gundersen (2012).

Figs 3*a* and 3*b* show the CCs and $T^{E^T} - \dot{E}^T$ diagram for the Brayton process. Notice that the cold stream (dashed line from 1 to 2) in Fig. 3*a* corresponds to the *cold pass* of the nitrogen through HX-100 (stream 5→1) in Fig. 2, which in turn corresponds to the exergy source curve (dashed line from 1 to 2) in Fig. 3*b*. One should have in mind that below ambient temperature, the cold streams are exergy sources and hot streams are exergy sinks. The segment from 1 to 2 of the hot composite curve represents the resulting heat surplus from the compression stages in COM-100-5. This heat surplus is located above ambient temperature, thus cooling water can be used as coolant in the intercoolers and aftercooler (AF-100). In terms of exergy (Fig. 3*b*) and because this heat surplus is situated above T_0 (hot stream), it is also is an exergy source (solid line from 1 to 2). Point number two in the exergy source curve has zero exergy (at ambient temperature) and is a meeting point for two exergy sources, one above and the other below T_0. The hot stream from 2 to 3 in Fig. 3*a* is the combination of the natural gas stream and the nitrogen *hot pass* through HX-100. In Fig. 3*b*, this section of the hot composite curve is the exergy sink.

The thermal energy surplus above ambient temperature is 46.9 MW (70 MW − 23.1 MW), while the exergy surplus is 4.44 MW.

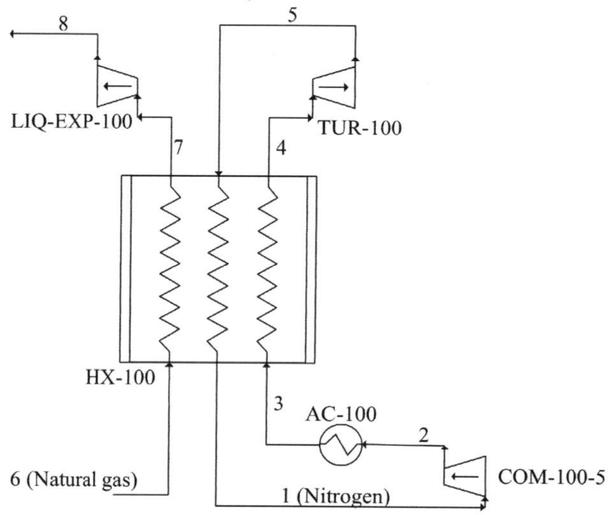

Fig. 2 Flowsheet of the Reverse Brayton process

Below ambient temperature, the cooling (refrigeration load) and heating requirements are zero (see left part of Fig. 3*a*). The exergy requirement and rejection are both zero, while the process-to-process exergy destruction due to driving forces is around 5.0 MW (quite large). Close to 35% of the exergy available in the exergy source (i.e. 18.8 − 4.4 = 14.4 MW) is destroyed (5.0 kW). Notice that there are two *heat pinches*: 25°C/22°C and -165°C/-168°C (ΔT_{min} = 3 K), while there is only one *exergy pinch* at 117.73 K for the exergy source curve and 112.35 K for the exergy sink curve.

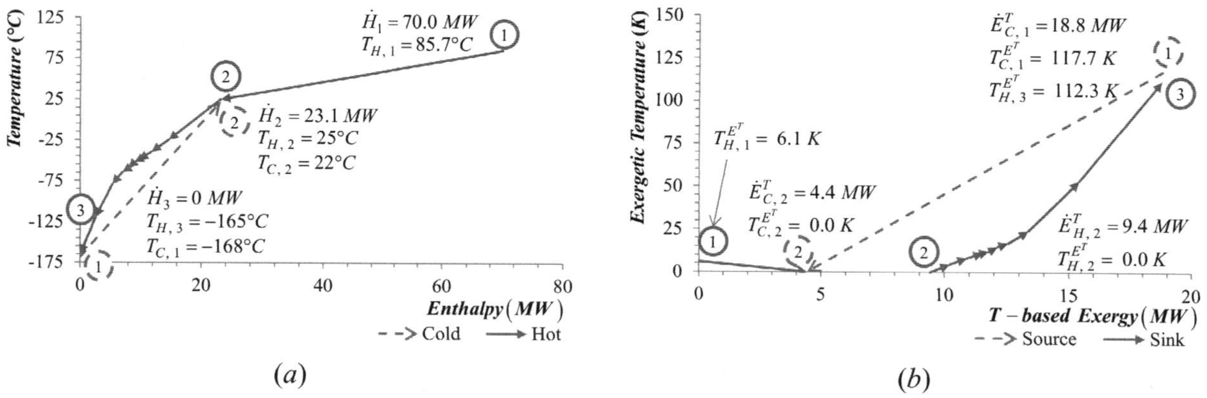

Fig. 3 Diagrams for the Reverse Brayton process: (a) CCs and (b) $T^{E^T} - \dot{E}^T$ *diagram*

Final Remarks

The paper presents a novel representation for exergy targeting which in combination with the heuristics presented by the Extended Pinch Analysis and Design (ExPAnD) method can be used to target the minimum thermal energy and exergy requirements as well as for designing low temperature processes. The new representation uses an alternative energy quality parameter called *exergetic temperature*. The Reverse Brayton process is used as case study for illustrating the application of this new exergy targeting representation.

References

Aspelund, A., D. O. Berstad and T. Gundersen (2007). "An extended pinch analysis and design procedure utilizing pressure based exergy for subambient cooling." Applied Thermal Engineering **27**(16): 2633-2649.

Linnhoff, B. and V. R. Dhole (1992). "Shaftwork targets for low-temperature process design." Chemical Engineering Science **47**(8): 2081-2091.

Marmolejo-Correa, D. and T. Gundersen (2012). A new procedure for the design of LNG processes by combining Exergy and Pinch Analyses. Proceedings of ECOS 2012 - The 25th International Conference on Efficiency, Cost, Optimization, Simulation and Environmental Impact of Energy Systems, Perugia, Italy, Elsevier (Submitted).

Venkatarathnam, G. (2008). Cryogenic mixed refrigerant processes. New York, Springer.

Wechsung, A., A. Aspelund, T. Gundersen and P. I. Barton (2011). "Synthesis of Heat Exchanger Networks at Sub-ambient Conditions with Compression and Expansion of Process Streams." AIChE Journal, **57**(8): 2090-2108.

CDU PREHEATING TRAIN RETROFIT KEEPING HEAT EXCHANGERS NETWORK TOPOLOGY

Antonio V. S. Castro, Claudio L. M. Kuboski, Danilo H. Biazi, Roberto A. Oliveira and Washington Geraldelli
CENPES – PETROBRAS Research and Development Center, Rio de Janeiro, Brazil

Abstract

As energy demand in the world is still rising, energy efficiency tools become more relevant for industry to meet the society expectations regarding energy consumption and greenhouse gas (GHG) emissions. In a refinery, Crude oil Distillation Unit (CDU) is one of the most significant energy consumers. However, retrofit projects focused on energy savings are not always easy to implement, as these projects are sensitive to payback and may result in several changes on the Heat Exchanger Network (HEN) topology. This work presents the methodology and results of a crude oil preheating train retrofit, considering that no HEN topology modification was allowed. This constraint made possible to implement the project in very short schedule, compared to traditional projects. In order to make the project feasible, the use of enhancements on heat exchangers (like the Twisted Tube bundles) were investigated. This work will also address the results achieved after project implementation regarding economics, GHG emissions and project schedule.

KEYWORDS: Heat Integration; Crude distillation, Retrofit

Introduction

Energy efficiency is recognized as one of the most important initiatives addressing GHG reduction.

PETROBRAS is committed in decreasing GHG emissions and improve energy efficiency of its process units. When considering downstream facilities, CDU is one of the most significant energy consumers in a refinery and the retrofit of CDU preheating train may contribute to meet this targets. However, projects considering heat exchanger network (HEN) retrofits focusing purely on energy efficiency improvement, excluding increase on throughput, are not always easy to implement, as the revenue obtained with energy savings may not be so relevant in order to result in fair economic results.

One of the major difficulties faced by designers when evaluating preheating train retrofits regards significant changes on network topology, thus requiring pipe work[1]. These topology changes incur in relevant capital costs increase and also may require additional time to the scheduled plant turnaround, which also contributes to turn the project economical evaluation more difficult.

In order to overcome these constraints, PETROBRAS developed a study to minimize the changes on the network topology in the preheating train retrofit at a CDU in one of its refineries. The use of Twisted Tube heat exchangers allowed tube bundles replacement with no HEN change on original topology.

The results achieved after design implementation will be presented and discussed.

Retrofit Methodology

The methods for HEN retrofit design are based on two main approaches[2]. The first approach uses pinch analysis and was first proposed by Tjoe and Linnhoff[3]. This method uses thermodynamics concepts to develop targets that can be achieved by applying rules in order to add or remove a heat exchanger on the HEN. The second approach is based on mathematical programming techniques which try to find the best solution in the expense of computer processing.

A discussion about the current methods and their drawbacks is presented by Asante and Zhu[2]. The main limitations pointed out for the pinch method are the potentially time consuming manual procedures, the dependence on the designer's experience and the difficulties to account the cost of structural changes, which might lead to complex designs.

The main drawbacks on the mathematical programming approach, besides the little scope for user interaction, are related to the high sensitivity of the results to the cost estimates used in the model.

There are also some methods that try to mix the benefits of both pinch and mathematical programming approaches[4] or minimize the shortcomings of both methods. However, considering the particular constraints regarding layout and schedule involved on this case, the application of these current methods could come up with either non-economic or unfeasible solutions, specially due to the high sensitivity to costs and heat transfer coefficients (HTC's) estimates. Therefore, it was developed a different approach to this particular retrofit design.

Basis And Constraints

This HEN retrofit design took place in a CDU of a PETROBRAS refinery. A schematic drawing of this crude oil preheating train is presented in Figure 1. The project's basis consisted on economically improve energy recovery without considering increase on throughput or product yields.

The major constraint to this retrofit design relied on the maintenance plant turnaround schedule. There were only about 15 months between the start of the Conceptual Project and the beginning of the plant turnaround. Because of this, it was stated that no modifications into the HEN topology were allowed, limiting the scope for energy recovery. The main constraints related to this short period for design and implementation were related to equipment procurement and delivery, detailing engineering and additional time to plant turnaround due to pipe and civil structures work. Pressure drop was not a major concern for this project. The HEN retrofit design was developed trying to minimize installation of new equipment, detail engineering work as well as plant turnaround work.

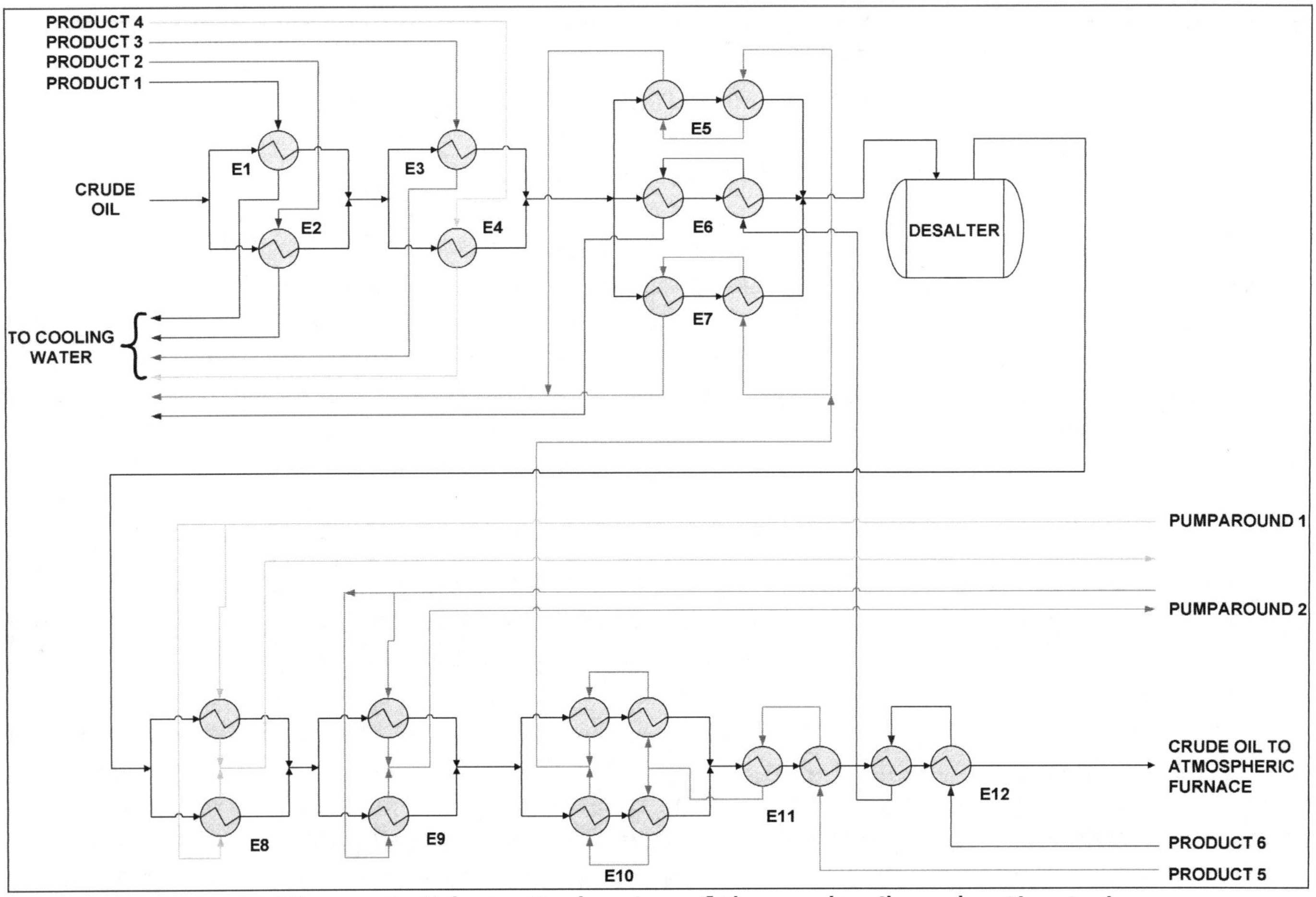

Figure 1. Schematic drawing of the crude oil pre heating train

As a result, the constraints limited the retrofit to few modifications in the HEN such as re-sequencing and re-piping (with caution to turnaround period), and other alternatives like heat transfer enhancements.

Methodology Applied

The combination of all constraints made this HEN retrofit design a peculiar challenge. Although the methodologies discussed above had been investigated, the results achieved by using them were not satisfactory. By tracking the root causes of the unsuccessful use of these methodologies, it was pointed out difficulties on establishing reliable data for costs related to pipe, civil and heat transfer enhancements estimates. Moreover, it was not possible to consider into the models the correct application of the several types of heat transfer enhancements.

Hence, it was applied a different approach in the search for a suitable solution, that is described in the following steps.

Plant Analysis

This is the first basic and crucial step in any retrofit study. A rigorous simulation of the CDU and the HEN with consistent and accurate data is critical. Data reconciliation between plant and simulation model is always a hard and time consuming work. The analysis of plant performance is important to determine many parameters such as fouling coefficients, pumparound (PA) duties, pressure drops, etc. Carefull understanding of these parameters can allow the designer to identify bottlenecks and major opportunities in energy recovery.

In this analysis, it was found some energy opportunities related to increasing PA duty, poor heat exchangers performance and poor product energy recovery in the preheating train.

The increase in PA duty was observed after rigorous simulation of the Atmospheric Distillation Column and analysis of the heat exchangers involved in these services. There was a significant opportunity for increasing 20% Pumparound 1 duty regarding column internal liquid flows while meeting product specifications. Furthermore, Pumparound 1 operational data tracking showed excessive fouling rates, indicating that equipment modification could be used in order to minimize this issue.

The coolers duties of all products leaving the preheating train were verified and are presented in Table 1.

Table 1. Coolers duties of products

	Cooler Duty Gcal/h
Product 1	1.56
Product 2	1.06
Product 3	3.22
Product 4	0.91

It can be observed that Product 3 has the greatest amount of energy lost to cooling water, highlighting scope for improving energy recovery.

The products 5 and 6 did not present any significant characteristic that could be conclusive in terms of energy recovery opportunity and so were not included in Table 1.

Another major constraint evaluated in this step was related to crude oil temperature. It was verified that an increase in the crude oil temperature at the end of the HEN greater than 10°C could induce vaporization of crude oil. This two phase flow could unbalance heater pass flows, leading to an uncontrollable atmospheric furnace[5]. Therefore, to avoid undesired layout changes by installation of a preflash drum or column[6], the maximum acceptable increase in crude oil temperature at the end of the HEN was defined as 10°C.

Defining Modifications

After screening potential services for modifications, it was necessary to analyze what type of changes in the heat exchangers were applicable and if the changes could contribute to raise the crude oil temperature to the target value.

At this point it was important a more detailed evaluation of each heat exchanger performance. It was observed that in E3, where Product 3 is placed in shellside and crude oil in tubeside, the heat transfer is controlled by the coldside conditions. The same fact was observed within exchanger E8, where the shellside product is Pumparound 1.

Thus, some alternatives were investigated to improve E3 and E8 energy recovery trying to keep the equipment original shells, for instance, new bundle designs, changes in number of tube passes, different tube types, tube internals, etc. It was observed that the best alternative in these two cases was to replace the original plain tube bundle by new bundles with twisted tubes technology.

One of the advantages observed on this alternative regards the possibility to convert an "E" shell to an "F" shell, without any changes in the original shell, nozzles or piping. Moreover, compact bundle layout allowed increase in heat transfer area and twisted tubes geometry incurred in a relevant increase in tubeside heat transfer coefficients.

Twisted tube characteristics may also be useful regarding fouling rates on Pumparound 1 service due to the higher turbulence within the heat exchanger, while pressure drop was not raised significantly.

One additional criterion used for evaluating twisted tube modification considered heat exchanger metallurgy. Some services demand the use of special metallurgy which incur in higher capital costs. In order to minimize capital costs, this project considered only modification on services requiring conventional metallurgy.

In order to keep system symmetry and allow some operational flexibility, it was also proposed to install the twisted tube bundle in exchanger E4.

At the end of this step, it was concluded that installing new twisted tubes bundles in exchangers E3, E4 and E8 could improve significantly energy recovery in each of these equipments. The next step was to evaluate the global effect over the HEN.

HEN Evaluation

The HEN performance related to the new twisted tubes bundles was evaluated by updating the rigorous simulation modeled in the Plant Analysis step with the heat exchangers modifications.

When evaluating an enhancement on an exchanger located at the beginning of the preheating train, it is important to confirm if the extra duty achieved with the enhancement will be sustained along the HEN. As the temperature approaches for the following exchangers will become smaller, the savings achieved by introducing an enhancement could become smaller than originally predicted.

The results of rigorous HEN simulation indicated an increase of 8°C in crude oil temperature at the end of the HEN.

The fact that most of the services at the end of the HEN were in series arrangement contributed to sustain their duties, even with lower temperature approaches. That supported the selection of the services to be modified.

Another benefit that shall be pointed out was the increase in desalters temperature. As the CDU also process heavy crude oils, which require higher temperatures for emulsion control, the increase on desalters temperature of about 10°C was considered beneficial. This is also another benefit of this methodology. Traditional methods consider target temperatures at the desalter. On a retrofit like this, the desalter temperature becomes a result of the modifications introduced at the preheating train and shall be carefully evaluated.

After deciding the project scope, the economic evaluation was performed using bundle supplier quotations. That pointed to a payback time of about three years, taking into account only the revenue due to fuel consumption decrease on fired heaters.

Even though the presented approach could not indicate if this was the best solution for the problem, this was considered a successful solution since no constraints were violated and no pipe work was required during plant turnaround period.

Project Implementation

The four new bundles were delivered as originally planned. There was very little scope for detailed engineering and no additional time was required in plant turnaround.

Figure 2. E3 Twisted tubes bundle

The twisted tubes bundles were easy to install, since the procedure was quite similar to the one used for conventional bundles replacement, simplifying the service.

Results after Start-Up and Discussions

After plant start-up, it was possible to verify the HEN performance with these new bundles.

The first evaluation was to compare operational and design data related to the new bundles. According to operational conditions for one period, clean and dirty heat transfer coefficients can be calculated via appropriate heat exchanger simulation software.

The clean and actual (dirty) heat transfer coefficients were compared for exchanger E8 and are presented in Figure 3. It can be observed for the period just after start-up, where clean conditions can be assumed, the global heat transfer coefficients U for the clean and dirty conditions are quite similar. It is possible to conclude by these results that the new bundles where performing like expected, since greater deviations from U clean and U dirty in start of run could indicate poor equipment performance.

The crude oil outlet temperature at the end of the HEN was also analyzed. In Figure 4, it can be seen an increasing of about 10ºC comparing the period before and after plant turnaround. However, it is not possible to determine via plant data how much of this temperature raise is related to the new bundles, since some heat exchangers were cleaned during plant turnaround.

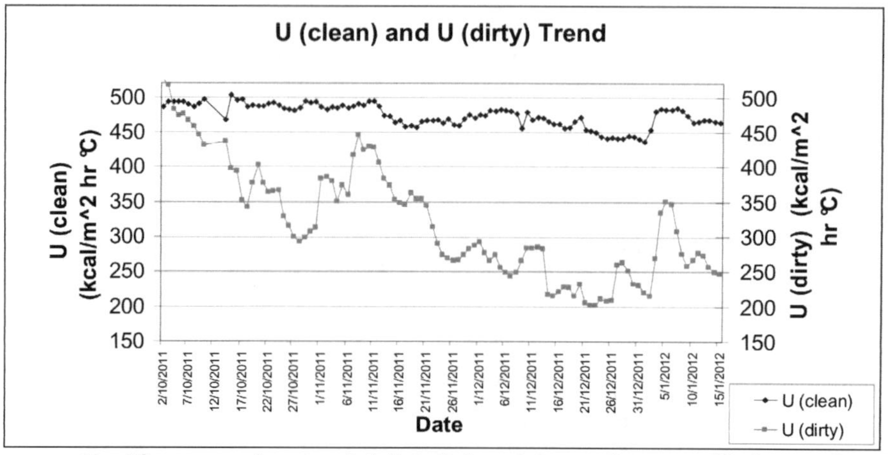

Figure 3. Clean and actual (dirty) heat transfer coefficients for E8

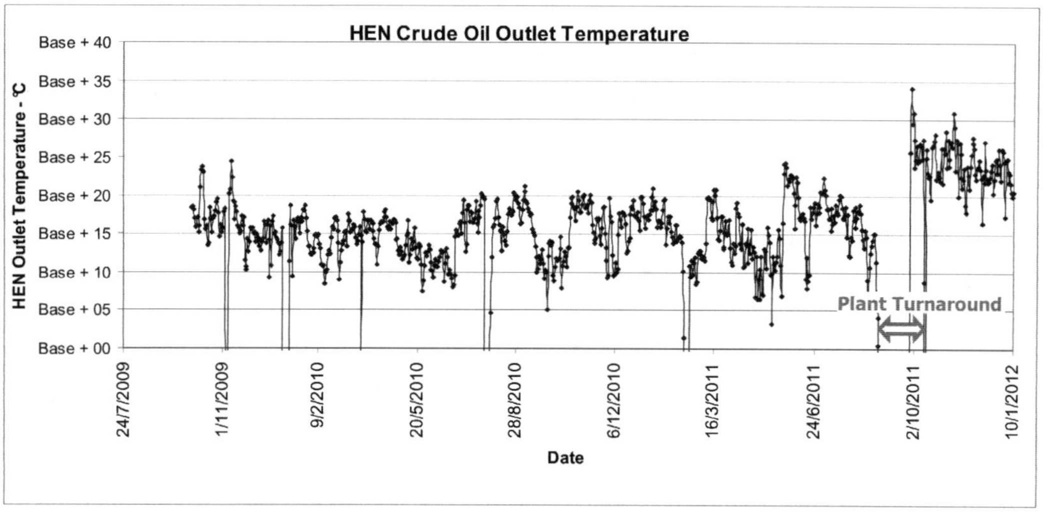

Figure 4. HEN crude oil outlet temperature

A qualitative investigation in plant data in order to overcome this difficulty was made comparing periods just after plant turnarounds. The analysis presented in Figure 5 tries to compare periods where the HEN could have similar fouling rates. Because of data confidentially, it was assumed a different basis temperature in each graph. The important information is to observe the temperature differences for each period considered (red and blue lines).

As it can be seen in Figure 5, it was possible to achieve a slightly higher crude oil temperature at the end of the HEN even with both atmospheric and vacuum furnaces operating at significant lower temperatures comparing the 2005 and 2011 runs. This was the major operational result achieved with this retrofit.

It is also important to verify in Figure 5 that the vacuum furnace outlet temperature has a high influence over HEN performance. When the vacuum furnace outlet temperatures on both periods are similar, the HEN outlet temperature for 2011 run is higher, showing the effect of the new bundles.

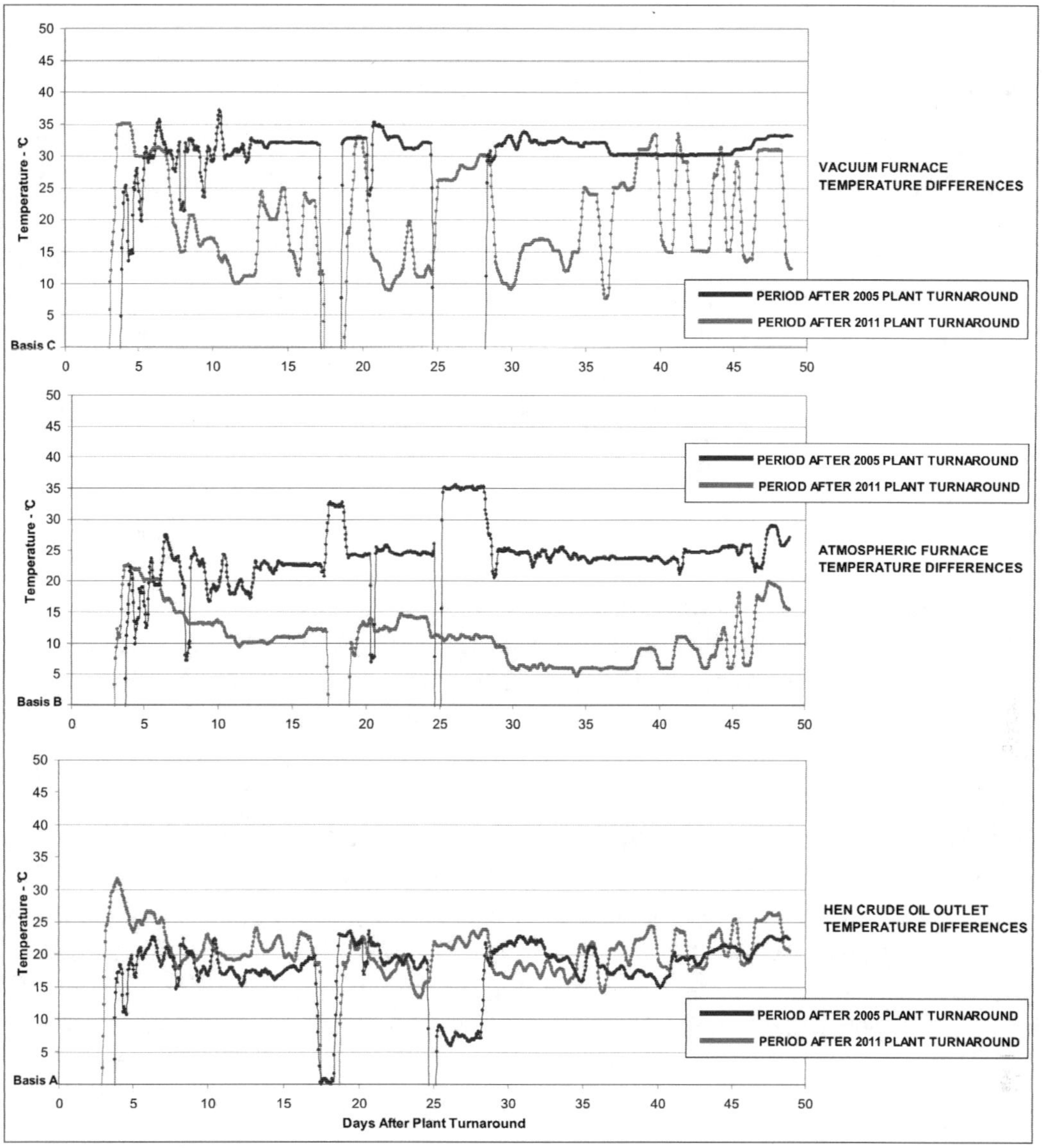

Figure 5. Plant data comparing periods after plant turnarounds

The quantitative determination of the benefits was only possible by rigorous simulation of the HEN. The simulation model was calibrated according to the conditions observed after start up. Then, it was simulated how the HEN could perform if the twisted tubes bundles were not installed. The simulation results showed that the crude oil temperature at the end of the HEN would decrease 7.5ºC. This result confirms the savings predicted during design phase. The additional energy recovery calculated in the simulation is equivalent to a reduction of 7% in atmospheric furnace duty. In terms of environmental benefit, this represents an estimated reduction of 5,000 tons per year in CO_2 emissions.

Besides furnace duty reduction, the additional energy recovery in the HEN can open scope for some process optimization. There is a possibility to keep the original furnace duty generating an increase in product yields or a reduction in the injection of stripping steam. The reduction in stripping steam causes some other beneficial effects like reduction of sour water production and reduction in atmospheric column pressure.

Conclusion

This HEN retrofit design was intended to economically improve energy recovery in a crude oil preheating train in a very short schedule that limited the scope for modifications. By using some alternative approaches and technologies, it was possible to reach an economically feasible solution.

Plant analysis was conducted indicating services in which enhancements could be investigated. Operational parameters such as fouling rates, products cooling requirements and Pumparound duties were used in this task. The present design focused on screening opportunities not only evaluating purely the HEN, but also the entire CDU performance (including crude and vacuum columns), making it possible to identify additional energy recovery opportunities.

Heat exchangers enhancements were investigated through rigorous simulation of overall HEN performance and the final solution came up with the installation of the twisted tubes bundles. The solution complied with the main project constraints, such as low capital cost required and fast and easy implementation.

Qualitative and quantitative analysis showed that the project objectives were achieved with an increase in crude oil temperature at the end of the HEN of 7.5°C, as predicted in conceptual design. The benefits can be resumed in a substantial reduction of about 7% in atmospheric furnace duty and an environmental saving of 5,000 tons of CO_2 emissions per year to atmosphere.

The economic evaluation pointed out to a payback time of about three years, taking into account only the revenue due to fuel consumption decrease.

According to observations during this retrofit design, the conventional methods could be used if the cost estimates for the many types of heat transfer enhancements were accurate. Unfortunately, this information was not available at the time of project design. The task of generate these cost estimates is highly time consuming, playing against the short project schedule. This highly constrained problem naturally indicated the solution for the retrofit design and the use of the engineer sense was the best tool in this case.

References

1. Smith, R.M. (2005), "Chemical Process: Design and Integration", *Ed. Wiley*
2. Asante, N.D.K. and Zhu, X.X. (1997), "An automated and interactive approach for heat exchanger network retrofit", *Trans IChemE,* Vol 75, Part A, March 1997,349-360
3. Tjoe, T. N. and Linnhoff, B. (1986), "Using pinch technology for process retrofit", *Chem Eng*, 28 April, 47-60
4. Briones, V. and Kokossis, A. (1996), " A new approach for the optimal retrofit of heat exchanger networks", *Comput and Chem Eng 20* (suppl): s7-s12
5. Waintraub, S., Lemos, A.J.F., Paschoal, L.C.M., Mainenti, M.R.M. (2001) "How to circumvent the shutdown of a preflash tower in an atmospheric distillation unit: An abnormal operation", 2001 AIChe Srping Meeting
6. Waintraub, S., Coutinho, R.C.C., Espósito, R.O. (2007), "Preflash drum when processing heavy oils: Paradox or reality?", 2007 AIChe Spring Meeting

(119f) An Innovative Design for Energy Saving by Self-Heat Recuperation in Crude Oil Distillation

Yasuki Kansha, Akira Kishimoto and Atsushi Tsutsumi, Institute of Industrial Science, The University of Tokyo, Tokyo, Japan

In our daily lives, we use many products that originate from oil, such as fuel and plastics. In fact, not only consumer products but also whole modern economies rely on petroleum. Currently, the amount of annual crude oil imports to Japan amounts to about 230 GL/y or about 5% of annual world crude oil production. It is reported that about 5% of this amount is used as fuel in oil refinery plants.

Recently, energy saving has attracted increased interest in many countries to minimize global warming, caused mainly by the consumption of fossil fuels. Although many heat integration techniques for process energy saving have been applied to oil refinery plants since the 1970s, oil refinery plants still consume large amounts of energy compared to the required values based on an exergy analysis for separation processes. In particular, crude oil distillation is an atmospheric distillation column using a furnace. It consumes about 50% of the energy required in an oil refinery plant. To reduce energy requirements, it is necessary to investigate crude oil distillation and to retrofit it with energy saving processes.

The authors developed an innovative process design technology, termed self-heat recuperation technology for saving energy. By using this technology, whole-process heat is recirculated within the process without heat addition, leading to large energy savings. In this research, crude oil distillation is analyzed and a crude oil distillation model for an energy saving design is developed. Furthermore, the feasibility of application of self-heat recuperation technology is investigated and an innovative design for energy saving by self-heat recuperation in crude oil distillation is proposed.

(121a) Hydrocracking - An Overview

Shankar Vaidyanathan, Fluor Corporation, Sugar Land, TX

The paper will present an overview of the Hydrocracking process as applied in the refineries today. It will cover Mild Hydrocracking, Partial Conversion Hydrocracking, Full Conversion Hydrocracking, Hydrocracker/FCC integration, Hydrocracker/Ethylene Steam Cracker integration and Production of Lube Oil Base stock. The paper will review several flow scheme options such as single stage once through, single stage recycle and two stage recycle.

Tutorial on Hydroprocessing: "Hydroprocessing Catalysts and Processes"

Paul R. Robinson, PhD
Criterion Catalysts & Technologies, LLC
Zeolyst International
Houston, Texas

Abstract

The products of petroleum refining must meet tight specifications, including limits on sulfur, nitrogen, olefins, aromatics, and other contaminants. Hydrotreating removes these contaminants from distilled crude oil fractions and intermediate process streams. Hydrocracking converts heavy oil fractions into lighter, more valuable products. Hydrotreating and hydrocracking processes share many common features, so they often are discussed together as "hydroprocessing."

Most hydroprocessing units employ specialized catalysts. As the name implies, they all consume hydrogen. Important chemical reactions include hydrodesulfurization (HDS), hydrodenitrogenation (HDN), hydrodeoxygenation, and the saturation of olefins and aromatics.

Most hydroprocessing catalysts are comprised of active metals and promoters on solid supports. For hydrotreating, the active component is molybdenum disulfide (MoS_2). MoS_2 is also used in some hydrocracking catalysts. Other hydrocracking catalysts employ either tungsten disulfide (WS_2), palladium, or platinum. Promoters for MoS_2 and WS_2 catalysts include nickel and/or cobalt. The main support for hydrotreating catalysts is γ-alumina. Hydrocracking requires highly acidic supports such as amorphous silica/alumina or synthetic zeolites.

Catalysts can be prepared in various ways. In some cases, the support is manufactured first and the metals are added afterwards. In other cases, the components are co-mulled prior to shaping, drying and calcination. Hydroprocessing configurations can include fixed-bed reactors, ebullating bed reactors, or slurry-phase reactors, each of which requires a different kind of catalysts or (in the case of slurry-phase processing) a non-catalytic additive.

Catalyst deactivation plays a major role in the economics of hydroprocessing.

(121c) Hydroprocessing Design - Tutorial

Stuart Shih, PetroTech, Gainesville, VA

Hydroprocessing or hydrotreating is the key technology to clean, to refine, and to convert crude oils into valuable products, such as gasoline and diesel. It covers a wide range of feed stocks and numerous configurations. The tutorial will discuss process configuration, material selection, and heat balance. Impacts of feed properties and catalyst selection also will be included.

LABORATORY HYDROTREATERS FOR MODELING AND SIMULATION OF INDUSTRIAL UNITS

Leonidas E. Kallinikos and Nikos G. Papayannakos
Chemical Engineering, National Technical University of Athens, Zografos, Athens, Greece

Introduction

Hydrotreatment of petroleum fractions is widely used in refineries for the improvement of fuel product characteristics and distribution. The interest on this three phase process is due to the need for ultra deep desulphurization of gas oils, e.g. all petrol and diesel sold in European area must contain less than 10 ppm after January 1st of 2009 (Directive 2003/17/EC), and for upgrading of heavy fractions to increase the production of middle distillates (Babich and Moulijn, 2003). To this end the main targets are the development of new and/or the improvement of existing processes for the catalytic hydrotreatment of the petroleum fractions by inventing new and more efficient catalysts, by optimizing the operating conditions of the existing units (Sie and Krishna, 1998; Schmitz et al., 2004), and by increasing the effectiveness and the operation life of the existing catalysts monitoring their activity online (Kallinikos et al., 2008).

Pilot, bench and small scale multiphase reactors are commonly used for the experimental simulation of industrial scale hydrotreaters (Sie and Krishna, 1998). However, the use of small scale reactors with total catalytic mass less than 10 gr (Bellos and Papayannakos, 2003; Kallinikos and Papayannakos 2007a,b) becomes more and more attractive in the scientific community due to their contribution to the reduction of experimentation cost, safer operation and minimization of experimental time. It is really important that an automated mini – unit is capable to operate without human attention to be required (Sie, 1996). These reactors appear to be a very promising and useful tool for future applications but their successful use will ultimately depend on their effective operation.

The operational problems of small scale reactors are mainly related to low gas and liquid mass flow rates which are needed to retain the same gas and liquid hourly space velocities (LHSV) as in commercial or pilot reactors. The flow rate of the liquid phase feeding a small scale reactor may be up to 10^7 times lower than the corresponding liquid flow feeding a common commercial hydrotreater and up to 300 times smaller than the liquid feed of pilot reactors. However, fluid dynamics and non – ideality problems may arise during operation of these small scale reactors like wall effects, incomplete catalyst particle irrigation, poor gas distribution, especially in cases of low gas/liquid ratios, as well as extended liquid axial dispersion and mass transfer resistances (Sie and Krishna, 1998). Although the problems arise when the ratio of the catalyst particles size to reactor diameter increases, the study of commercial catalyst extrudates must be performed by using them in their original dimensions and not to be crushed because in commercial operation the catalyst effectiveness factor is usually less than one and crushing would disguise any size and skin effect that the commercial catalyst might display (Gierman, 1998).

Concurrent gas and liquid downflow and upflow modes have been widely attempted in the operation of small scale three phase reactors. Upflow operation improves catalyst wetting and utilization (Khadilkar et al., 1996) but extended liquid phase mixing problems in non-diluted beds affect the reactor performance (Papayannakos et al., 1992).

Many investigators use the dilution technique to avoid problems related with wall effects, low catalyst utilization, incomplete wetting and axial back mixing for downflow or upflow operation in bench or micro reactors (Bellos and Papayannakos, 2003; Klinken et al., 1980), putting small diameter inert fines to fill the voids between the catalyst extrudates. The improved performance of the diluted beds operating in downflow mode is attributed to restricted liquid axial dispersion and the much higher liquid hold-up (Klinken et al., 1980) that mainly leads to improved wetting and utilization of the catalyst particles. For upflow operation the improvement in reactor performance is mainly attributed to diminished liquid phase dispersion.

The main problem of the diluted beds, operating in upflow or downflow mode, is the reproducibility of the loading procedure. Al – Dahhan et al. (1995) have presented techniques to reproducibly build diluted beds in bench scale reactors; however the loading of diluted beds is a stochastic procedure and the voids created cannot be totally controlled. Moreover, for low gas and liquid flow rates distribution problems of both gas and liquid phases may arise.

In many cases fluid dynamics drawbacks and mass transfer limitations can not be neglected, thus it is necessary to take them into account in the mathematical simulation of the reactor. To this end it is indispensable to quantitatively estimate their effects on the reactor performance.

To address the downscale problem of three phase reactors, a different type structured bed reactor has been presented in some of our previous works. This type of reactor is called string bed spiral reactor and it is a structured bed reactor that can be used in any two or three phase reactions catalyzed by solid particles. It comprises a tube with a diameter a little bigger than the diameter of the catalyst particles used, in which the solid particles are successively introduced one by one building a bed with all the particles being in a line. Thus, a 'string bed' of catalyst particles is created. Because the reactor tubes containing a satisfactory catalyst mass (1–5 g) are long enough, they can be coiled to form a spiral reactor. The gas and liquid phases flow over all the catalyst particles resulting in avoidance of by-passing, channeling and poor gas distribution. A certain advantage in using spiral reactors in scaling down is the operation with higher liquid and gas superficial velocities than in common small scale reactors due to the small internal reactor diameter.

In this work, the procedure followed for the experimental simulation of deep catalytic desulphurization process of gasoil with the string bed spiral reactor is presented. Special features of the spiral reactor operation obtained by following this procedure are discussed.

Discussion

The procedure for the experimental study and simulation of three phase hydrotreatment of hydrocarbon mixtures with small scale laboratory reactors is based on two parallel experimental methods that have been employed for the investigation of the phenomena involved in the operation of the reactors (Trambouze, 1990). The schematic approach of the procedure is illustrated in Fig. 1. The first method aims at the study of the phenomena with mock up experiments (Trambouze, 1990) (route a, Fig. 1) while the second one focuses on them through reacting systems, isolating the influence of the reaction rates (route b, Fig. 1). The fluid dynamics (step 1, Fig. 1) (Kallinikos and Papayannakos 2007a,b), the gas - liquid mass transfer (Metaxas and Papayannakos, 2006) (step 2, in Fig. 1), the liquid - solid mass transfer (Stüber et al., 1996; Saroha, 2010) (step 3, in Fig. 1) and the intra – particle mass transfer resistances (Schmitz et al., 2004) (step 4, in Fig. 1) are considered as the process controlling phenomena and therefore they attract the interest of this work. After these phenomena are studied and the controlling mechanisms established the kinetics of the main reactions of interest like deep desulphurization of gasoil fraction are determined by carrying out experiments at the appropriate reaction conditions, (step 5, Fig. 1).

In the framework of the first method, the fluid dynamic characteristics of the spiral reactor (1a) were studied with mock up tests performed at ambient conditions. For these experiments simple gas – liquid systems (water or heptane as liquid and nitrogen as gas phase) were used. PTFE tubes of several lengths were loaded with non porous extrudates to simulate the spiral reactor concept. The effect of the gas and liquid superficial velocities, the reactor internal diameter to catalyst diameter ratio and the length of the reactor on the flow regime, the axial dispersion of the liquid phase, liquid hold-up and pressure drop have already been presented (Kallinikos and Papayannakos 2007a).

Following the second method, the pertinent mechanisms were investigated performing experiments at reaction conditions (1b). In this case the major problem is the interference of mass transfer in the measured reaction rates. In order to deal with this problem reaction experiments were carried out at the same reaction conditions, LHSV, temperature, pressure and gas to liquid ratio, but with several catalyst masses and spiral reactor lengths. Moreover, the performance of the spiral reactor was compared with a convectional small scale diluted bed reactor as well as with a pilot reactor (Kallinikos and Papayannakos 2007b). The test reaction was the gasoil hydrodesulphurization. Some representative results are presented in Fig. 2 as far as the exit sulfur concentration of the hydrotreated products vs. LHSV are concerned for experiments performed with spiral reactors with different lengths and compared with some conventional experimental reactors. As it is shown, the performance of the spiral reactor is not affected by its length and the fluid velocities. This implies that down-scaling effects like wall effects, axial backmixing and incomplete catalyst particle irrigation are not important in its operation (Kallinikos and Papayannakos 2007b). Moreover, the performance of a spiral reactor was found to be comparable with that of a convectional small scale diluted bed reactor and of a pilot scale reactor. Thus, the operation of the spiral reactors can effectively represent the operation of the pilot reactor which is considered to operate as an ideal reactor since the impact of problems like axial backmixing, poor irrigation and wall effects on its operation is minimized (Gierman, 1988).

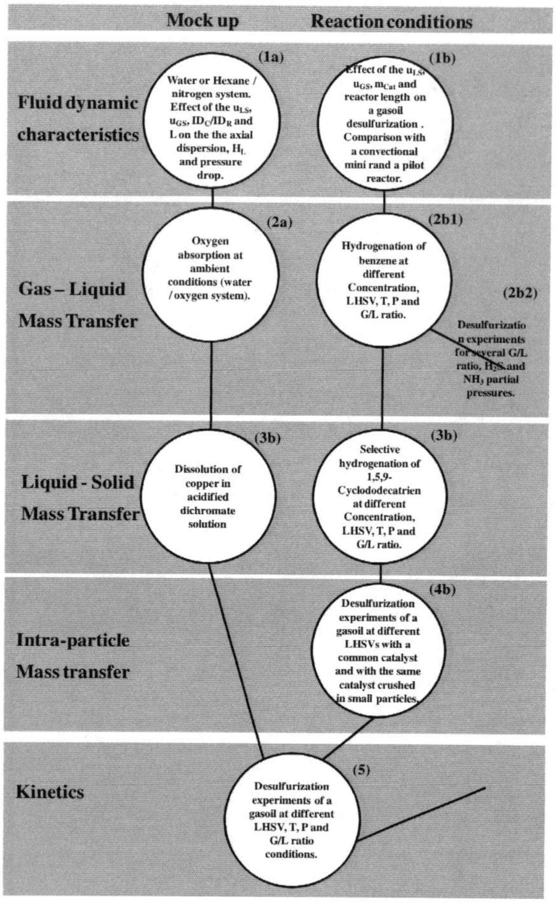

Fig. 1. Procedure followed for the experimental simulation of deep desulphurization of gasoil with a small scale laboratory reactor

In the framework of the study of the controlling mass transfer phenomena, the interfacial gas-liquid mass transfer limitations at ambient conditions are quantified (2a). For this study several PTFE tubes and non porous extrudates were used and oxygen absorption or desorption experiments were performed for the estimation of the mass transfer coefficients.

The gas - liquid mass transfer limitations of the system can also be studied by performing experiments with a model reaction (2b1), in our case benzene hydrogenation in hexane, while the reaction kinetics was obtained from kinetic experiments in a batch vessel in the absence of mass transfer limitations (Metaxas and Papayannakos, 2006). The results obtained with the model reaction are compared with gasoil hydrotreatment experiments for the confirmation of the gas - liquid mass transfer limitations effects. Desulphurization experiments of a gasoil were performed at industrial reaction conditions of LHSV, temperature, pressure but for several gas and liquid superficial velocities. The effects of hydrogen sulphide and ammonia on reaction kinetics for several gas to liquid ratios were also studied by adding ammonia and hydrogen sulphide in the gas phase (2b2).

In Fig. 3, the overall mass transfer coefficients ($k_{LG}a$) for hydrogen vs gas and liquid superficial velocities are presented. As it is shown, the mass transfer coefficients estimated for the spiral reactor are directly related to the gas to liquid superficial velocity ratios. These results are in agreement with results obtained with HDS reaction of gasoil (Kallinikos and Papayannakos 2007b) which show that the performance of the spiral reactor is not affected by the individual gas and liquid superficial velocities but it is mainly affected by the ratio of these velocities.

The liquid–solid mass transfer generally increases with the liquid flow rate due to the improved wetting of catalyst; however some investigators claim that they also increase with the gas flow rate. The mass transfer decreases with the increase of particle diameter. For the investigation of these limitations, systems without reaction can be used (3a) like the dissolution of copper in acidified dichromate solution (Saroha, 2010), while for reacting systems the selective hydrogenation of 1,5,9-Cyclododecatriene to Cyctododecene has been used as model reaction (Stüber et al., 1996). As referred above, the influence of fluid velocities on the performance of the spiral reactor concept for the same gas to liquid ratio is negligible indicating limited liquid – solid mass transfer resistances taking into account that these mainly are affected by the liquid phase velocity.

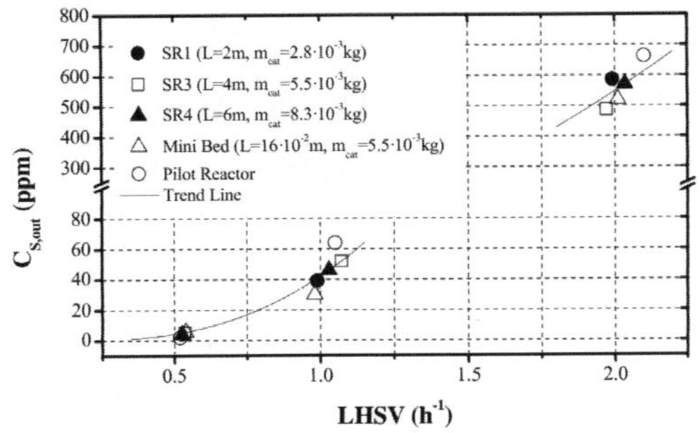

Fig. 2. Exit sulfur concentration vs. LHSV for the experiments performed with spiral reactors with different lengths and with the conventional experimental reactors

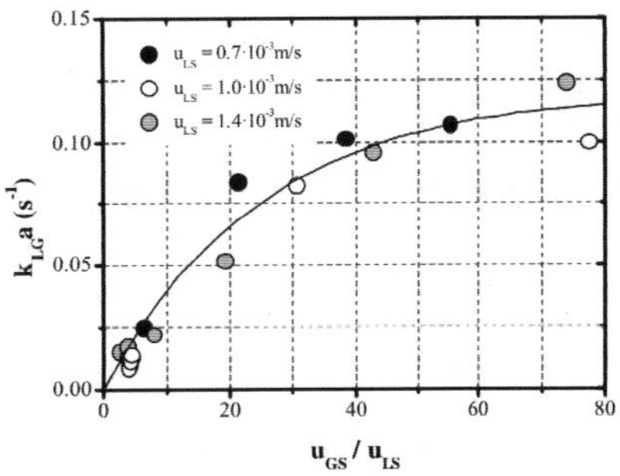

Fig. 3. Overall mass transfer coefficient for hydrogen vs gas and liquid superficial velocities obtained with benzene hydrogenation experiments

Finally, the intra-particle mass transfer limitations on the catalyst effectiveness of a three phase process can be studied with reaction experiments using non – crushed and crushed catalyst particles as it has already been presented by several investigators (Schmitz et al., 2004) (4b). From the results it was indicated that the catalyst effectiveness factor is higher than 0.85. Similarly, Schmitz et al. (2004) showed that the reaction rate constant calculated from the results of DBT desulphurization at common reaction conditions using a commercial catalyst with an equivalent diameter of 1.7 mm appeared similar to those calculated from experiments using the same catalyst crushed in small particles with mean diameter 0.2 mm.

References

1. Babich, I., Moulijn, J., Fuel 82 (2003) 607.
2. Bellos, G., Papayannakos, N., Cat. Tod. 79–80 (2003) 349.
3. Directive 2003/17/EC of the European Parliament and of the Council of 3 March 2003 amending Directive 98/70/EC relating to the quality of petrol and diesel fuels.
4. Gierman, H., Appl. Catal. 43 (1988) 277.
5. Kallinikos, L.E., Bellos, G.D., Papayannakos, N.G., Fuel 87 (2008) 2444.
6. Kallinikos, L.E., Papayannakos, N., Chem. Eng. Sci. 62 (2007a) 5979.
7. Kallinikos, L.E., Papayannakos, N., Ind. Eng. Chem. Res. 46 (2007b) 5531.
7. Khadilkar, M., Wu, Y., Al-Dahhan, M., Duducovic, M., Colakyan, M., Chem. Eng. Sci. 51 (1996) 2139.
9. Klinken, J., Dongen, R., Chem. Eng. Sci. 35 (1980) 59.
10. Metaxas, K., Papayannakos, N., Ind. Eng. Chem. Res. 45 (2006) 7110.
11. Papayannakos, N., Galtier, P., Bigeard, P., Kasztelan, S., Chem. Eng. Sci. 47 (1992) 2275.
12. Saroha, A.K., Chem. Eng. Res. and Des., 88 (2010) 744.
13. Schmitz, C., Datsevitch, L., Jess, A., Chem. Eng. Sci. 59 (2004) 2821.
14. Sie, S., AIChE J. 15 (1996) 3498.
15. Sie, S.T., Krishna, R. , Rev. Chem. Eng. 14 (1998) 203.
16. Stüber, F., Wilhelm, A. M., Delmas, H., Chem. Eng. Sci. 51 (1996) 2161.
17. Trambouze, P., Chem. Eng. Prog. 23 (1990).

Isomalk-2: A Low-Temperature, Light Naphtha Isomerization Process

by

Dr. Alexander Shakun, General Director, JSC SIE Neftehim, Krasnodar, Russia

Ilya Aranovich, Sales Engineer, GTC Technology, Houston, Texas, USA

Emanuel Reis, Licensing Manager, GTC Technology, Houston, Texas, USA

Over the last ten years, stronger government requirements and a very competitive market have pushed refiners to produce higher-quality automobile fuels. With the introduction of new mandated fuel regulations throughout the world, refiners must produce gasoline that meets the quality standards set by Euro-IV and Euro-V throughout the majority of the world. These new standards require a significant decrease in sulfur, benzene, and olefin species in the gasoline pool.

In view of these requirements, isomerate is a prime component for the gasoline pool. It brings several benefits, including low sulfur, very low benzene, aromatics, and olefins, a boiling range of 80-90°C, and an octane number that can reach 92-93 RON. The C_5-C_6 hydrocarbon fraction is abundant in the straight run light naphtha and secondary light fractions. This stream has a low octane value with great potential for improvement with the isomerization process. The refiner must choose the right isomerization process and technology partner for such units.

The start-up of a world-scale isomerization unit at the Omsk Gazpromneft Refinery (Figure 1) in November 2010 drew attention to alternatives for conventional chlorinated aluminum oxide-based technology. The SI-2 selective isomerization catalyst yields the same amount of branched paraffins as the chlorinated aluminum catalysts. The unit in Omsk produces an isomerate with the maximum achievable octane value for this stream of 92RON. The success of this unit utilizing the new Isomalk-2 isomerization technology is significant not only for the Russian market, but for the refining industry as a whole. A diligent analysis of this case will help refiners make a fact-based decision on planning the further development of their facilities.

Figure 1. Omsk Gazpromneft Refinery

Isomerization technology has been applied in refineries for over 50 years, but the catalyst types and performances are still misunderstood today. Until the beginning of this century, the choice of isomerization technology using the light naphtha C_5-C_6 fraction was between two types – chlorinated aluminum oxide or zeolite. Each of these technologies had its advantages and limitations.

The increasingly stringent requirements of automobile gasoline quality made the technology based on the use of zeolite catalyst less attractive, because of the higher temperatures required to facilitate the isomerization reaction. These high temperatures are less favorable for the production of branched C_5-C_6 isomers as seen in Figure 2. The licensing partnership of GTC Technology/Neftehim has experience in three revamp cases of the old zeolite technology to a new application of the SI-2 selective isomerization catalyst, which operates at lower temperature and yields higher octane.

Figure 2 shows the temperature ranges in which different catalytic systems are active, and the equilibrium of the main branched isomer of n-hexane: 2,2-dimethylbutane. Historically, the first light naphtha isomerisation catalytic systems were based on the use of platinum on fluorinated aluminium oxide catalysts, which operated at very high temperatures and yielded only very small amounts of branched paraffins. On the other end of the spectrum, researchers have been methodically studying the possibility of applying highly fluorinated metal super acid catalytic systems that operate at a lower temperature and would yield up to 70% 2,2-dimethylbutane. However, there is no stable, safe example of such a technology at the present time.

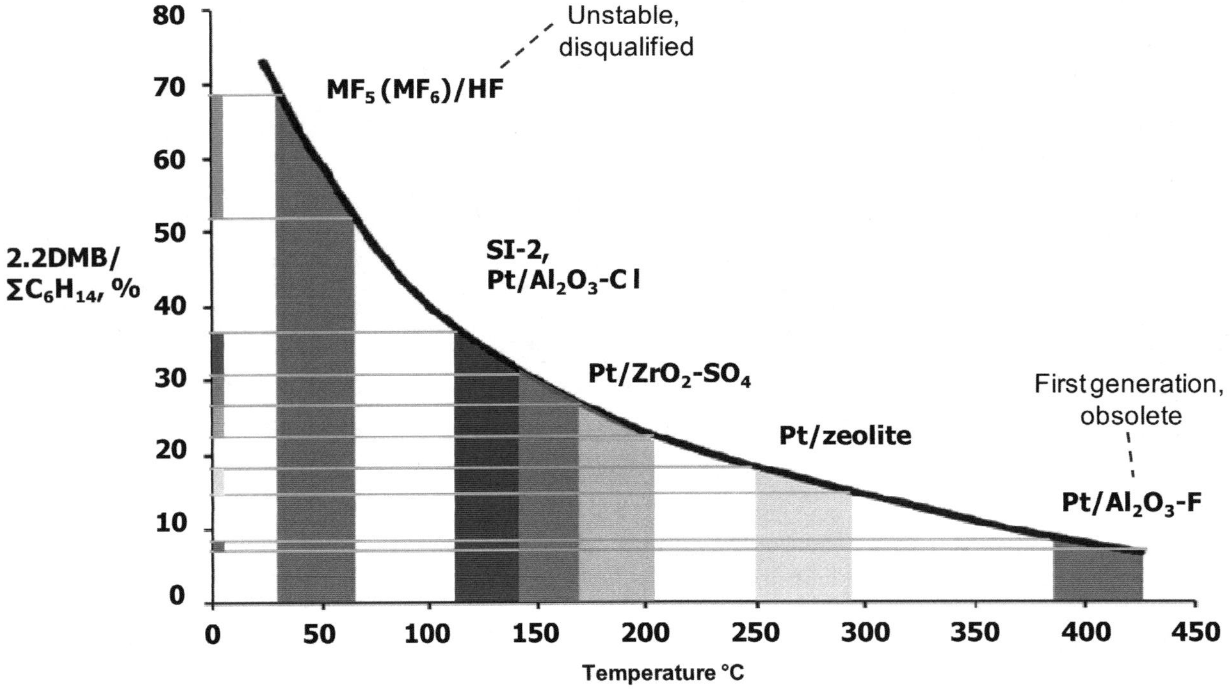

Figure 2. Temperature range of the activity for n-hexane isomerization of different catalysts

In the contemporary marketplace, there are only two types of proven, commercial isomerisation technologies that work in the low temperature range and can yield up to 36% 2,2-dimethylbutane. These are the chlorinated aluminum oxide and the SI-2 sulfated zirconium mixed metal oxide catalysts. Other commercially available sulfated zirconium catalysts operate in a 140 – 200 °C range and have a lower yield of the valuable branched paraffins.

As the market becomes more competitive and governed by buyer requirements, refiners are choosing process technology that:

- Offers simple operation.

- Avoids of the use of corrosive materials.
- Does not generate toxic waste.
- Has regenerable catalysts.
- Tolerates process contaminants.
- Does not require complicated analytical equipment to monitor.
- Has lower operating and initial capital cost.

The Isomalk-2 technology meets all of these requirements; others do not.

A broad comparison of the process parameters of the Isomalk-2 technology to other conventional technologies is shown in Table 1.

Table 1. Process parameters comparison of commercial isomerization technology

Parameter	Chlorinated Catalysts Pt/Al_2O_3-Cl	Zeolite Catalysts Pt/Zeolite	Pt/Sulfated Oxides	Isomalk-2 Pt/ZrO_2-M_xO_y-SO_4
Operating temperature (°C)	120-180	250 – 280	140-200	120-180
Pressure (kg/cm^2)	30 – 40	20 – 30	30 – 35	25 – 30
H_2/HC	0.2 – 0.5	1 – 2	2 – 3	2 – 3
Octane Number (RON)				
Once-through	82 – 85	78 – 79	80 – 81	82 – 85
DIH	86 – 88	81 – 82	84	86 – 88
DIP + DIH	89 – 90	83 – 86	86 – 87	89 – 90
DIP + DP + DIH	91 – 93	84 – 87	89 – 90	91 – 93
Trace Contaminant Limits				
Sulfur (ppm)	< 0.5	< 10	< 1.0	< 2.0
Nitrogen (ppm)	< 0.1	< 2.0	< 0.5	< 1.0
Water (ppm)	< 0.1	< 50	< 3.0	< 10.0

Capability to Regenerate	**No**	**Yes**	**Yes**	**Yes**
Water Poisoning	**Irreversible**	**Recoverable**	**Recoverable**	**Recoverable**

The SI-2 catalyst provides a higher conversion rate of the naphthenic hydrocarbons than chlorinated aluminum oxide systems. For example, at 160 ˚C the naphthenes conversion is twice the conversion attained with the chlorinated aluminum oxide catalyst. Isomalk-2 also produces less low octane C7+ hydrocarbons. This effectiveness allows the Isomalk-2 technology to operate at a lower recycle to feed rate, thereby maintaining a lower operating cost and allowing higher initial benzene content in the light naphtha feed. A detailed comparison of these technologies in terms of equipment requirements and operation conditions is shown in Table 2.

Table 2. Operating comparison of Isomalk-2 and Chlorinated Aluminum systems

Parameter	Chlorinated Systems	Isomalk-2
Preliminary Hydrotreatment	**yes**	**yes**
Adsorption Feed Treatment	**Required**	**Not required**
H_2 Consumption (wt. % per feed)	**0.6 – 0.8**	**0.2 - 0.3**
Columns	**2**	**2**
Recycle (% to feed)	**80-100**	**40 – 70**
Irreversible Contaminant Poisoning	**Yes**	**No**
Constant Chlorine Supply	**Yes**	**No**
Spent Caustic	**Yes**	**No**

Regeneratibility	**No**	**Yes**
Service life, years	**4-5**	**10 - 12**
Relative Operating Costs	**110**	**100**

The life cycle costs of Isomalk-2 are also far lower than chlorinated aluminium oxide catalyst systems.

Case Study – Gazpromneft Refinery, Isomalk-2 Technology with SI-2 Catalyst

The SI-2catalyst was developed by JSC Neftehim, a renowned scientific and engineering company, historically successful in the development of catalysts and refining technologies for isomerization and reforming processes. The Isomalk-2 light naphtha technology is offered by the partnership of GTC Technology and Neftehim, and has been implemented in nine different units with six more units at a stage of either design or construction.

The major technical requirements for the Gazpromneft refinery were: the overall octane number of the isomerate to be not less than 91 RON; liquid yield of the isomerate not less than 97.0 (wt%); catalyst operating cycle not less than five years; overall catalyst lifetime not less than ten years; minimal recycle of normal pentane and hexane; and absence of any solid, liquid or gaseous wastes. The process flow diagram for the Isomalk-2 technology at the Omsk Refinery is shown in Figure 3.

Figure 3. Isomalk-2 process flow diagram at Omsk Gazpromneft Refinery

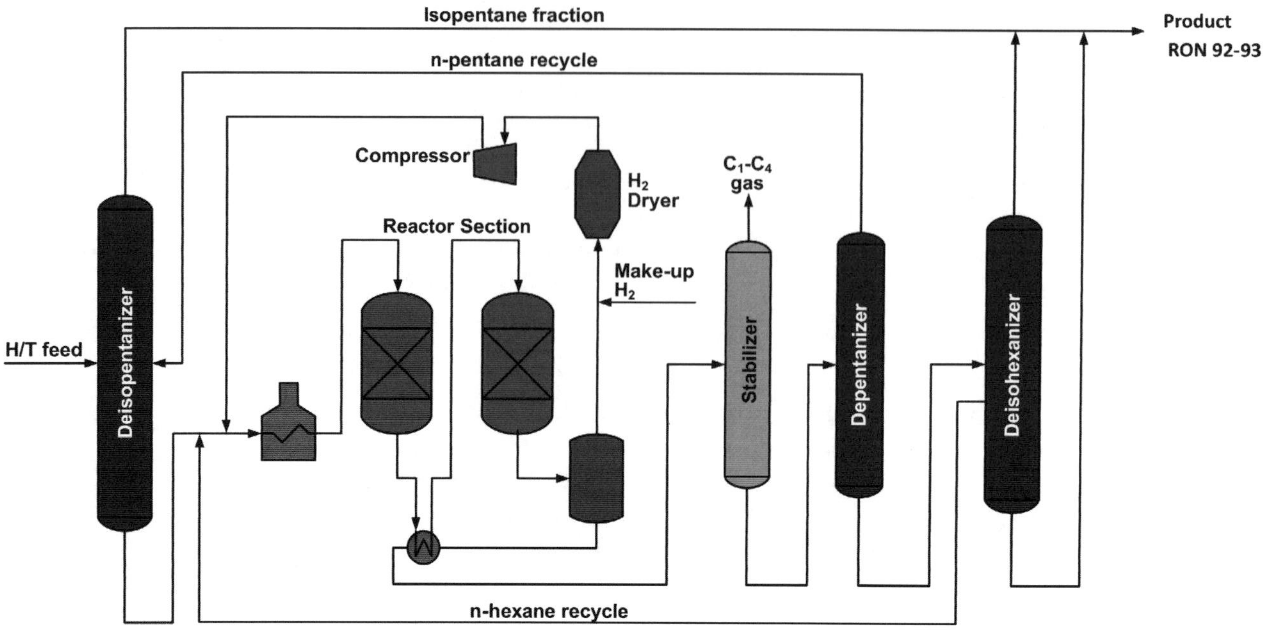

As the plant's goal was to achieve the maximum octane value of the isomerate product, a process scheme with full recycle of low-octane components was made. The configuration of the isomerization block includes a deisopentanizer; two reactors operating with the SI-2 catalyst; a stabilizer column; two columns for the recycle of n-pentane and n-hexane; and the separation of the final iso-component blend stocks. The feedstock is a blend of straight run light naphtha and by-product C_5-C_6 fractions with overall benzene content of 1 wt. % and C7+ hydrocarbons 1 wt. %. The sulfur content in the feed is in the range of 200-300 ppm.

A light naphtha feed hydrotreatment unit is used to achieve an optimum composition of the feed for the isomerization reactor, to reduce the contaminant level of the sulfur. The required sulfur content in the feed should not be more than 2 wppm, and nitrogen not more than 1 wppm (both requirements highly exceed the amount which should be controlled by the chlorinated aluminum oxide systems). The water amount in the feed should be less than 10 wppm (100 times more tolerant than the chlorinated aluminum systems).

The principal configuration of the light naphtha hydrotreating unit for the Isomalk-2 technology is the same as used to hydrotreat feeds to reforming units, and may in fact be the same unit. The isomerization section also uses the classical hydroprocessing flow scheme. Thus this technology is very convenient for the revamp of existing HDS/semi-regenerative reforming units or other hydroprocessing units into Isomalk-2 technology. The GTC Technology/Neftehim partnership has experience in providing technical solutions for such revamps at three commercial units with capacities ranging from 300 to 500 KTA of light naphtha. We see high interest in such revamp possibilities.

The isomerization block is a very simple arrangement that includes two fixed-bed reactors applying the SI-2 selective isomerization catalyst, a heat exchanger between the reactors for heat integration, a hydrogen recycle scheme with a gas compressor for the effective utilization of plant hydrogen, and hydrogen gas dryers. The performance parameters of the Isomalk-2 isomerization block correspond to the design requirements and are shown in Table 3.

Table 3. Performance parameters of the SI-2 catalyst Reactor Block

Parameter	Unit performance
Inlet Pressure, MPa	**3.0**
Inlet temperature first reactor, °C	**130-140**
Inlet temperature second reactor, °C	**130-140**
WHSV, hour^{-1}	**2.0-3.0**
Conversion and selectivity	
$iC_5/\sum C_5$, wt%	**70-77**
2,2-DMB/$\sum C_6H_{14}$, wt%	**30-36**
Stable Isomerate Yield. wt%	**97.5-98.5**

The depentanizer column performs two very important tasks: recycles the equilibrium amount of n-pentane left in the isomerate and provides the possibility of separating the i-hexane fraction from the dehexanizer column top with the maximum achievable RON of 92.

The deisohexanizer column produces the final isohexane gasoline pool blend stock. The light naphtha isomerization unit can be efficiently used only if a thorough hexane isomerization is achieved in the reactor block and the deisohexanizer column is operated satisfactorily.

Exclusion of any of these conditions will lower the process's effectiveness and will not allow a maximum isomerate product RON to be achieved.

The Isomalk-2 technology has a successful track record in achieving more than satisfactory results. The start-up and operation of the Isomalk-2 unit at the Omsk Refinery marks the appearance of a stable and environmentally safe alternative to chlorinated aluminum systems in a world-scale unit. The Isomalk-2 technology allows refiners to achieve the maximum isomerate octane value of 92 RON, and is a clear winner in every category of merit.

(138b) Light Naphtha Isomerization Chlorided Alumina Catalyst Deactivation

Dana K. Sullivan, R&D, UOP LLC, Des Plaines, IL

Light naphtha isomerization has been used for over fifty years to increase the octane of light naphtha streams and to manage benzene. It continues to be important today as refiners limit gasoline benzene content to comply with MSAT II regulations, and must offset the resulting octane loss. Chlorided alumina type catalysts are most commonly used for light naphtha isomerization, and their deactivation can have a serious effect on a refiner's ability to produce gasoline.

This paper will focus on optimizing chlorided alumina catalyst performance, particularly stability. The key data for tracking catalyst performance will be identified, followed by a discussion of the causes of both temporary and permanent catalyst deactivation and unit monitoring to detect loss of catalyst activity.

Impact of Silica on the Performance of Reforming Catalysts

Manuela Serban, Mark P. Lapinski, Sesh Prabhakar and Steve Bradley

UOP LLC, A Honeywell Company.
Gasoline Development
25E Algonquin Road, Des Plaines, IL, 60017

Summary

Carbon rejection technologies such as delayed cokers are a popular means to improve the profitability of processing heavy crudes. Given the growing trend to process heavy sour crudes, more refiners are hydrotreating coker naphtha with high contents of sulfur, olefins, nitrogen and Si-compounds.

Reforming catalysts become contaminated with Si if the naphtha hydrotreating (NHT) catalyst is operated beyond its Si capacity. The cyclic trimer and tetramer Si-containing compounds originate from the decomposition of the polydimethylsiloxane de-foaming agent injected in the coke drums to control the foam-over to the coker fractionator. These Si compounds are transferred in the hydrotreated naphtha feeding the reformer unit and will irreversibly poison the reforming catalyst.

Several refiners operating UOP CCR Platforming™ process units started to experience both production and catalyst performance issues that were linked to the heavily Si-contaminated reforming catalysts. We will show that Si-poisoned aged catalysts are less stable, active and selective than fresh reforming catalysts. Si contamination leads to poor chloride retention and poor Pt dispersion. EDXRF (Energy Dispersive X-Ray Fluorescence) and SEM (Scanning Electron Microscopy) analyses of the aged catalyst pills confirmed the extent of Si-contamination and ^{29}Si-MASNMR (Magic Angle Spinning Nuclear Mass Resonance spectrometry) indicated that Si is present as silica, several layers thick anchored to the alumina surface, present as a shell or as islands on the outer edges of the catalyst pills.

Introduction

Carbon rejection technologies such as delayed cokers are a popular means to improve the profitability of processing heavy crudes. Given the growing trend to process heavy sour crudes, more refiners are hydrotreating coker naphtha with high contents of sulfur, olefins, nitrogen and Si-compounds.

This paper focuses on the origins and impact of silicon/silica on the chemistry and performance of reforming catalysts. Reforming catalysts become contaminated with Si if the naphtha hydrotreating (NHT) catalyst is operated beyond its Si breakthrough. The presence of Si in the naphtha feed originates from the decomposition products of the polydimethylsiloxane de-foaming agent (Figure 1) used in the coke drums.

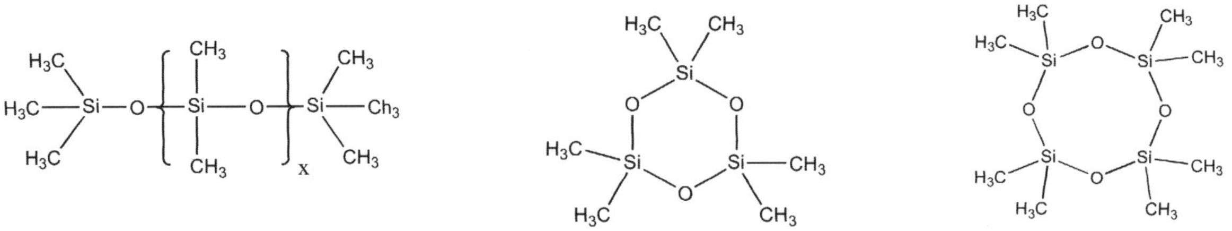

Polydimethylsiloxane (PDMS) anti-foaming agent

Trimer
BP=134oC

Tetramer
BP=175oC

PDMS decomposition products

Figure 1: Polydimethylsiloxane de-Foaming Agent and its Decomposition Products.

The de-foaming agent is injected in the coke drums in order to control the foam-over to the coker fractionator. Under the coker operating conditions, the PDMS de-foaming agent decomposes to the cyclic trimer and tetramer Si-containing compounds and will be carried downstream in the coker products. The amount of silicon in the naphtha leaving the coker is significant, such that it is crucial that the naphtha hydrotreater (NHT) catalysts are Si tolerant. In fact, more than 50% of the Si from the PDMS ends up in the coker naphtha. The NHT catalysts have a maximum total capacity of ~10 wt% Si; however, the Si breakthrough usually occurs when the catalyst reaches ~50% of its Si capacity. If the refiner operates the NHT catalyst beyond its Si breakthrough, the Si-containing compounds transfer in the hydrotreated naphtha to the reforming unit and end up on the reforming catalyst. Figure 2 shows that Si built up on a Platforming CCR catalyst after Si breakthrough on the NHT catalyst.

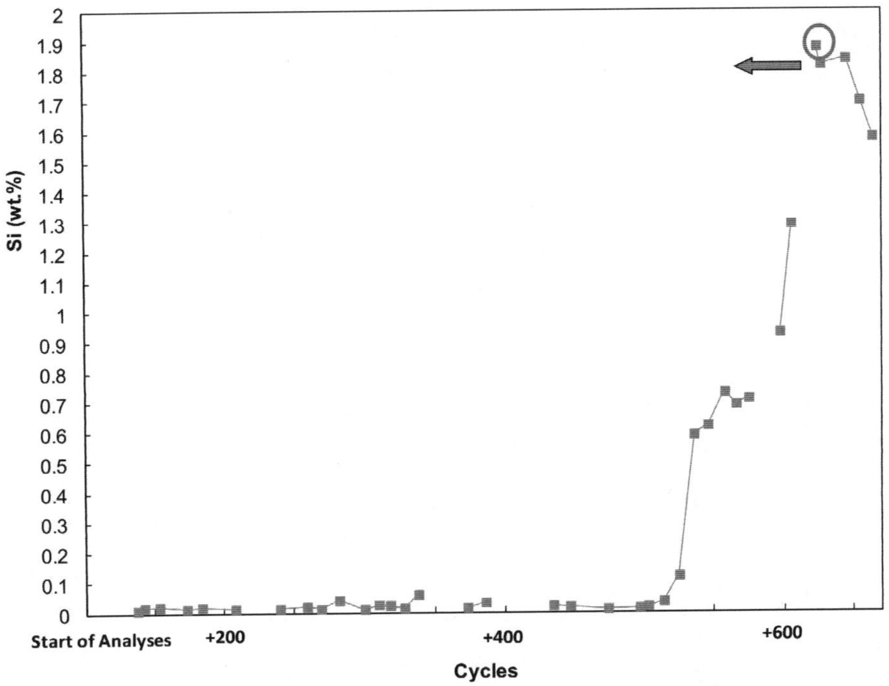

Figure 2: Silicon (wt.%) level on the Platforming CCR catalyst after 500+ cycles.

Experimental

The catalytic performance of an aged Si-contaminated reoxychlorinated/reduced reforming catalyst (catalyst nr. 1 in Table 1) and fresh, reference reforming catalyst (catalyst nr. 2 in Table 1) was compared in pilot plant studies using a paraffinic naphtha feed with EP = 165°C (329F) in an Accelerated Stability Test (AST) done at 75 psig, H_2/HC = 2.1 and constant RON of 103.3. The temperature was increased continuously during the run to maintain constant RON.

Table 1: Selected Properties for the Aged and Fresh Reforming Catalysts

Nr.	Catalyst	Reduced Cl (wt.%)	Pt Dispersion	Si (wt.%)
1.	Aged reforming catalyst, lab. oxychlorinated and reduced	0.7	0.85	1.88
2.	Fresh, reference reforming catalyst	1.0	1.00	~0.01

In order to understand and assess the extent of damage caused by the presence of Si and to elucidate the impact on the chemistry of the catalyst, the samples were also analyzed using advanced analytical techniques such as EDXRF (Energy Dispersive X-Ray Fluorescence) for metal concentrations in the catalyst pills, SEM (Scanning Electron Microscopy) for metal profiles in the catalyst pills, ^{29}Si-MASNMR (Magic Angle Spinning Nuclear Mass Resonance spectrometry) for Si coordination numbers and nearest neighbors identification, and STEM (Scanning Tunneling Electron Microscopy) for metal cluster sizes and chemistry.

Results and Discussions

Figure 3 shows that the aged Si-contaminated catalyst has significantly lower activity than the fresh reference catalyst and is more than 3 times less stable. At 10 bpcf, for example, the Si-contaminated aged catalyst is 21°C less active than the fresh catalyst and the C5+ yield (Figure 4) is much poorer than that of the fresh catalyst.

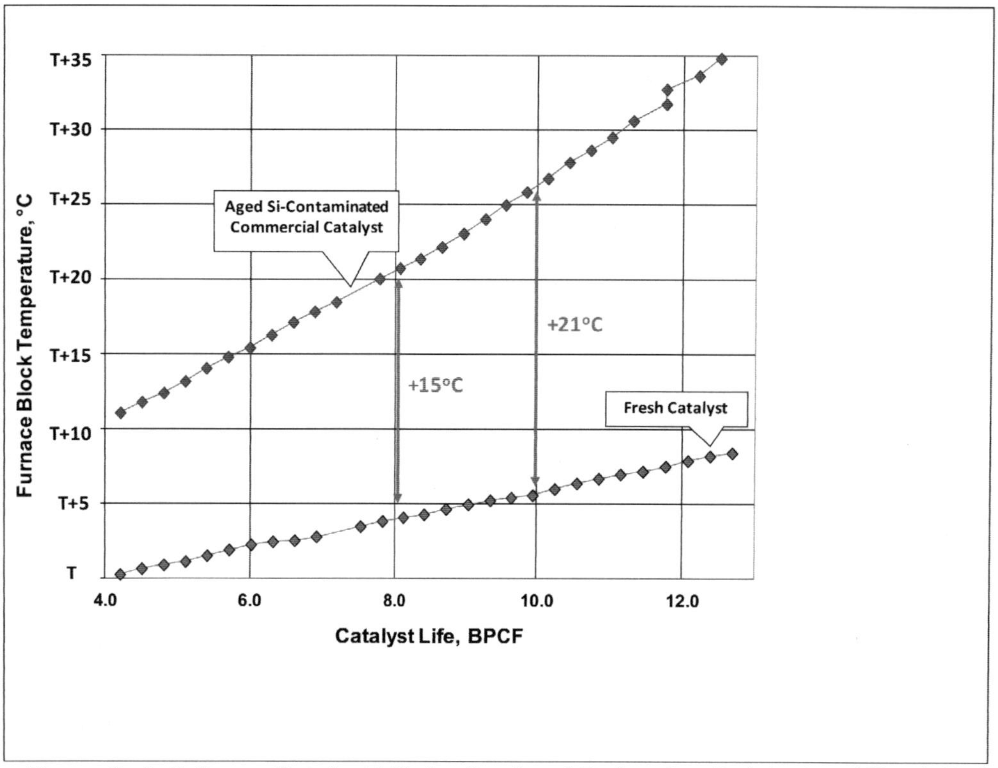

Figure 3: Activity vs. Catalyst Life for the Aged Reforming Catalyst and Fresh Reforming Reference Catalyst

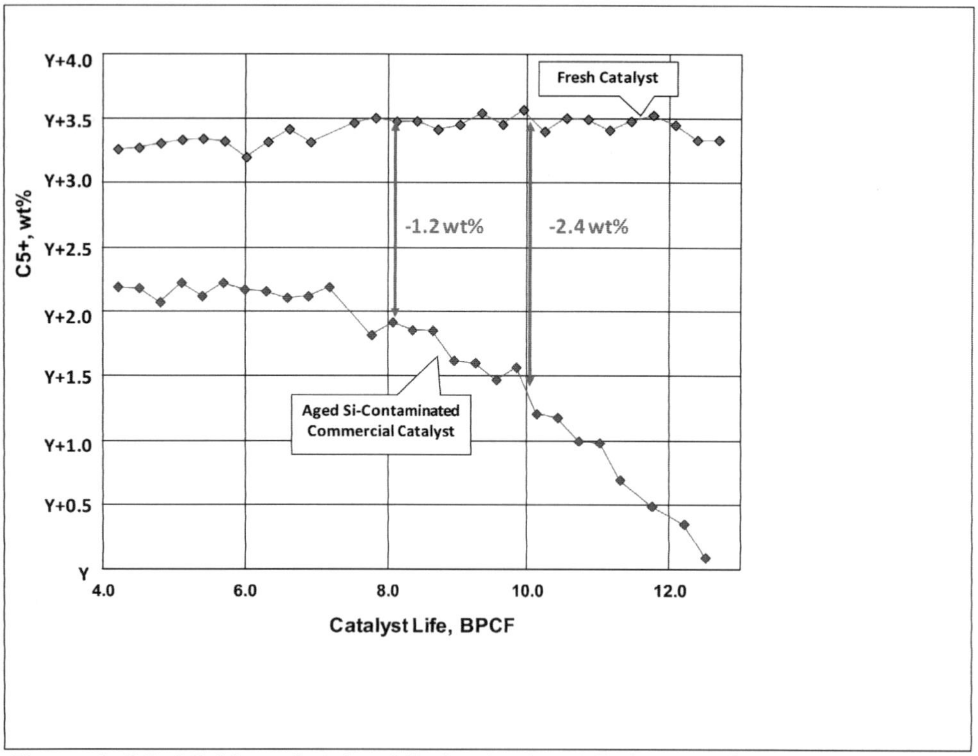

Figure 4: C5+ Yield vs. Catalyst Life for the Aged Reforming Catalyst and Fresh Reforming Reference Catalyst

A summary of the catalytic performance at 10 bpcf (Table 4) indicates the extensive poisoning effect caused by the Si contamination. The inactivity of the Si-contaminated aged catalyst translates into a higher temperature needed to reach the set octane point, which leads to high yields of cracked products and hence low C5+ yields. The high temperature and increased cracking also leads to high coke make. The activity offset is significantly higher than expected for the lower catalyst chloride level.

Table 4: Performance Comparison at 10 BPCF. Aged Si-Contaminated vs. Fresh Reforming Catalysts.

Catalyst	Δ T (°C)	Δ C5+ (wt%)	Δ H2 Yield (wt%)	Δ Coke Make (%)
Fresh Reference Catalyst, 1.0 wt.% Cl	Base	Base	Base	Base
Aged Si-Contaminated Catalyst, 0.7 wt.% Cl	+ 21	- 2.2	- 0.20	+ 44

The Si-contaminated aged catalyst was characterized by EDXRF 96 pills elemental analysis and 6 pills SEM metal profiles. The EDXRF analysis detected Si on every pill analyzed with a significant variation in the Si level across the population. The 6 pills SEM metal analysis clearly showed the presence of distinct profiles for Si and Cl. Keeping in mind that the weight % values reported are semi-quantitative as standards were not prepared for the EDXRF analysis, there is still a fairly good agreement between the EDXRF and SEM results.

Figure 5 shows the Si concentration on the outer shells of the 96 aged catalyst pills analyzed, while Figure 6 shows the Si metal profile via SEM analysis. Both analyses agree on the extent of Si contamination. As it can be seen from Figure 6, the edges of catalyst pills are highly contaminated with Si, with the concentration varying between 5-12 wt% and decreasing continuously from the outer layer towards the center of the pills. It is apparent that the Si contamination is ~100 micron deep forming a shell around the catalyst pills.

Figure 7 shows the lack of chloride on the edges of the catalyst pills ~100 micron deep, corresponding to the same pill depth highly contaminated with Si. Silicon (converted to silica as evidenced by ^{29}Si-MASNMR data) anchors strongly and irreversibly on the alumina surface via the $[OH]^-$ sites which play a key role in the ability of the catalyst to retain Cl^-. Since the terminal alumina $[OH]^-$ groups are tied up by the silicon/silica deposited on the alumina surface and since most probably SiCl4 is not formed, the catalyst's 100 μm outer shell is completely depleted of Cl-.

The poor Cl- retention ability of the aged catalyst caused mainly by the high Si concentration causes an inefficient Pt re-dispersion during the regeneration step. The decreased Pt dispersion as the Si contamination increased is evident in Figure 9.

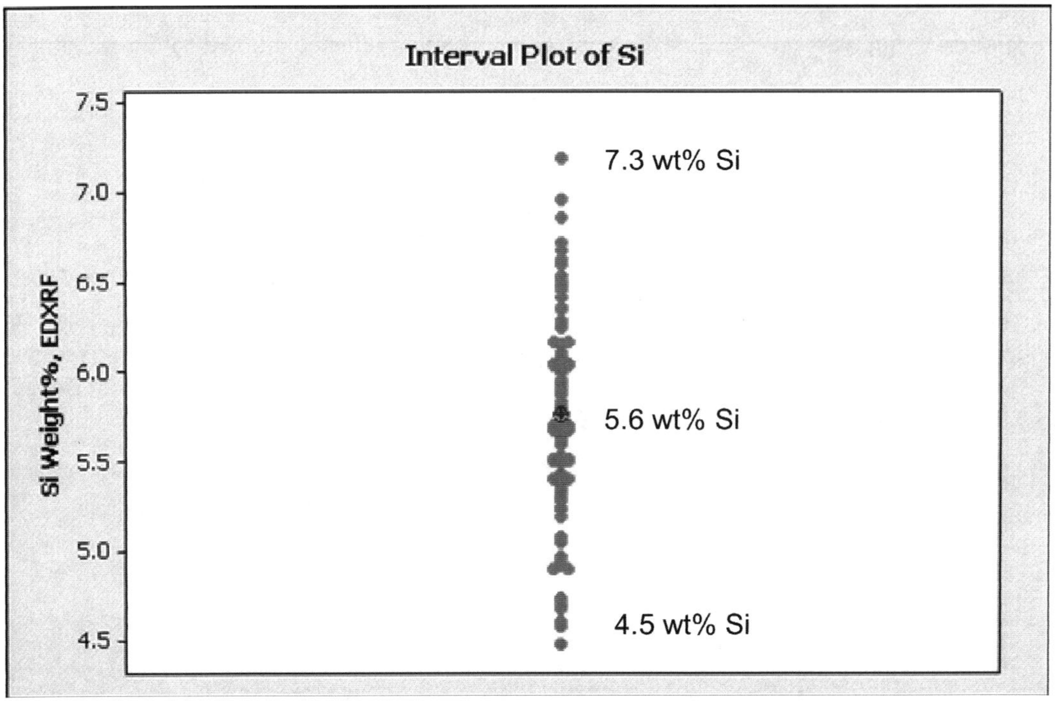

Figure 5: EDXRF Si concentration in the outer shell of the Si-contaminated aged catalyst pills. 96 catalyst pills were analyzed.

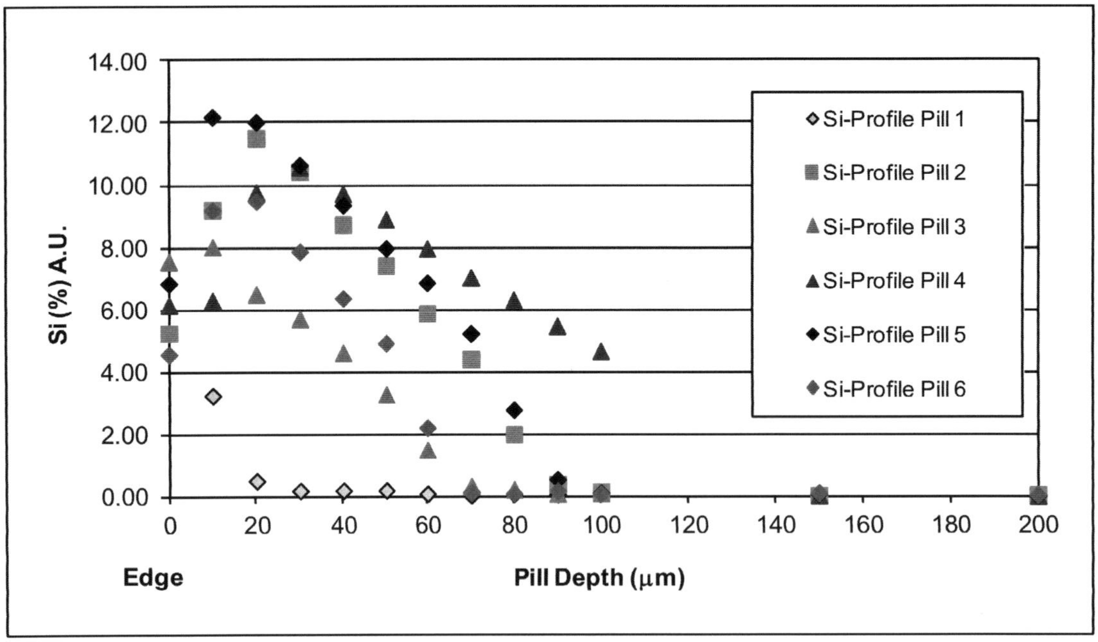

Figure 6: SEM Si profile in the Si-contaminated aged catalyst pills. 6 catalyst pills were analyzed.

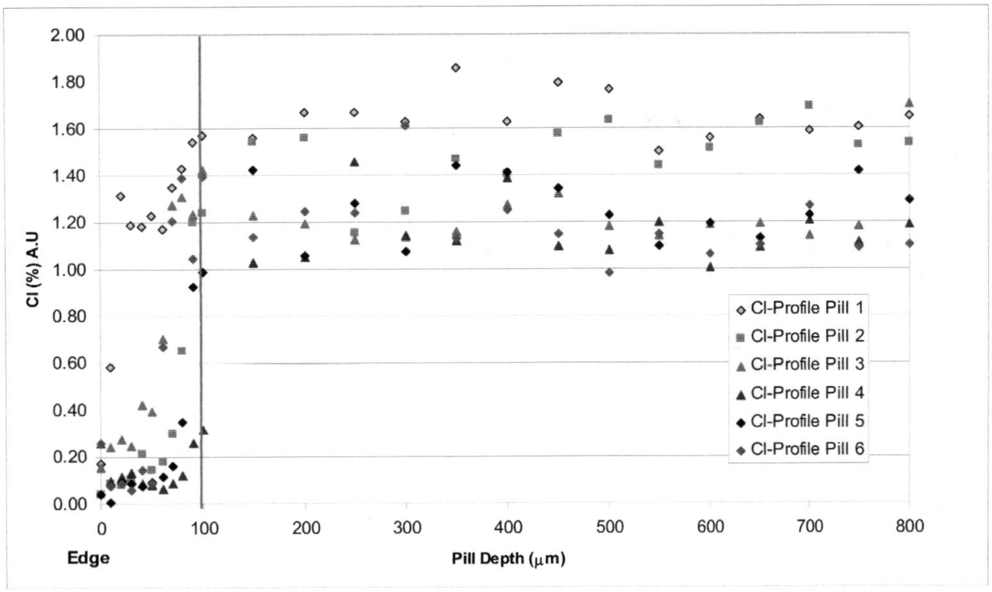

Figure 7: SEM Cl- Profile in the Si-contaminated aged catalyst pills. 6 catalyst pills analyzed. Cl levels were not calibrated to absolute levels; trends are relative.

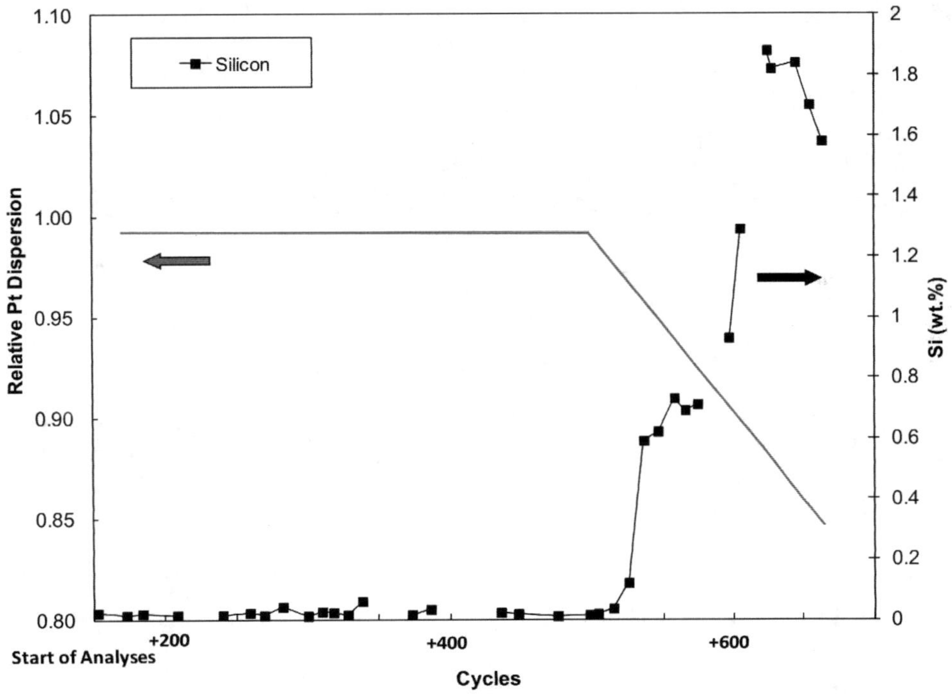

Figure 9: Pt dispersion for the aged Si-contaminated catalyst.

The ^{29}Si-MAS-NMR analysis of the aged Si-contaminated catalyst showed evidence of silicon with 4, 3, and 2 bridging oxygen atoms (Figure 10) which suggests that there could be several layers (as evidenced by the compound with 4 bridging oxygens) of silica anchored to the alumina surface, most probably present as islands mainly on the edges of the catalyst pills. Based on SEM metal profiles, Si forms a shell ~100 μm deep.

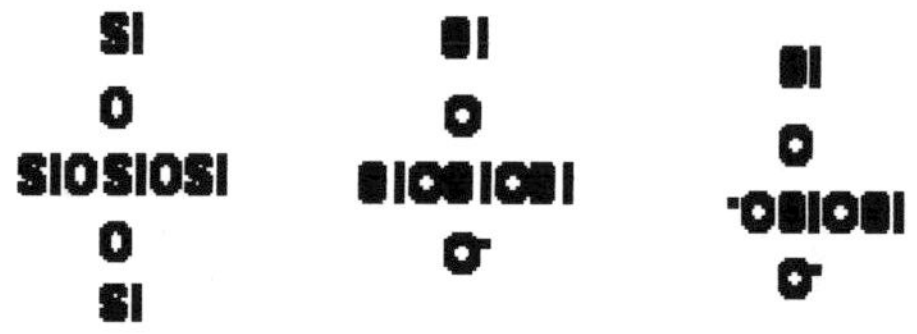

Figure 10: Silicon Atoms with 4, 3, and 2 Bridging Oxygen Atoms

The ^{29}Si-MAS-NMR data showed no evidence of Si-C bonds, such that the compound pictured in Figure 11 postulated to be formed during the silicon deposition on a hydrotreating catalyst must decompose to SiO_2 on a reforming catalyst during the regeneration cycle.

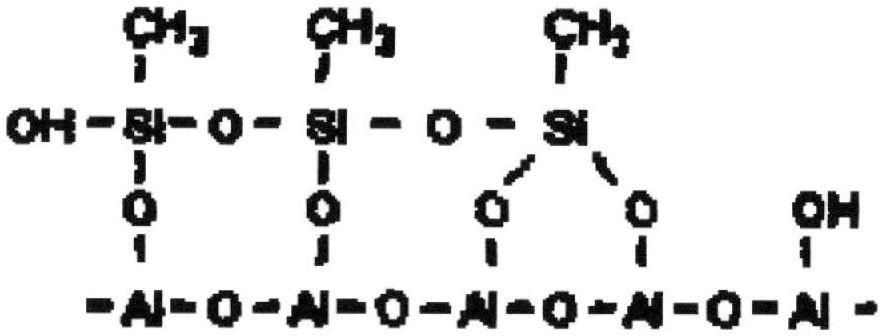

Figure 11: Compound Formed during the Silicon Deposition on a Hydrotreating Catalyst

The scenario for the Si contamination mechanism is proposed in Figure 12. The existence of the Si atoms present in the final proposed structure and labeled in red is confirmed by the NMR data. This scenario is also consistent with the fact that silicon/silica depletes substantially the $[OH]^-$ concentration from the surface of the alumina support, leading thus to a much poorer chloride retention ability by a Si-contaminated catalyst and to poor Pt dispersion.

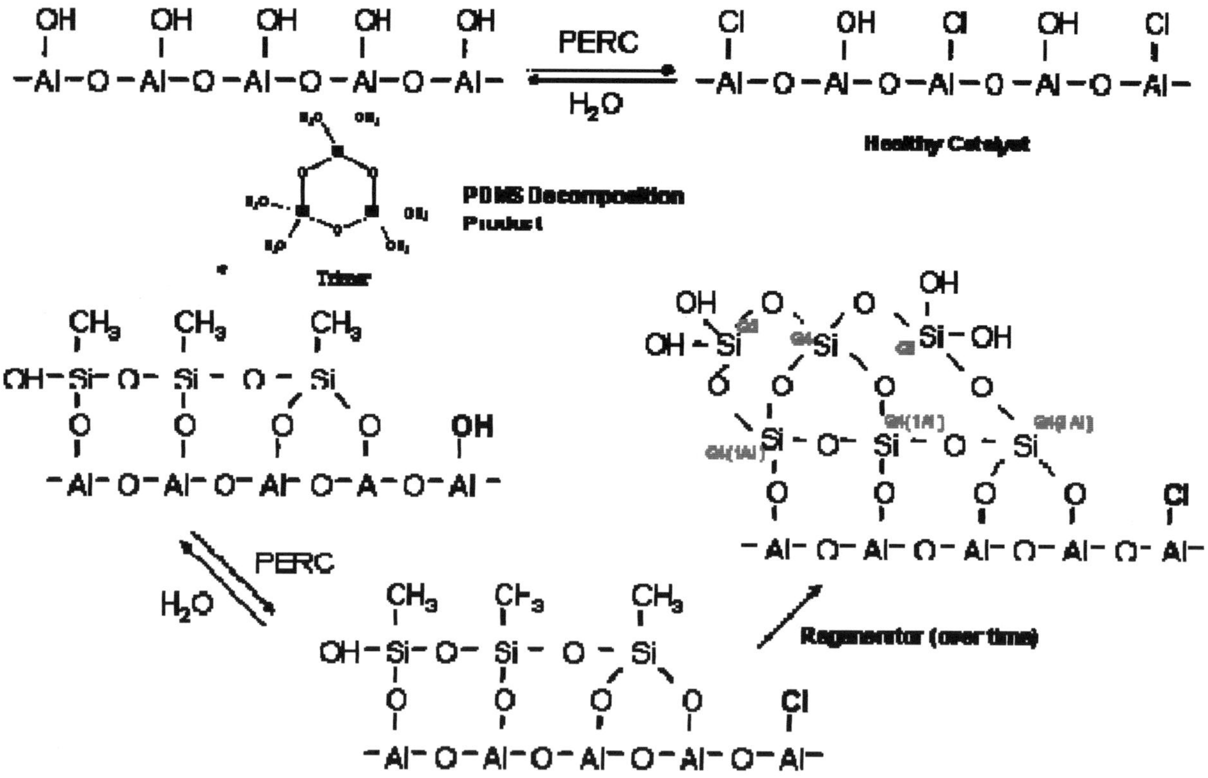

Figure 12: Proposed mechanism for Si poisoning of a reforming alumina catalyst at high levels of contamination.

Conclusions

The catalytic performance of the aged Si-contaminated reforming catalyst was significantly worse than that of the fresh reference reforming catalyst. At 10 bpcf, the aged catalyst was 21°C less active, deactivated 3 times faster, and had 2.4 wt.% lower C5+ yield. The low activity of the aged catalyst is mainly linked to the high level of Si poisoning. The presence of Si hinders the chloride retention ability. Aggressive chloriding treatments are needed to maintain the required Cl- wt%. High temperatures of operation are required to reach the target octane. The performance of a reforming catalyst will continuously decline as the Si content increases. The Si poisoning effect is irreparable and irreversible.

The EDXRF and SEM analyses confirmed the extent of Si-contamination suffered by the aged catalyst. The edges of the catalyst pills are highly contaminated with Si, with the concentration varying between 5-12 wt% and decreasing continuously from the outer layer towards the center of the pills. ^{29}Si-MASNMR analysis indicated that Si is present as silica, several layers thick anchored to the alumina surface, present as a shell or as islands on the outer edges of the catalyst pills. The thickness of the silica shell is ~100 micron deep around the catalyst pills.

Acknowledgments

The authors would like to acknowledge the UOP colleagues from the Pilot Plant and Analytical groups.

References

Franck, J-P., Martino, G.P., *Deactivation of Reforming Catalysts.*
Murthy, et al., *Structure and Performance of Reforming Catalysts.*

STRUCTURAL CHARACTERIZATION OF LANTHANIDE PROMOTED SULFATED ZIRCONIA CATALYSTS

Dana K. Sullivan, Shelly D. Kelly, Ally S.Y. Chan, George E. Mickelson, Simon R. Bare, and Kurt Vanden Bussche

UOP LLC, a Honeywell Company, Des Plaines, IL, USA

Abstract

Increasing the octane value of light naphtha streams through isomerization is accomplished commercially by using a range of catalysts including chlorided alumina, sulfated zirconia, and zeolites. Sulfated zirconia (SZ) catalysts are stable, can be used at low temperature, and have activities significantly higher than zeolitic catalysts but are less active than chlorided alumina catalysts. The activity of sulfated zirconia can be enhanced with the addition of first row transition metal cations such Fe and Mn. It has been shown that these promoters are in part incorporated into the zirconia lattice, leading to stabilization of the tetragonal (or cubic) ZrO_2 structure.

It is well known that the addition of lanthanides into zirconia affects many properties including cation diffusion, defect ordering, and oxygen diffusion. In this paper the structural characterization of a Tm-promoted SZ catalyst (as representative lanthanide) is presented with the aim of determining the structural promotion of the lanthanide. EXAFS data are presented, together with synchrotron XRD, XANES, and XPS data.

Introduction

Isomerization of straight chain C5 – C6 alkanes to their branched isomers is of great industrial importance. By forming the branched products the octane value of the molecule is greatly enhanced, increasing their value in the hydrocarbon pool for gasoline blending.

The desired skeletal isomerization reaction is difficult as alkanes are relatively unreactive molecules, and a strong acid is needed. Over the years three classes of solid acid have been developed that are used in commercial catalysts ranging from zeolites to amorphous chlorided alumina (Figure 1). Sulfated zirconia is intermediate in acid strength between chlorided alumina and zeolites, and its activity can approach that of chlorided alumina. UOP was the first company to introduce a sulfated zirconia-based isomerization catalyst into the marketplace in 1996.

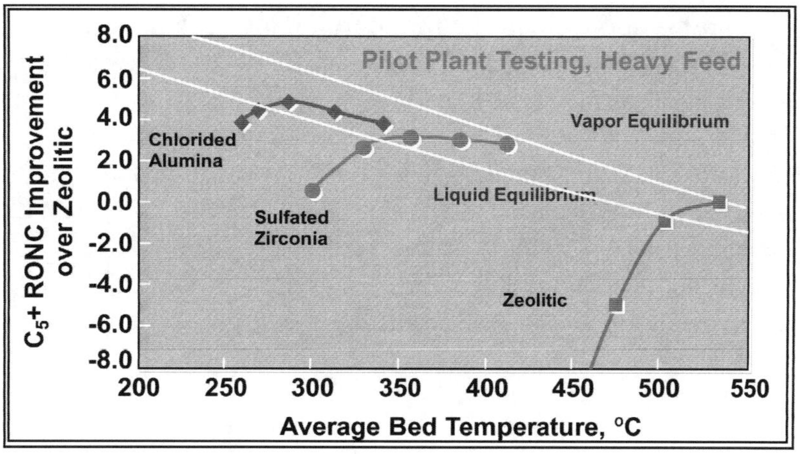

Figure 1. Isomerization Catalyst Types

Sulfated Zirconia Catalysts

Addition of Promoters

The base activity of sulfated zirconia (SZ) can be improved by the addition of promoters. For example in 1990 Sun Refining patented a modified catalyst containing Mn and Fe as promoters which boosted its activity over the unpromoted catalyst (1). Much work has been published as to the origin of the promotion effect of the Mn and Fe, and while there is still debate in the literature, it is believed that both promoters are incorporated into the zirconia lattice leading to a stabilization of the tetragonal phase (2-3). These promoters thus create oxygen vacancies which, in cooperation with the sulfate groups, produce a site or sites responsible for the increased isomerization activity.

It is also well-known that the addition of lanthanide elements into zirconia affects many properties including cation diffusion (4), defect ordering (5), and oxygen diffusion (6). It was therefore of interest to prepare sulfated zirconia catalysts with added lanthanide elements and to investigate the effect of the promoter on the paraffin isomerization activity.

An evaluation of lanthanide additives showed that addition of thulium improved the conversion at a given temperature. This paper focuses on the characterization of a 3.6 wt% thulium promoted sulfated zirconia catalyst, which had a higher activity than a non-promoted SZ, or SZ samples with 1.4 wt% or 5.5 wt% Tm added.

Zirconia Phases

At standard pressure zirconia is known to exhibit the monoclinic, tetragonal, or cubic phase with increasing temperature. At room temperature only the monoclinic phase stable but the tetragonal phase can be stabilized by the addition of cationic dopants, in addition to other factors including the small particle size (7), increased strain (8), and the Zr:O stoichiometry (9).

During the preparation of sulfated zirconia catalysts mixtures of monoclinic and tetragonal phases may be obtained, but with the addition of sulfate the tetragonal phase usually dominates (10-12). Given the complexity of the phase diagram it is of interest to be able to quantitatively determine the relative amount of each phase in the catalyst. However, it is not straightforward to use X-ray Diffraction (XRD) to distinguish the amount of cubic and tetragonal phases. Additional advanced characterization methods were therefore employed.

The distribution of atoms about Zr in the monoclinic, tetragonal, and cubic phase of zirconia are shown as a function of radial distance in Figure 2. The Zr distribution is shown on the left in green and the oxygen distribution is shown on the right in red. The red circles in the ball and stick models of the structures represent the distances to the oxygen atoms in the first coordination shell and illustrate that the coordination of oxygen about Zr is distinct for each phase. The monoclinic structure has three shells, close in distance. The tetragonal and cubic phases have two shells, but they differ in the number of oxygen atoms. The tetragonal phase is equally split into 4 and 4 while the cubic phase is split into 6 and 2. Extended X-ray absorption fine structure (EXAFS), a local atomic probe, can be used to distinguish these phases, as will be shown here. There is also similarity in the Zr distribution for the tetragonal and cubic phases.

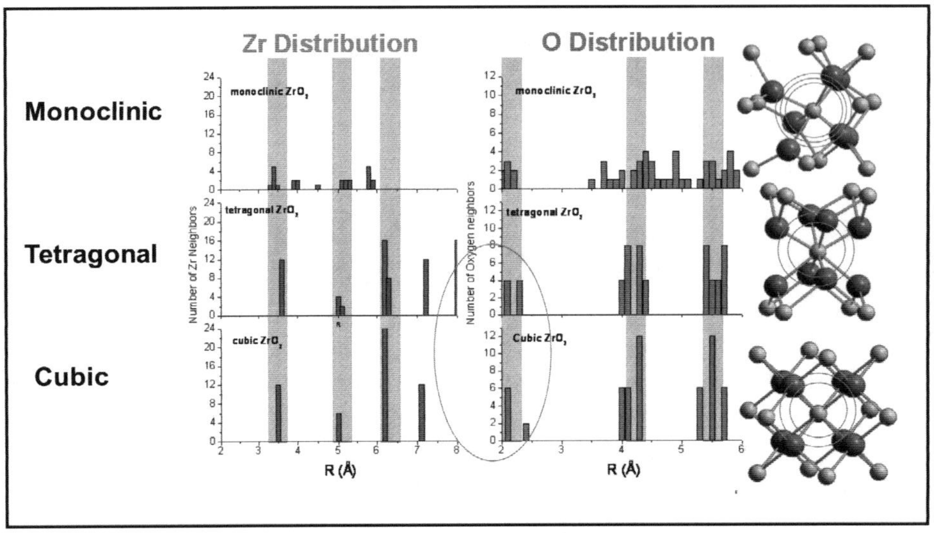

Figure 2. Radial Distribution of Atoms about Zr

EXAFS Modeling

Extended X-ray Absorption Fine Structure (EXAFS) is a short range local atomic coordination probe. As EXAFS spectra can be independently recorded from each element in a sample separate Zr and Tm EXAFS data were collected from the same sample. The two combined datasets provide a more complete analysis of the sample.

As an illustration of EXAFS modeling, the magnitude of the Fourier transform (FT) of the Tm L_3-edge EXAFS data and EXAFS model for bulk thulium(III) oxide, Tm_2O_3, are shown in Figure 3. The EXAFS data are shown as black symbols and the results of the model as the solid red line. The first peak in the FT at approximately 2Å corresponds to the first shell oxygen atoms (red in the ball and stick inset). The second peak at approximately 3.5 Å is from the six Tm atoms in the next nearest shell of atoms (blue in the ball and stick diagram).

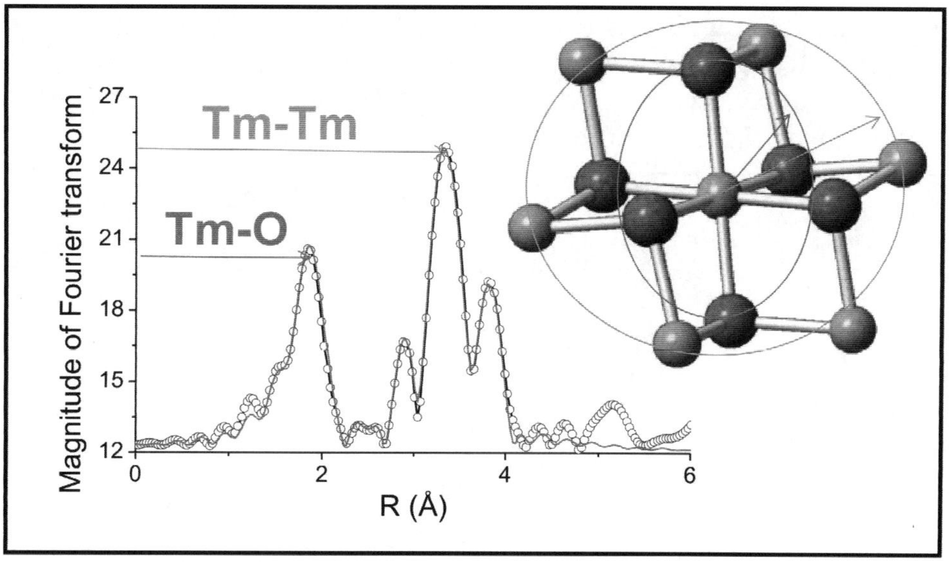

Figure 3. Magnitude of the FT of the Tm L_3-edge EXAFS of Tm_2O_3

Characterization of Thulium Promoted Sulfated Zirconia

Zr and Tm EXAFS

The Zr K-edge EXAFS spectrum of the calcined 3.6wt% thulium promoted sulfated zirconia sample (Tm/SZ) is shown by the symbols and the model is shown as the solid line in Figure 4. The model is based on the tetragonal structure of zirconium oxide as shown by the ball and stick model. The excellent agreement between the model and the data all the way out to 7Å is strong evidence of tetragonal ZrO_2 being the predominant phase. This indicates that the Tm is likely playing a role in stabilizing the tetragonal phase at room temperature. As was shown previously one major difference between the cubic and tetragonal phases is in the first oxygen shell. The tetragonal phase has four oxygen atoms at 2.1Å and four oxygen atoms at 2.4Å.

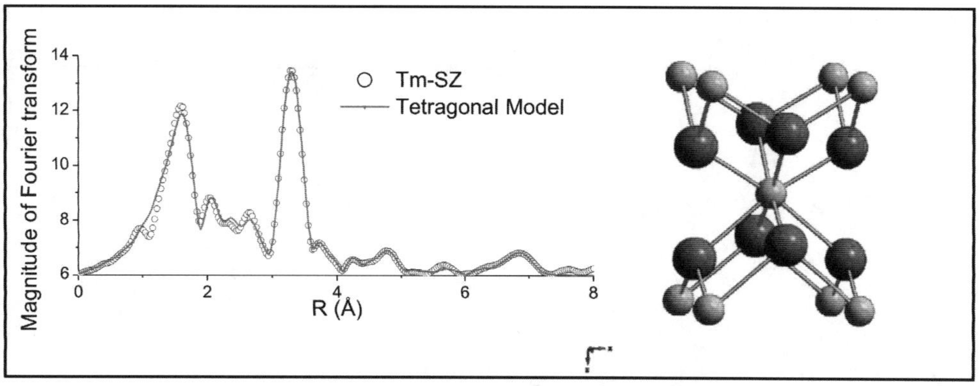

Figure 4. Zr K-edge EXAFS Spectrum of Tm/SZ (symbols) Modeled with Tetragonal ZrO_2 (red line)

Figure 5 shows a comparison of the magnitude of the FT of Zr K-edge EXAFS (black line), that has already been modeled as tetragonal ZrO_2, and the magnitude of the FT of the Tm L_3-edge EXAFS (blue line). There is a the large shift in the radial distance of the first peak in the FT, assigned to the Zr-O first shell oxygen signal, as shown by the red arrows in Fig. 5, but the peaks in the FT assigned to the scattering from the cations are at similar distances, as shown by the blue arrows. If the Tm directly substitutes for Zr into the tetragonal ZrO_2 lattice, then the EXAFS signals from both Zr and Tm would be similar, as they would be occupying the same lattice positions but they are clearly different, especially with regards to the first oxygen shell.

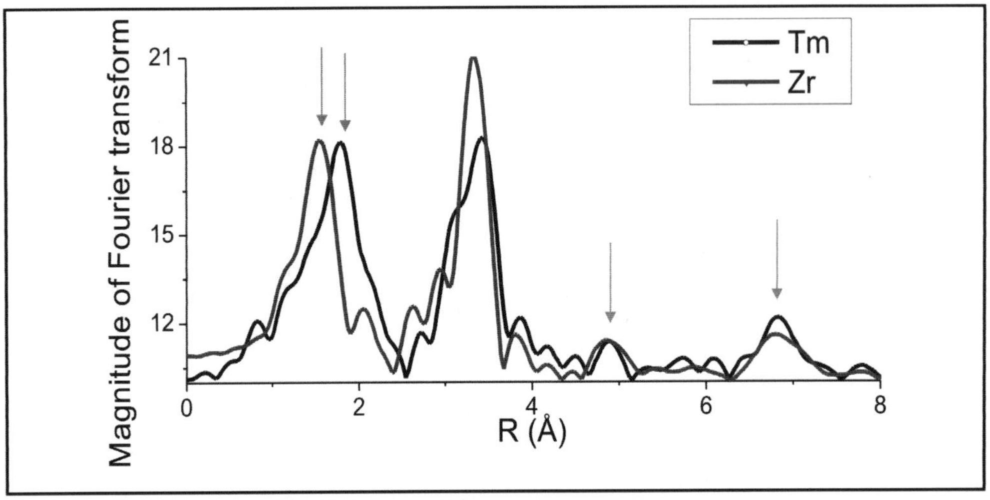

Figure 5. Comparison of magnitude of the FT of the Tm L3-edge (black) and Zr K-edge (blue) EXAFS from Tm/SZ

As a first attempt to interpret the measured Tm EXAFS spectra, the same tetragonal ZrO_2 model that was shown to correctly model the Zr data was applied to the Tm EXAFS data. The fitting was constrained to a tetragonal unit cell and the unit cell parameters were optimized by the model fit to the measured spectrum. The model is able to reproduce the main features of the measured spectrum but resulted in extreme distortion to the tetragonal unit cell. The distorted unit cell has similar a, b, and c parameters of about 4.0Å. This resulting structure from this model is similar to cubic ZrO_2. Thus, a cubic ZrO_2 structure model was applied to the data.

Figure 6 shows the resulting Tm EXAFS data (symbols) and the cubic unit cell model (red line). The inset shows a ball and stick model of the best fit cubic structure. There is excellent agreement between the data and the model out to 7Å, implying strong evidence of a cubic-like structure about the Tm in the calcined Tm/SZ sample.

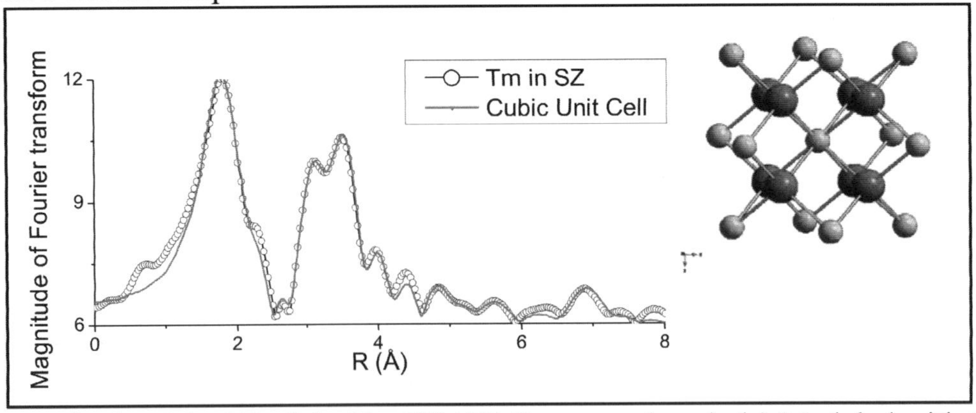

Figure 6. Magnitude of the FT of the Tm EXAFS Spectrum (symbols) Modeled with Cubic ZrO_2 Structure (red line)

The question then is how to reconcile these data: the local structure surrounding the zirconia is tetragonal while that around the Tm is cubic. The proposed structure is to have the cubic unit cells of Tm form a superlattice within the tetragonal zirconia. The cubic Tm unit cells are constrained to face share, forming rows of Tm atoms within the tetragonal zirconia. Based on this proposed structure, that is consistent with the measured EXAFS spectra for both Tm and Zr in the samples, each Tm atom stabilizes 30 Zr atoms. The 3.6wt% Tm sample studied has one Tm atom for every 35 Zr atoms which is in excellent agreement with the proposed structure.

A possible reason for the fact that the Tm can be present as cubic (as a super lattice) within the tetragonal ZrO_2 structure is because of the similarity in the Zr cation distribution for both the cubic and tetragonal structures, as shown in Figure 7.

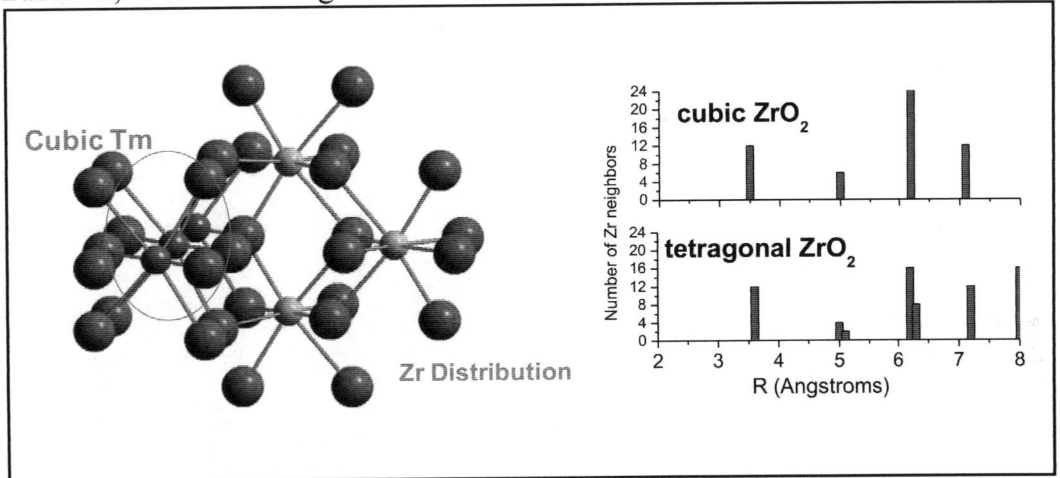

Figure 7. Cation distribution in the tetragonal and cubic structures, and the proposed superlattice of Tm (blue) in the tetragonal zirconia (green)

Synchrotron XRD

As previously mentioned it can be difficult using standard laboratory XRD methods to reliably determine the phase composition of the zirconia for the sulfated zirconia catalysts. However, if the XRD data are recorded with increased resolution at a synchrotron (by minimizing the instrumental broadening) then the quality of the fitting is improved. Such data were collected for the thulium-promoted sulfated zirconia catalyst sample. The resulting data show that a model using both tetragonal and cubic zirconia results in a statistically better fit to the diffraction data than a tetragonal-only model, validating the EXAFS-derived model discussed earlier.

Sulfur K-edge XANES

Important information regarding the local structure of the sulfate group present in the catalyst can be obtained using sulfur K-edge X-ray Absorption Near Edge Structure (XANES).

Figure 9 shows a comparison of the sulfur XANES of aqueous sulfate to that from the in situ dried Tm-promoted sulfated zirconia catalyst. The spectra are quite different, and can be interpreted considering the local symmetry of the sulfate group. In aqueous sulfate the sulfate group has tetrahedral symmetry (T_d), and the resultant transition from the 1s to 2p orbitals results in a narrow white line. If the sulfate group is distorted, lowering the symmetry, (e.g. to C_{3v}), then the final state will be split resulting in a broader white line, as observed. Thus, we conclude that the local geometry of the sulfate group on the catalyst is not simply tetrahedral but is distorted.

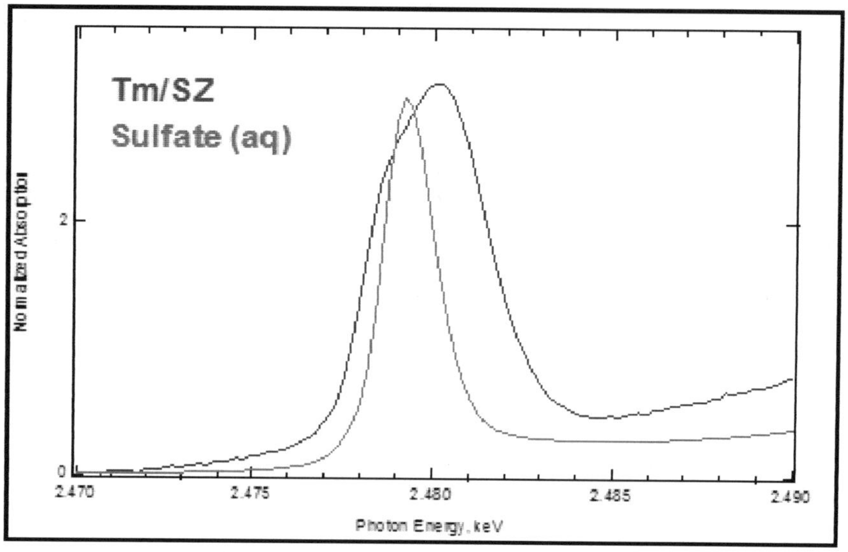

Figure 8. Sulfur K-edge XANES of Tm/SZ and aqueous sulfate

XPS

The binding energy of the Tm 4d peak in XPS suggests that the Tm is in the Tm^{3+} oxidation state, as shown in Figure 9. There is, however, a small shift of ~0.6 eV to higher binding energy in the Tm 4d peak, compared to bulk Tm_2O_3. This is consistent with the oxidized Tm in SZ being present in a more dispersed form than bulk-like crystallites of Tm_2O_3.

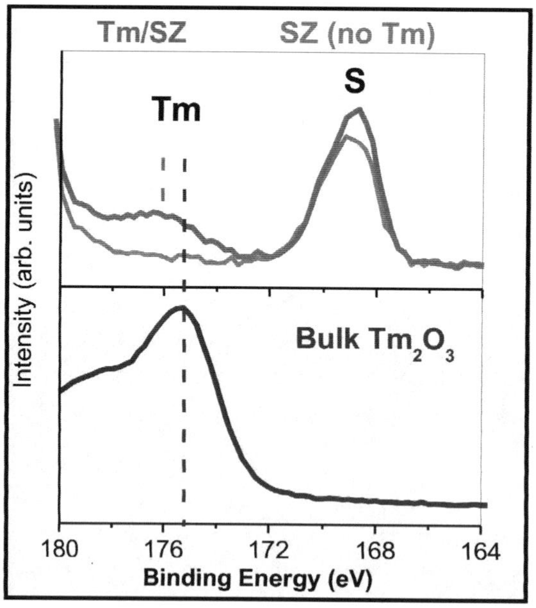

Figure 9. Tm 4d XPS spectra of the Tm/SZ sample compared to SZ and Tm_2O_3

Argon ion sputter depth profiling combined with XPS of the Tm/SZ provided information on the distribution of the sulfur and thulium in the catalyst. The intensity of the S 2p peak decreases dramatically after sputter-removal of ~2 nm (approx. 5 atomic layers) of the Tm/SZ catalyst surface, as shown in Figure 10. This indicates that the sulfate groups are primarily localized at the surface of the zirconia and not distributed evenly in the bulk. At the same time, the intensity of the Tm 4d peak remains essentially constant, implying that the Tm is distributed throughout the bulk of the zirconia.

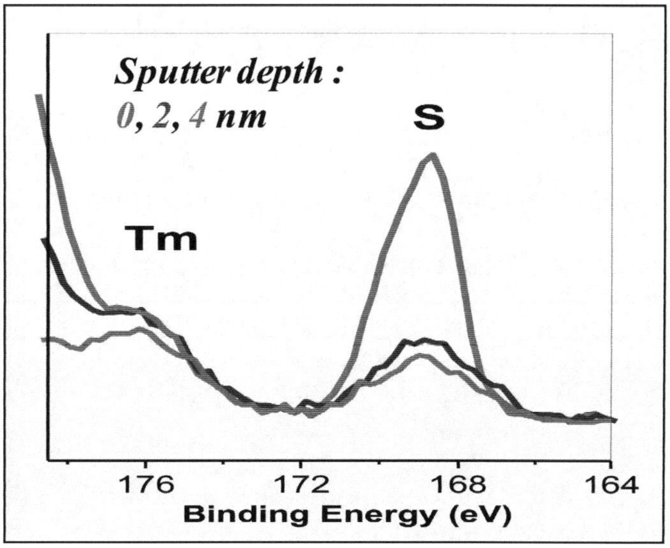

Figure 10. Sulfur 2p and Tm 4d XPS spectra before and after argon ion sputtering

Summary

A combination of characterization techniques (EXAFS, XANES, synchrotron XRD, and XPS) have been used to provide insight regarding the promoting effect of the addition of thulium to a sulfated zirconia paraffin isomerization catalyst. The data show that the sulfate groups are primarily on the surface of the zirconia while the thulium is in the +3 oxidation state and is distributed throughout the bulk. The sulfate groups are distorted from tetrahedral symmetry, likely as a result of the bonding with the zirconia surface. The local coordination environment around the thulium is cubic, whilst that around the zirconium is tetragonal. A model consistent with these data is proposed where the thulium is present as a superlattice within the tetragonal zirconia lattice. Clearly the addition of a +3 promoter cation into the zirconia lattice must create vacancies to compensate the charge but the techniques used in this study were not sensitive to the presence of the defects. It is likely that the improved catalytic performance is a result of complex interplay between the structural modifications, which affect the bonding of the sulfate group on the zirconia surface, and thus its acidic properties.

References

1. E.J. Hollstein, et al.; U.S. Patent 4,918,041; assigned to Sun Refining and Marketing Company (1990)

2. C.Y. Hsu, C.R. Heimbuch, C.T. Armes, and B.C. Gates, J. Chem. Soc. Chem. Comm. 1645 (1992).

3. F.C. Jentoft, A. Hahn, J. Krohnert, G. Lorenz., R.E. Jentoft, T. Ressler, U. Wild, R. Schlogl, C. Hassner, and K. Kohler, J. Catal., 124, 224 (2004)

4. M.A. Taylor, C. Argirusis, M. Kilo, G. Borchardt, K.D. Luther, and W. Assmus, *Solid State Ionics* 51, 173 (2004).

5. A. Bogicevic, C. Wolverton, G.M. Crosbie, and E.B. Stechel, *Phys. Rev. B* 6401 (2001).

6. R. Krishnamurthy, and B.W. Sheldon, *J. Am. Ceram. Soc.* 1549, 88 (2005).

7. R. C. Gravie, J. Phys. Chem., 82, 218 (1978).

8. H.-C. Wang and K.-L. Lin, J. Mater. Sci., 26, 2501 (1991).

9. B. Djuricic, S. Pickering, D.McGarry, P. Glaude, P. Tambuyser, and K. Schuster, Ceram. Int. 21, 195 (1995).

10. J.A. Navio, M. Macias, C. Real, and G. Colon, Mater. Lett., 20, 345 (1994).

11. R. Srinivasan, B.H. Davis, O. Burl Cavin, and C.R. Hubbard, J. Am. Ceram. Soc., 75, 1217 (1992).

12. M.A. Ecormier, K. Wilson, and A.F. lee, J. Catal., 215, 57 (2003).

MOLECULAR MODELING AND NEW NAPHTHA REFORMING CATALYSTS

Nancy Iwamoto, Molecular Modeling, Honeywell International, Ramona, CA
Gregory Gajda, R&D Catalysis, Honeywell UOP, Des Plaines, IL
Mark Lapinski, Catalyst Research & Development, Honeywell UOP, Des Plaines, IL

Abstract

Continued research and development in catalytic naphtha reforming has led to new catalysts that increase reformate, hydrogen and aromatic yields for the transportation and petrochemical industries. In order to make the next significant improvement in yield selectivity, advanced characterization and molecular modeling tools have been employed to understand the current catalyst systems and to design improved catalysts.

The presentation will discuss how molecular modeling is being used to understand the coordination and bonding of aluminum, hydrogen and chloride as a function of alumina crystal face, water and HCl addition. The partial charge of the hydrogen atom can be either positive or negative depending upon the local interaction conditions which will impact the hydrocarbon reactions at these sites. Work is in progress to investigate the effect of aluminum defect sites, impurities and other elements on the alumina surface.

Introduction

Naphtha reforming catalysts catalyze reactions such as hydrogenation/dehydrogenation, dehydrocyclization, paraffin and ring isomerization, hydrocracking, hydrogenolysis and dealkylation. The catalyst system consists of a noble metal on an alumina carrier in which the alumina acidity is promoted with Cl. The role of the metal is to first dehydrogenate the paraffins to olefins which then undergo cyclization reactions on the acid function to form naphthenes. The naphthenes undergo a series of dehydrogenation and ring isomerization reactions to create the desired aromatic and H_2 products. The metal and acid functions need to be balanced to provide high paraffin selectivity to aromatics with minimal cracking and dealkylation to the less desired C_1 to C_4 products.

One method to study the selectivity of the acid function of the chlorided alumina support is to test model compounds such as 1-heptene which forms naphthenes and cracked products. Recent testing with 1-heptene (Figure 1) showed that the selectivity to naphthenes increased with increasing Cl up to about 0.5 wt%, peaked over the 0.5-1 wt% Cl range, and then decreased above about 1.2 wt% Cl. The Lewis acidity of the chlorided alumina was then characterized with pyridine IR. The Lewis acidity was found to steadily increase to about 1.1 wt% Cl and then stayed approximately constant up to 1.4 wt% Cl. Since these and other acidity characterization methods did not explain the 1-heptene selectivity trends, fundamental modeling studies were employed to examine the nature of acidity based upon Al coordination, charge distribution and chloride population of the alumina.

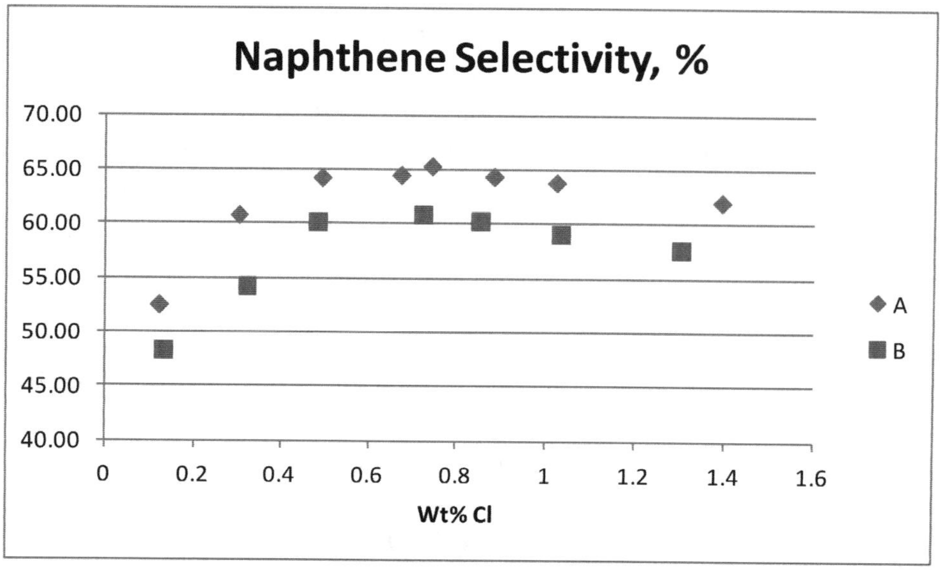

Figure 1: Selectivity for conversion of 1-heptene to naphthenes at constant temperature (space velocity was varied to maintain constant conversion) for two gamma alumina catalysts at different chloride levels.

Molecular Model Basis

In order to establish general charge and bonding trends, surface models were set-up using primarily the [100] alumina cut in order to represent γ alumina as described by Knozinger[1]. The model of the [100] cut was represented by an initial oxide (stoichiometric Al_2O_3), and hydration to represent approximately 10-11 OH/nm^2 which is similar to experimental estimates of OH content. Chloride was added to the surface by addition of HCl to the surface. The initial models made use of cleaved unit cells with a two unit cell depth ($Al_{24}O_{36}$ stoichiometry for the oxide). The bottom half of the unit cell was fixed in order to increase the computation speed.

Because of the bottom atom constraints, only the top surface of the unit cell (the "free" surface) was used to determine trends. Since the small Al_{24} models could only represent a lower limit of 1.3% Cl, later models increased the size of the unit cell (ranging from Al_{48} to Al_{92}) so that chloride levels below 1% could be studied.

All models made use of the DMol3 program using Materials Studio 5.5 from Accelrys[2]. All models were geometry optimized using a general gradient approximation in the Beck-Perdue (GGA/BP) functional with a double numerical polarization (DNP) basis set (basis file 3.5); thermal smearing was applied. When converged, each model was further used in a population analysis in which the relative Hirshfeld partial charges were calculated in order to obtain an indication of charge distribution at the surface. It is cautioned that the partial charge calculation was meant as a trend comparison only and does not reflect actual charges in the system. The surface aluminum atoms were analyzed for both coordination and partial charge.

Modeling Results and Discussion

The OH Surface -- The current models are distinctly different than those reported by Sohlberg[1] which use full unit cells. Both show that H_2O can easily be incorporated in the material, although Sohlberg postulates bulk H incorporation, whereas the current models concentrate on surface OH formation.

It was found that the surface coordination of the oxide form (i.e., no surface OH) averaged 5.13, but the coordination changed with OH introduction. The type of surface coordination expected from Digne[3] was 5.5 corresponding roughly to 25% tetrahedral coordination. However, in the cut case that was used here, an average coordination of 5 was found. Addition of non-bonded water to the smaller models indicated that a coordination of 5.5 could be achieved only if a surface layer of water was present. If all the water was removed, the coordination dropped to 5.25. These model experiments suggested that measured coordination will be dependent upon the hydration state of the material, which then brings into question the true surface state of the material at high process temperatures.

The partial Al charge also changed with OH introduction to the surface. In general, the partial charge started at around 0.5 but dropped with the introduction of OH to the surface. The charge was as low as 0.42 depending upon the cell size and whether any additional water was added to the surface.

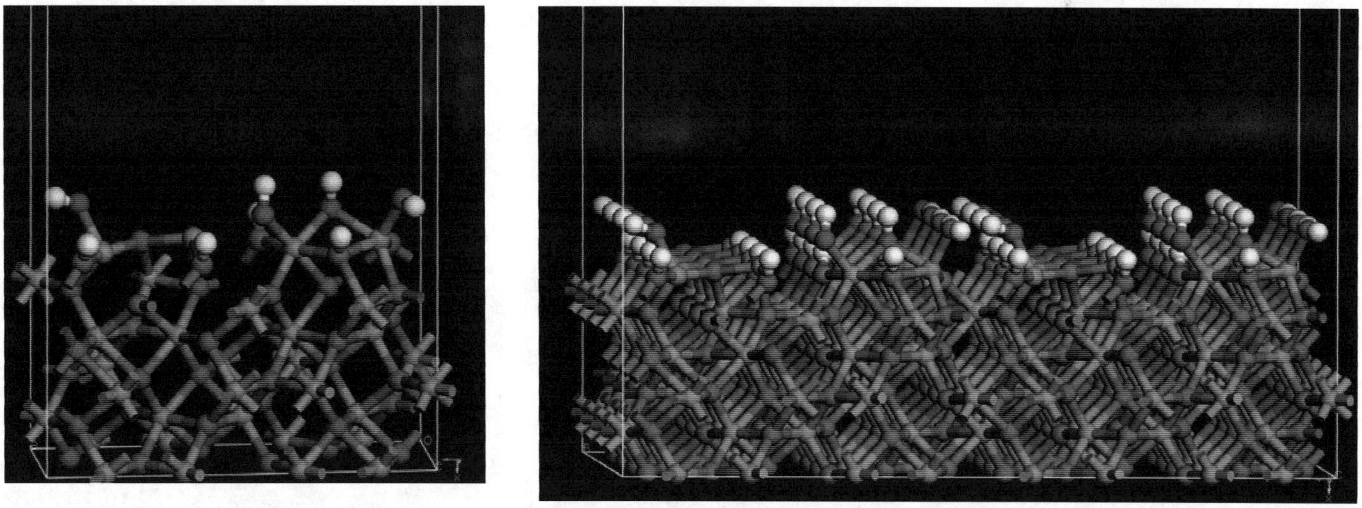

Figure 2: Example unit cells showing OH surface. Left: smaller Al_{24} example; Right: large surface Al_{92} example.

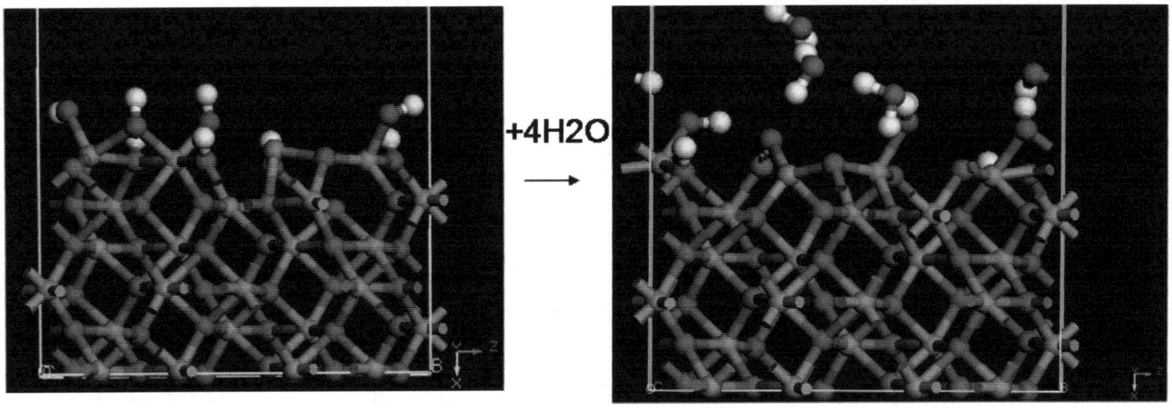

Figure 3: OH Surface with water layer.

Charge and Coordination Trends with Cl Addition -- Upon HCl addition to the OH surface, not every HCl molecule reacted to the surface; however, the surface charges did change depending upon the amount of HCl added. The data was combined for all the surfaces calculated for the [100] cut.

It is important to note that the model size did change the Al charge trends slightly. The smaller Al_{24} models before addition of 1 HCl had a partial charge of 0.437. Upon addition of 1 HCl to the surface, the charge increased to 0.457 (+5%). One of the largest models, using a 1x2x4 replicate of the smaller model (Figure 4), started with an average Al charge of 0.429. Upon addition of 1 HCl, the charge increased to 0.435 (only +1.5%), whether or not the HCl bonded to the surface. In both cases, a rise in acidity of the Al center (or increase in the partial positive charge) was found. Because the smaller cell represented a larger surface density of HCl, a larger impact was found. This may not be surprising, but this demonstrates that the large models are necessary in order to discern trends at the low chloride content found in actual catalysts.

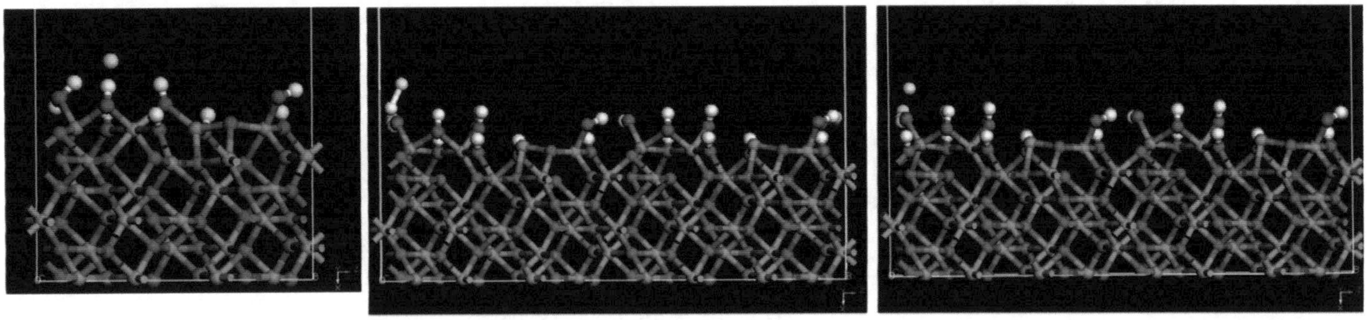

Figure 4: Addition of 1 HCl to a Al_{24} model (far left) and Al_{92} models (center: HCl placed on surface, right: HCl bonded)

Figure 5 shows that overall, the Al partial charge initially decreased with Cl, then increased, mirroring the coordination trends with increase and then decrease (Figure 6). In general, the coordination increased but then reached a maximum at greater than 5. This inverse trend of coordination and charge is in agreement with the general supposition that if the surface starts from a low coordination state (such as the dehydrated oxide state), and coordination increases with process, then the charges should also decrease as the Al center becomes coordinate saturated (more like the full coordinate bulk structure). At some point the coordination reaches it's maximum and the charges begin to increase again. As with the Al charges, the coordination reaches its optimum somewhere above an HCl/nm^2 level of about 0.5 (where the HCl/nm^2 level is roughly similar to the weight percent of Cl).

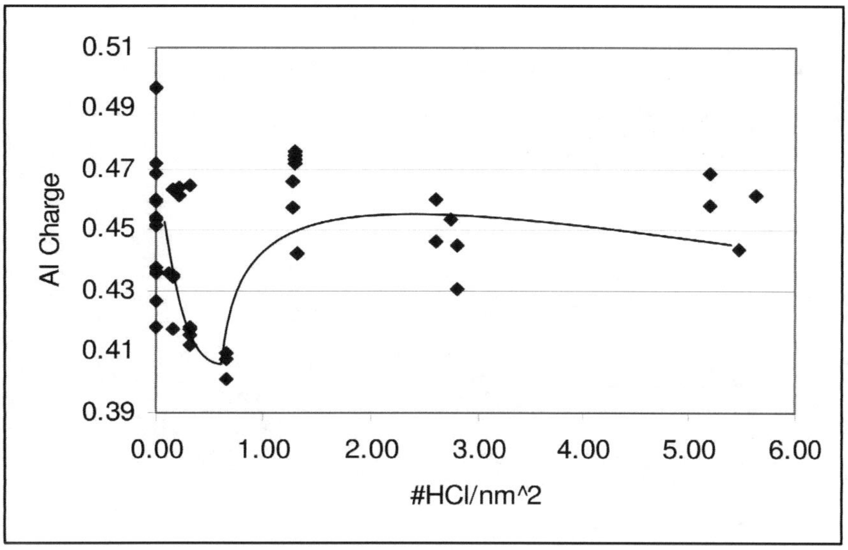

Figure 5: Al charge trends with HCl

The most interesting comparison comes when looking at the energy changes and comparing them to the charge and coordination. The energy trends for both the small and large models appear to generally increase and plateau with HCl introduction suggesting that there is a destabilization occurring. The large scale models (Figure 7, right) at Al_{48} and above, however, show a slight optimization of the energies at low HCl concentration (~0.3 HCl/nm^2) before the energies increase. This suggests that early HCl introduction is favorable. At these levels, the coordination is also increasing, moving toward the optimum coordination. The charges then start to increase, i.e. the acidity is increasing. This is consistent with the experimental observation of the more severe conditions required to remove chloride from a chlorided alumina to a chloride concentration of 0.2 wt% or lower.

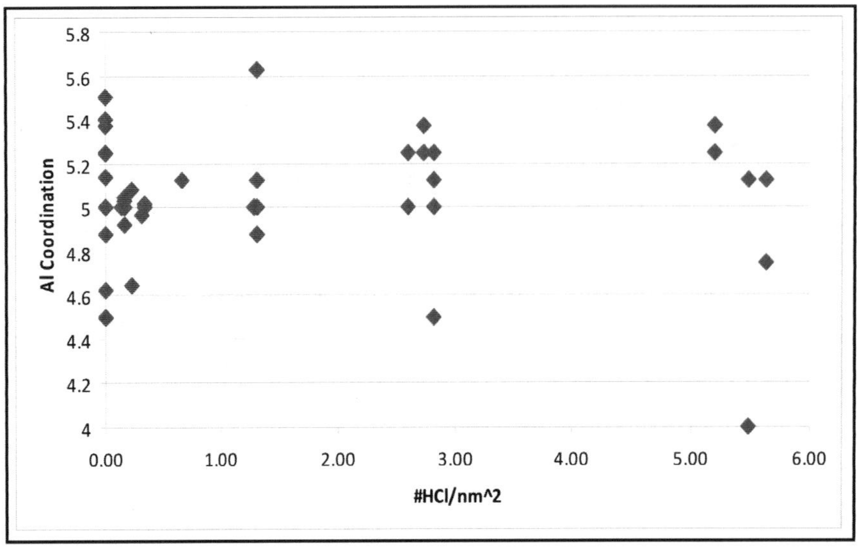

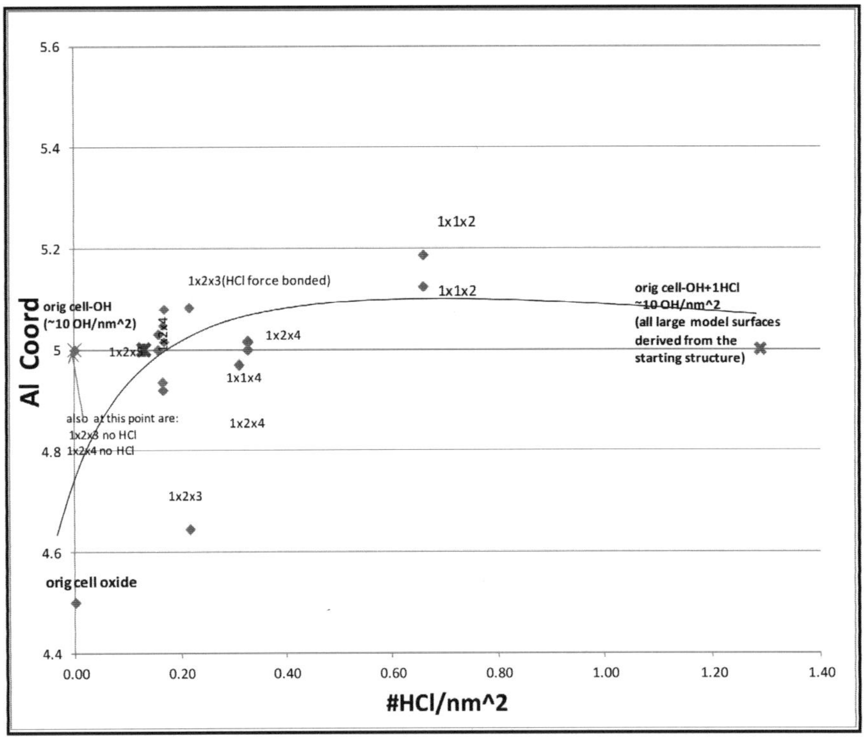

Figure 6: Al coordination trends. Top: all data using Al_{24} to Al_{92}; Bottom: large models with some small model Al_{24} comparisons added at 0 HCl axis and highest 1.3 HCl/nm^2 point at right to show distribution effects.

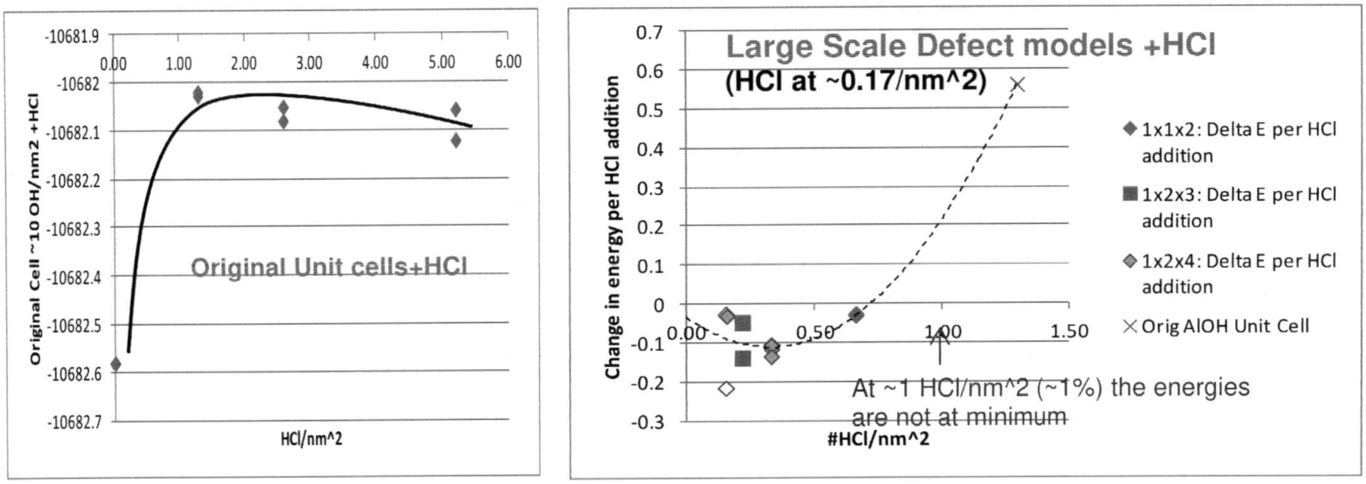

Figure 7: Energy trends of the models.
(Original cell sizes are represented in Figure 2 left, and Figure 3)

One additional interesting trend can be seen with the H partial charges in Figures 8 and 9. In general, the proton charge derived from HCl increases with HCl introduction. However, if the data are plotted by coordination, both positive and negative charges are derived at the coordination number of 5. This is the point where the system energies are already high. Normal positive H charges are resumed when the coordination is higher than 5, and the energy has plateaued indicating some coordination stabilization.

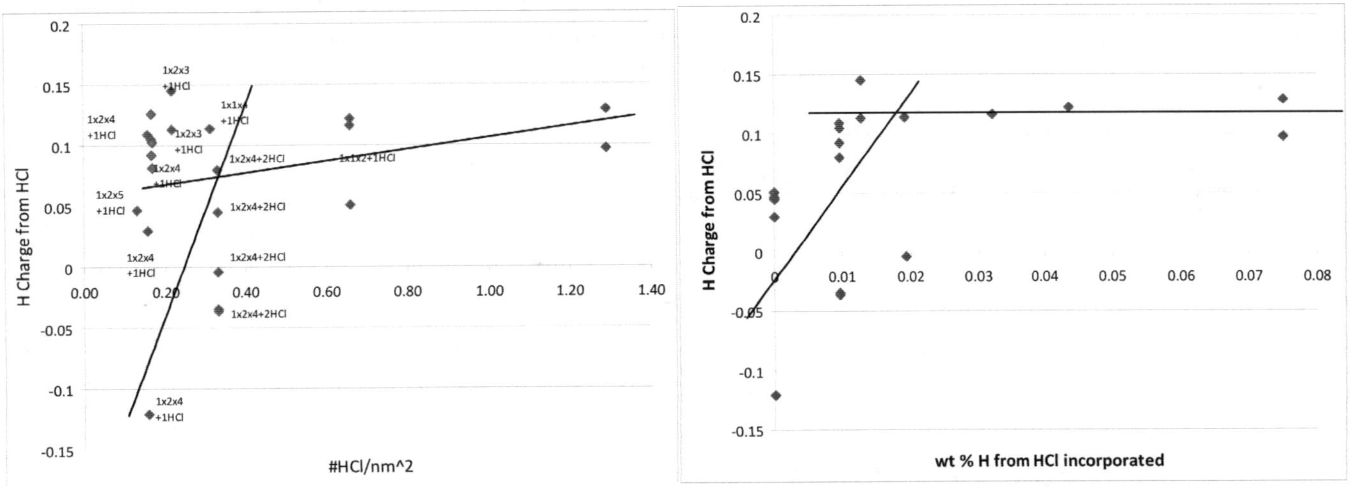

Figure 8: Partial charge of H center derived from HCl.
(red-filled points represent surfaces with surface defects, 1Al removed)

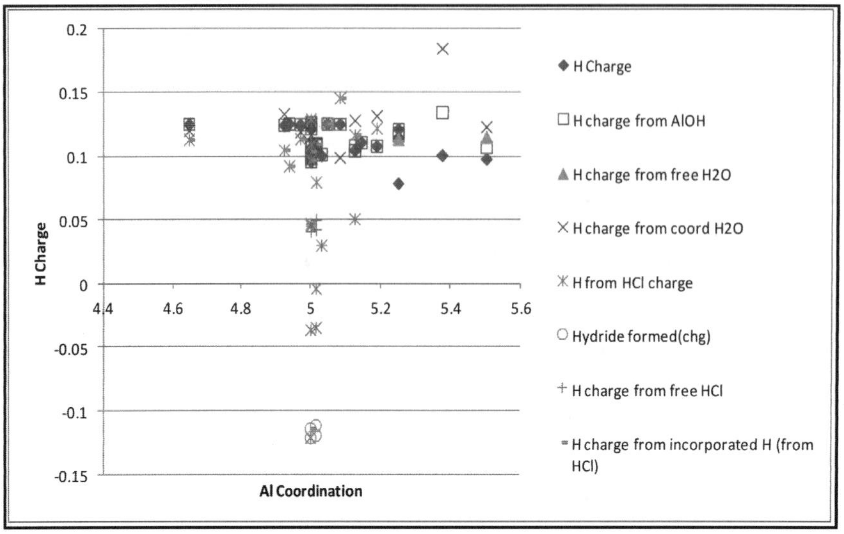

Figure 9: Partial charge of H center. All H centers plotted by Al coordination.

Summary and Conclusions

The molecular modeling demonstrated that larger surface models are needed to understand the underlying coordination and charges at the low chloride levels normally used in catalysis.

At low Cl introduction, the acidity increases because the Al charges are increasing and the coordination is maximized. The Al center is coordination deficient and the energy change (upon HCl addition) is also decreasing suggesting stabilization. This region becomes a sweet spot for Al-Cl incorporation stability, but becomes an undesirable region for acidity since a small increase in Cl content leads to a large change in charge.

Above a general region of ~0.4 HCl/nm^2, the energy change increases again, suggesting that the system is destabilizing which infers higher reactivity. At the same time, the average Al coordination as well as the average Al charge increase toward their maximum.

The coordination and charge appear to optimize in the range between 0.5 and 1 HCl/nm^2. Before maximum coordination, the hydrogen can form either protons or hydrides, suggesting a reactive surface forms at just the correct conditions. These interpretations suggest that the Al center is very sensitive to the coordination environment. Therefore, the catalyst activity is also very sensitive to the Cl reacted to the surface.

References

[1] H. Knozinger and P. Ratnasamy, "Catalytic Aluminas: Surface Models and Characterization of Surface Sites", Catal. Rev. Sci. Eng 17(1), 31-70(1978).

[2] Accerlys, Inc. San Diego.

[3] Karl Sohlberg, Stephen J. Pennycook and Sokrates T. Pantelides, "Hydrogen and the structure of the Transition Aluminas", JACS 1999, 121, 74930-7499.

[4] M. Digne, P.Sautet, P. Raybaud, P. Euzen and H. Toulhoat, "Use of DFT to Achieve a Rational Understanding of Acid-base Properties of γ–Alumina Surfaces", J. Catal. 226 (2004) 54-68.

Disclaimer

Innovative Hydrocavitation Technology Process to Create Composite Fuels

Harold N. Knickle[1], Oleg Kravchenko[2], Suvorova Iryna[2] and Igor Baranov[2], (1)Chemical Engineering, University of Rhode Island, Kingston, RI knickle@egr.uri.edu, (2)Podgorny Institute, Ukraine

Keywords: composite fuels, liquid fuels, cavitation, combustion

Goal of the Project

The primary goal of the project is the creation of composite fuels for combustion processes containing biomass, coal, coal residues, crude oil processing residues, and other combustible material for effective firing in power plants

The principal technological approach is hydrodynamic cavitation application to hydrocarbon fuels. The components of composite fuels are initially mixed and then processed in special cavitation devices where homogenization, reactions and partial changing of fractional compositions of the end product are realized. The hydrocavitation transformation development for composite fuels (including biofuels) ultra-dispersed spray and homogenization is an effective and inexpensive method to prepare waste materials for combustion

Because these composite fuels did not work well with current furnace spray nozzles, improved combustion nozzles were developed.

Existing Technologies

Other technologies to make use of biomass, coal, coal residues, crude oil processing residues, and other combustible material include:

Pyrolysis: This technology is good for fuel oil formation but is expensive from an energy consumption point of view. Biogas production: The long term duration of biochemical reactions, energy expenses current control for reactor conditions, formatted in reactor residues as the result of biogas generation can be considered as disadvantages (although the residues can be utilized like good fertilizer). Biodiesel: Biodiesel can be a substitute for diesel fuel. Biodiesel needs an expensive biomass (raps for example). The production of biodiesel is expensive and there is substantial waste from its production. Bioethanol: Bioethanol can be substituted for the use of gasoline. But bioethanol needs an expensive biomass (sugar, corn) and in production a great deal of energy is lost.

New Process: PHCR Process – Introduction

An innovative hydrocavitation technology for creation and effective burning of composite liquid fuels of new types based on technology of a new principle of cavitation transformations obtained by a cascade aggregate of cavitators with different functional performance capabilities. During the process of hydro-cavitation, liquid composite fuel emulsions and slurries are obtained and activation of physical and chemical transformations of hydrocarbons is observed. Solid waste of coal production, waste from the oil-refining industry, tank-washes, waste from production of different oils, different wastes of organic and inorganic origin, peat, silt, biomass, and wood can be used in this process.

There should be great interest in developing and exploiting this new reactor process called the pump-homogenizer-cavitator-reactor or PHCR for short. This process has the potential to enhance fuels by increasing their octane or cetane number and to formulate composite fuels. The process is also applicable to oil cleanup in ships or tanks and the commercial use of the recovered oil. The process will provide an important contribution to environmental cleanup and recycling of oils. The objective is to help bring this reactor to near commercialization. This process can be of importance to the United States gas and oil industry as well as environmental firms.

PHCR Process

• Pump: The pump moves the fluid through the recycle pipe and back to the PHCR and builds pressure and temperature in the system.

• Homogenizer: This part of the PHCR causes mixing of the fuel and subsequent enhancement of the homogenization of the fuel.

• Cavitation: The introduction of cavitation in a small volume is an important feature of the PHCR. The introduction of liquid water in the mix contributes to the success of the PHCR. A small cavitation space focuses energy on a water bubble and breaks its bonds introducing Hydrogen in the PHCR.

• Reactor: The Hydrogen reacts with the hydrocarbons increasing the quality of the fuel as measured by the octane or cetane number. Concurrently the viscosity is reduced. In addition the sulfur bonds can appear in the discard stream.

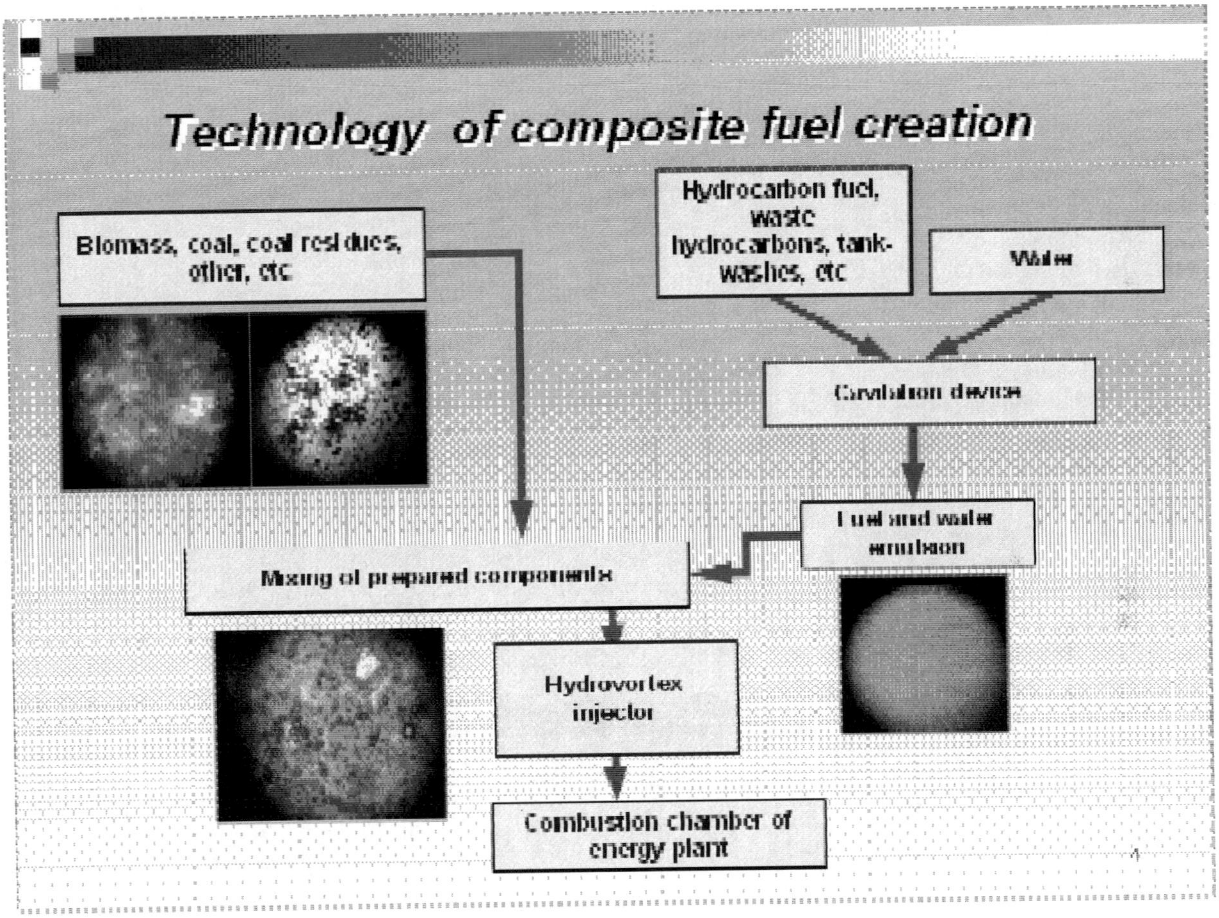

Figure 1 Flow Schematic of Process for Creating Composite Fuels

Composite Fuels Examples

Example 1

• Input: 70% Waste Heavy Oil, 15% Diesel Fuel, 15% Water

• *Apply the Cavitation Process*

• Result: Form of Diesel Fuel with No Free Water Observed

Example 2

• Input: Bitumen Tar, and Water

• **(bitumen** and asphalt are mostly interchangeable words and is a mixture of organic liquids that are highly viscous, black and sticky, composed primarily of highly condensed polycyclic aromatic hydrocarbons.)

• *Apply the Cavitation Process*

• Gross Heat of Combustion 133,577 Btu/gal

• **(Compared to #2 Heating Oil 137,000 Btu/gal)**

• Sediment and Water 0.05 vol%

• Asphaltenes 3.23 wt%

• Ash 0.021 wt%

• Kinematic Viscosity at 50C 6.972 cSt

(Compared to #2 Heating Oil of 3+cSt)

Example 3

• Input: 20% Ground Rubber Tires and 80% #6 Fuel Oil

• *Apply the Cavitation Process*

• Results: Clean Paste with No Trace of Hard Rubber

• Combustion of Paste is good.

Problem: Reduce Viscosity with additive?

Devices used in PHCR Process

Three key components of the process include a ***ROTOR-PULSATION DISPERSER*** for producing medium and high-viscosity emulsions and producing suspensions with an option of adding up to 30 % of a not very abrasive solid phase, a ***ROTOR CAVITATOR*** for producing low and medium-viscosity emulsions and a ***RECESSES-TYPE CAVITATOR*** for producing suspensions with an option of adding up to 70 % of a solid phase.

Figure 2 Experimental equipment used for creation of composite fuels based on rotor-pulsation disperser.

Figure 3 Experimental device for emulsion fuel creation – ROTOR CAVITATOR

Figure 4 Experimental Device for suspension - composite fuel creator – RECESSES CAVITATOR

Experiments to Determine Fuel Combustion Quality

Experiments to determine combustion quality included three key studies. The first process included water-carbon-black oil suspension droplets, in which pulverized coal is activated by model activator substances such as; NH_4NO_3, $NaNO_3$, and NH_4ClO_4. A second examined brown coal with the use of activator substances. A third study involved the estimation of activators influence on pre-flaming processes is carried out. The possibility of complete combustion increase of coal conglomerate is shown. A products combustion balance of activation slurries has been determined and the decrease of some substances in products combustion balances have been calculated.

Hydro-Vortex Fuel Injector

Tests of hydro-vortex injectors on the composite fuel in boiler combustion of composite slurry fuels with use of standard types of injectors is impossible since the presence of a dispersed condensed phase in the fuel results in rapid blocking of the fuel pipelines and calibrated orifices used for composite slurry fuels combustion. This is a distinct challenging problem of using suspension composite fuels for effective combustion because typical burner nozzles are very sensitive to fuel purity. Thus high fuel viscosity and the presence of a solid fraction is the cause of low spraying quality and rapid nozzle contamination. This issue was addressed by developing a methodology of designing hydro-vortex nozzles whose operation is based on the principle of vortex dispersion and mixing. These nozzles allow for a solid phase and form an air-fuel mixture. Based on R-functions-based computational research on hydrodynamic flows in the flowing parts of - nozzles, laboratory specimens of such nozzles were made. They demonstrated high effectiveness, even for very viscous fuels and fuels with a high percentage content of a solid phase and made it possible to have a relatively homogeneous ultra-dispersed spray of slurry fuel.

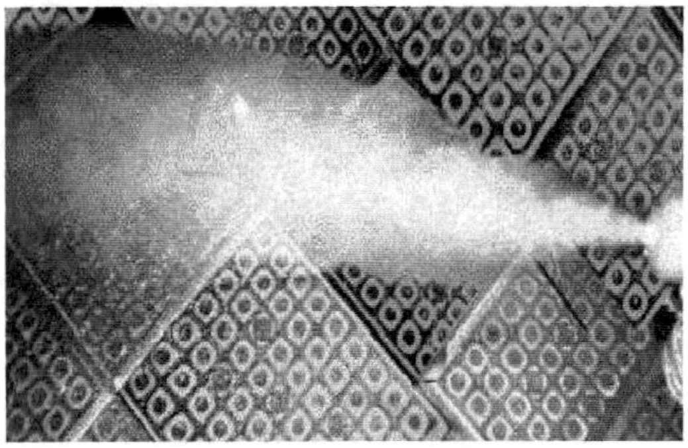

Figure 5 Model of the developed and tested sprayers

Figure 6 Laboratory testing of fuel atomization

Results and Summary

Dramatic improvement of both energy and ecological indices of the production and combustion of these new composite liquid fuels has been obtained with the use of this innovative hydrocavitation technology (PHCR).

This process is developed to substitute for natural gas with inexpensive composite liquid fuels at different power plants based on solid waste of coal production, waste from oil-refining industry, waste from the production of different oils, different waste of organic and inorganic origin such as peat, silt, biomass, and wood.

Algorithms for the Automated Generation of Reaction Networks in the Hydroprocessing of Oil Fractions

Jorge M. Martinis

Bryan Research & Engineering, Inc. P.O. Box 4747, Bryan, Texas 77805

The field of characterization and estimation of physical properties of complex mixtures has achieved a significant progress with the expansion of computing power and higher resolutions in analytical tools. As never before, scientists and engineers now have broader access to more accurate descriptions of mass transfer phenomena and reaction kinetics in commercial process simulators, resulting in faster and more reliable engineering workflows in the chemical processing industry.

Historically, the standard approach to the problem of the exhaustingly large number of species conforming oil fractions has been to split each fraction into pseudo species with narrower boiling intervals. Physical properties for the pseudo species are then obtained from matching the average boiling point of the pseudo species to a model molecule within homologous series of species. For instance, normal alkanes have been widely used as model molecules for pseudo species by correlating their critical properties to the average properties of petroleum fractions such as specific gravity, molecular weight and boiling point. Most recent applications of these methods have been extended to series of cycloalkanes, alkenes and aromatics.

The use of pseudo species (a.k.a. pseudo components, oils, lumps, Etc.) has enabled the simulation of a great variety of unit operations in chemical process plants where the conservation of species holds. However, the lack of information about the chemical structure of the species within lumps prevents their application to unit operations carrying chemical reactions. Though recent advances in the field of continuous kinetics have shown promising results in predicting the conversion of oil fractions based on their true-boiling-point representation, only the fundamental kinetic information coming from the chemistry of the process at the elementary step level has proven predictive capabilities when dealing with complex feed stocks.

In the present work, a series of algorithms have been formulated to solve the formidable problem of generating detailed reaction networks for a given set of species (NP-complete problem) following well-established rules governing the underlying chemistry of the process. In the particular case of hydroprocessing, computer science fundamentals such as object-orientation, functional programming and recursive operations on relational sets are applied to generate consistent reaction networks starting from the fundamentals of transition-metal catalysis and carbenium ion chemistry. As a result, reaction networks at the elementary step level are generated for catalytic reforming, hydroisomerization, hydrodewaxing and hydrocracking.

(141c) HDS and HDA Characteristics of Pt-Encapsulated Zeolite Catalysts

Jinwen Chen[1], Song Chen[2], Rafal Gieleciak[3]and Craig Fairbridge[3], (1)Natural Resources Canada, Devon, AB, Canada, (2)CanmetENERGY Upgrading Program, Natural Resources Canada, Devon, AB, Canada, (3)Natural Resources Canada

Noble metal catalysts (e.g. Pt, Pd) are commonly used in hydrogenation processes. They have higher hydrogenation capability than base catalysts (e.g. Co, Mo, Ni), but can be easily poisoned by sulphur compounds and lose their activity. This paper presents a specially prepared noble metal catalyst which uses a Pt encapsulated zeolite material to prevent the Pt particles from sulphur poisoning. A Mo-Ni catalyst was also prepared using the same procedure for comparison. A set of pilot plant tests was conducted with naphthalene as a hydrodearomatization/hydrogenation model compound and dibenzothiophene as a sulphur model compound. Two-dimensional GCxGC analysis was performed to identify and quantify the model compounds and their hydrotreated products.

The prepared Pt-encapsulated catalyst exhibited naphthalene conversion comparable to that obtained using the Mo-Ni catalyst. However, the Pt catalyst yielded significant amounts of the deep naphthalene hydrogenation product decalin in the low-sulphur poisoning environment. In this case, the relative amount of decalin increased concomitantly with reaction temperature. In the high-level sulphur poisoning environment, the Pt catalyst gave much more intermediate hydrogenation product tetralin than decalin. These observations indicate that sulphur poisoning has an impact on the hydrogenation selectivity. At the same time, the Mo-Ni catalyst gave similar amounts of tetralin and decalin regardless of the presence of sulphur poisoning. Based on the analysis of HDS products of dibenzothiophene, it was observed that the HDS of dibenzothiophene over the Pt- encapsulated catalyst proceeded mainly via the direct desulphurization pathway and yielded biphenyl as the main product. This observation is contrary to the commonly accepted understanding that the hydrogenation pathway is preferred for noble metal catalysts. The Mo-Ni catalyst, on the other hand, yielded cyclohexylbenzene as the main product. The implication is that use of the Pt catalyst leads to lower hydrogen consumption than the Mo-Ni catalyst while retaining comparable HDS conversion.

(141d) Design Considerations for Diesel Hydrotreaters

R.E. (Ed) Palmer, Mustang Engineering, Houston, TX

In 2006, the United States EPA mandated the use of 15 ppmw ultra-low sulfur diesel (ULSD) for on-road diesel fuel. At that time several refiners either revamped existing facilities or constructed grass roots facilities to meet the specification. Some facilities did not implement ULSD process capacity since they were located such that they could still easily sell off-road diesel, No.2 Fuel Oil, and/or diesel fuel to other countries that did not require ULSD. However, the number of locations that will accept higher sulfur fuel is quickly being reduced as states in the northeast set time tables for Fuel Oil desulfurization requirements, and developing nations begin to implement similar ULSD regulations. Additionally, as the market for gasoline continues to have limited growth, refiners are looking for opportunities to expand their diesel processing capacity.

In this paper, critical process design issues for hydrotreating diesel fuel to meet ULSD requirements will be updated based on the knowledge that has been gained since ULSD regulations were put in place.

(141e) Hydrotreating of Biooils to Jet Fuels: An Overview

Maoqi Feng, Chemical Engineering Dept., Div. of Chemistry and Chemical Engineering, Southwest Research Institute, San Antonio, TX and Jimell Erwin, Chemical Engineering Department, Southwest Research Institute, San Antonio, TX

This paper will give an overview of the current status of jet fuels production from hydrotreating of biooils, including vegetable oils and microalgae oils. Hydrotreating of biooils is different from the conventional hydrotreating process; there are two reasons:

1) Different routes. There are at least four different kinds of chemical reactions in the hydrotreating of biooils: hydrodeoxygenation, hydrodecarboxylation, hydrocracking and hydrogenation of unsaturated chemical bonds.

2) Different heteroatoms contaminants in the feedstock. This effect includes the inhibitory actions of the byproducts of the reactions with hydrogen. Sulfur concentration is much lower in biooils than that in conventional fossil fuels, while phosphorus contaminant in biooils is much higher. Phosphorus originates from phospholipids in biomass feedstock. Phospholipids removal is required in biofuels production.

Traditional hydrodesulfurization catalysts can still be used for hydrotreating of biooils, however, the catalyst lifetimes are much shorter. Future catalysts for biooils hydrotreating should be more selective for hydrodeoxygenation and be able to tolerate phosphorus contaminants. This paper will address the bioils pretreatment requirement and the direction for the development of biooils hydrotreating catalyst as well as product property adjustment.

INFLUENCE OF BLENDING VEGETABLE OIL INTO HVGO ON HYDROTREATING/MILD HYDROCRACKING PERFORMANCE

Jinwen Chen, Song Chen, Hena Farooqi
CanmetENERGY, Natural Resources Canada
One Oil Patch Drive, Devon, AB, Canada T9G 1A8

1. Introduction

The production of Canadian bitumen is expected to increase significantly in the near future, from current 1.5 million barrel per day (MBPD) to 3.5 MBPD by 2025 (CAPP, 2010). It is known that the carbon footprints related to heavy oil and bitumen production, upgrading and refining are higher than those related to conventional crudes (Burkhard et al., 2011; Keesom et al., 2009; Sheehan et al., 1998). Meanwhile biomass derived fuels from renewable sources are considered sustainable and have the potential to help reducing green house gas (GHG) emission and other environmental impact (Keesom et al., 2009; Sheehan et al., 1998). Currently the renewable biofuel makes up about 8% of total fuels consumed in US (EPA, 2010). Co-processing petroleum with biomass derived feedstocks is one of the most attractive and practical options for petroleum refiners, and heavy oil and bitumen upgraders (Templis, et al., 2011). It offers a wide range of advantages, such as little or no extra capital investment, no need to change existing technologies and catalysts, potential synergy between aromatic petroleum feedstocks and paraffinic biomass feedstocks, and improved fuel quality. However, technical challenges are still present and need to be addressed in commercial implementation and operation.

There have been a number of studies published on co-processing petroleum and biomass feedstocks through hydroprocessing. In the majority of these studies two different types of biomass feedstocks are used, pyrolysis oils (Mercader et al., 2011; Pinheiro et al., 2011, 2009; Bui et al., 2009) and vegetable oils (Templis et al., 2011; Šimáček and Kubička, 2010; Tóth et al., 2010; Bezergianni et al., 2009; Walendziewski et al., 2009; Watkins et al., 2009; Huber et al., 2001; Viljava et al., 2001; Furimsky, 2000; Wong and Hong, 1999). However the influence of biomass derived components in the co-processed feedstocks on hydroprocessing performance is not conclusive, especially on hydrodesulphurization (HDS) and hydrodenitrogenation (HDN). Some researchers observed inhibition effects while others reported no inhibition by these components.

The objective of this study is to investigate the impact of low-grade raw canola oil on the hydrotreating/mild hydrocracking of Canadian bitumen derived heavy vacuum gas oil (HVGO).

2. Experimental

A bench-scale reactor system was used for the experiments. The system consisted of mainly a single tubular reactor, a high-pressure phase separator, and a product stabilizer.Hydrogen gas and liquid oil co-currently flowed upwards through the tubular reactor. The reactor tube, 100 cm in length and 9 mm in diameter (ID), was heated with a four-zone electric furnace. The liquid product from the stabilizer went to a collector while the gases leaving the HPS and the stabilizer were combined and directed to the vent, from which a small slip-stream was analyzed by an online gas analyzer (HP5880 GC).

A commercial $NiMo/Al_2O_3$ hydrotreating catalyst (extrudes of 1.6 mm in diameter and 5 mm in averaged length) was used. A total of 20 ml of catalyst, with 20 ml of 0.2 mm diluting glass beads, were loaded in the reactor. Prior to the planned runs, the fresh catalyst was sulphided and then stabilized by processing the feed for about 170 hours to get relatively stable activity. Slight catalyst deactivation was observed and monitored by conducting check-back runs periodically at the same operating conditions. However, all the experimental data presented in this paper were not corrected for catalyst deactivation.

The feeds used in this study were blends of bitumen derived HVGO and low-grade raw canola oil (Milligan Bio-Tech, SK, Canada) with 3 different mass blending ratios (HVGO/canola oil): 95/5, 90/10, and 80/20. For comparison, pure HVGO was also tested under the similar operating conditions. The properties of the feeds are listed in Table 1. The feeds and the products were analyzed for simulated distillation (SimDis), density, sulphur and nitrogen contents, and hydrocarbon type composition using ASTM methods and other in-house developed methods (Chen et al., 2003).

The operating conditions used for the experiments were typical for hydrotreating/mild hydrocracking and covered a relatively wide range: temperature from 360°C to 395°C, pressure from 80 bars to 110 bars; liquid hourly space velocity from 1.0/h to 2.5/h; gas/oil ratio fixed at 8000 NL/L.

3. Results and Discussion

3.1 Total Liquid Product

The objective of this study is to investigate the catalyst performance and product quality of hydrotreating/mild hydrocracking of bitumen derived HVGO with and without the addition of canola oil. SimDis of the total liquid product (TLP) obtained from each mass balance run was measured to determine the product (360°C- material) yield and feed conversion (360°C+ material). Figures 1 and 2 show the product yield and feed conversion, respectively, obtained with different feeds.

As seen in Figures 1 and 2, product yield and feed conversion increase with the increase in temperature for all the 4 feeds investigated in this study. The trends show typical exponential effect of temperature on reaction conversion when the conversion is relatively low (less than 50%). Note that only 2 experiments were performed at temperatures of 385°C and 395°C with the 80/20 HVGO/canola oil feed due to reactor plugging caused by the high content of canola oil, which has high contents of metals and phosphorus. Since a hydrotreating catalyst was used in this study, the product yield and feed conversion are generally moderate (mild hydrocracking), varying from 25wt% to 47wt% for yield and from 15wt% to 40wt% for conversion. Figures 1 and 2 also show that with the increase in canola oil content in the feed, product yield and feed conversion increase, with the 80/20 HVGO/ canola oil feed having the highest while the pure HVGO has the lowest. This simply confirms that canola oil is easier to crack than HVGO. In another word, to achieve the same product yield or feed conversion, canola oil containing feed requires lower reaction temperature, indicating potential energy savings during commercial operation. Similar observations have been reported by Bezergianni et al. (2009), while opposite observations were made by Lappas et al. (2009) who observed inhibition effect of sunflower oil on hydrocracking of vacuum gas oil (VGO).

3.2 Hydrodesulphurization and Hydrodenitrogenation

Another important objective of this study is to investigate the impact of canola oil on HDS and HDN of the HVGO. Figures 3 and 4 illustrate the HDS and HDN conversions at different temperatures with different feeds. As seen in these figures, HDS and HDN conversions increase with the increase in temperature, but are not affected much by the addition of canola oil. Even with the 80/20 HVGO/canola oil feed, at temperature of 395°C, the HDS conversion is about 99.5wt%, similar to that obtained with pure HVGO (varying between 99wt% and 100wt%) as seen in Figure 3. Similar observations can be made with HDN in Figure 4. Therefore it can be concluded that in the present study, canola oil does not affect the HDS and HDN performance. In published studies, there are inconsistent conclusions regarding the effect of adding vegetable oil in petroleum on HDS and HDN. Reduced HDS efficiency was reported due to competitive adsorption of sulphur and oxygen molecules on the same catalyst active sites (Mercader et al., 2011; Tóth et al., 2010; Pinheiro et al., 2009). However, other researchers reported no significant HDS and HDN inhibition by vegetable oils (Lappas et al., 2009; Huber et al., 2007). Furthermore it has been reported that the inhibition effect of vegetables oils or pyrolysis bio oils on HDS and HDN could be significant or insignificant, depending on the operating conditions or the vegetable/bio oil contents in the feed (Templis et al., 2011; Bui et al.; 2009)).

3.3 Hydrocarbon Type Composition

Figures 5 and 6 show the contents of saturates and aromatics in the TLPs obtained at different temperatures with different feeds. It is seen in the figures that the saturates contents in the TLPs obtained with canola oil containing feeds are about 10% higher than those obtained with the pure HVGO feed at all temperatures investigated. Correspondingly, the aromatics contents in the TLPs obtained with canola oil containing feeds are about 10% lower than those obtained with the pure HVGO feed. This is expected since hydrotreating/hydrocracking of canola oil always yields paraffinic hydrocarbons, effectively increasing the content of saturates in the product. With the increase in canola content in the feed, the content of saturates/aromatics in the TLPs obtained at the same temperature also increases/decrease. It is also noted that the content of saturates increases slightly with the increase in reaction temperature, indicating increased aromatics hydrogenation conversion.

3.4 Yields and Properties of Gasoline and Diesel

Spinning band distillation was performed with the TLP from each mass balance run to obtain 3 fractions: gasoline (IBP – 200°C), diesel (200 – 343°C) and unconverted oil (343°C +). Figures 7 and 8 show the gasoline and diesel yields, respectively, obtained from distillation with different feeds at different temperatures. As seen in these figures, gasoline yield increases with the increase in temperature, but decreases with increasing canola oil content in the feed, indicating that the hydrotreating and hydrocracking of canola oil does not favor gasoline production, but favors diesel production as will be discussed shortly. Generally the gasoline yield is relatively low (less than 7%) at any temperature with any feed. Therefore, co-processing of canola oil (or other vegetable oils) with petroleum feedstocks through hydroprocessing might not be the best option if the target product is gasoline. Instead, it could be suitable for diesel production since the increase in canola oil content in the feed favors diesel yield as observed in Figure 8. Figure 8 also demonstrates that the diesel yields ranging between 20 and 40wt% are higher than those of gasoline, supporting the conclusion that co-processing through hydroprocessing favors diesel production.

Figures 9 to 11 show the sulphur, nitrogen, and aromatics contents, respectively, of the gasoline fractions obtained at different temperatures with different feeds. As seen in these figures, the sulphur, nitrogen, and aromatics contents decrease with the increase in temperature as expected due to increased hydrotreating/hydrogenation. However, among the different feeds, it seems that there are no significant differences. Figure 12 shows the corresponding research octane numbers of the gasoline fractions. The research octane number remains relatively constant regardless of the feed and temperature as evident in Figure 12.

Figures 13 and 14 show the sulphur and nitrogen, respectively, of the diesel fractions obtained at different temperatures with different feeds. Similar to those observed with the gasoline fractions discussed above, no significant differences in sulphur and nitrogen contents are present with the diesel factions obtained with different feeds as shown in Figures 13 and 14. Figure 15 and Figure 16 illustrates the corresponding aromatics contents and cetane index (an indication of cetane number) of the corresponding diesel fractions, respectively. As observed in these figures, increased canola oil content in the feed results in reduced aromatics content (Figure 15), which consequently increases the cetane index (Figure 16).

From the experimental findings discussed above, a preliminary conclusion can be drawn that while the addition of canola oil into bitumen derived HVGO slightly decreases the gasoline yield but significantly increases the diesel yield. At the same time the properties of the gasoline and diesel obtained are not significantly affected by the addition of canola oil, or even are improved.

4. Conclusions

In this study, the hydrotreating/mild hydrocracking of feeds consisting of HVGO and canola oil with different blending ratios were experimentally investigated using a pilot plant unit with a commercial hydrotreating catalyst. The experimental results showed that:

1) Feed conversion and TLP yield increase with the increase in canola oil content in the feed. Co-processing requires lower temperature than processing pure HVGO under similar conversion or product yield.
2) HDS and HDN are not affected by the addition of canola oil into HVGO. The TLPs obtained from canola oil co-processed feeds contain less aromatics and more saturates than those from pure HVGO.
3) With the increase in canola oil content in the feed, gasoline yield decreases slightly but diesel yield increases significantly, indicating that co-processing bitumen feedstocks with canola oil favors diesel production.
4) The sulphur, nitrogen and aromatics contents, and octane number of the gasoline fractions do not change significantly with the change in canola oil content in the feed.
5) While the sulphur and nitrogen contents of the diesel fractions remain relatively constant with the change in canola oil content in the feed, the aromatics content decreases and therefore the cetane index increases with the increase in canola oil content in the feed.

Acknowledgments

Partial funding for this study was provided by the Canadian Interdepartmental Program of Energy Research and Development (PERD 1.1.3). The authors are grateful to the CanmetENERGY pilot plant staff for conducting the experiments and to the analytical lab staff for performing the analyses. Dr. Edward Little's valuable inputs and suggestions on revising this paper are greatly appreciated.

References

1) Bezergianni, S., Kalogianni, A., Vasalos, "Hydrocracking of Vacuum Gas Oil-Vegetable Oil Mixtures for Biofuels Production", Bioresource Technology, 100, 3036-3042, 2009.
2) Bui, V. N., Toussaint, G., Laurenti, D., Mirodatos, C., Geantet, C., "Co-processing of Pyrolisis Bio Oils and Gas Oil for New Generation of Biofuels: Hydrodeoxygenation of Guaïacol and SRGO Mixed Feed, Catal. Today 143, 172-178, 2009.
3) Burkhard, J., Forrest, J., and Gross S., "Oil Sands, Greenhouse Gases, and European Oil Supply: Getting the Numbers Right", IHS CERA Special Report, April 2011.
4) Canadian Association of Petroleum Producers (CAPP), "2010-2025 Canadian Crude Oil Forecast and Market Outlook". http://www.capp.ca/aboutUs/mediaCentre/NewsReleases/Pages/2010-Oil-Forecast.aspx#A7lJAtIkyfHq, June, 2010.
5) Chen, J., Ring, Z., and Briker, Y., "Building a Hydrotreating Process Model for Modern Refineries", proceedings of 2003 AIChE Spring Conference, Paper 52e, March 30 –April 3, 2003, New Orleans, LA, USA.
6) Environmental Protection Agency (EPA), "EPA Finalizes 2011 Renewable Fuel Standards", http://www.epa.gov/oms/fuels/renewablefuels/420f10056.htm, November 2010.
7) Furimsky, E., "Catalytic Hydrodeoxygenation", Appl. Catal. A: Gen. 199, 147-190, 2000.

8) Huber, G.W., O'Connor, P., Corma, A., "Processing Biomass in Conventional Oil Refineries: Production of High Quality Diesel by Hydrotreating Vegetable Oils in Heavy Vacuum Oil Mixtures", Appl. Catal. A: Gen. 329, 120-129, 2007.

9) Kalnes, T. N., Marker T., Shonnard, D., R., and Koers, K., P., "Green Diesel Production by Hydrorefining Renewable Feedstocks", Biofuels Technology, Issue 4, 7-11, 2008.

10) Keesom, W. H., Unnasch, S., and Moretta, J., "Life Cycle Assessment Comparison of North American and imported Crude Oils". Report by Jacobs Consultancy for Alberta Energy Research Institute (AERI), July 2009.

11) Lappas, A. A., Bezergianni, S., Vasalos, I. A., "Production of Biofuels via Co-Processing in Conventional Refining Processes", Catal. Today 145, 55-62, 2009.

12) Mercader, F., M., Groeneveld, M. J., Kersten, S. R. A., Geantet, C., Toussaint, G., Way, N. W. J., Schaverien, C. J., Hogendoorn, J. A., "Hydrodeoxygenation of Pyrolysis Oil Fractions: Process Understanding and Quality Assessment through Co-Processing in Refinery Units", Energy Environ. Sci. 4, 985-997, 2011.

13) Pinheiro, A., Dupassieux, N., Hudebine, D., Geantet, C., "Impact of the Presence of Carbon Monoxide and Carbon Dioxide on Gas Oil Hydrotreatment: Investigation on Liquids from Biomass Cotreatment with Petroleum Cuts", Energy Fuels 25, 804-812, 2011.

14) Pinheiro, A., Hudebine, D., Dupassieux, N., Geantet, C., "Impact of Oxygenated Compounds from Lignocellulosic Biomass Pyrolysis Oils on Gas Oil Hydrotreatment", Energy & Fuels, 23, 1007-1014, 2009.

15) Sheehan, J., Camobresco, V., Duffield, J., Graboski, M., Shapouri, H., "Life Ccycle Inventory for Biodiesel and Petroleum Diesel for Use in an Urban Bus", Final Report by US Department of Agriculture and US Department of Energy, May 1998.

16) Šimáček, P., Kubička, D., "Hydrocracking of Petroleum Vacuum Distillate Containing Rapeseed Oil: Evaluation of Diesel Fuel", Fuel, 89, 1508-1513, 2010.

17) Templis, Ch., Vonortas, A., Sebos, I., and Papayannakos, N., "Vegetable Oil Effect on Gasoil HDS in Their Catalytic Co-hydroprocessing", Applied Catalysis B: Environmental, 104 (3-4), 324-329, 2011.

18) Tóth, C., Baladincz, P., Kovács, S., Hancsók, J., "Producing Diesel Fuel by Co-hydrogenation of Vegetable Oil with Gas Oil", Chem. Eng. Trans., 21, 1219-1224, 2010.

19) Viljava, T, R., Saari, E. R. M., Krause, A. O., I., "Simultaneous Hydrodesulfurization and Hydrodeoxygenation: Interactions between Mercapto and Methoxy Groups Present in the Same or in Separate Molecules", Appl. Catal. A: Gen. 209, 33-43, 2001.

20) Walendziewski, J., Stolarski, M., Łużny, R., Klimek, B., "Hydroprocessing of Light Gas Oil – Rape Oil Mixtures", Fuel. Process. Tech., 90, 686-691, 2009.

21) Watkins, B., Olsen, C., Sutovich, K., Deady, J., Petti, N., Wellach, S., "New Opportunities: for Coprocessing Renewable Feeds in the Refinery Process", Hydrocarbon Eng. 14(1), 49-58, 2009.

22) Wong, A., Hong, E., "Biomass: A growth Opportunity in Green Energy and Value-added Products", the Fourth Biomass Conference of the Americas, Elsevier, Oakland, CA, 1999, p.837.

Table 1. Feed properties

Tests	Units	100% HVGO	100% canola oil	95/5 HVGO/ canola oil	90/10 HVGO/ canola oil	80/20 HVGO/ canola oil
Density	g/ml	0.9678	0.9237	0.9645	0.9628	0.9586
Carbon	wt%	84.57	77.41	84.90	84.72	83.30
Hydrogen	wt%	10.74	11.46	11.15	11.09	11.22
Sulfur	wt%	3.59	0.0208	3.36	3.09	2.81
Nitrogen	wppm	1800	192.7	1800	1718	1425
Oxygen	wt%	0.37	8.31	0.74	1.42	2.48
Hydrocarbon types	wt%					
Saturates		31.16	0.15	59.53	29.05	19.96
Aromatics		59.68	7.67	55.39	52.99	53.71
Polars		8.38	92.18	13.07	17.24	25.94
Simulated distillation	°C					
IBP		243.8	360	250.6	240.8	240.2
10wt%		337.4	541.2	339.6	338.8	342.8
30wt%		390.8	609.8	392.6	393.6	395.6
50wt%		427.4	614.2	430.4	432.8	437.4
70wt%		468	617.2	474.2	479.8	493.2
90wt%		530.6	619.8	547.6	570	604.4
FBP		622.8	715.6	653.2	643.8	657

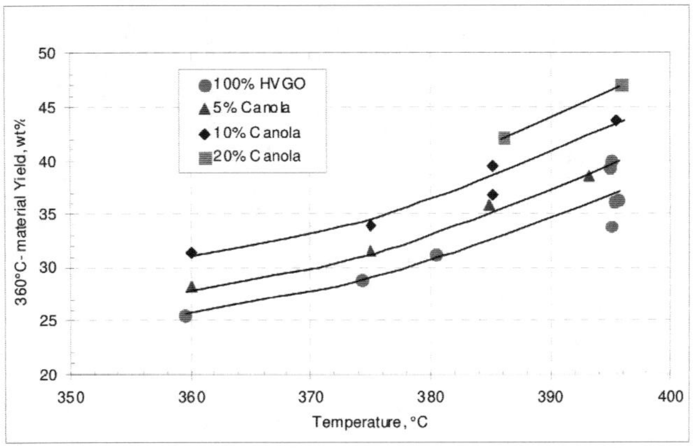

Figure 1. Product (360°C-) yields obtained with different feeds at different temperatures

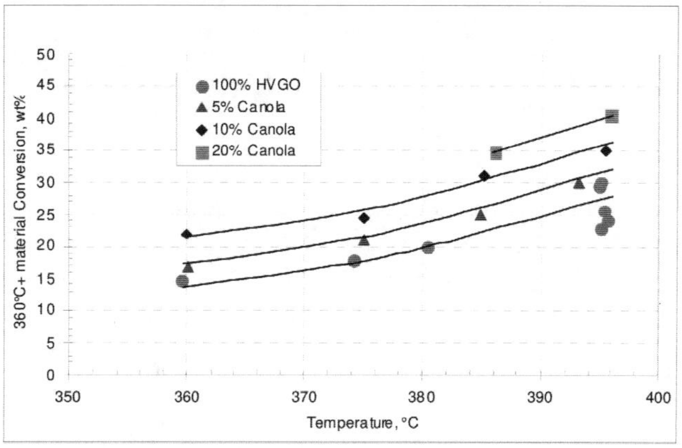

Figure 2. Feed (360°C+) conversions obtained with different feeds at different temperatures

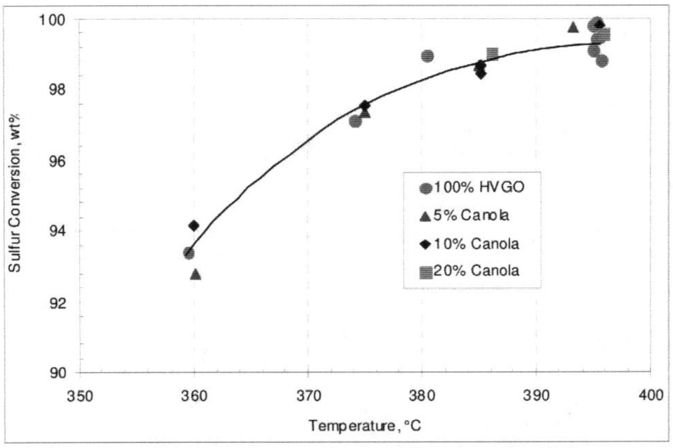

Figure 3. HDS conversions obtained with different feeds at different temperatures

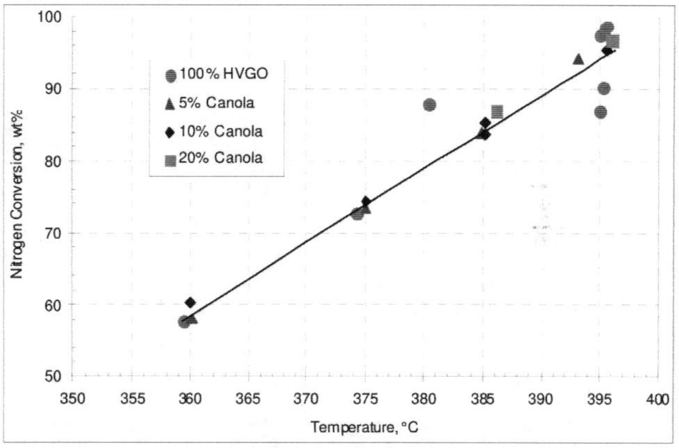

Figure 4. HDN conversions obtained with different feeds at different temperatures

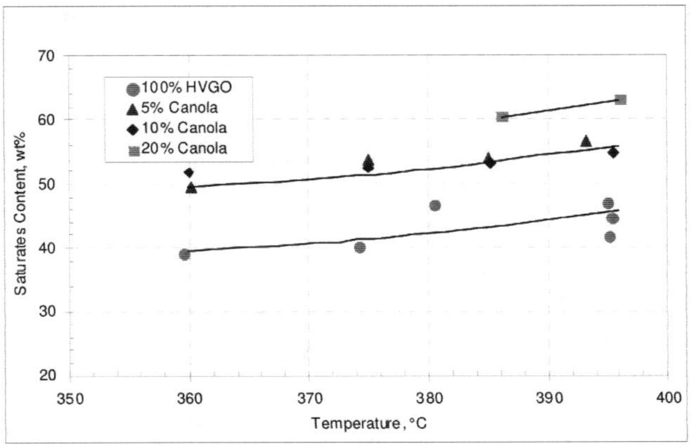

Figure 5. Saturates content in TLPs obtained with different feeds at different temperatures

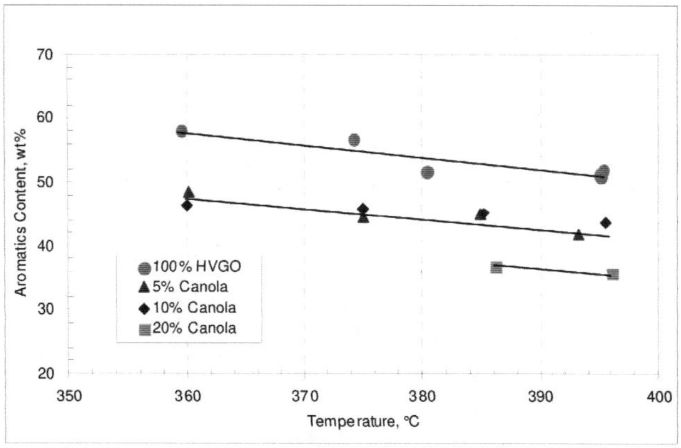

Figure 6. Aromatics content in TLPs obtained with different feeds at different temperatures

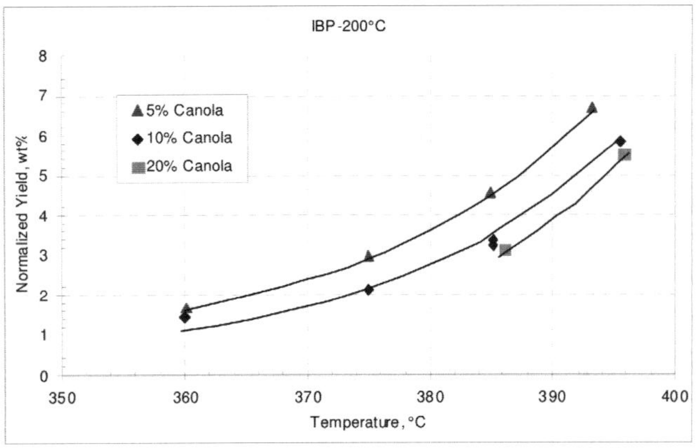

Figure 7. Gasoline yields obtained with different feeds at different temperatures

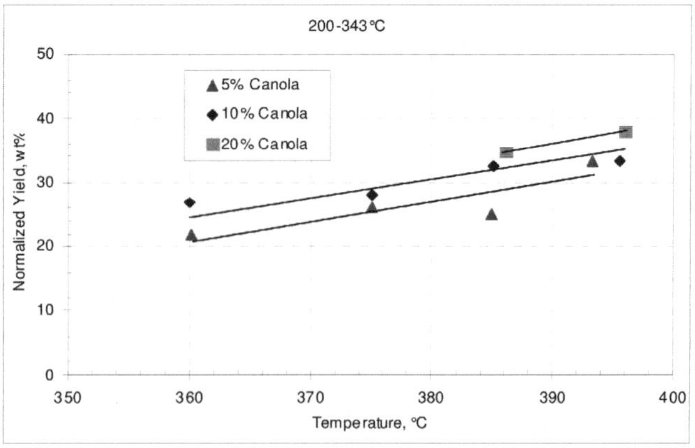

Figure 8. Diesel yields obtained with different feeds at different temperatures

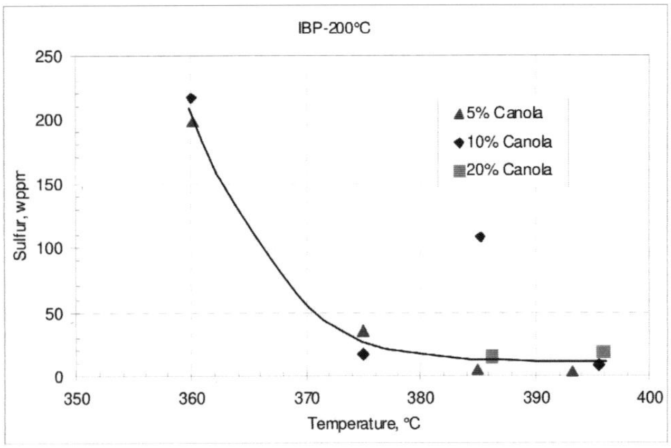

Figure 9. Sulphur content of gasoline fractions obtained with different feeds at different temperatures

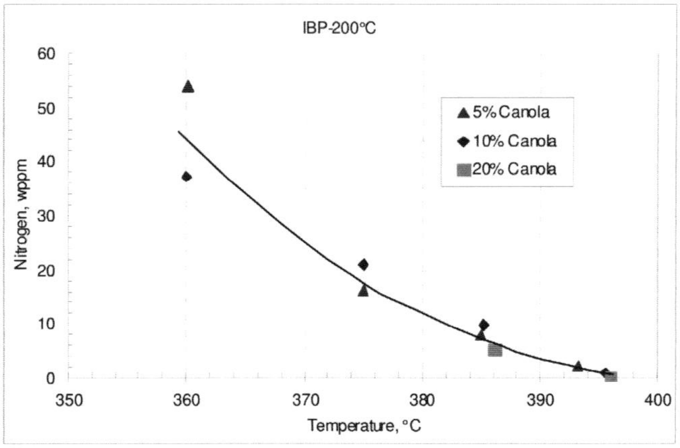

Figure 10. Nitrogen content of gasoline fractions obtained with different feeds at different temperatures

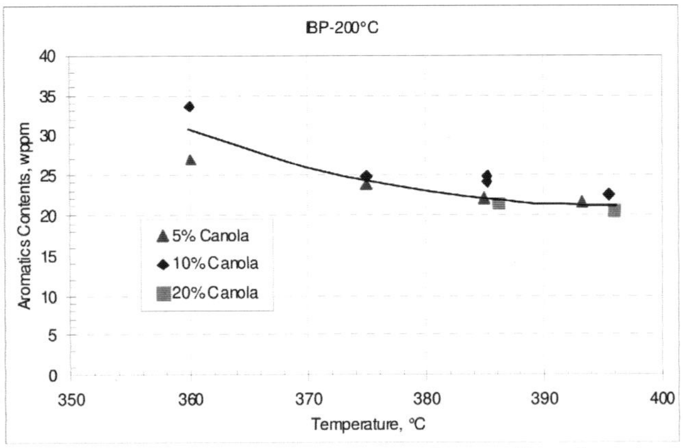

Figure 11. Aromatics content of gasoline fractions obtained with different feeds at different temperatures

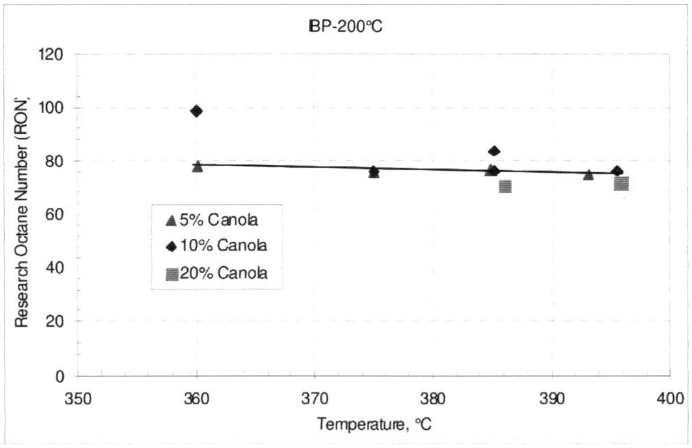

Figure 12. Octane number of gasoline fractions obtained with different feeds at different temperatures

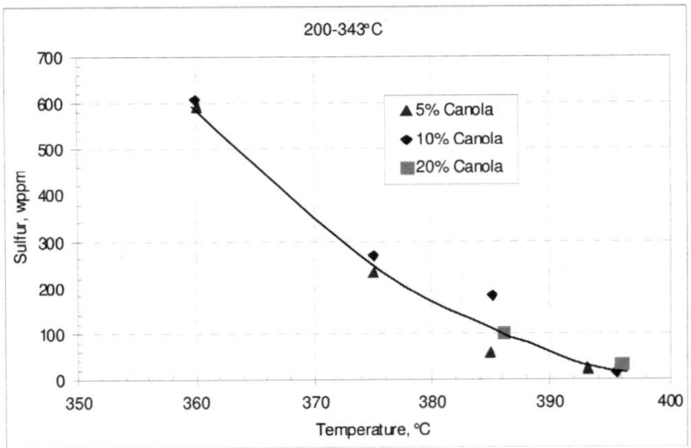

Figure 13. Sulphur content of diesel fractions obtained with different feeds at different temperatures

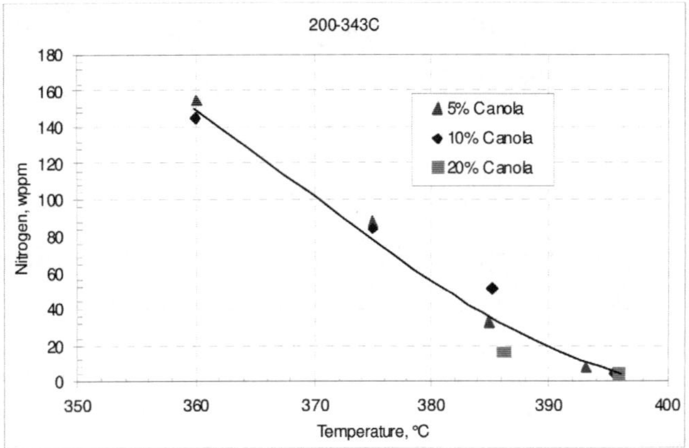

Figure 14. Nitrogen content of diesel fractions obtained with different feeds at different temperatures

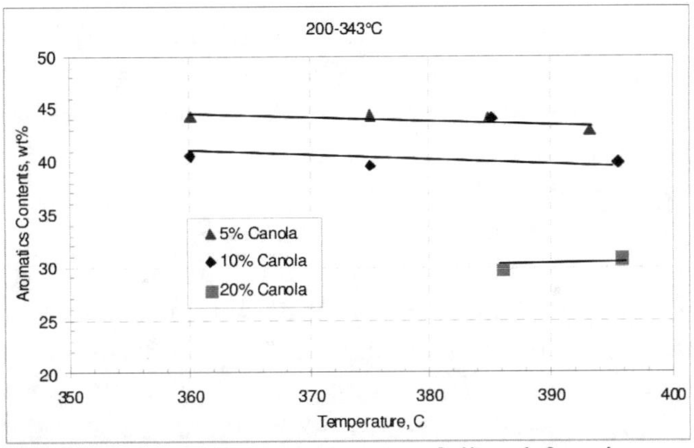

Figure 15. Aromatics content of diesel fractions obtained with different feeds at different temperatures

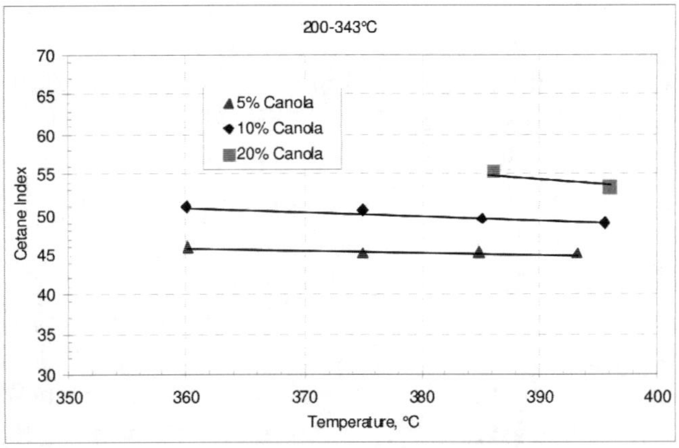

Figure 16. Cetane index of diesel fractions obtained with different feeds at different temperatures

Revamping an Existing Flare Network to Meet Increased Flaring Rates

Dennis Aguirre, Process Engineer, WorleyParsons

In attempts to expand refinery throughput, by debottlenecking existing facilities or installing additional capacity onsite, often offsite limitations are overlooked, particularly flare systems, which can potentially result in costly solutions jeopardizing capital investment and production revenue. In this paper, a baseline refining facility is examined with the intent of generating a cost effective strategy for both mitigating existing sources of refinery relief while at the same time expanding flaring capacity in order to meet relieving requirements from new sources. In order to obtain valid solutions, Process Protection Manager™ was used as the hydraulic solver for the flare network assessments; note that this paper will forgo discussion regarding the particulars of the solver methodology and instead focus on the design strategy.

Initially, the baseline refinery was observed to have overpressure issues with the backpressure at various Unit flare headers being relatively high, in addition to the flare network equipment being marginally able to safely process relief material. Within the confines of a pseudo project proposed scope to expand refining capacity, an initial assessment deemed additional flaring capacity was necessary with a select number of relief valves in the existing refinery being relocated to an alternate flare network. This in effect added an additional stack and a new independent flare header with a balance line to equalize pressures and process excess material. However, although this did provide an initial solution to proceed with the pseudo project, later it was determined that new sources being added would overwhelm the flare network resulting in radiation isopleth levels, at designated areas, being the acceptable limits. In addition, the backpressure at various points in the flare network was still too high during the governing relief scenario. Furthermore, the cost of the new flare and flare network were well beyond the allocated budget. As such, it became evident that the initial proposed solutions fell short and the refinery would need to be re-examined to find a better solution to the problem.

Rigorous models, in addition to detail assessments of refinery utility distribution systems, were generated to tackle this problem with the most viable option to mitigate the largest relieving sources in the refinery through use of a High Integrity Process Protection Systems (HIPPS). Although this item is briefly discussed in API 520/521 for averting overpressure contingencies, a meticulous understanding of the shutdown logic confirmed by dynamic modeling was required to validate the solution. Needless to say, the solution proved efficient with a very low probability of failure which allowed the pseudo project to proceed. Notably, the cost of the added HIPPS was justified by reducing the flare stack size as well as the piping size required.

Design Challenges in Tankage Modernization

James Kong, PE, Principle Process Engineer, WorleyParsons

Environmental regulations on refined petroleum products have become more stringent over the years. For example, a change in the requirement of sulfur level in highway diesel fuel from 500 ppm to 15 ppm has phased in between 2006 and 2010 in the United States. The same trend of lowering the sulfur content of diesel products is gradually being seen around the world. This means that different countries are currently in different phases of implementing low sulfur diesel requirements and to supply this market a refinery needs to be capable of producing different grades of diesel products, such as Regular Diesel, Off-road Grade, and Ultra Low Sulfur Diesel (ULSD). This presentation discusses the design challenges for revamping an existing tank farm to accommodate the different grades of diesel products.

The existing tankage allocation often needs to be changed to accommodate the different diesel grades. The design conditions of the existing lines need to be re-evaluated and possibly re-rated to the higher conditions. In the refineries where more than one grade of diesel product is produced, product segregation is critical. Additional means of isolation are often required at any junction points where two different grades of diesel meet. Double block and bleed is often utilized for product segregation. When additional isolation valve are installed on an existing line, it is important to check the thermal relief requirements.

To produce lower sulfur diesel grades requires a redesign of the refinery diesel hydrotreating facilities. This generally results in feed and product rates that are different from the existing operation. The hydraulic design of the feed and product lines has to be checked, and any deficiencies have to be remedied so as to achieve the required design results. A common problem is that the existing pumps do not have enough discharge pressure to deliver the required flow rate to the required destination. In this case, the remediation options can include removing high-pressure drop instruments, increasing the size of the existing line(s), increasing the size of existing impeller(s), and installing new additional pumps.

When the feed and products rates increase, it is also important to verify that the feed tanks and the product tanks are fully equipped to handle the higher flow rates. Specifically, the tank breather valves have to be checked to make sure they have sufficient capacity to handle the higher vapor inflows and outflows that result from the higher filling and higher draw rates. API 2000 is commonly used as the standard for breather valve designs. For revamping existing tanks, the sixth edition of API 2000 also includes an appendix, which offers an alternative design method.

In some special occasions, the function of a tank would change. Other modifications to the tank, such as the installation of a nitrogen blanketing system may be required. The diesel tanks are generally cone roof tanks that operate at atmospheric pressure. Adding a nitrogen blanketing system on existing diesel tanks can be challenging as the set pressure of the control valve or regulator needs to be high enough to maintain a positive pressure in the tank, but low enough to avoid the release of nitrogen through the breather valves during normal operations.

(146c) Real-Time Utilities Optimization for Refineries Incorporating Hydrogen Supply Systems

Carlos Ruiz, Soteica Ideas & Technology LLC, Houston, TX

Hydrogen management can have a significant effect on refinery utility supply through the integration with the rest of the systems. Optimization of hydrogen production in conjunction with steam, power, and fuels can yield significant savings opportunities for the refinery. Examples of actual benefits that have been captured at refineries will be shown.

(146d) Reduction of Refinery CO2 Emissions Via Utilities & Offsites

Jason L. Stahlman, UOP, Des Plaines, IL

The importance of reducing CO2 Emissions in process units has become more important to refiners. The presentation will focus on optimizing and identifying the Utilities and Offsites related equipment, which can be modified to reduce CO2 admissions and save energy. Topics covered will include steam generation, steam usage, cogeneration, flares, tankage, electricity, and cooling water. The CO2 emissions from the entire facility may be reduced by: reducing the amount of fuel that is consumed in the Utility Boilers and by controlling the production and CO2 throughout the complex.

(146e) Reducing CO2 Emissions in Plant Flaring Operations

Christopher Ferguson, Callidus Technologies, Honeywell, Tulsa, OK and Brian Duck, Callidus technologies by Honeywell, Tulsa, OK

The owners of an ethylene plant under construction in China faced pressure from neighborhood facilities and regulatory agencies to reduce flare emissions, visible flame, and noise. A new system that reduces flaring, CO2 emissions, noise, radiation, maintenance cost, operating costs and increases energy efficiency and flare tip life without sacrificing safety was required. To enable installation during start-up of the plant in 2009, it was necessary to design, fabricate, and test a new system that included flare gas recovery system on a fast track schedule.

Production tests and operating data confirm the new flare and flare gas recovery system achieves the required increase in smokeless capacity, reduction in noise levels, decreased flame visibility, enhanced operational efficiency and reduced CO2 emissions. This paper will discuss how these results were obtained, and estimate results that could be obtained with a similar system in a refinery application.

(146f) Applied Aspects of Forward Osmosis Desalination

Raafat Alnaizy[1], Ahmed A. Aidan[1] and M. Qasim Rafiq[2], (1)Chemical Engineering, American University of Sharjah, Sharjah, United Arab Emirates, (2)AUS, Sharjah, United Arab Emirates

Forward Osmosis (FO) uses an osmotic driving force to extract water from water containing dissolved species including salt water. This experimental work evaluates the performance of the FO process in seawater desalination using various membranes to study the permeability and productivity of the FO process. Moreover, the work investigates the effect of temperature, pH, membrane surface area, and saline water concentration on the permeability, productivity and the overall performance of the FO desalination process. Since permeability increases with temperature theoretically, the optimum temperature for the FO operation is determined. Also, the surface area of the membrane is varied to study permeability and fouling of the membrane. The concentration of saline water is also varied to investigate for the operating range of the FO process and study the performance of the process at high saline water concentrations. Preliminary results shows that water flux is relatively low due to the weaker driving force induced by the seawater draw solution; however, feed water quality has a minimal effect on water flux through the FO membrane. The bench-scale batch experiments show minimal fouling of the FO membrane. Results indicated that draw solution concentration has minimal effect on the flux but flow velocity affects water flux in both sides of the membrane. Work in progress is directed at testing the performance of the process with several membrane and draw solutions.

AUTHOR INDEX

AUTHOR INDEX